T0181895

Lecture Notes in Computer Science 13756

Founding Editors

Gerhard Goos
Karlsruhe Institute of Technology, Karlsruhe, Germany

Juris Hartmanis
Cornell University, Ithaca, NY, USA

Editorial Board Members

Elisa Bertino
Purdue University, West Lafayette, IN, USA

Wen Gao
Peking University, Beijing, China

Bernhard Steffen
TU Dortmund University, Dortmund, Germany

Moti Yung
Columbia University, New York, NY, USA

More information about this series at https://link.springer.com/bookseries/558

Hujun Yin · David Camacho · Peter Tino (Eds.)

Intelligent Data Engineering and Automated Learning – IDEAL 2022

23rd International Conference, IDEAL 2022
Manchester, UK, November 24–26, 2022
Proceedings

 Springer

Editors
Hujun Yin
University of Manchester
Manchester, UK

Peter Tino
University of Birmingham
Birmingham, UK

David Camacho
Technical University of Madrid
Madrid, Spain

ISSN 0302-9743   ISSN 1611-3349 (electronic)
Lecture Notes in Computer Science
ISBN 978-3-031-21752-4   ISBN 978-3-031-21753-1 (eBook)
https://doi.org/10.1007/978-3-031-21753-1

© The Editor(s) (if applicable) and The Author(s), under exclusive license
to Springer Nature Switzerland AG 2022
This work is subject to copyright. All rights are reserved by the Publisher, whether the whole or part of the
material is concerned, specifically the rights of translation, reprinting, reuse of illustrations, recitation,
broadcasting, reproduction on microfilms or in any other physical way, and transmission or information
storage and retrieval, electronic adaptation, computer software, or by similar or dissimilar methodology now
known or hereafter developed.
The use of general descriptive names, registered names, trademarks, service marks, etc. in this publication
does not imply, even in the absence of a specific statement, that such names are exempt from the relevant
protective laws and regulations and therefore free for general use.
The publisher, the authors, and the editors are safe to assume that the advice and information in this book are
believed to be true and accurate at the date of publication. Neither the publisher nor the authors or the editors
give a warranty, expressed or implied, with respect to the material contained herein or for any errors or
omissions that may have been made. The publisher remains neutral with regard to jurisdictional claims in
published maps and institutional affiliations.

This Springer imprint is published by the registered company Springer Nature Switzerland AG
The registered company address is: Gewerbestrasse 11, 6330 Cham, Switzerland

Preface

The International Conference on Intelligent Data Engineering and Automated Learning (IDEAL) is an annual international conference dedicated to emerging and challenging topics in intelligent data analytics and associated machine learning paradigms and systems. The conference provides a sample of current trends and a unique and stimulating forum for presenting and discussing the latest theoretical advances and real-world applications.

After last year's virtual event, the 23rd edition, IDEAL 2022, was held in Manchester, UK, during November 24–26 in hybrid mode. Given the uncertainties and many travel difficulties, a hybrid event appeared to be practical and best for the time being, like many other similar conferences.

For the past two decades, IDEAL has served an important role in the data analytics, machine learning, and AI communities. The conference aims to bring together researchers and practitioners to present their latest findings, disseminate state-of-the-art results, and share experiences and forge alliances on tackling many real-world challenging problems. During the difficult and turbulent past two years, the IDEAL conference continued to play its role in these communities. The core themes of IDEAL 2022 included big data challenges, machine learning, deep learning, data mining, information retrieval and management, bio-/neuro-informatics, bio-inspired models, agents and hybrid intelligent systems, and real-world applications of intelligence techniques and AI.

In total, 52 papers were accepted and presented at IDEAL 2022, which were selected from nearly 80 submissions. Our Program Committee did a sterling job in providing timely and useful feedback and peer reviews for all the submissions, with each paper receiving three-blind reviews. In addition to the IDEAL 2022 main track, there were a number of special sessions as listed on the IDEAL 2022 website. We would also like to thank the special sessions chairs, Antonio J. Tallón-Ballesteros, Teresa Gonçalves, Vítor Nogueira and Fernando Nuñez Hernandez, and the publicity chairs, Bing Li, Guilherme Barreto, Jose A. Costa, and Yimin Wen, for their great efforts. Our distinguished keynote speakers, Paolo Rosso, Ioannis (Yiannis) Kompatsiaris and Barbara Hammer were also greatly appreciated for their outstanding lectures.

We would like to thank the IEEE UK and Ireland CIS Chapter for their technical co-sponsorship. We would also like to thank all the people who devoted so much time and effort to the successful running of the conference, in particular the members of the Program Committee and reviewers and organisers of the Special Sessions, as well as the authors who contributed to the conference.

Finally, we are grateful for the hard work by the local organising team at the University of Manchester, especially Yating Huang, and our event management collaborators,

Magnifisense, in particular Lisa Carpenter and Gail Crowe. Continued support, collaboration, and sponsorship for the best paper awards from Springer LNCS are also greatly appreciated.

October 2022

Hujun Yin
David Camacho
Peter Tino
Richard Allmendinger
Antonio J. Tallón-Ballesteros
Ke Tang
Sung-Bae Cho
Paulo Novais
Paulo Quaresma

Organization

General Chairs

Hujun Yin — University of Manchester, UK
David Camacho — Technical University of Madrid, Spain
Peter Tino — University of Birmingham, UK

Program Chairs

Richard Allmendinger — University of Manchester, UK
Antonio J. Tallón-Ballesteros — University of Huelva, Spain
Ke Tang — Southern University of Science and Technology, China
Sung-Bae Cho — Yonsei University, South Korea
Paulo Novais — University of Minho, Portugal
Paulo Quaresma — Universidade de Evora, Portugal

Special Session Chairs

Antonio J. Tallón-Ballesteros — University of Huelva, Spain
Teresa Gonçalves — University of Évora, Portugal
Vítor Nogueira — University of Évora, Portugal
Fernando Nuñez Hernandez — University of Seville, Spain

Steering Committee

Hujun Yin — University of Manchester, UK
Colin Fyfe — University of the West of Scotland, UK
Guilherme Barreto — Federal University of Ceará, Brazil
Jimmy Lee — The Chinese University of Hong Kong, Hong Kong, China
John Keane — University of Manchester, UK
Jose A. Costa — Federal University of Rio Grande do Norte, Brazil
Juan Manuel Corchado — University of Salamanca, Spain
Laiwan Chan — The Chinese University of Hong Kong, Hong Kong, China
Malik Magdon-Ismail — Rensselaer Polytechnic Institute, USA
Marc van Hulle — KU Leuven, Belgium
Ning Zhong — Maebashi Institute of Technology, Japan

Peter Tino	University of Birmingham, UK
Samuel Kaski	Aalto University, Finland
Vic Rayward-Smith	University of East Anglia, UK
Yiu-ming Cheung	Hong Kong Baptist University, Hong Kong, China
Zheng Rong Yang	University of Exeter, UK

Publicity and Liaisons Chairs

Bin Li	University of Science and Technology of China, China
Guilherme Barreto	Federal University of Ceará, Brazil
Jose A. Costa	Federal University of Rio Grande do Norte, Brazil
Yimin Wen	Guilin University of Electronic Technology, China

Program Committee

Hector Alaiz Moreton	University of León, Spain
Richardo Aler	Universidad Carlos III de Madrid, Spain
Romis Attux	University of Campinas, Brazil
Carmelo Bastos Filho	University of Pernambuco, Brazil
Riza T. Batista-Navarro	University of Manchester, UK
Lordes Borrajo	University of Vigo, Spain
Vicent Botti	Universitat Politècnica de València, Spain
Federico Bueno de Mata	Universidad de Salamanca, Spain
Robert Burduk	University of Wroclaw, Poland
Anne Canuto	Federal University of Rio Grande do Norte, Brazil
Roberto Carballedo	University of Deusto, Spain
Joao Carneiro	Polytechnic of Porto, Portugal
Mercedes Carnero	Universidad Nacional de Rio Cuarto, Argentina
Pedro Castillo	University of Granada, Spain
Luís Cavique	University of Aberta, Potugal
Songcan Chen	Nanjing University of Aeronautics and Astronautics, China
Xiaohong Chen	Nanjing University of Aeronautics and Astronautics, China
Stelvio Cimato	Università degli Studi di Milano, Italy
Manuel Jesus Cobo Martin	University of Cádiz, Spain
Leandro Coelho	Pontifícia Universidade Católica do Parana, Brazil
Carlos Coello Coello	CINVESTAV-IPN, Mexico
Roberto Confalonieri	Free University of Bozen-Bolzano, Italy
Paulo Cortez	University of Minho, Portugal
Jose Alfredo F. Costa	Federal University of Rio Grande do Norte, Brazil
Carlos Cotta	Universidad de Málaga, Spain

Raúl Cruz-Barbosa	Universidad Tecnológica de la Mixteca, Mexico
Ernesto Damiani	University of Milan, Italy
Andre de Carvalho	University of São Paulo, Brazil
Dalila A. Durães	Universidade do Minho, Portugal
Bruno Fernandes	University of Minho, Portugal
João Ferreira	ISCTE, Portugal
Joaquim Filipe	EST-Setubal/IPS, Portugal
Felipe M. G. França	COPPE-UFRJ, Brazil
Pedro Freitas	Universidade Católica Portuguesa, Portugal
Dariusz Frejlichowski	West Pomeranian University of Technology, Poland
Hamido Fujita	Iwate Prefectural University, Japan
Marcus Gallagher	University of Queensland, Australia
Isaias Garcia	University of León, Spain
María José Ginzo Villamayor	Universidad de Santiago de Compostela, Spain
Teresa Goncalves	University of Evora, Portugal
Anna Gorawska	Silesian University of Technology, Poland
Marcin Gorawski	Silesian University of Technology, Poland
Manuel Graña	University of the Basque Country, Spain
Maciej Grzenda	Warsaw University of Technology, Poland
Pedro Antonio Gutierrez	Universidad de Cordoba, Spain
Barbara Hammer	Bielefeld University, Germany
J. Michael Herrmann	University of Edinburgh, UK
Wei-Chiang Hong	Jiangsu Normal University, China
Jean-Michel Ilie	Sorbonne Université, France
Dariusz Jankowski	Wrocław University of Technology, Poland
Vicente Julian	Universitat Politècnica de València, Spain
Jason Jung	Chung-Ang University, South Korea
Rushed Kanawati	Université Paris 13, France
Bin Li	University of Science and Technology of China, China
Victor Lobo	Universidade Nova de Lisboa, Portugal
Wenjian Luo	Harbin Institute of Technology, Shenzhen, China
Jesús López	Tecnalia Research & Innovation, Spain
José Machado	University of Minho, Portugal
Rui Neves Madeira	Instituto Politécnico de Setúbal, Portugal
José F. Martínez-Trinidad	INAOE, Mexico
Cristian Mihaescu	University of Craiova, Romania
José M. Molina	Universidad Carlos III de Madrid, Spain
Paulo Moura Oliveira	UTAD University, Portugal
Tatsuo Nakajima	Waseda University, Japan
Susana Nascimento	Universidade Nova de Lisboa, Portugal

Grzegorz J. Nalepa	AGH University of Science and Technology, Poland
Antonio Neme	Universidad Nacional Autonoma de Mexico, Mexico
Vitor Nogueira	Universidade de Évora
Fernando Nuñez	University of Seville, Spain
Eva Onaindia	Universitat Politècnica de València, Spain
Eneko Osaba	Tecnalia Research & Innovation, Spain
Jose Palma	University of Murcia, Spain
Carlos Pereira	Instituto Superior de Engenharia de Coimbra, Portugal
Radu-Emil Precup	Politehnica University of Timisoara, Romania
Héctor Quintián	University of A Coruña, Spain
Izabela Rejer	University of Szczecin, Poland
Matilde Santos	Universidad Complutense de Madrid, Spain
Richardo Santos	Polytechnic of Porto, Portugal
Jose Santos	University of A Coruña, Spain
Fábio Silva	University of Minho, Portugal
Ivan Silva	University of São Paulo, Brazil
Dragan Simic	University of Novi Sad, Serbia
Marcin Szpyrka	AGH University of Science and Technology, Poland
Murat Caner Testik	Hacettepe University, Turkey
Qing Tian	Nanjing University of Information Science and Technology, China
Stefania Tomasiello	University of Salerno, Italy
Alexandros Tzanetos	University of the Aegean, Greece
Eiji Uchino	Yamaguchi University, Japan
José Valente de Oliveira	Universidade do Algarve, Portugal
Alfredo Vellido	Universitat Politècnica de Catalunya, Spain
Gianni Vercelli	University of Genoa, Italy
Tzai-Der Wang	Cheng Shiu University, Taiwan
Dongqing Wei	Shanghai Jiao Tong University, China
Michal Wozniak	Wroclaw University of Technology, Poland
Xin-She Yang	Middlesex University, UK

Special Session on Intelligent Techniques for Real-World Applications of Renewable Energy and Green Transport

Organizers

| J. Enrique Sierra García | University of Burgos, Spain |
| Matilde Santos Peñas | Complutense University of Madrid, Spain |

Fares M'zoughi   University of the Basque Country, Spain
Payam Aboutalebi   University of the Basque Country, Spain

Special Session on Computational Intelligence for Imbalanced Classification

Organizers

Wenbin Pei   Dalian University of Technology, China
Bing Xue   Victoria University of Wellington, New Zealand
Antonio J. Tallón-Ballesteros   University of Huelva, Spain

Contents

**Special Session on Intelligent Techniques for Real-World Applications
of Renewable Energy and Green Transport**

**Special Session on Computational Intelligence for Imbalanced
Classification**

Main Track

Ensemble Stack Architecture for Lungs Segmentation from X-ray Images

Asifuzzaman Lasker[1(✉)], Mridul Ghosh[2], Sk Md Obaidullah[1],
Chandan Chakraborty[3], Teresa Goncalves[4], and Kaushik Roy[5]

[1] Department of Computer Science and Engineering, Aliah University,
Kolkata 700160, India
`asifuzzaman.lasker@gmail.com`
[2] Department of Computer Science, Shyampur Siddheswari Mahavidyalaya,
Howrah 711312, India
[3] Department of Computer Science and Engineering, National Institute of Technical
Teachers' Training and Research, Howrah 700106, India
[4] Computer Science Department and ALGORITMI Center, University of Évora,
Évora, Portugal
`tcg@uevora.pt`
[5] Department of Computer Science, West Bengal State University,
Barasat 700126, India

Abstract. In healthcare, chest X-rays are an inexpensive medical imaging diagnostic tools. The lung images segmentation from chest X-rays (CXRs) is important for screening and diagnosing diseases. The lungs are opacified in many patients' CXRs, making it difficult to segment them. A segmentation algorithm based on U-Net is proposed in this paper to address this problem. The proposed architecture was developed using three pre-trained models: MobileNetV2, InceptionResNetV2, and EfficientNetB0. In this architecture, we designed a ensemble stacked framework which is based on the pre-trained models to improve segmentation performance. Compared with the conventional U-Net model, our method improves by 3.02% dice coefficient and 3.43% IoU experimenting on the three public lung segmentation datasets.

Keywords: Deep learning · Lung segmentation · Unet · Medical imaging · Chest X-rays

1 Introduction

Modern healthcare systems typically use medical imaging; X-rays are one of the most commonly used diagnostic tools for respiratory diseases like asthma, influenza, pneumonia, COVID-19, tuberculosis, lung cancer, etc. [1]. Clinical decision support systems that analyze pulmonary images rely heavily on lung segmentation, a process of identifying lung regions and boundaries from surrounding tissue. It is possible to extract clinically relevant features from lung

© The Author(s), under exclusive license to Springer Nature Switzerland AG 2022
H. Yin et al. (Eds.): IDEAL 2022, LNCS 13756, pp. 3–11, 2022.
https://doi.org/10.1007/978-3-031-21753-1_1

fields [2] and train a deep learning algorithm for detecting disease and anomalies in infected lung regions [3]. Using computational methodologies can help physicians provide their patients with a timely, accurate medical diagnosis, improving both qualities of care and outcomes.

Research in artificial intelligence has helped localize lung infection using a deep learning-based segmentation model [4,5]. Lesion assessment can be performed by identifying regions of interest through semantic segmentation. Therefore, segmentation results can assist clinicians to find missing lesions, prevent diseases, and improve therapy planning and treatment. Obtaining the desired algorithmic goal for automated medical image segmentation involves many high-quality labeled and annotated datasets, which is a significant challenge. The pixel-level lungs lesion segmentation task often results in noise, and clean annotations are difficult to collect. The annotation standards of different annotators may lead to inter-observer variability, and high intra-observer variability may also exist. An annotation with these variabilities is likely to demonstrate disagreement between two annotators. Some researchers use a human-in-the-loop strategy to reduce the annotation effort [6], where the annotators refine the algorithm-generated labels for annotation. Therefore, the annotations can contain noise at the pixel level because they are biased towards the algorithm's results. Using the annotation, deep learning has proven its efficiency compared with traditional models, which are reflected in literature [7–10]. For this reason, we have used deep learning for segmenting lung form X-ray images. Here we have designed three ensemble stacked architecture which is based on the deep learning models to segregate the lungs from the X-ray images efficiently.

In this paper, we proposed a stacked ensemble framework that is based on U-Net architecture and pretrained architectures namely MobileNetV2, Inception-ResNetV2, and EfficientNetB0 for binary-class semantic segmentation of lung structures from chest radiography. In this stacked ensemble framework, the pretrained models are input to the encoder module of U-Net which are stacked and ensembled to produce a final prediction from the meta learner of the system.

2 Related Works

The semantic segmentation of X-ray, CT scan, and MRI images is widely used to diagnose lung diseases. The segmentation process often uses the therapeutic domain or stage of disease development (usually exclusively biomedical) as criteria for differentiation. Depending on the disease type, individuals will receive the same additional support and service in addition to the appropriate medical treatment. A diagnosis can often be made by segmenting different organs and providing doctors with essential lung disease information. Several convolutional neural network (CNN) supported deep learning architectures were used for segmentation, including Fully Convolutional Networks (FCN), U-Net, Mask RCNN, and other combinations of these architectures.

Wu et al. [11] introduced a joint diagnostic system of classification and segmentation of COVID-19 diseases. They prepared and fine-grained pixel labeled large scale COVID-19 dataset. The segmentation features are encoded using the backbone features of the classification model. Xu et al. [12] applied a two-stage classification framework using segmentation method to solve the COVID-19 diagnosis problem. In the first stage, segmentation models are applied to predict masks for all CXR images. Using CNN, a second stage of classification was conducted based on the preserved lung regions. Segmentation was done with U-Net and ResNet, and classification was done using four state-of-the-art deep learning models. Ter-Sarkisov et al. [13] Implemented a lightweight segmentation model for chest CT scans with Ground Glass Opacity and consolidation. The model uses truncated versions of ResNet18 and ResNet34 networks as its backbone in order to reduce the number of parameters. The lightweight version of the model requires significantly less training time than its original version.

Maxime et al. [14] discussed three important aspects of radiographic imaging, which include image characterization, deep learning, and statistical analysis. A deep learning model was used in this article for lung segmentation and image opacity detection. The authors applied several statistical analyses to identify the type of lung images. Singh et al. [15] implemented preprocessed and segmentation techniques for chest X-rays to enhance overall accuracy compared with the original dataset. To increase accuracy, five techniques were applied: weight initialization, training class distribution, preprocessing, segmentation, and ensemble learning. In addition to the five techniques, they also proposed a meta-learner for generalizing model complexity. The results were visualised for medical communities to prove the model's performance.

3   Methodology

The proposed stacked ensemble architecture consists of MobileNetV2, Inception-ResNetV2, EfficientNetB0 along with U-Net model. The U-Net model is based on CNN technique that was developed for biomedical image segmentation [16]. There are three components of the U-Net model. The first is downsampling, the second is bottle necking between encoder and decoder, and the third is expanding or up sampling.

In the proposed architecture, the encoder part of U-Net is associated with the three different individual pretrained models, which are stacked and ensembled and fed to the meta learner to produce the final prediction. In the meta learner, there are 3 dense layers having sizes 128, 256, and 512. The dropout value of 0.3 was considered throughout the experiment after experimental trial. A batch normalization layer was also considered before the output dense layer. The design of the meta learner was set after a few trial runs. At the beginning, we found out the predictions from the individual U-Net, U-Net with MobileNetV2, U-Net with EfficientNetB0, and U-Net with InceptionResNetV2. Then to increase the performance of the system, we developed the stacked ensemble framework with these individual models. We set the alpha parameter to 0.35 for MobileNetV2, in order to reduce the complexity of the pretrained architecture. From the activation layer of the pretrained encoder, a skip connection was prepared. In the decoder block, four different filter sizes are used such as 16, 32, 64, and 128. As the output layer of the encoder uses a filter size of 16, we used the same filter size for the first layer of the decoder as well. To maintain the similarity in the skip connection with the same dimensions of stacked architecture, InceptionResNetV2 was adjusted using zero padding in the encoder output of 127×127, 62×62, 29×29, and 14×14 dimensions to 128×128, 64×64, 32×32, and 16×16 dimensions. The input images were resized to $256 \times 256 \times 3$. Figure 1 represents the architecture of the proposed model.

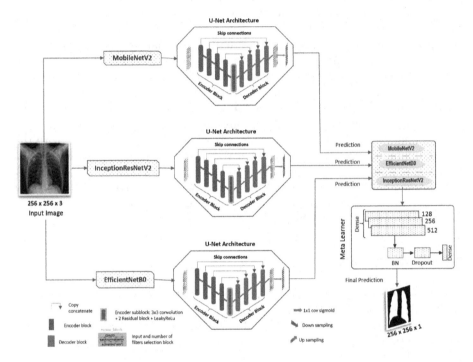

Fig. 1. Proposed ensemble stack architecture for lungs segmentation from X-ray images.

4 Experiment

4.1 Evaluation Protocols

To assess segmentation performance, Dice Coefficient (DC) and Intersection over Union (IoU) are most significant. Along with the dice coeff and IoU, we considered three other performance evaluation metrics.

Dice Coefficient (DC): To calculate the DC, divide the overlap area between the predicted segmentation and the ground truth by the sum of the overlap areas. It is formulated as

$$Dice\ Coeff = \frac{2.area(seg_{pred} \bigcap seg_{g.true})}{area(seg_{pred} + seg_{g.true})} \qquad (1)$$

Intersection Over Union (IoU): A relationship between the predictions and the ground truths is calculated by using the IoU. It is the ratio between the size of the intersection and the size of the union based on the predicted outcome and labels. It is depicted as

$$IoU = \frac{area(seg_{pred} \bigcap seg_{g.true})}{area(seg_{pred} \bigcup seg_{g.true})} \qquad (2)$$

The other three evaluation metrics are loss, precision, and recall. A simple cross-category entropy loss function is used for semantic segmentation.

4.2 Dataset

We used three publicly available datasets [17] namely, Japanese Society of Radiological Technology (JSRT), Shenzhen Hospital (SH), and Montgomery County (MC) in this work. There are 247, 138, and 662 X-ray and associated mask images in JSRT, MC, and SH, respectively. There are 247 X-ray images in the JSRT dataset, including 93 normal images and 154 abnormal images from 14 medical centers. In SH dataset, there are 138 frontal chest images, 80 normal and 58 tuberculosis images, which were captured with a stationary Eureka x-ray machine in 12-bit grayscale. Using a Philips DR digital diagnose system, the MC's datasets include 326 normal images and 336 tuberculosis images. These datasets contain images having different resolutions, such as 2048×2048, 4892×4020, and 3000×3000 pixels, respectively. To maintain the homogeneity of the input, all images are resized to 256×256 dimensions to improve training performance.

4.3 Training Regime

The proposed architecture was tuned using different hyperparameters to achieve better results. The parameters in these cases were batch size, learning rate, and the number of epochs, and their corresponding values are 32, 0.00001, and 50, respectively. In comparison with U-Net model, there are 75.89%, 7.74%, and 0.01% fewer trainable parameters for pretrained U-Net+MobileNetV2, U+InceptionRes- NetV2, and U-Net+EfficientNetB0 architecture. The inference time (value) for U-Net, U-Net+MobileNetV2, U+InceptionResNetV2, and U-Net+EfficientNetB0 in 187 ms, 140 ms, 157 ms, and 177 ms. Based on these parameter values, four different learning curves are depicted in Fig. 2.

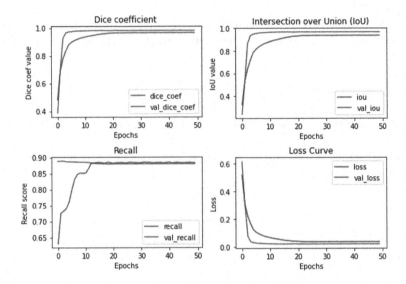

Fig. 2. Learning curve of U-Net with four stack model.

4.4   Results

To evaluate our segmentation experiment's performance, we present five metrics for comparative analysis with various architectures. The results of the experiments are shown in the Table 1. The first row shows the performance of the U-Net architecture, followed by MobileNetV2, InceptionResNetV2, EfficientNetB0, and stacked framework, respectively. We observed that in comparison to only U-Net models, the concatenated stack architecture produced almost 2% better dice coeff and IoU scores on Montgomery, JSRT, and Shenzhen datasets. Also, it was observed that the DC, IoU, and Recall values of U-Net with three pre-trained models such as MobileNetV2, InceptionResNetV2, and EfficientNetB0 were improved in the respective three experiments. In comparison with the other experiments, the loss values decreased significantly in the final results. Similarly, in other four metrics dice coeff, IoU, Recall, and Precision values are 0.9832, 0.9669, 0.9296, and 0.9976, respectively.

After evaluating the proposed model, Fig. 3 depict the X-ray image and its corresponding ground truth, binary mask, and segmented lungs. The column (Binary Mask) demonstrates that the suggested model performs effectively with test data.

5   Comparison with State-of-the-Arts

We have compared our results with different state-of-the-art techniques. To show the efficiency of the proposed methods, we have compared Liu et al. [18], Frid-Adar et al. [19], Munawar et al. [20], and Cao et al. [21] since they have used the same datasets that we used in our work. It is seen that the dice coefficients are

Table 1. Results of U-Net model with different state-of-the-art architectures.

	Loss	Dice coeff	IOU	Recall	Precision
U-Net	0.0259	0.9580	0.9373	0.8694	0.9866
U-Net + MobileNetV2	0.0246	0.9673	0.9519	0.8948	0.9985
U-Net + InceptionResNetV2	0.0202	0.9712	0.9602	0.9101	0.9980
U-Net + EfficientNetB0	0.0275	0.9654	0.9464	0.9154	0.9968
U-Net + EfficientNetB0 + MobileNetV2 + InceptionResNetV2	0.0102	0.9882	0.9716	0.9356	0.9986

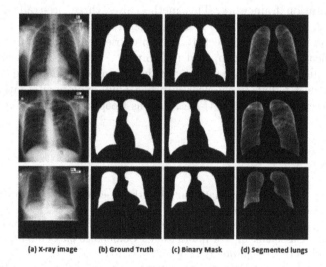

(a) X-ray image (b) Ground Truth (c) Binary Mask (d) Segmented lungs

Fig. 3. Lungs segmentation on X-ray image.

increased by 1.02%, 2.72% and 1.42%, compared with [18,19] and [20]. Similarly, the IoU score is improved by 1.51%, 2.86% 5.15% and compared with [18,20] and [21]. In this experiment, the recall score did not improve, but the precision values showed significant improvements (Table 2).

Table 2. Comparison of U-Net model and U-Net stacked model with some state-of-the-art models.

Author's	Dice coeff	IOU	Recall	Precision
Liu et al. [18]	0.9780	0.9565	0.9795	0.9890
Frid-Adar et al. [19]	0.9610	–	–	–
Munawar et al. [20]	0.9740	0.9430	–	–
Cao et al. [21]	–	0.9201	0.9504	0.9693
Ours	**0.9882**	**0.9716**	**0.9356**	**0.9986**

6 Conclusion

On the basis of the U-Net architecture, the paper proposes a reliable and efficient lung segmentation framework. This method uses three pre-trained networks: MobileNetV2, InceptionResNetV2, and EfficientNetB0 and further stacked them to develop an ensemble architecture. The proposed ensemble architecture performs significantly better compared to individual models. A variety of public datasets was used to test the efficiency of the proposed architecture and impressive outcome was found. In the near future, we will explore alternative segmentation architectures other than U-Net, such as DeepLab, SegNet, mask R-CNN, etc.

References

1. Mukherjee, H., et al.: Automatic lung health screening using respiratory sounds. J. Med. Syst. **45**(2), 1–9 (2021)
2. Lasker, A., et al.: Deep features for COVID-19 detection: performance evaluation on multiple classifiers. In: 4th International Conference on Computational Intelligence in Pattern Recognition (CIPR-2022)
3. Mukherjee, H., et al.: Deep neural network pneumonia detection using chest X-Rays. In: 5th International Conference on Computer Vision & Image Processing (CVIP 2020)
4. Saad, M.N., et al.: Image segmentation for lung region in chest X-ray images using edge detection and morphology. In: 2014 IEEE International Conference on Control System, Computing and Engineering (ICCSCE 2014). IEEE (2014)
5. Abedalla, A., et al.: Chest X-ray pneumothorax segmentation using U-Net with EfficientNet and ResNet architectures. PeerJ Comput. Sci. **7**, e607 (2021)
6. Shan, F., et al.: Lung infection quantification of COVID-19 in CT images with deep learning. arXiv preprint arXiv:2003.04655 (2020)
7. Ghosh, M., et al.: Understanding movie poster: transfer-deep learning approach for graphic-rich text recognition. Visual Comput. **38**(5), 1645–1664 (2022)
8. Ghosh, M., et al.: A deep learning-based approach to single/mixed script-type identification. In: Advanced Computing and Systems for Security, vol. 13, pp. 121–132. Springer, Singapore (2022)

9. Ghosh, M., et al.: Text/non-text scene image classification using deep ensemble network. In: Proceedings of International Conference on Advanced Computing Applications. Springer, Singapore (2022)

10. Ghosh, M., et al.: LWSINet: a deep learning-based approach towards video script identification. Multimed. Tools Appl. **80**(19), 29095–29128 (2021)

11. Wu, Y.-H., et al.: Jcs: an explainable covid-19 diagnosis system by joint classification and segmentation. IEEE Trans. Image Process. **30**, 3113–3126 (2021)

12. Xu, Y., Lam, H.-K., Jia, G.: MANet: a two-stage deep learning method for classification of COVID-19 from Chest X-ray images. Neurocomputing **443**, 96–105 (2021)

13. Ter-Sarkisov, A.: Lightweight model for the prediction of COVID-19 through the Detection and Segmentation of lesions in chest ct scans. medRxiv (2020)

14. Blain, M., et al.: Determination of disease severity in COVID-19 patients using deep learning in chest X-ray images. Diagnostic and interventional Radiol. **27**(1), 20 (2021)

15. Singh, R.K., Pandey, R., Babu, R.N.: COVIDScreen: explainable deep learning framework for differential diagnosis of COVID-19 using chest X-rays. Neural Comput. Appl. **33**(14), 8871–8892 (2021)

16. Ronneberger, O., Fischer, P., Brox, T.: U-Net: convolutional networks for biomedical image segmentation. In: Navab, N., Hornegger, J., Wells, W.M., Frangi, A.F. (eds.) MICCAI 2015. LNCS, vol. 9351, pp. 234–241. Springer, Cham (2015). https://doi.org/10.1007/978-3-319-24574-4_28

17. Tuberculosis chest X-ray datasets. National Libray of Medicine, 30 Mar 2021. lhncbc.nlm.nih.gov/LHC-downloads/dataset.html

18. Liu, W., et al.: Automatic lung segmentation in chest X-ray images using improved U-Net. Sci. Rep. **12**(1), 1–10 (2022)

19. Frid-Adar, M., et al.: Improving the segmentation of anatomical structures in chest radiographs using u-net with an ImageNet pre-trained encoder. In: Image Analysis for Moving Organ, Breast, and Thoracic Images, pp. 159–168. Springer, Cham (2018)

20. Munawar, F., et al.: Segmentation of lungs in chest X-ray image using generative adversarial networks. IEEE Access **8**, 153535–153545 (2020)

21. Cao, F., Zhao, H.: Automatic lung segmentation algorithm on chest X-ray images based on fusion variational auto-encoder and three-terminal attention mechanism. Symmetry **13**(5), 814 (2021)

Synonym-Based Essay Generation and Augmentation for Robust Automatic Essay Scoring

Tsegaye Misikir Tashu[1,2](✉) 🆔 and Tomáš Horváth[1,3] 🆔

[1] ELTE - Eötvös Loránd University, Faculty of Informatics, Department of Data Science and Engineering, Telekom Innovation Laboratories, Pázmány Péter sétány 1/C, Budapest 1117, Hungary
{misikir,tomas.horvath}@inf.elte.hu

[2] Wollo University, Kombolcha Institute of Technology, College of Informatics, 208 Kombolcha, Ethiopia

[3] Pavol Jozef Šafárik University, Faculty of Science, Institute of Computer Science, Jesenná 5, 040 01 Košice, Slovakia

Abstract. Automatic essay scoring (AES) models based on neural networks (NN) have had a lot of success. However, research has shown that NN-based AES models have robustness issues, such that the output of a model changes easily with small changes in the input. We proposed to use keyword-based lexical substitution using BERT that generates new essays (adversarial samples) which are lexically similar to the original essay to evaluate the robustness of AES models trained on the original set. In order to evaluate the proposed approach, we implemented three NN-based scoring approaches and trained the scoring models using two stages. First, we trained each model using the original data and evaluate the performance using the original test and newly generated test set to see the impact of the adversarial sample of the model. Secondly, we trained the models by augmenting the generated adversarial essay with the original data to train a robust model against synonym-based adversarial attacks. The results of our experiments showed that extracting the most important words from the essay and replacing them with lexically similar words, as well as generating adversarial samples for augmentation, can significantly improve the generalization of NN-based AES models. Our experiments also demonstrated that the proposed defense is capable of not only defending against adversarial attacks, but also of improving the performance of NN-based AES models.

Keywords: Adversarial attack · Data augmentation · Automatic essay scoring

1 Introduction

Systems for automatic essay scoring (AES) are used to reduce the workload of examiners, improve the feedback cycle in the teaching-learning process, save

© The Author(s), under exclusive license to Springer Nature Switzerland AG 2022
H. Yin et al. (Eds.): IDEAL 2022, LNCS 13756, pp. 12–21, 2022.
https://doi.org/10.1007/978-3-031-21753-1_2

time and costs in grading [7]. The task of AES was regarded as a supervised learning [7,23] problem that learns to approximate the assessment process using handcrafted features. Recently, AES systems have achieved success by moving from linear models over sparse and handcrafted feature inputs to nonlinear deep neural network (NN) models over dense inputs [7,8]. Despite their success, models based on NNs are not inherently robust. In particular, it has been shown that the addition of imperceptible deviations to the input, known as adversarial perturbations, can cause NNs to make incorrect predictions with high confidence [3,22]. The most common adversarial perturbations in text-based models, like AES, are adding distracting text to the input [13], paraphrasing the text [3], replacing words with similar words [3], or inserting character-level typos [5]. These unwanted glitches can significantly affect the performance of a model. To improve the resilience against these perturbations, several defense strategies have been proposed. Among them, data augmentation methods, which extend the training set to include adversarial examples, are widely used in the training process (fine-tuning) to improve the robustness of the model. These methods, playing a special and important role, achieve generalization without affecting the representation capacity of the model and without re-tuning its other hyperparameters [12,19].

In this work, we proposed to use a synonym-based word replacement to generate an adversarial essay set to attack NN-based AES models for checking their vulnerability and robustness. Furthermore, we proposed to use synonym-based word replacement for essay generation and augmentation as a defense strategy to train a more robust AES model against these perturbations. We extended the work of Alshemali Basemah and Kalita Jugal [1] for automated essay scoring using BERT. As it was pointed out by Zhou et al. [26], approaches that use WordNet fails to consider the substitution's influence on the global context of the sentence; and are likely to overlook good substitute candidates that are not synonyms of the target words in the lexical resources [26]. We restrict the generation of adversarial essays to certain important words, which leads to high-quality results that are semantically similar to the input essay. We conduct experiments with three NN-based model architectures and five unique prompts. The results show that the proposed approach is sound and relevant. The rest of the paper is organized into six sections. Section 2 describes related works relevant to our research. Section 3 presents the proposed approach. AES scoring models used in the study are presented in Sect. 4. Section 5 presents the overall experimental settings, implementation and evaluation of the proposed system. Section 7 concludes the study.

2   Related Work

Most AES models treat scoring as a supervised learning task [15], where the most recent NN-based approaches range from finding the hierarchical structure of documents [8] across using attention over words [9] to modeling coherence [24]. AES systems have had some success with this transition. Despite the success of

AES systems, [17,20] have shown that NN-based models are vulnerable to adversarial perturbations. Moreover, in the context of Deep Learning models, so-called adversarial examples/attacks are increasingly generated by small changes in the input data that lead the algorithms to make incorrect predictions [10].The most common and popular attacks in natural language processing (NLP) systems are character-level attacks [11], token-level attacks [2] and sentence-level attacks [14]. Adversarial attacks based on synonym substitution are becoming more common [3,18]. In attacks of this type, the attacker substitutes tokens with synonyms. Since synonym substitution attacks aim to satisfy all lexical, grammatical and semantic constraints, they are difficult to detect both with automatic syntax checking and by humans.

Using adversarial training, generators of adversarial examples can be used as methods for data augmentation [3,25], making NN-based models less susceptible to such simple changes [4]. The most extensive research on data augmentation is in computer vision. This is due to the intuitive construction of simple, label-preserving transformations. Data augmentation methods in computer vision include geometric transformations [25], neural style transfers, interpolation of images, and generative adversarial network data generation [4,10]. Alshemali Basemah and Kalita Jugal [1] proposed a method combining word importance ranking, synonym extraction, word embedding, averaging, and majority voting to mitigate interference from synonym substitution. Zhou et al. [26] proposed an end-to-end BERT-based lexical substitution approach that can propose and validate substitute candidates without using any annotated data or manually curated resources. Our work extends the work of Alshemali Basemah and Kalita Jugal [1] using BERT for synonym extraction and replacement for essay generation. The generated data will be augmented with the original data as a defense strategy against word-level synonym replacement attacks.

3   Proposed Methodology

In Fig. 1, we present an overview of the proposed approach, where our objective is to build a model that is robust to label-preserving perturbations. In this work, we focus on perturbations where important keywords of the input are substituted with lexically similar words. Then, adversarial data augmentation is used as a defense strategy that augments the training set by the adversarial examples in the training process to enhance the robustness of the model.

3.1   Synonym Replacement and Essay Generation

Given an essay with a sequence of terms, only a few important "content bearing" terms has influence on the model's prediction. To enhance the model robustness, the proposed approach consists of the following steps, described in more detail later.

1. Extract the most important or keywords from each essay using term frequency/inverse document frequency (TF-IDF);

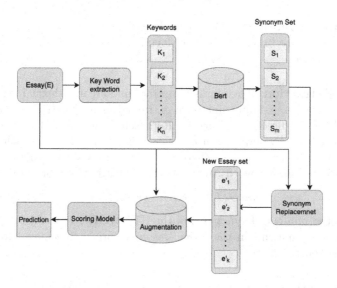

Fig. 1. The architecture of the proposed synonym-based replacement adversarial generation and augmentation.

2. For each keyword or important terms in each essay, extract its synonyms using BERT (build a synonym set);
3. Generate new essays (adversarial examples) by replacing the Keywords in each essay using their synonyms;
4. Augment the training set by the adversarial examples in the training process.

Keyword Extraction. Given a sequence of words, only a few words affect the prediction of the model. To extract the most important terms from the sequence of terms in each essay, we used a mechanism called TF-IDF to select the terms with the greatest influence. Formally, the TF-IDF score for the word t in essay e from the essay set E is calculated as

$$TFIDF_{t,e,E} = tf_{t,e} \cdot idf_{t,E} \tag{1}$$

where

$$tf_{t,e} = log(1 + freq(t,e)) \tag{2}$$

and

$$idf_{t,E} = log \left(\frac{N}{count(e \in E : t \in e)} \right) \tag{3}$$

while $freq(t,e)$ is the frequency of occurrence of a word t in an essay e and N is the total number of essays in E. The mechanism based on the TF-IDF weighting scheme decreases the effect of globally frequent terms and increases the effect of locally frequent terms in the dictionary. It assigns a relatively higher value to those terms that are important in the document but at the same time rare in the whole corpora.

Synonym Extraction. The aim is to extract and create a lexical synonym set for a given keyword or important term extracted using TF-IDF. In this study, we used BERT (Bidirectional Encoder Representations from Transformers) [6] for extracting lexically similar words for each Keyword [26]. For each keyword, we used the same technique as Zhao et al. [26] to create a synonym set that contains lexically similar words to the given keyword. More formally,

$$synset(T_i) = BERTs_1, s_2, \dots s_m \tag{4}$$

where m is the number of synonyms of the KEYWORD that exist in BERT. If a token has no synonyms in the lexical resource, processing moves to the next important token.

Essay Generation. Once the important terms and the synonyms for each important term have been extracted, for each essay, a new essay sentence is created from each essay by replacing the synonyms. In this way, a new essay corpus is built, expanding the original essay corpus. For every important term T_i in the given essay input E, the set of allowed synonym substitutions are, $synset(T_i)$ including the term T_i itself. We chose $\hat{E}$ to denote the perturbed version of E, where each word of $\hat{E}$ is, including T_i, in $synset(T_i)$. We choose to use the synonym set so that $\hat{E}$ is more likely to be grammatically and lexically similar and have the same label as E.

3.2 Data Augmentation

Given the set of original training instance pairs $D_{or} = \{(E_i, Y_i)\}_{i=1}^n$ where E_i contains a set of training essay and Y_i contains the corresponding essay scores, we use word synonym substitution-based adversarial essay generation to create a score-preserving adversarial training $D_{ad} = \{(\hat{E}_i, Y_i)\}_{i=1}^n$. Then we train the model with the augmented training dataset $D_{aug} = D_{or} \bigcup D_{ad}$.

4   Scoring Models

To analyze the potential benefits of the proposed adversary generation and augmentation, we implemented three different NN-based scoring models using convolutional NNs (CNN) and recurrent NNs (RNN). We used bidirectional RNN (BiRNN), bidirectional gated recurrent unit (BiGRU) and bidirectional long short-term memory (BiLSTM) variants of RNN-based models, A CNN-BiLSTM/BiGRU/BiRNN architecture is a hybrid architecture that combines CNN layers for feature extraction on input data and BLSTM/BGRU/BRNN layers for sequence prediction. The Relu function was used to predict the score, and the mean squared error was used to compute the loss. To minimize the mean squared error (MSE) loss, we used the RMSProp optimization algorithm.

5   Experiment

5.1   Data Sets

Experiments were conducted using the dataset provided as part of the AES contest on the Kaggle[1]. The dataset contains student essays with eight different prompts (discussion questions). Four data sets contained essays of traditional writing genres such as persuasive, expository, or narrative. The training set has been evaluated by at least two human experts and their average evaluation is used as the final score, i.e. the label.

5.2   Essay Pre-processing

We tokenize the essays using the NLTK[2] tokenizer and lowercase text. To learn the representation of each essay, the freely available Glove[3] word embedding, an unsupervised learning algorithm for obtaining vector representations for words, was used. Training is performed on aggregated global word-word co-occurrence statistics from a corpus, the resulting representations of which showcase interesting linear substructures of the word vector space [16].

5.3   Evaluation Methodology

We split the data into training set, validation set, and test set. 60% of the data was used as a training set, 10% as the validation set, and 30% as the test set. We train all models for 40 epochs and select the best model based on the performance on the validation set. The evaluation is conducted in a prompt-specific fashion (essay topic). We also use dropout regularization to avoid overfitting [21]. The input of the dropout is the output of the recurrent layer. Dropout randomly selects some entries in its input with probability p and resets them by assigning 0 (zero) to their value. The rest of the entries will be copied to the output.

We use the most widely used evaluation metric called quadratic weighted kappa (QWK). QWK is an error metric that measures the degree of agreement between the automated scores and the resolved human scores, and is analogous to the correlation coefficient.

6   Results and Discussion

To test the robustness of the NN-based AES, we implemented and trained three NN-based scoring models. First, we trained these models with the original training set and then evaluated their performance with the original test set, presented as "without attack" in Table 1. Second, we evaluate the robustness of these models (trained with the original data) using the test set generated using

---

[1] https://www.kaggle.com/c/asap-aes.
[2] http://www.nltk.org.
[3] https://nlp.stanford.edu/projects/glove/.

synonym-based replacement and data generation (with adversarial attack) presented as "with attack" in Table 1. Finally, we trained NN-based AES models by augmenting the original data with the adversarial set generated using synonym-based replacement and data generation, presented as "With augmentation" in Table 1.

Table 1. Test Accuracy of the various NN-based models on non-modified data (without attack); on modified data with synonym replacement (with attack) and adversarial data augmentation.

Datasets	Without attack			With attack			With augmentation		
	BiLSTM	BiGRU	BiRNN	BiLSTM	BiGRU	BiRNN	BiGRU	BiLSTM	BiRNN
Set 1	0.764	0.721	0.678	0.654	0.632	0.615	0.858	0.889	0.838
Set 2	0.699	0.719	0.556	0.549	0.529	0.475	0.886	0.883	0.790
Set 3	0.713	0.702	0.581	0.627	0.650	0.611	0.864	0.865	0.774
Set 4	0.738	0.761	0.705	0.581	0.608	0.596	0.876	0.866	0.774
Set 5	0.813	0.799	0.810	0.612	0.690	0.692	0.918	0.896	0.849

The experimental results show that simple attacks based on synonym substitution significantly reduce the robustness of the model, as shown in Table 1. We observe a drop in performance when we test the model trained with the original data with the adversarial test set. This shows that these models can be easily fooled by simple attacks. The classification accuracy of the models decreased significantly due to the attacks of the adversaries. For the BiLSTM-based model, the accuracy deteriorated by 0.11, 0.15, 0.09, 0.16, and 0.20 for test sets one to five, respectively. For the BiGRU-based model, the accuracy deteriorated by 0.09, 0.19, 0.05, 0.15, and 0.11 for essay set one through five, respectively. For the BiRNN-based model, the accuracy decrease is 0.06, 0.08, 0.11, and 0.12 for essay sets one, two, four, and five, respectively, but the accuracy increased by 0.03 for data set three. From the results, the classification accuracy decreased for almost all five datasets used in the experiment. The worst deterioration was when the classification accuracy decreased by 0.20 for BiLSTM in essay set five, 0.19 for BiGRU in essay set two, and 0.12 for simple BiRNN in essay set five.

6.1   Improving Robustness with Adversarial Data Augmentation and Training

Since the aim of this paper is to build models that are robust to synonym-based perturbations, we proposed synonym-based attacks and synonym-based data augmentation as a defense strategy. The comparison and results on the influence of synonym-based perturbations on the robustness of the model were presented above (Table 1). The next step is to examine how synonym-based adversarial data augmentation and training affect the robustness of the model against synonym-based adversarial attacks. Table 1 shows the results of augmented adversarial training, which is a typical method for using adversarial examples to improve models. Specifically, we augmented the generated adversarial examples with the original dataset and split the augmented dataset into two subsets, one of these set is used for training model and the other set is for testing model. Using the augmented training set, we re-train three NN-based AES models and test them with the test set. As we can see in 1, augmenting the original set with the synonym-based adversarial generation improves the performance much better than the models trained with the original data only.

The results show that augmenting the original set with adversarial examples and using them as a defense strategy can improve the performance of AES models. The classification accuracy of the BiLSTM-based model was increased by 0.13, 0.18, 0.15, 0.13, and 0.08 for essays one through five, respectively. Similarly, the classification accuracy of the BiGRU-based model was also increased by 0.14, 0.17, 0.16, 0.12, and 0.12 for essays one through five, respectively, compared to the models without attack. As shown in Table 1, the classification accuracy of all models decreased on average by 0.14, 0.12, and 0.07 for BiLSTM-, BiGRU-, and BiRNN-based AES models, respectively, when the models were attacked. However, when the model was trained by augmenting examples generated by replacing synonyms, the performance of each model increased by 0.28, 0.26, and 0.21 on average for BiLSTM-, BiGRU-, and BiRNN-based models, respectively. The most interesting part is an increase in accuracy between models trained and evaluated with the original set and the augmented set. For BiLSTM, BiGRU, and BiRNN, an average performance increase of 0.13, 0.14, and 0.41, respectively, was observed between the models trained and evaluated with the original data and the models trained and evaluated with the augmented data. The results suggest that defense is not only beneficial in adversarial situations, but also in situations without adversarial attacks to train more robust scoring models.

7   Conclusions

In this work, we proposed a synonym-based essay generation to evaluate the vulnerability of NN-based AES systems. Moreover, we proposed a synonym-based essay generation and augmentation method as a defense strategy capable of improving the performance and robustness of NN-based AES models. Our results show that extracting the most important words from the essay and replacing important words in the input samples with their synonyms, and generating adversarial samples for augmentation can significantly improve the generalization of NN-based AES models. Our experimental results show that the proposed defense is not only able to defend against adversarial attacks, but also improve the performance of NN-based AES models. On average, the proposed defense improved the classification accuracy of BiLSTM-, BiGRU-, and BiRNN-based models by 0.28, 0.26, and 0.21, respectively, when tested against adversarial attacks. The experimental results also show that our defensive method can improve the robustness of the models with simple data augmentation operations and achieve an average classification accuracy of the BiLSTM-, BiGRU-, and BiRNN-based models of 0.13, 0.14, and 0.14, respectively, when the models are not attacked. We believe that our work can establish a reasonable evaluation protocol and provide a competitive basis for future work to improve the robustness of NN-based AES models. The future direction of our work will be to extend our approach to other types of adversarial perpetuation in deep learning-based scoring models.

Acknowledgements. This research is supported by the ÚNKP-21-4 New National Excellence Program of the Ministry for Innovation and Technology from the source of the National Research, Development and Innovation Fund.

References

1. Alshemali, B., Kalita, J.: Generalization to mitigate synonym substitution attacks. In: Proceedings of Deep Learning Inside Out (DeeLIO): The First Workshop on Knowledge Extraction and Integration for Deep Learning Architectures, pp. 20–28 (2020)
2. Alshemali, B., Kalita, J.: Improving the reliability of deep neural networks in NLP: a review. Knowl.-Based Syst. **191**, 105210 (2020)
3. Alzantot, M., Sharma, Y., Elgohary, A., Ho, B.J., Srivastava, M., Chang, K.W.: Generating natural language adversarial examples. In: Proceedings of the 2018 Conference on Empirical Methods in Natural Language Processing, pp. 2890–2896. Association for Computational Linguistics, Brussels, Belgium (2018)
4. Bayer, M., Kaufhold, M., Reuter, C.: A survey on data augmentation for text classification. CoRR abs/2107.03158 (2021)
5. Belinkov, Y., Bisk, Y.: Synthetic and natural noise both break neural machine translation. In: 6th International Conference ICLR. OpenReview.net (2018)
6. Devlin, J., Chang, M.W., Lee, K., Toutanova, K.: BERT: pre-training of deep bidirectional transformers for language understanding. In: Proceedings of the 2019 Conference of the North American Chapter of the Association for Computational Linguistics, pp. 4171–4186. Association for Computational Linguistics (2019)
7. Alikaniotis, D., Yannakoudakis, H., Rei, M.: Automatic text scoring using neural networks. In: Proceedings of the 54th Annual Meeting of the Association for Computational Linguistics, pp. 715–725 (2016)
8. Dong, F., Zhang, Y.: Automatic features for essay scoring–an empirical study. In: Proceedings of the 2016 Conference on Empirical Methods in Natural Language Processing, pp. 1072–1077. Association for Computational Linguistics (2016)
9. Dong, F., Zhang, Y., Yang, J.: Attention-based recurrent convolutional neural network for automatic essay scoring. In: Proceedings of the 21st Conference on Computational Natural Language Learning, pp. 153–162. Association for Computational Linguistics (2017)
10. Ebrahimi, J., Rao, A., Lowd, D., Dou, D.: HotFlip: white-box adversarial examples for text classification. In: Proceedings of the 56th Annual Meeting of the Association for Computational Linguistics (Volume 2: Short Papers), pp. 31–36 (2018)
11. Gao, J., Lanchantin, J., Soffa, M.L., Qi, Y.: Black-box generation of adversarial text sequences to evade deep learning classifiers. In: 2018 IEEE Security and Privacy Workshops (SPW), pp. 50–56. IEEE (2018)

12. Hernández-García, A., König, P.: Data augmentation instead of explicit regularization (2020)
13. Jia, R., Liang, P.: Adversarial examples for evaluating reading comprehension systems. In: Proceedings of the 2017 Conference on EMNLP, pp. 2021–2031. Association for Computational Linguistics (2017)
14. Jia, R., Liang, P.: Adversarial examples for evaluating reading comprehension systems. arXiv preprint arXiv:1707.07328 (2017)
15. Ke, Z., Ng, V.: Automated essay scoring: a survey of the state of the art. In: Proceedings of the Twenty-Eighth International Joint Conference on Artificial Intelligence, pp. 6300–6308 (2019)
16. Pennington, J., Socher, R., Manning, C.D.: Glove: global vectors for word representation. In: Empirical Methods in Natural Language Processing (EMNLP), pp. 1532–1543 (2014)
17. Pham, T., Bui, T., Mai, L., Nguyen, A.: Out of order: How important is the sequential order of words in a sentence in natural language understanding tasks? In: Findings of the Association for Computational Linguistics: ACL-IJCNLP 2021, pp. 1145–1160. Association for Computational Linguistics (2021)
18. Ren, S., Deng, Y., He, K., Che, W.: Generating natural language adversarial examples through probability weighted word saliency. In: Proceedings of the 57th Annual Meeting of the Association for Computational Linguistics, pp. 1085–1097 (2019)
19. Shorten, C., Khoshgoftaar, T.M.: A survey on image data augmentation for deep learning. J. Big Data **6**(1), 1–48 (2019). https://doi.org/10.1186/s40537-019-0197-0
20. Singla, Y.K., Bhatia, M., Kabra, A., Li, J.J., Jin, D., Shah, R.R.: Calling out bluff: attacking the robustness of automatic scoring systems with simple adversarial testing. ArXiv abs/2007.06796 (2020)
21. Srivastava, N., Hinton, G., Krizhevsky, A., Sutskever, I., Salakhutdinov, R.: Dropout: a simple way to prevent neural networks from overfitting. J. Mach. Learn. Res. **15**, 1929–1958 (2014)
22. Szegedy, C., et al.: Intriguing properties of neural networks. In: Bengio, Y., LeCun, Y. (eds.) 2nd International Conference on Learning Representations, ICLR 2014, Banff, AB, Canada, 14–16 April 2014, Conference Track Proceedings (2014)
23. Tashu, T.M., Horváth, T.: Smartscore-short answer scoring made easy using sem-LSH. In: 2020 IEEE 14th International Conference on Semantic Computing (ICSC), pp. 145–149 (2020). https://doi.org/10.1109/ICSC.2020.00028
24. Tay, Y., Phan, M.C., Tuan, L.A., Hui, S.C.: Skipflow: Incorporating neural coherence features for end-to-end automatic text scoring. In: Proceedings of the Thirty-Second AAAI Conference on Artificial Intelligence, pp. 5948–5955. AAAI Press (2018)
25. Taylor, L., Nitschke, G.: Improving deep learning using generic data augmentation. CoRR (2017), http://arxiv.org/abs/1708.06020
26. Zhou, W., Ge, T., Xu, K., Wei, F., Zhou, M.: BERT-based lexical substitution. In: Proceedings of the 57th Annual Meeting of the Association for Computational Linguistics, pp. 3368–3373 (2019)

Characterizing Cardiovascular Risk Through Unsupervised and Interpretable Techniques

Hugo Calero-Díaz⬤, David Chushig-Muzo$^{(\boxtimes)}$⬤, and Cristina Soguero-Ruiz⬤

Department of Signal Theory and Communications, Telematics and Computing
Systems, Rey Juan Carlos University, Madrid 28943, Spain
h.calero.2017@alumnos.urjc.es, {david.chushig,cristina.soguero}@urjc.es

Abstract. In the last decades, the prevalence of noncommunicable diseases has significantly increased, causing millions of deaths worldwide.
Among them, cardiovascular diseases (CVDs) have become a global
health problem due to the high rates of mortality and morbidity. This
issue is exacerbated in patients with type 1 diabetes (T1D) since they
have a higher risk of developing CVD and death. Machine Learning (ML)
methods have reached remarkable results in both industry and academia.
In the clinical setting, these methods are promising to extract clinical knowledge and help to prevent acute events caused by CVDs. This
paper aims to identify clusters of patients with different CVD risks and
interpret the clinical variables that play a significant role in CVD risk.
Towards that end, unsupervised and interpretable ML techniques were
considered, in particular k-prototypes and the agglomerative hierarchical
clustering technique with Gower's distance. We used a dataset that collects information from 677 adult Danish patients with T1D. Experimental results showed that albuminuria, smoking, and gender are crucial for
identifying T1D patients at different risks of suffering CVD. Our paper
contributes to the identification of CVD risk factors for T1D patients,
paving the way for clinical decision-making. This identification of individuals at higher CVD risk would allow physicians to make early interventions and adequate treatments to prevent the onset of diseases.

Keywords: Cardiovascular disease · Diabetes · Unsupervised
methods · k-prototypes · Hierarchical clustering · Interpretability

1 Introduction

In recent decades, noncommunicable diseases (NCDs) have become a health
problem of global importance. According to the World Health Organization,
these diseases are the leading causes of disability and mortality worldwide [1].
Among these diseases, cardiovascular diseases (CVD) and diabetes are the most
prominent, causing 1,9 and 1,5 million deaths per year, respectively [1]. The
prevalence of these diseases has dramatically increased due to heterogeneous factors, including longer life expectancy, smoking, harmful use of alcohol, unhealthy

© The Author(s), under exclusive license to Springer Nature Switzerland AG 2022
H. Yin et al. (Eds.): IDEAL 2022, LNCS 13756, pp. 22–30, 2022.
https://doi.org/10.1007/978-3-031-21753-1_3

diet, physical inactivity, among others [2]. NCDs are not only a health issue but also an economic issue. Patients diagnosed with NCDs frequently visit practitioners, and in general, they require long-term supervision and care, causing high consumption of health resources. People with type 1 diabetes (T1D) are at elevated risk of CVD and mortality compared to healthy populations [2].

Thanks to clinical research, several CVD risk factors including age, hypertension, family history of CVD, lack of physical activity, smoking, and abdominal adiposity have been identified and extensively studied. The clinical evidence and understanding of these factors have led to the development of prediction models for assessing the risk of developing CVD. This risk stratification becomes key to conducting an appropriate preventive strategy, aiming to make early interventions and prevent the onset of CVD. In the state-of-the-art, several CVD risk engines have been created, including the Framingham risk score [3], United Kingdom prospective diabetes study risk engine. However, most of these approaches are built using information from general populations. In the last decade, three risk scores have been specifically developed for estimating CVD risk in people with T1D [4]. Among them, the Steno Type 1 Risk Engine (ST1RE) has been internationally used in multiple clinical studies [4].

Data-driven approaches have become a reliable option to identify patients with a higher risk of developing CVD and provide an efficient way to identify risk factors. The implementation of these methods could potentially have important implications on the health care and patient quality of life [5]. In particular, ML methods have shown great performance to analyze and find patterns from large and heterogeneous data [5]. The potential use of ML models is not only limited to the development of predictive models, since they have also been successfully used for identifying clusters of patients with specific clinical characteristics [6]. In this work, the aim is to find clusters of T1D patients with different CVD risks and identify risk factors involved in the development of CVD by using unsupervised and interpretable techniques based on ML.

2   Materials and Methods

2.1   Dataset Description

In this paper, we work with a dataset that is publicly available and collects information from 677 Danish patients diagnosed with T1D [4]. It is composed of 10 variables, 2 demographics (age, gender), 2 variables linked to patients' lifestyle (smoking, exercise), and 6 clinical variables, including: diabetes duration (in years), systolic blood pressure (SBP), low-density lipoprotein (LDL), glycosylated hemoglobin (Hba1c), estimated glomerular filtration rate (EGFR) and albuminuria. All clinical variables are continuous, except albuminuria which is composed of three categories: normoalbuminuria, microalbuminuria, and macroalbuminuria. These types of albuminuria differ in the urinary albumin creatinine ratio, with $<30\,mg/g$, 30–$299\,mg/g$, and $\geq 300\,mg/g$ for normoalbuminuria, microalbuminuria, and macroalbuminuria, respectively. Albuminuria was transformed using one-hot encoding. The ST1RE [4] was used to calculate the 10-year CVD risk and the 5-year risk of end-stage kidney disease (ESKD). For 10-year CVD risk and ESKD, the dynamic range is between $[0.021, 0.909]$, $[0.001, 0.049]$, respectively. As observed, the ESKD risk of the patients in the studied cohort is quite low. The clinical relationship between albuminuria and ESKD is also worthy to mention. According to clinical guidelines [7], kidney disease is caused by different factors, with albuminuria being one the most relevant.

2.2   Unsupervised and Interpretable Methods

Unsupervised techniques are used to discover underlying data structure and identify hidden patterns. Clustering is an unsupervised learning task that aims to group observations into k_c clusters by maximizing the similarity between observations (similar characteristics) [8]. Many clustering algorithms have been proposed and applied to multiple problems and domains [8]. Since our data is composed of categorical and continuous features, it is necessary to choose methods that adequately work with mixed-type data. In this paper, k-prototypes and the agglomerative hierarchical clustering (AHC) were considered [8]. The former is a variant of k-means designed to handle mixed data [9], and the latter was combined with Gower's distance [10] to cluster mixed-type data. The k-prototypes algorithm relies on 3 stages [9]. Firstly, k prototypes are randomly selected, and then, each observation is assigned to one of the k clusters according to its nearest prototype. Next, new k prototypes are obtained by computing the similarity of observations against the current prototypes. These last 2 steps are repeated until the difference between the new and previous prototypes is not significant. For the AHC's algorithm, all observations are initially treated as a unique cluster, which are then merged successively until a single cluster gathers all observations [8]. Several linkage criteria are available as merging strategies and thus determine the observations present in each cluster at every iteration. The average linkage was considered, which works through the minimization of the sum of squared differences within all the clusters [8]. It is worth to mention that a min-max scaling procedure has been carried out to scale continuous variables.

The selection of the most appropriate number of clusters is key in clustering applications. The cluster validity indices (CVIs) are extensively used in the literature for this purpose [11]. Generally, CVIs aim to measure separation and compactness [11]. The separation measures the distance between clusters, and compactness quantifies the distance of the observations within the cluster [11]. The silhouette coefficient (SC) (ranged between $[-1, 1]$) and the Davies-Bouldin score (DBS) were considered as CVIs in this study. Higher values in SC and smaller values of DBS indicate a better clustering performance [11]. Both CVIs were selected because they quantify both the intra-cluster and inter-cluster distances [11]. Thus, they allow us to measure how compact are samples within their corresponding clusters, and how separated are the clusters between them.

In recent years, the search for explainable ML approaches has been an important research area. Interpretability is defined as the degree to which a human can understand the causes of decisions reached [16], which is crucial in certain fields such as medical research. Most approaches for gaining interpretability have mainly focused on supervised learning methods [12], aiming to understand the way how predictions are made. A few works have been proposed for unsupervised learning. Recently, some authors have focused on the generation of tree-based rules for gaining interpretability in unsupervised and clustering approaches. One of the most remarkable works was the expanding explainable k-means clustering (ExKMC) [12]. This method builds a tree-based schematic by explaining each cluster assignment by a small number of thresholds, each depending on a single feature (single-variable thresholds) [12]. The algorithm leverages the Mixed Integer Optimization techniques [12] to generate an interpretable tree-based schematic. Each partition of the feature space is represented as a new node in the tree-like structure. Nodes that are further away from leaves can be understood as the most relevant, since they partition the greatest number of samples. ExKMC only works with numerical variables, and to work with mixed-type data, the algorithm was modified by using prototypes.

3   Experimental Results

In this section, we present the results of applying k-prototypes and the AHC with Gower's distance, as well as the characterization of clusters found. Additionally, a statistical analysis was performed to check the differences between clusters.

3.1   Characterization of Clusters and CVD Risk Analysis

For the clustering methods, the aforementioned CVIs were computed to select the most appropriate number of clusters. The SC and the DBI pointed out $k_c = 3$ and $k_c = 6$ as the number of clusters for the AHC and k-prototypes, respectively. The clustering methods were trained considering these values. To search for differences between the clusters found, a statistical approach was considered. For binary variables, the two-proportion Z test was used, which studies if the difference between the proportions of a specific parameter taken from two independent samples is significantly different. For continuous variables, a two-step procedure was followed. Firstly, the Shapiro-Wilks test was applied to choose between a parametric and nonparametric test. Then, to determine if the means of two populations are equal, the t-test (for variables that are normally distributed) and the Mann-Whitney U test were considered. It is important to consider that the level of significance imposed on statistical tests is 0.05. Additionally, to complement the cluster characterization, a correlation analysis between the variables of each cluster and the 10-year CVD risk was carried out. The point-biserial correlation coefficient (PBCC) (for binary against continuous variables) and the Spearman correlation coefficient (SPCC) (for continuous variables) were considered. Both measure linear correlation and are ranged between $[-1, 1]$, where higher absolute values indicate high linear correlation.

Firstly, we analyze the AHC with $k_c = 3$. Figure 1 shows the barplots and normalized histograms linked to each categorical/continuous variable for each cluster. To identify the difference between clusters, these were depicted in different colors. Figure 1 revealed that albuminuria significantly contributes to clustering results, obtaining clusters that gather patients with a single type of albuminuria: cluster 0 (microalbuminuria), cluster 1 (normoalbuminuria) and cluster 2 (macroalbuminuria). Table 1 summarizes the mean and standard deviation for age, 10-year CVD risk, and ESKD risk. As observed, cluster 2 presented the highest 10-year CVD risk, followed by cluster 0 and cluster 1. Regarding the ESKD risk, cluster 0 and cluster 1 presented similar values (see Table 1) and the t-test provided a p-value = 0.192 (no significant risk differences between these clusters). The main statistical differences between macroalbuminuric and microalbuminuric patients were observed in the variables age and Hba1c, with a p-value of 0.026 and 0.044, respectively. Only adult patients presented macroalbuminuria (over 28 years). Patients in cluster 2 (macroalbuminuria) presented significantly higher levels of Hba1c than people in cluster 0 (microalbuminuria). HbA1c is an indicator of blood glucose levels and is highly linked with a long-term CVD risk in people with T1D [13].

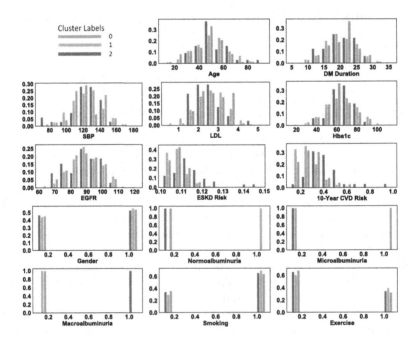

Fig. 1. Barplots and histograms of variables with AHC and $k_c = 3$.

Table 1. Number of patients, and mean ± standard deviation of age, ESKD risk and 10-year CVD risk for each cluster (considering AHC and $k_c = 3$).

Cluster	# patients	Albuminuria	Age	10-year CVD risk	ESKD risk
0	72	Microalbuminuria	47.063 ± 12.713	0.197 ± 0.109	0.009 ± 0.004
1	573	Normoalbuminuria	50.264 ± 13.734	0.163 ± 0.101	0.008 ± 0.005
2	32	Macroalbuminuria	53.387 ± 14.041	0.333 ± 0.167	0.015 ± 0.010

Now, we perform the characterization of clusters found when considering the k−prototypes technique and $k_c = 6$. In Fig. 2, we show the barplots and histograms associated with the categorical/continuous variables. As in the previous analysis, 3 clusters were obtained, each one with patients suffering from a type of albuminuria: cluster 0 (normoalbuminuria), cluster 1 (microalbuminuria), and cluster 2 (macroalbuminuria). The remaining clusters were composed of patients with mixed types of albuminuria (see Table 2). It is worth noting that cluster 2 (macroalbuminuria), and cluster 4 (normoalbuminuria and microalbuminuria) present the highest values for 10-year CVD risk (see Table 2). These patients also present a greater age mean (over 53 years) compared to other clusters. Similarly to the scenario with the AHC and $k_c = 3$, the 3 clusters with a single type of albuminuria were significantly different in CVD risk, with a p-value≃0.0 for the t-test. ESKD risk differs in all cases except when comparing cluster 0 (normoalbuminuria) and cluster 1 (microalbuminuria). Note that patients of these clusters present a similar age mean. Moreover, differences in smoking, exercise, age, gender and Hba1c were the most recurrent ones. It is remarkable to observe that CVD risk was statistically different in all cases except when comparing the microalbuminuric cluster against the mixed ones (mainly composed of patients with normoalbuminuria). Even though the previous analysis proved that albuminuria itself is useful for the stratification of CDV risk, insights obtained here suggest that when pairing it with information from other variables, a more robust stratification is achieved. To gain more interpretability, the tree-based schematic obtained with ExKMC is depicted in Fig. 3. As shown, albuminuria is crucial to distinguish clusters 0, 1, and 2, while the variables smoking, exercise and gender help to identify other clusters. Several studies have identified both microalbuminuria and macroalbuminuria as relevant risk factors that contribute to developing CVD [14]. In [15], clinical evidence showed that macroalbuminuria increases the risk of CVD independently of other factors such as age, diabetes duration, and HbA1c. Smoking is a well-studied modifiable risk factor that contributes to the development of acute CVD events [4]. These risk factors identified in our work are supported by clinical evidence.

In the clinical setting, several studies have identified age as an important risk factor for developing CVD [13]. To complement this work, a cluster characterization using demographic variables was conducted. First, we split the population by gender (male and female), obtaining a significant statistical difference between the clusters in ESKD risk and 10-year CVD risk with p-value<0.01 for the t-test. Regarding the correlation of the variables with the 10-year CVD risk, SBP

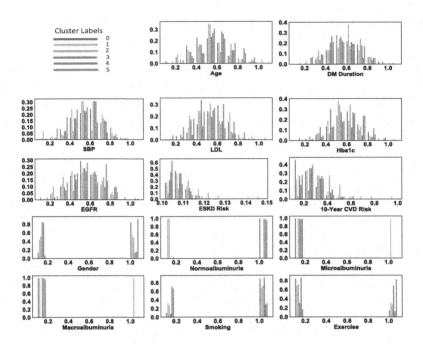

Fig. 2. Barplots and histograms when considering k−prototypes and $k_c = 6$.

(SPCC $= 0.062, 0.149$ for males and females, respectively), microalbuminuria (PBCC $= 0.015, 0.125$ for males and females, respectively), smoking (PBCC $= 0.0220, 0.057$ for males and females, respectively), and ESKD risk (SPCC $= -0.074, -0.173$ for males and females, respectively) result significantly different. According to correlation values, the variables in the female population appear to be more related to 10-year CVD risk than the group of male patients. Since there are significant differences between men and women in the CVD risk and the ESKD risk this can indicate that the pathophysiology of CVD and T1D differs by gender due to other factors not studied in this work. It opens the way for further clinical studies on the pathogenesis of CVD by gender.

By analyzing the differences by age, four clusters split into the following age ranges were studied: young patients (YP) with individuals younger than 25 years; middle-aged patients (MP) aged between 25 and 50 years; adult patients (AP) aged between 50 and 60 years; and elderly patients (EP) that gather individuals than 60 years. These clusters gathered 23, 315, 190, and 149 patients, respectively. According to the statistical tests, only a few variables show significant differences between pairs of clusters. This was the case for: *(i)* smoking between MP and EP (p-value = 0.0043); *(ii)* microalbuminuria between YP and EP (p-value = 0.017), and between AP and EP (p-value = 0.096); and *(iii)* diabetes duration comparing MP and EP (p-value = 0.052). Through the correlation analysis, it was observed that significant differences appear when studying either YP or EP against the rest of the clusters. For the remaining clusters, the correlation

Table 2. Number of patients, and mean ± standard deviation of age, ESKD risk, and 10-year CVD risk for each cluster (considering k−prototypes and $k_c = 6$.)

Cluster	# patients	Albuminuria	Age	10-year CVD Risk	ESKD Risk
0	222	Normoalbuminuria	45.871 ± 13.389	0.136 ±0.089	0.009 ± 0.005
1	72	Microalbuminuria	46.258 ± 12.501	0.190 ± 0.104	0.009 ± 0.004
2	32	Macroalbuminuria	53.701 ± 14.702	0.340 ± 0.175	0.016 ± 0.010
3	178	Normo & Micro	50.887 ± 12.659	0.177 ± 0.099	0.009 ± 0.005
4	76	Normo & Micro	56.217 ± 13.505	0.209 ± 0.125	0.007 ± 0.003
5	97	Normo & Micro & Macro	53.013 ± 12.865	0.166 ± 0.090	0.006 ± 0.004

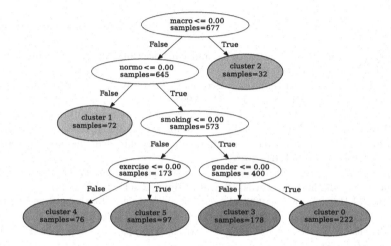

Fig. 3. Explainable clustering results (with k-prototypes and $k_c = 6$) through the decision tree of ExKMC. The number of samples in each leaf/cluster is shown.

of macroalbuminuria increases with age. Finally, Hba1c, SBP and LDL were relevant for characterizing patients belonging to the YP.

4   Conclusions

In this work, we analyze the effectiveness of unsupervised ML techniques to identify T1D patients with significant differences in CVD risk. ML techniques, in particular, k-prototypes and AHC allowed us to distinguish clusters of patients with significant differences in CVD risk. Regarding risk factors, albuminuria, smoking, and gender were identified as relevant in the development of CVD in T1D patients. In particular, patients with macroalbuminuria present higher risks than those with microalbuminuria and normoalbuminuria. It is also important to note that only adult patients presented macroalbuminuria, suggesting a link between age and abnormal blood albumin levels. Our paper contributes to the identification of individuals with higher CVD risk, supporting physicians to make early interventions and adequate treatments to prevent the onset of CVDs.

Acknowledgements. This work was partly funded by the Spanish Government (grant AAVis-BMR PID2019-107768RA-I00); by Project Ref. 2020-661, financed by Rey Juan Carlos University and the Community of Madrid; and by the Young Researchers R&D Project Ref. F861 (AUTO-BA-GRAPH) funded by the Community of Madrid and Rey Juan Carlos University.

References

1. Harris, R.E.: Epidemiology of chronic disease: global perspectives. Jones Bartlett Learn. (2019)
2. Schnell, O., et al.: Type 1 diabetes and cardiovascular disease. Cardiovasc. Diabetol. **12**(1), 1–10 (2013)
3. Wilson, P.W., et al.: Prediction of coronary heart disease using risk factor categories. Circulation **97**(18), 1837–1847 (1998)
4. Vistisen, D., et al.: Prediction of first cardiovascular disease event in type 1 diabetes mellitus: the steno type 1 risk engine. Circulation **133**(11), 1058–1066 (2016)
5. Shameer, K., et al.: Machine learning in cardiovascular medicine: are we there yet? Heart **104**(14), 1156–1164 (2018)
6. Chushig-Muzo, D., et al.: Learning and visualizing chronic latent representations using electronic health records. BioData Mining **15**(1), 1–27 (2022)
7. de Boer, I.H., et al.: Kdigo 2020 clinical practice guideline for diabetes management in chronic kidney disease. Kidney Int. **98**(4), S1–S115 (2020)
8. Rodriguez, M.Z., et al.: Clustering algorithms: a comparative approach. PLoS ONE **14**(1), e0210236 (2019)
9. Hsu, C.-C., Lin, S.-H., Tai, W.-S.: Apply extended self-organizing map to cluster and classify mixed-type data. Neurocomputing **74**(18), 3832–3842 (2011)
10. Foss, A.H., Markatou, M., Ray, B.: Distance metrics and clustering methods for mixed-type data. Int. Stat. Rev. **87**(1), 80–109 (2019)
11. Arbelaitz, O., et al.: An extensive comparative study of cluster validity indices. Pattern Recogn. **46**(1), 243–256 (2013)
12. Frost, N., Moshkovitz, M., Rashtchian, C.: Exkmc: expanding explainable k-means clustering, arXiv preprint arXiv:2006.02399 (2020)
13. Vergès, B.: Cardiovascular disease in type 1 diabetes: a review of epidemiological data and underlying mechanisms. Diabetes Metabolism **46**(6), 442–449 (2020)
14. Gerstein, H., et al.: Albuminuria and risk of cardiovascular events, death, and heart failure in diabetic and nondiabetic individuals. JAMA **286**(4), 421–426 (2001)
15. Cederholm, J., et al.: A new model for 5-year risk of cardiovascular disease in type 1 diabetes; from the swedish national diabetes register (ndr). Diabet. Med. **28**(10), 1213–1220 (2011)
16. Montavon, G., Samek, W., Müller, K.R.: Methods for interpreting and understanding deep neural networks. Digit. Sig. Proc. **73**, 1–15 (2018)

Identification of Sedimentary Strata by Segmentation Neural Networks of Oblique Photogrammetry of UAVs

Daniel Theisges dos Santos[1]([⊠])[iD], Mauro Roisenberg[1], and Marivaldo dos Santos Nascimento[2]

[1] Department of Statistics and Computer Science, Federal University of Santa Catarina, Florianopolis, SC 88034-400, Brazil
`daniel.theisges@posgrad.ufsc.br`, `mauro.roisenberg@ufsc.br`
[2] Department of Geology, Federal University of Santa Catarina, Florianopolis, SC 88034-400, Brazil
`marivaldo.nascimento@ufsc.br`

Abstract. The determination of the shape and extent of strata is one of the most important steps in the classification of architectural elements in an outcrop. However, mapping and accessing outcrops with broad lateral and vertical continuity can be difficult due to their geographic position. In this work, UAV-captured outcrop images are used to apply seeding methods to identify and measure strata size. A computational system based on Segmentation Neural Networks was used as an effective method to automatically identify outcrop strata. The results show that the DeepLabV3 semantic segmentation neural network can achieve interesting results when combined with the PointRend technique. The different combinations of architectures were compared with UAV images from outcrops with turbidite systems of the Itararé Group (Paraná Basin, Brazil). The results show that machine learning approaches are alternative and efficient techniques that contribute to the improvement of traditional semi-supervised segmentation methods.

Keywords: Semantic segmentation · PointRend · DeepLab · Turbidite system

1 Introduction

The identification of the strata architecture is one of the main challenges in the analysis of sedimentary basins. Outcrops allow the investigation and identification of each stratum and are still the main data collected today [5]. The integration of outcrop studies can enhance the interpretation of complex reservoirs in deep-water depositional settings. The strata classification process is considered time-consuming and laborious, and the accuracy of stratum identification often depends on the experience and knowledge of experts in the field. In addition, the detailed identification of strata can be difficult when the outcrops are very

© The Author(s), under exclusive license to Springer Nature Switzerland AG 2022
H. Yin et al. (Eds.): IDEAL 2022, LNCS 13756, pp. 31–41, 2022.
https://doi.org/10.1007/978-3-031-21753-1_4

extensive or are in terrain with difficult to access. Several approaches have been used applying remote sensing techniques, which allow the acquisition of data from an object without being in direct contact with it. From this technology, image segmentation methods can be applied to differentiate strata based on field photos [12, 22, 33], using three-dimensional images behind devices such as LiDAR to build a 3D model of outcrop strata and use geological information to differentiate strata and classify lithology [4, 6, 14, 23].

Digital photogrammetry tools and Unmanned Aerial Vehicles (UAVs) are methods that have attracted much attention for application of sensing in geosciences [13, 31]. UAVs have been adopted by the geoscience community due to their ability to capture high-resolution data remotely, quickly, and inexpensively in mountainous areas, where in many cases accessing rocky outcrops is difficult [15, 34]. Commercial models of UAVs [26] generally acquire Red-Green-Blue, RGB [11], infrared and NIR images [8]. However, this volume of data makes manual analysis of captured images time consuming. Some studies use different approaches to image segmentation. While some use semi-supervised algorithms, where the user must demarcate the area with characteristics to be grouped, other studies use deep learning for identification in a supervised way, without human intervention.

New Deep Learning architectures (e.g. Convolutional Neural Networks/ CNN) were developed to perform semantic instance or panoptic segmentation tasks [18], object detection [21] and instance segmentation [3]. The use of such architectures has highly accurate results and they are commonly used in many disciplines within the field of computer vision. The use of supervised strata identification methods can contribute a lot to the area of geology, since once a neural network is correctly trained, its use does not depend on the interpretation of specialists.

In this work we chose to evaluate methods of semantic segmentation of deep neural networks, especially architectures that currently reach high metrics. The performance of the U-net and DeepLab topologies are compared using a custom dataset created from images captured from UAVs of an outcrop belonging to a turbidite system. These neural network architectures identify discontinuities or changes in visual cues such as color and texture (grain size), thus allowing to predict mineralogical changes between stratigraphic units.

2   Theoretical Foundations and Related Works

One of the main procedures used on digital image processing is segmentation, where the image is split into its constituent parts or objects. In the literature, there are different well-known methods used for segmentation, such as clustering, thresholding, segmentation using neural network and segmentation using region growing.

The approaches known as clustering are also known as unsupervised, which do not need prior training on the data, but need the intervention of the expert during their use [29]. Several semi-supervised segmentation techniques were used to detect lithological boundaries of remote sensing data, using superpixel and region growing approaches from user-selected colors. Other works have similar approaches, such as the rotation variant model matching (RTM) algorithm [25, 27] and the Walsh transform [19].

A pilot study by [20] shows that, although the automated technique detects some of the boundaries well, it produced over-segmentation in some lithologies. [32] proposed a method using a superpixel algorithm [1], merging each superpixel with neighboring superpixels based on their color similarity and in the next step applying the growth of regions specified by the user-marked lines to determine different lithological units.

Semi-supervised methods are dependent on human interaction, which can lead to different segmentation results for different application users. In this way, supervised machine learning segmentation can be an effective way to reduce this type of problem. Recent advances in deep learning and convolutional neural architectures, along with the availability of large-scale annotated datasets, are useful for the impressive performance of recent semantic segmentation algorithms [35].

As image segmentation technology based on deep learning methods has become more and more mature in the field of earth sciences, studies comparing traditional image segmentation methods with machine learning based methods for rock analysis have been carried out [35]. Andrew [2] mentioned that machine learning methods may change our ability to extract information from earth science images. In recent years, many researchers have applied the advantages of deep learning to lithology recognition [16,28,36], fracture identification in geological disasters such as landslides [17,37], fracture identification in auxiliary logging [10,30] and automatic outcrop cavity extraction method based on the Mask R-CNN architecture [35].

The purpose of this work is to evaluate the use of CNNs structures to perform the outcrop segmentation task with images captured from UAVs. Previous works use CNNs in similar tasks or use other approaches to identify lithofacies in outcrops. Correct identification of the facies transition is important in this problem and ways to improve this process are analyzed.

3   Data and Methods

3.1   Segmentation Architecture

Stratigraphic structures can be described at different scales, from pore/plug scale to system stack scale passing by lithology bed stack and depositional architectural elements identification. The depositional process occurs over time, making it possible to visually identify the different vertical sequences and their horizontal continuities. In this study we are considering the ability of different segmentation neural network architectures to be able to identify these patterns. It is known that for object segmentation, where there are large differences in contrast at the edges of each object, such methods present good results. Here we evaluated the outcrops, where the transition of colors and texture is smoother between the elements.

Different architectures for segmentation have been proposed over the years. The U-net has a similar design of an encoder and a decoder [24]. The former is used to extract features by downsampling, while the latter is used for upsampling the extracted features using the deconvolutional layers. Unlike U-net, which uses features from every convolutional block and then concatenates them with their corresponding deconvolutional block, DeepLab uses features yielded by the last convolutional block before upsampling. Features are extracted from the backbone network (VGG, DenseNet, ResNet) [7].

The DeepLabV3 architecture brought important advances to perform the semantic segmentation task. The aim of DeepLabV3 was to capture sharper object boundaries. This was achieved by adopting the encoder-decoder architecture with atrous convolution in the last few blocks of the backbone. This method modifies the kernel by inserting gaps/spaces between the elements. On top of extracted features from the backbone, an Atrous Spatial Pyramid Pooling (ASPP) network is added to classify each pixel corresponding to their classes. The output from the ASPP network is passed through a 1×1 convolution to get the actual size of the image which will be the final segmented mask for the image [7].

Segmentation models can tend to generate over-smooth boundaries which might not be precise for objects or scenes with irregular boundaries. To get a crisp segmentation boundary, a point-based rendering neural network module called PointRend was proposed. A PointRend module consists of three main components: (i) A point selection strategy chooses a small number of real-value points to make predictions. (ii) For each selected point, a point-wise feature representation is extracted. (iii) A point head: a small neural network trained to predict a label from this point-wise feature representation, independently for each point.

The architecture used in this work combining DeepLab V3 and PointRend can be seen in Fig. 1. The network receives images in the size of 512×512 with 3 channels as input. The encoder backbone is a ResNet network [16], which uses residual blocks and downsampling blocks at each layer. Contextual information at multiple scales is acquired by Atrous Spatial Pyramid Pooling (ASPP). Based on the original idea of PointRend, the decoder consists of two parts: upsampling layers (for 4 times) and PointRend which will retrieve the learned features to the detailed road boundary. Output labels are predicted by a point-head network from point-to-point resources.

3.2 Dataset

The database used in this work was built from images of an outcrop of a Permian turbidite system of the Itararé Group (Paraná Basin; Fig. 2A), witch records permocarboniferous sedimentation in Western Gondwana Supercontinent. The choice of this outcrop for this work was motivated by the importance of tur-bidite systems studies for petroleum geology. In addition, the outcrop presents wide lateral and vertical continuity, with depositional structures very well pre-served (Fig. 2B-C). A DJI Phantom 4 Advanced UAV from Basin and Reservoir

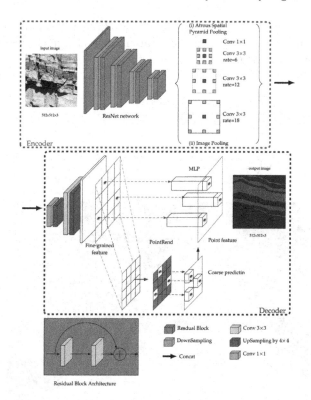

Fig. 1. Neural network implemented in research based on DeepLab V3, and PointRend.

Analysis Laboratory (LABAC) was used to collect 380 photographs at an average altitude of 30m outcrop. There was over 70% overlap between neighboring photographs, and the drone camera view was orthogonal to the target surface at most shooting points. After all the images were captured, these were used to build a high-resolution 3D model within EasyUAV version 2.1.

The EasyUAV software processed the image datasets according to the following procedures: (1) spatially align all images using information from the global positioning system (GPS); (2) optimize the alignment by applying a correction to the drone's camera lens parameters; (3) set point clouds, digital elevation model (DEM) and orthophoto output options; (4) build dense point clouds with SfM photogrammetry; (5) build a dense cloud mesh; and (6) generate a high resolution DEM and orthophoto.

From the oblique photogrammetry, a 3D mosaic of the outcrop was produced (Fig. 2B). The sedimentary facies of this turbidity system were classified into three categories (Fig. 2C): (1) medium to fine-grained laminated sandstones (sand-rich lobe facies); (2) massive and medium-grained sandstones (sand-rich channel facies); and (3) massive silt and black laminated shale (light facies). To train the machine learning models in different architectures, a dataset was built from the 3D mosaic, with a resolution of 23865 × 11600 pixels. Crop images with a resolution of 512×512 pixels were created, resulting in a dataset of 334 images. Each image resulting from this training was classified according to previously interpreted geological data from outcrop.

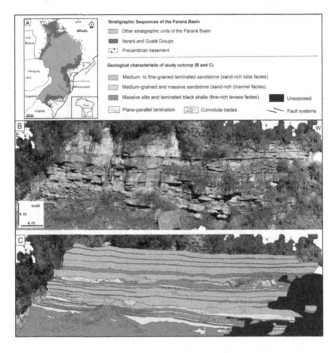

Fig. 2. (A) Study outcrop location in the Paraná Basin; (B) digital model west-east oriented view using photogrammetry; and (C) facies classification based on conventional mapping methods including a posteriori correlation.

The generated dataset was classified using the labeling tool for image segmentation LabelMe. LabelMe generates a json file counting the coordinates of each segment present within the image. From this, two scripts were created to convert the jsons into the COCO Database and Cityscape formats, formats expected by the instance threading architectures and semantic threading accordingly. The generated dataset with the corresponding classes can be seen in the Fig. 3.

4   Experiment and Discussion

4.1   Experiment

Several frameworks were used. The generated frameworks were used during the development of the project. However it was decided to implement the neural network proposed in Pytorch 1.9 and Python 3.7. In addition to allowing the implementation of neural network architectures, Pytorch has a library of known

Fig. 3. Images belonging to the generated dataset with the respective classes.

segmentation neural network architectures to carry out transfer learning. In this sense, the proposed neural network was compared with a standard U-net neural network. The neural network was trained on a GPU computer with an RTX 2060 GPU.

In addition to comparing the results with the U-net as a baseline, a study was carried out comparing the DeepLabV3 with and without the use of the PointRend structure. As a backbone, that is, a neural network structure to extract features, the ResNet50 structure was used, which uses residual blocks and downsampling blocks at each layers.

We chose not to use data augmentation mechanisms, following the same app-roach used in the original DeepLabV3 and PointRend publications to compare the feasibility of the methods. As evaluation metric we used Mask Intersection-over-Union (Mask IoU) and Boundary IoU [9]. Boundary IoU allows a more granular analysis of segmentation-related errors for the complex multifaceted tasks. To survey the results, five images were used to evaluate those metrics. As a training configuration, we used the Adadelta optimizer with the default settings and a learning rate of 0.0001. The training was done for 100 epochs.

Table 1 presents the results obtained for each architecture analyzed, using mIoU in five dataset images and three classes. It can be observed that the pro-posed architecture showed good results in relation to the U-net network, chosen as the baseline architecture. The application of the PointRend structure helped to improve the quality of the result, proving to be an interesting approach to the proposed problem. The Fig. 4 presents visual comparison between the three generated models for the five images.

Table 1. Metrics of extracted five images via different models.

Method	Mask IoU	Boundary IoU
U-net	80.0	69.0
DeepLabV3	84.6	75.1
DeepLabV3 + PointRend	85.5	82.0

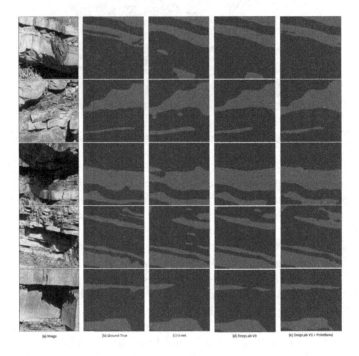

Fig. 4. Segmentation result for the 3 generated models.

4.2 Discussion

The identification of outcrop strata is an important field within sedimentary geology, where traditional methods based on semi-supervised segmentation take a lot of specialist time. Compared with existing purely manual or semi-automatic methods, our proposed model structure guarantees a high accuracy rate and greatly improves efficiency and meets the objective of segmenting outcrops.

The combination of the PointRend method with the DeepLabV3 architecture has improved the accuracy in identifying areas of facies change, allowing the method to be extended to other similar cases. The depth of the ResNet50 convolution layers plays a very important role in extracting the high-dimensional features.

To investigate the effects of the PointRend algorithm, we built two encoder-decoder models: the decoder of the first has only upsampling layers and the decoder of the second has both upsampling layers and the PointRend module. We can observe that the adoption of the method slightly surpasses the metric value, showing a smoother and clearer lithological alteration. The use of metrics such as Boundary IoU contributes to a better analysis of the facies transition effect.

5 Conclusion

Photogrammetry in combination with a drone-based remote sensor to study and model geological outcrops is a solution in terms of cost, accuracy, and efficiency. In this paper, we presented a Convolutional Neural Network architecture based on Semantic Segmentation for the identification of strata in turbidite system outcrops using photogrammetric data based on UAVs. The dataset was based on an outcrop of Permian turbidite systems from the Itararé Group (Paraná Basin).

The proposed architecture has the encoder-decoder structure. The encoder employs ResNet50 network and ASPP technique, and the decoder uses the upsampling layer and PointRend algorithm to recover the road boundary. Experimental results show that the generated model outperforms the U-net model used as baseline. The PointRend method makes the extracted facies have a smooth and sharp edge with better connectivity. The approach adopted in this work proved to be promising, as it combined remote sensing techniques with recent neural network architectures. This information can be used to evaluate and improve the outcrop classification process in future work.

References

1. Achanta, R., Shaji, A., Smith, K., Lucchi, A., Fua, P., Süsstrunk, S.: Slic superpixels compared to state-of-the-art superpixel methods. IEEE Trans. Pattern Anal. Mach. Intell. **34**(11), 2274–2282 (2012)
2. Andrew, M.: A quantified study of segmentation techniques on synthetic geological xrm and fib-sem images. Comput. Geosci. **22**(6), 1503–1512 (2018)
3. Bai, M., Urtasun, R.: Deep watershed transform for instance segmentation. In: Proceedings of the IEEE Conference on Computer Vision and Pattern Recognition, pp. 5221–5229 (2017)
4. Becker, I., Koehrer, B., Waldvogel, M., Jelinek, W., Hilgers, C.: Comparing fracture statistics from outcrop and reservoir data using conventional manual and t-lidar derived scanlines in ca2 carbonates from the southern permian basin, germany. Mar. Pet. Geol. **95**, 228–245 (2018)
5. Boggs, S., et al.: Principles of sedimentology and stratigraphy (2012)
6. Casini, G., Hunt, D., Monsen, E., Bounaim, A.: Fracture characterization and modeling from virtual outcrops. AAPG Bull. **100**(1), 41–61 (2016)

7. Chen, L.C., Papandreou, G., Schroff, F., Adam, H.: Rethinking atrous convolution for semantic image segmentation. arXiv preprint arXiv:1706.05587 (2017)

8. Chen, S.C., Hsiao, Y.S., Chung, T.H.: Determination of landslide and driftwood potentials by fixed-wing uav-borne rgb and nir images: a case study of Shenmu area in Taiwan. In: EGU General Assembly Conference Abstracts, p. 2491 (2015)

9. Cheng, B., Girshick, R., Dollár, P., Berg, A.C., Kirillov, A.: Boundary IoU: improving object-centric image segmentation evaluation. In: CVPR (2021)

10. Cruz, R.A.Q., Cacau, D.C., dos Santos, R.M., Pereira, E.J.R., Leta, F.R., Clua, E.G.: Improving accuracy of automatic fracture detection in borehole images with deep learning and gpus. In: 2017 30th SIBGRAPI Conference on Graphics, Patterns and Images (SIBGRAPI), pp. 345–350. IEEE (2017)

11. Endres, F., Hess, J., Sturm, J., Cremers, D., Burgard, W.: 3-d mapping with an rgb-d camera. IEEE Trans. Rob. **30**(1), 177–187 (2013)

12. Gong, X., Liu, J.: Rock detection via superpixel graph cuts. In: 2012 19th IEEE International Conference on Image Processing, pp. 2149–2152. IEEE (2012)

13. Harwin, S., Lucieer, A.: Assessing the accuracy of georeferenced point clouds produced via multi-view stereopsis from unmanned aerial vehicle (uav) imagery. Remote Sensing **4**(6), 1573–1599 (2012)

14. Jacquemyn, C., Huysmans, M., Hunt, D., Casini, G., Swennen, R.: Multiscale three-dimensional distribution of fracture-and igneous intrusion-controlled hydrothermal dolomite from digital outcrop model, latemar platform, dolomites, northern italy. AAPG Bull. **99**(5), 957–984 (2015)

15. Joyce, K.E., Samsonov, S., Levick, S.R., Engelbrecht, J., Belliss, S.: Mapping and monitoring geological hazards using optical, lidar, and synthetic aperture radar image data. Nat. Hazards **73**(2), 137–163 (2014)

16. Karimpouli, S., Tahmasebi, P.: Segmentation of digital rock images using deep convolutional autoencoder networks. Comput. Geosci. **126**, 142–150 (2019). https://doi.org/10.1016/j.cageo.2019.02.003. https://www.sciencedirect.com/science/article/pii/S0098300418303911

17. Li, Y., Liu, P., Chen, S., Jia, K., Liu, T.: The identification of slope crack based on convolutional neural network. In: Sun, X., Zhang, X., Xia, Z., Bertino, E. (eds.) ICAIS 2021. CCIS, vol. 1423, pp. 16–26. Springer, Cham (2021). https://doi.org/10.1007/978-3-030-78618-2_2

18. Long, J., Shelhamer, E., Darrell, T.: Fully convolutional networks for semantic segmentation. In: Proceedings of the IEEE Conference on Computer Vision and Pattern Recognition, pp. 3431–3440 (2015)

19. Maiti, S., Tiwari, R.: Automatic detection of lithologic boundaries using the walsh transform: A case study from the ktb borehole. Comput. Geosci. **31**(8), 949–955 (2005)

20. Ngcofe, L., Minnaar, H.: A study on automated segmentation for object-based image analysis for geological mapping in the northern cape province, South Africa. In: Proceedings of the 4th GEOBIA, pp. 7–9 (2012)

21. Papageorgiou, C.P., Oren, M., Poggio, T.: A general framework for object detection. In: Sixth International Conference on Computer Vision (IEEE Cat. No. 98CH36271), pp. 555–562. IEEE (1998)

22. Perez, C.A., Saravia, J., Navarro, C., Castillo, L., Schulz, D., Aravena, C.: Lithological classification based on gabor texture image analysis. In: 2012 International Symposium on Optomechatronic Technologies (ISOT 2012), pp. 1–3. IEEE (2012)

23. Phelps, R.M., Kerans, C., Scott, S.Z., Janson, X., Bellian, J.A.: Three-dimensional modelling and sequence stratigraphy of a carbonate ramp-to-shelf transition, permian upper san andres formation. Sedimentology **55**(6), 1777–1813 (2008)

24. Ronneberger, O., Fischer, P., Brox, T.: U-Net: convolutional networks for biomedical image segmentation. In: Navab, N., Hornegger, J., Wells, W.M., Frangi, A.F. (eds.) MICCAI 2015. LNCS, vol. 9351, pp. 234–241. Springer, Cham (2015). https://doi.org/10.1007/978-3-319-24574-4_28

25. van Ruitenbeek, F.J., van der Werff, H.M., Hein, K.A., van der Meer, F.D.: Detection of pre-defined boundaries between hydrothermal alteration zones using rotation-variant template matching. Comput. Geosci. 34(12), 1815–1826 (2008)

26. Sadeghipoor, Z., Lu, Y.M., Süsstrunk, S.: Gradient-based correction of chromatic aberration in the joint acquisition of color and near-infrared images. In: Digital photography XI, vol. 9404, p. 94040F. International Society for Optics and Photonics (2015)

27. Salati, S., van Ruitenbeek, F.J., van der Meer, F.D., Tangestani, M.H., van der Werff, H.: Lithological mapping and fuzzy set theory: automated extraction of lithological boundary from aster imagery by template matching and spatial accuracy assessment. Int. J. Appl. Earth Obs. Geoinf. 13(5), 753–765 (2011)

28. Saporetti, C.M., da Fonseca, L.G., Pereira, E., de Oliveira, L.C.: Machine learning approaches for petrographic classification of carbonate-siliciclastic rocks using well logs and textural information. J. Appl. Geophys. 155, 217–225 (2018). https://doi.org/10.1016/j.jappgeo.2018.06.012. https://www.sciencedirect.com/science/article/pii/S092698511630667X

29. Taye, W.: Lithological boundary detection using multi-sensor remote sensing imagery for geological interpretation. Master's thesis, University of Twente (2011)

30. Tian, M., Li, B., Xu, H., Yan, D., Gao, Y., Lang, X.: Deep learning assisted well log inversion for fracture identification. Geophys. Prospect. 69(2), 419–433 (2021)

31. Turner, D., Lucieer, A., Watson, C.: An automated technique for generating georectified mosaics from ultra-high resolution unmanned aerial vehicle (uav) imagery, based on structure from motion (sfm) point clouds. Remote sensing 4(5), 1392–1410 (2012)

32. Vasuki, Y., Holden, E.J., Kovesi, P., Micklethwaite, S.: Semi-automatic mapping of geological structures using uav-based photogrammetric data: An image analysis approach. Comput. Geosci. 69, 22–32 (2014)

33. Vasuki, Y., Holden, E.J., Kovesi, P., Micklethwaite, S.: An interactive image segmentation method for lithological boundary detection: a rapid mapping tool for geologists. Comput. Geosci. 100, 27–40 (2017)

34. Vollgger, S.A., Cruden, A.R.: Mapping folds and fractures in basement and cover rocks using uav photogrammetry, cape liptrap and cape paterson, victoria, australia. J. Struct. Geol. 85, 168–187 (2016)

35. Wu, S., Wang, Q., Zeng, Q., Zhang, Y., Shao, Y., Deng, F., Liu, Y., Wei, W.: Automatic extraction of outcrop cavity based on a multiscale regional convolution neural network. Computers & Geosciences p. 105038 (2022)

36. Xu, Z., Ma, W., Lin, P., Shi, H., Pan, D., Liu, T.: Deep learning of rock images for intelligent lithology identification. Comput. Geosci. 154, 104799 (2021). https://doi.org/10.1016/j.cageo.2021.104799. https://www.sciencedirect.com/science/article/pii/S009830042100100X

37. Yang, Y., Mei, G.: Deep transfer learning for identifications of slope surface cracks. arXiv preprint arXiv:2108.04235 (2021)

Detection of False Information in Spanish Using Machine Learning Techniques

Arsenii Tretiakov[✉], Alejandro Martín, and David Camacho

Department of Computer System Engineering, Universidad Politécnica de Madrid, Calle de
Alan Turing, 28031 Madrid, Spain
a.tretiakov@alumnos.upm.es, {alejandro.martin,
david.camacho}@upm.es

Abstract. This research focuses on the detection of false claims in Spanish
through the use of machine learning techniques. Most of the current work related
to automated, or semi-automated, fake news detections are carried out for the
English language, however, there is still a large room for improvement in other
languages such as Spanish. The detection of fake news content and its dissemination (spread) in online platforms is an open and hard problem, this work is focused
on the detection of false and misleading information spreading during the election campaign in the Spanish Parliament and Catalonia crisis in 2019, migration
crisis, COVID-19 pandemic, and hate speech against minorities. We propose the
use of a machine learning model adapted for dealing with human language understanding tasks, called BERT, which has been trained and experimentally tested.
We have collected a corpus of different types of false information and claims
such as articles, posts on Facebook, WhatsApp's messages, tweets, and others.
The results evidence how usage of machine learning techniques can help in the
identification of false statements with more than 88% accuracy, and in collecting
samples of false information. The experiments, with a comparison between different machine learning methods, have also been carried out using previous datasets,
providing a comparison between different approaches.

Keywords: Fake news · False information · Mass media · Social media ·
Machine learning · Media literacy

1 Introduction

On 28 April 2019 (28A) and 10 November (10N), parliamentary elections were held
in Spain. Before and after the elections, political parties actively used and spread false
information in mass media and social media against each other. The day before 28A
social networks and messaging platforms such as WhatsApp increased "social pressure
to mobilize the undecided and dissuade the unbelievers" [28]. During the two election
campaigns in 2019, the Ciudadanos, center-right political party, placed the largest number of advertisements on Facebook containing elements of disinformation, while Partido
Popular (traditional center-right party), Podemos (left-wing party) and VOX (far-right

© The Author(s), under exclusive license to Springer Nature Switzerland AG 2022
H. Yin et al. (Eds.): IDEAL 2022, LNCS 13756, pp. 42–53, 2022.
https://doi.org/10.1007/978-3-031-21753-1_5

party) [35] did it to a lesser extent [8]. It demonstrated how social media groups changed as the election approached, the Facebook page "The love of God is always with me" changed name and became the "Vox Affiliates" [28]. At the same time, the media are facing a post-Twitter era because platforms such as WhatsApp (84,3% of user penetration in Spain) [22], or Telegram, transformers direct messages into the news so that the audience can condition the agenda.

Due to the active development in Spain of initiatives to identify false news and socio-political processes such as parliamentary elections and the crisis in Catalonia, hate speech against minorities, there are many opportunities to study this topic and collect data. The work carried out by Elías and Catalan-Matamoros [12] shows that one of the reasons explaining the dissemination of false information about the COVID-19 is related to the state's policy when official sources give journalists confusing data so they do not have time to look for independent sources and check them.

Another opportunity for this study is a weak representation of fake news detection studies in other languages, including Spanish. There is a high representation of media projects and research on methods of detecting false information in English and Chinese, as the most common on the Internet. However, according to Internet World Stats [21], Spanish is the third most common worldwide language on the Internet (the penetration level is 7.9%, English - 25.9%, Chinese - 19.4%).

Despite plenty of false information on the internet, the same publication is being disseminated, copied and rewritten on any digital media platform. Strategies to detect fake news have traditionally been based either on analysis of news content and context-oriented methods that use syntactic and semantic features to classify between real-life articles and articles based on a trigram language model that Della Vedova et al. [9] implemented in English Facebook posts.

MIT's Computer Science and Artificial Intelligence Lab (MIT MediaLab), in collaboration with the Qatar Computing Research Institute, conducted a project to determine information sources and identify whether it is politically or ideologically biased [5]. Web sites with more complex URLs are less trustworthy than with simpler ones. Moreover, the reliability of the original news source is achieved by blockchain (Distributed Ledger Technology) and its features such as cryptographic hashing, digital signatures and distributed consensus. The blockchain system for media is applied in cases where online platforms retroactively change news content or publication dates. The source of origin news can be traced by recording a timestamp using a blockchain-based approach for decentralized distributed storage [34].

ClaimBuster, the claim spotter, provides a priority ranking that helps fact-checkers focus on the top-ranked sentences without sifting through a large number of sentences [16]. It gives each sentence a score indicating how likely it contains a factual claim. Project Fabula investigates how the stories are shared to recognize patterns of dissemination that can only correspond to fake news [2]. FacTeR-Check provides semi-automated verification of claims and tracking known hoaxes [29], and Google News provided fact-check tags to its users to identify news information verified by mass media and fact-checking initiatives [24].

Thus, this work analyses the use of machine learning (ML) strategies to detect false claims. A comparison between different strategies is presented, aimed at detecting both false and misleading information in Spanish which are based on automated analysis of stylistic features and context of the unique corpus of false information samples. To achieve this, all the major ML methods were adapted to the specificities of the Spanish language and were tested.

2 Background and Related Work

There is no all-purpose model or algorithm suitable for all cases of detecting false information, so the model using a combination of deep learning and pre-trained algorithms is one of the methods still under study. Classical ML methods for fake news detection, such as Bayesian classifier, natural language processing (NLP), support vector methods (SVM), Deep learning (DL), k-nearest neighbor (KNN) method, sentiment analysis, random forest (RF) and others are used in developing models of such systems. Decision tree algorithms are simple to understand and implement, but they are not effective when the number of distinctive features between documents is significant; KNN is applicable to various tasks, but it requires more computing resources for a large corpus [15]. SVM is able be used as a discriminating document classifier and it demontrates high accuracy with a small corpus [23]. Among DL methods, the following types of neural networks have been used: recurrent (RNN) (Long short-term memory block, LSTM) and convolutional neural networks (CNN).

Boididou, et al. [6] obtained a high classification accuracy 98.01% by applying a logistic regression (LG) model and 91.78% with a random forest by using a dataset MediaEval VMU 2016 with 1103 Spanish tweets for training. Also, they got less accuracy (93.80%) in English with a much larger dataset of English tweets (12229 items).

Silva et al. performed a comprehensive analysis for fake news detection in Portuguese with a corpus of 3600 false and 3600 true news [38]; the stacking classifier performed 97.8% accuracy, superior to the best outcomes obtained with the bag-of-words model's linguistic-based features. The study of stance classification for a dataset of Russian tweets (700 items) and news comments from multiple sources performed accuracy over 90% using classifiers without tuning [26], tweets showed less variability than arbitrary texts. Yang et al. achieved test set accuracy of 88.06% in Chinese-Mandarin [40] with the implementation of natural language inference, BERT (Bidirectional Encoder Representations from Transformers), and the Decomposable Attention models.

Posadas-Durán et al. gathered 480 samples of fake news and trained classification algorithms on lexical features Bag-of-Words, Part-of-Speech tags, n-grams combination, which allowed to achieve the best result (77.28%) [33]. Huertas-García et al. [18] studied how different tuning methods affect performance on semantic-aware tasks and dimensional reduction techniques interact with the high-dimensional embeddings computed for the Semantic Textual Similarity and NLP tasks. The embedding reduction improves the performance by extending their evaluation in contextual-based models from a multilingual approach.

Villar-Rodríguez et al. demonstrated the result of analysis of a coronavirus-related corpus of tweets using NLP techniques to focus on high impact tweets, on all posts that promote false information or deny it, regardless of their virality [39]. It was determined that although attention is given to relevant tweets in terms of their type of reaction (retweets, likes, quotes). Most tweets about misinformation do not have interactions beyond clicking on those posts because users only give their massive support to a minimal number of tweets [39].

Furthermore, Almatarneh et al. [3] obtained 85% deployed TF-IDF (Term Frequency Inverse Document Frequency) method for predicting hateful information in Spanish tweets against women or immigrants in the framework of SemEval 2019. The analysis of 150 real and fake news articles using an AI application demonstrated that titles of false publications are more negative [31].

Huertas-Tato et al. [20] developed a Siamese Inter-Lingual Transformer architecture for efficiently aligning multilingual embeddings and processing unmatched language pairs, where the two input sentences attend to each other to later be combined through a matrix alignment method. The multilingual semantic-aware Transformer model across 15 languages performed increasing the average Spearman correlation coefficient from 81.89% to 83.60% [17] with applying this approach in Covid-19 MLIA @ Eval Multilingual Semantic Search Task.

Huertas-Tato et al. [19] presented BERTuit for profiling fake news spreaders on Twitter and detecting this type of user by generating a representation of each author by averaging the embedding vector of each tweet of that user. BERTuit was trained in a corpus composed of 230M Spanish tweets and was validated in tasks, such as hate speech detection or sentiment analysis [19].

The fundamental objective for creating an efficient fake news detector is collecting the correct data to train the neural network [13]. Spanish fake news researchers commonly have used Twitter, and various dialects of Spanish [36] can be used in one case and the socio-political background of any Spanish-speaking country is individual. In this study the task is to test ML models only for sources of information from Spain and on different types of text and sources, not only on Twitter.

3   Data and Resources

The data array represents most of the recorded false and true news messages and their variations till 01.02.2021. Through this method, 2742 samples of false information were collected (Table 1). Among the samples of false information are news, headlines, tweets, posts in Facebook, Instagram, and Telegram channels, YouTube captions, messages and transcripts of voice messages in WhatsApp, transcription of video fragments with false statements, photo captions, fake government documents, texts on images.

Only Spanish Castilian texts have been taken to avoid distinctions in different types of Spanish languages (Catalan, Basque, Galician). In the study, fake news samples were collected from feeds by Maldito Bulo (https://maldita.es/malditobulo/), Newtral (https://www.newtral.es/zona-verificacion/fakes/), AFP (https://factual.afp.com/). Maldita, Newtral, AFP Factual have a status as fact-checking agencies by the Poynter Institute, U.S. nonprofit fact-checking, media literacy and journalism ethics institute.

Staff authors of those initiatives provide checked news as observed articles why the information is false and write evidence or truthful information about the event or statement. An observed article contains general information about the situation, quotes of the false statements and claims themselves or links to them, and even transcripts of WhatsApp voice messages. As a result, the quotes of false statements and texts that were indicated in reviews were collected, a news article usually contains a mix of true and false information. Also, upon request to a Maldita staff member, a database of 657 publications containing false information was obtained.

The corpus was collected manually (especially extracting false statement quotes) and automatically using MyNews service (https://mynews.es/), library of news of Spanish mass media, and using parser. Regarding non-textual disinformation formats, a text-recognition service was used to extract false statements from pictures. To increase the dataset, a corpus of false news from open sources was added from other studies [4, 14, 33], non-Spanish sources (e.g. Mexican, Venezuelan) were removed. Thus, the result is 2742 samples.

Samples of the true news, posts and tweets were collected from El Mundo, El Pais, ABC, RTVE and their social media pages. When collecting data for the corpus of truthful publications, a proportion by topic, type and size of publications were maintained. Thus, the whole true news dataset contains 2750 samples. In this study only text information was used, at the same time, the corpus contains metadata of news (link, media source, date, author, headline, false statement, full text), which can be used for future studies.

Table 1. Media sources of misinformation in the current dataset.

Media source	Number of samples
Twitter	549
Facebook	319
WhatsApp	269
Mediterraneo Digital	72
Hay Noticia	35
Alerta digital	35
OkDiario	31
Caso Aislado	29
Periodista digital	16
YouTube	15
Somatemps	14
12minutos	13
Others	517
No labeled	828

The dataset was separated into 16 topics which were based on Magallón's study [27]. The most common topics of false information are politics (25,6% of the dataset) and COVID-19 (23,4%). Any unclassifiable false information was categorized as Society (10,3%); Migration (9,6%), authors note the nationality and migration status; Religion (5,8%), religious affiliation stands out. Machismo (4,6%), fake news about sexism, feminism and harassment. Science (4,2%), false research, technology news and health advice;

False alerts (3%) about hazards and natural disasters. Crime (2,8%) includes unclassifiable false information about criminal incidents; Celebrities (2,5%); Business and economics (2,5%); fake news about enacted laws (3,8%); statements against LGBTQ + community (1,6%); Racism (1,2%); disinformation about terrorist attacks (0,5%); false sport news (0,4%).

4 Methodology

4.1 Linguistic Features

The method determines the credibility of the news in the text [5] as structure (linguistic characteristics based on the use of specific words) and complexity (lexical diversity coefficient, readability, cognitive process word count, inconsistency detection).

Psycholinguistic features are word categories that represent psycholinguistic processes (e.g., positive emotions, perceptual processes, functional words), composite categories (words in a sentence, analytical thinking, emotional tone) and categories of parts of speech, linguistic processes (operative words, pronouns). Readability indicates the effectiveness of text perception. These include the number of characters, the presence of compound words, the length of words, the number of syllables, the types of words, and the number of paragraphs [37]. Syntax extracts a set of derived rules based on context-free grammar trees, whose derived functions consist of all lexical production rules (including child nodes) combined with their parent node [32].

4.2 The Conceptual Architecture of the Fine-Tuned Model

Since the research task contains textual information from social networks and news sites, and the BERT model is pre-trained on common cases, it is crucial to analyze the contextual information extracted from the pre-trained layers of BERT transformers [30]. Different levels of syntactic and semantic information are encoded in different layers of the BERT model [10]. The lower layers of the BERT model may contain more general information, while the upper layers contain task-specific information. The fine-tuning method was performed with BETO [7]; a pre-trained BERT model was trained by making use of the task-specific corpus. BERT is capable of defining named entities, it is an essential subtask for information extraction and context identification.

The processed fake news samples are filed into the pre-trained BERT model to be fine-tuned according to task-specific modification. Using the trained classifiers, the labels are predicted of the test set and evaluate the results. The technical essence of the model is realized by recognizing and analyzing the task presented as a system of facts in text format. BertForSequenceClassification was used, a BERT model with a line layer on top of the model to provide classification and fine-tuning [11]. For fine-tuning BERT on a specific task, the model obtains a batch size of 16, 12 self-attention layers and 110M parameters; the special '[CLS]' classifier token was put at the beginning of the array; the 768-dimensional embeddings were extracted. The documents' length of 500 was selected because the corpus contains samples of the length of the characters. Learning weight is 1e-12 as it offers the lowest training and validation loss.

The model's training process selects a layer from which a certain number of neurons are randomly excluded, which in turn can be turned off from further calculations. More trained neurons in the network get more weight, so this method allows us to improve the learning efficiency and quality of the result, despite the reduced accuracy of the definition. The model will be trained with a more nuanced representation of patterns in the data and thus will be able to summarize and make accurate predictions when it sees completely new data. Thus, the accuracy of the model is increased by raising the corpus and combining different methods and technical solutions.

4.3 The Technological Implementation

A Google Colab Research notebook was used for fine-tuning BERT implementation; it allowed for high-RAM GPU processing. The data was separated into Texts and Classes ("TRUE"/"FALSE"). In the process of data preprocessing, BERT tokenizes each tweet (as an input sentence) in such a way that invalid characters are removed, and all words are converted to lowercase.

Each "batch" contains three PyTorch tensors as input ids, attention masks, labels. The number of iterations for which the model was to be trained was 4 epochs. AdamW (Adam algorithm with weight decay) was utilized as the fine-tuning optimizer with a learning rate of $2e-5$ to minimize the Cross-Entropy loss function; furthermore, the dropout probability is set to 0.1 for all layers.

For classic machine learning methods, the Scikit-learn library was used; Keras and Tensorflow libraries were used to create deep learning models.

4.4 Evaluation Metrics

In classification problems, preference is given to classifiers with higher precision and recall rates. The performance of the models was generalized in a macro-averaged F-measure, which is the geometric mean of precision and recall and gives better results of the performance characteristics of each classification model.

5 Results

Spanish false information detection models were performed in F1-measure when the amount of labeled data is restricted. The ML strategies were tested over a specific concentration range $[0.1-1.0]$. The training is performed using 80% of the data and 20% for testing. Experiments were conducted with neural network models and other classical ML methods (Table 2), such as LG, SVM with TF-IDF technique, RF, multinomial Naive Bayes (NB), decision tree (Tree), KNN, AdaBoost algorithm (ABoost), Gradient Boosting Classifier (GBoost). CNN, LSTM and an integrated method that combines CNN and LSTM architectures were also tested among deep learning methods. In addition, experiments with various pre-trained BERT models, in combination with LG, CNN, and LSTM layers, were carried out (Table 3).

These strategies were chosen for testing on a corpus of news because they are the most widely employed for text analysis, which in turn is applied in studies of misinformation and analysis of hate speech detection in the mass media, as described in Related work.

Regarding DL models, LSTM captures contextual information better and is ideal for implementing the semantics of long texts [1]. The LSTM is considered to be a biased model because last words in the text have more weight than earlier words because it is used to capture the semantics of a whole document, since key components can appear anywhere in the document, not just at the end [25]. Hence, CNN is able to capture text semantics better than RNN, and such a model can reserve a more extensive range of word order for learning, which is necessary when dealing with different formats and volumes of text. To minimize the disadvantages of both presented DL models, a combination of CNN and LSTM is used for text analysis (LSTM + CNN). The key feature of this model is a sequential fact input system, which allows using an unlimited number of conditions and rules to describe the final system.

Table 2. Experimental results with comparable approaches.

Method	SVM	LG	RF	NB	Tree	KNN	ABoost	GBoost
Accuracy	0.872	0.860	0.848	0.763	0.815	0.785	0.840	0.838
F-measure	0.863	0.852	0.844	0.805	0.810	0.794	0.830	0.824
Precision	0.926	0.904	0.873	0.684	0.833	0.764	0.888	0.905
Recall	0.809	0.805	0.816	0.978	0.788	0.827	0.779	0.757

Table 3. Experimental results using Deep Learning (CNN and LSTM) methods and a combination of DL methods and pretrained NLP-DL architectures.

Method	CNN	LSTM	LSTM + CNN	BERT + LG	BERT + CNN	BERT + LSTM	Fine-tuned BERT
Accuracy	0.866	0.849	0.844	0.785	0.810	0.680	0.891
F-measure	0.863	0.851	0.842	0.784	0.810	0.664	0.889
Precision	0.882	0.842	0.859	0.788	0.811	0.702	0.895
Recall	0.846	0.860	0.825	0.781	0.808	0.630	0.883

An experiment was conducted with 2742 samples of fakes and 2750 of true news information. As a result, the fine-tuned BERT model obtains the highest possible value of F-measure, 88.97%. The loss validation value is 32%. SVM demonstrates a high F-measure too (86,3%), but there is a higher number of errors (Figs. 1). The fine-tuned BERT model predicted 472 cases of false information, and it made only 62 errors. The SVM model predicted 436 fake news cases and made 103 prediction errors. Thus, with almost equal accuracy, the fine-tuned BERT model performed fewer errors in predicting false information publications.

Since fine-tuned BERT performs better at defining named entities, especially for automated categorization of false news topics or discovering themes (subjects) within each topic. Also, although SVM is a "lighter" ML method, when the dataset size is increased, BERT will perform better, reducing the number of errors when validating the trained model.

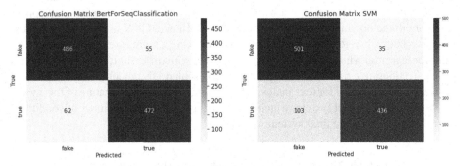

Fig. 1. Confusion Matrices of the most effective models: fine-tuned BERT and SVM.

Thus, the proposed fine-tuned model is able to analyze the style of writing, context, taking into account the named entities, as well as to determine the relationship between the characteristics, and on this basis to identify false news. Such a model contributes with a high degree of accuracy to identify the characteristics that are not inherent in the true news, and also it is adaptive for other languages and various tasks. Therefore, this led to the decision to develop a fine-tuned model.

6   Conclusions and Future Work

In this study, a wide number of ML methods were selected and experimentally tested, and the most efficient strategy among them was selected to solve the task of detecting false information in Spanish. The fine-tuned BERT model on a unique corpus of Spanish fake news publications has allowed it to obtain a detection accuracy of 88.97%. This method is based on the analysis of stylistic characteristics, context, and complexity of the presented corpus.

The methodology of the data collection used is described, and the linguistic input parameters of the fake news samples are analyzed. An experience of training neural networks has been presented on varied types of fake information such as messages on WhatsApp, posts on Facebook, tweets, news articles, video subtitles, and others.

Even though the proposed method does not solve the problem of dissemination of false news, it contributes with a high degree of reliability to detect non-true news writing style, which in combination with other available methods, such as crowdsourcing, classification of sources and authors, fact-checking, has the potential to provide the highest level of accuracy results at a low cost of resources.

This study allows designing a fine-tuned ML model to develop a qualimetric evaluation of the quality of publications in mass media according to a range of linguistic features on specific topics, such as politics or migration. Furthermore, the more profound analysis of texts and metadata of misleading texts could help us comprehend the nature and patterns of misleading messages.

The proposed approach can be scaled for other dialects and languages of Spain, for instance, in Catalan. The ML model itself will learn a more nuanced representation of patterns in the data and thus will be able to summarize and make accurate predictions when it sees completely new data. Moreover, the proposed method is able to show greater efficiency when training the model on more data. While other machine learning models analysed in this study will not be able to show the same performance in terms of accuracy and data processing speed as Fine-tuned BERT.

References

1. Aghakhani, H., Machiry, A., Nilizadeh, S., Kruegel, C., Vigna, G.: Detecting deceptive reviews using generative adversarial networks. In: IEEE Security and Privacy Workshops (SPW), pp. 89–95. IEEE (2018). https://doi.org/10.1109/spw.2018.00022
2. Alarcon, N.: Fabula AI develops a new algorithm to stop sake sews. Nvidia Developer. News Center. https://developer.nvidia.com/blog/fabula-ai-develops-a-new-algorithm-to-stop-fake-news/. Accessed 15 Apr 2022
3. Almatarneh, S., Gamallo, P., Pena, F.J.R.: CiTIUS-COLE at semeval-2019 task 5: combining linguistic features to identify hate speech against immigrants and women on multilingual tweets. In: Proceedings of the 13th International Workshop on Semantic Evaluation, pp. 387–390. Association for Computational Linguistics, Minneapolis, Minnesota, USA (2019). https://doi.org/10.18653/v1/s19-2068
4. Aragón, M.E., Jarquín-Vásquez, H.J., Montes-y-Gómez, M., Escalante, H.J., Pineda, L.V., Gómez-Adorno, H., et al.: Overview of MEX-A3T at IberLEF 2020: fake news and aggressiveness analysis in Mexican Spanish. In: IberLEF@ SEPLN, pp. 222–235 (2020). https://doi.org/10.29057/mjmr.v8i16.3926
5. Baly, R., Karadzhov, G., Alexandrov, D., Glass, J., Nakov, P.: Predicting factuality of reporting and bias of news media sources. arXiv preprint arXiv:1810.01765 (2018). https://doi.org/10.48550/arXiv.1810.01765
6. Boididou, C., Papadopoulos, S., Zampoglou, M., Apostolidis, L., Papadopoulou, O., Kompatsiaris, Y.: Detection and visualization of misleading content on Twitter. Int. J. Multimed. Inf. Retrieval 7(1), 71–86 (2017). https://doi.org/10.1007/s13735-017-0143-x
7. Canete, J., Chaperon, G., Fuentes, R., Ho, J.H., Kang, H., Pérez, J.: Spanish pre-trained bert model and evaluation data. Pml4dc at iclr, pp. 1–10 (2020)
8. Cano-Orón, L., Calvo, D., García, G.L., Baviera, T.: Disinformation in Facebook ads in the 2019 Spanish general election campaigns. Media Commun. 9(1), 217–228 (2021). doi: https://doi.org/10.1080/13216597.2019.1634619
9. Della Vedova, M.L., Tacchini, E., Moret, S., Ballarin, G., DiPierro, M., de Alfaro, L.: Automatic online fake news detection combining content and social signals. In: 2018 22nd Conference of Open Innovations Association (FRUCT), pp. 272–279. IEEE. Jyvaskyla, Finland (2018). https://doi.org/10.23919/fruct.2018.8468301
10. Devlin, J., Chang, M.W., Lee, K., Toutanova, K.: Bert: pre-training of deep bidirectional transformers for language understanding. arXiv preprint arXiv:1810.04805 (2018). https://doi.org/10.48550/arXiv.1810.04805
11. Dukic, D., Kržic, A.S.: Detection of hate speech spreaders with BERT. CLEF (Working Notes) (2021)
12. Elías, C., Catalan-Matamoros, D.: Coronavirus in Spain: Fear of "official" fake news boosts WhatsApp and alternative sources. Media Commun. 8(2), 462–466 (2020). https://doi.org/10.17645/mac.v8i2.3217

13. Flores Vivar, J.M.: Artificial intelligence and journalism: diluting the impact of misinformation and fakes news through bots. Doxa. Comunicación n° 029, 197–212 (2019)
14. Gómez-Adorno, H., Posadas-Durán, J.P., Enguix, G.B., Capetillo, C.P.: Overview of FakeDeS at IberLEF 2021: fake news detection in Spanish shared task. Procesamiento del Lenguaje Natural **67**, 223–231 (2021). https://doi.org/10.26342/2021-67-19
15. Han, E.-H., Karypis, G., Kumar, V.: Text categorization using weight adjusted k-nearest neighbor classification. In: Cheung, D., Williams, G.J., Li, Q. (eds.) PAKDD 2001. LNCS (LNAI), vol. 2035, pp. 53–65. Springer, Heidelberg (2001). https://doi.org/10.1007/3-540-45357-1_9
16. Hassan, N., Zhang, G., Arslan, F., Caraballo, J., Jimenez, D., Gawsane, S., et al.: ClaimBuster: the first-ever end-to-end fact-checking system. Proc. VLDB Endowm. **10**(12), 1945–1948 (2017). https://doi.org/10.14778/3137765.3137815
17. Huertas-García, Á., Huertas-Tato, J., Martín, A., Camacho, D.: Countering misinformation through semantic-aware multilingual models. In: Yin, H., et al. (eds.) IDEAL 2021. LNCS, vol. 13113, pp. 312–323. Springer, Cham (2021). https://doi.org/10.1007/978-3-030-91608-4_31
18. Huertas-García, Á., Martín, A., Huertas-Tato, J., Camacho, D.: Exploring Dimensionality Reduction Techniques in Multilingual Transformers. arXiv preprint arXiv:2204.08415 (2022). https://doi.org/10.48550/arXiv.2204.08415
19. Huertas-Tato, J., Martin, A., Camacho, D. BERTuit: understanding Spanish language in Twitter through a native transformer. arXiv preprint arXiv:2204.03465 (2022). https://doi.org/10.48550/arXiv.2204.03465
20. Huertas-Tato, J., Martín, A., Camacho, D.: SILT: Efficient transformer training for interlingual inference. Expert Syst. Appl. **200**, 116923 (2022). https://doi.org/10.1016/j.eswa.2022.116923
21. Internet world stats. INTERNET WORLD USERS BY LANGUAGE. Top 10 Languages. https://www.internetworldstats.com/stats7.htm. Accessed 02 Apr 2022
22. Iqbal, M.: WhatsApp Revenue and Usage Statistics (2022). Business of Apps. https://www.businessofapps.com/data/whatsapp-statistics/. Accessed 30 June 2022
23. Isa, D., Lee, L.H., Kallimani, V.P., Rajkumar, R.: Text document preprocessing with the Bayes formula for classification using the support vector machine. IEEE Trans. Knowl. Data Eng. **20**(9), 1264–1272 (2008). https://doi.org/10.1109/tkde.2008.76
24. Kosslyn, J., Yu, C.: Fact check now available in google search and news around the world. Google Blog. https://blog.google/products/search/fact-check-now-available-google-search-and-news-around-world/. Accessed 03 Apr 2022
25. Lai, S., Xu, L., Liu, K., Zhao, J.: Recurrent convolutional neural networks for text classification. In: Twenty-Ninth AAAI Conference on Artificial Intelligence, pp. 2267–2273 (2015)
26. Lozhnikov, N., Derczynski, L., Mazzara, M.: Stance prediction for Russian: data and analysis. In: Ciancarini, P., Mazzara, M., Messina, A., Sillitti, A., Succi, G. (eds.) SEDA 2018. AISC, vol. 925, pp. 176–186. Springer, Cham (2020). https://doi.org/10.1007/978-3-030-14687-0_16
27. Magallón-Rosa, R. Nuevos formatos de verificación. El caso de Maldito Bulo en Twitter. Sphera publica **1**(18), 41–65 (2018). https://doi.org/10.6084/m9.figshare.6142808
28. Magallón-Rosa, R.: Desinformación en campaña electoral. Telos. https://telos.fundaciontelefonica.com/desinformacion-en-campana-electoral/. Accessed 25 Mar 2022
29. Martín, A., Huertas-Tato, J., Huertas-García, Á., Villar-Rodríguez, G., Camacho, D.: FacTeR-Check: Semi-automated fact-checking through Semantic Similarity and Natural Language Inference. arXiv preprint arXiv:2110.14532 (2021). https://doi.org/10.48550/arXiv.2110.14532

30. Mozafari, M., Farahbakhsh, R., Crespi, N.: Hate speech detection and racial bias mitigation in social media based on BERT model. PLoS ONE **15**(8), e0237861 (2020). https://doi.org/10.1371/journal.pone.0237861

31. Paschen, J.: Investigating the emotional appeal of fake news using artificial intelligence and human contributions. J. Prod. Brand Manage. **29**(2), 223–233 (2019). https://doi.org/10.1108/JPBM-12-2018-2179

32. Pérez-Rosas, V., Kleinberg, B., Lefevre, A., Mihalcea, R.: Automatic detection of fake news. arXiv preprint arXiv:1708.07104 (2017)

33. Posadas-Durán, J. P., Gómez-Adorno, H., Sidorov, G., Escobar, J. J. M.: Detection of fake news in a new corpus for the Spanish language. J. Intell. Fuzzy Syst. **36**(5), 4869–4876 (2019). https://doi.org/10.48550/arXiv.1708.07104

34. Qayyum, A., Qadir, J., Janjua, M.U., Sher, F.: Using blockchain to rein in the new post-truth world and check the spread of fake news. IT Professional **21**(4), 16–24 (2019). https://doi.org/10.1109/mitp.2019.2910503

35. Rama, J., Cordero, G., Zagórski, P.: Three Is a Crowd? Podemos, Ciudadanos, and Vox: The End of Bipartisanship in Spain. Front. Political Sci. **95** (2021). https://doi.org/10.3389/fpos.2021.688130

36. Reyes, J., Palafox, L.: Detection of fake news based on readability. RIIAA 2019 Conference Submission (2019). https://openreview.net/forum?id=ByxTOnokxr. Accessed 30 May 2022

37. San Norberto, E.M., Gómez-Alonso, D., Trigueros, J.M., Quiroga, J., Gualis, J., Vaquero, C.: Readability of surgical informed consent in Spain. Cirugía Española (English Edition) **92**(3), 201–207 (2014). doi:https://doi.org/10.1016/j.cireng.2013.02.010

38. Silva, R.M., Santos, R.L., Almeida, T.A., Pardo, T.A.: Towards automatically filtering fake news in Portuguese. Expert Syst. Appl. **146**, 113199 (2020). https://doi.org/10.1016/j.eswa.2020.113199

39. Villar-Rodríguez, G., Souto-Rico, M., Martín, A.: Virality, only the tip of the iceberg: ways of spread and interaction around COVID-19 misinformation in Twitter. Commun. Soc., 239–256 (2022). https://doi.org/10.15581/003.35.2.239-256

40. Yang, K.C., Niven, T., Kao, H.Y.: Fake news detection as natural language inference. In: WSDM '19 Proceedings of the Twelfth ACM International Conference on Web Search and Data Mining 2019 arXiv preprint arXiv:1907.07347 (2019). https://doi.org/10.48550/arXiv.1907.07347

An Approach to Authenticity Speech Validation Through Facial Recognition and Artificial Intelligence Techniques

Hugo Faria$^{(\boxtimes)}$, Manuel Rodrigues$^{(\boxtimes)}$, and Paulo Novais

University of Minho, Braga, Portugal
a81283@alunos.uminho.pt, {mfsr,pjon}@di.uminho.pt

Abstract. Since all times, humans tend to adapt their speech, in terms of authenticity, according to the moments specific needs, making some statements or claims about something that do not correspond to reality, in short, lying about some matter. Identifying such moments has always been a challenging task, not at all times successful, and requiring external artefacts, with scarce availability and inducing stressful situations. With recent advances in hardware technology, making enormous computational power available in our hands through smartphones and fast network technologies, and with Artificial Intelligence evolution, namely Deep Learning, it seems possible to achieve the goal of validating speech authenticity, with smartphones, using facial recognition to detect signs of untruthful speech. This paper presents a framework to achieve this goal.

Keywords: Artificial intelligence · Deep learning · Facial recognition · Deception detection

1 Introduction

Deception is very common in daily activities. Some lies prove to be inoffensive, while others can lead to severe consequences and can become dangerous. Detecting when a person is lying, is a difficult task, not always successful, but extremely needed nowadays. Being able to detect untruthful speech is a much-needed feature in practically every domains and situations. Such a feature would induce major outcomes if widely available, and with simple usage. Adding the possibility of doing so, using daily objects that are present in practically any person's pocket nowadays (smartphones), and with some degree of accuracy, would constitute an enormous leap towards facts verification. Since the beginning of 20th century, the standard for lie detecting has been the polygraph, which relies on physiological sensors to measure heart rate, blood pressure, breathing rate and breathing rhythm. Shifting to a non-intrusive method, with no need for specialist and special machines, is then a great improvement. The process to detect lies relies on the fact that there are contradictory actions, true and false, that generate internal conflict in the individual that lies [1]. The task of detecting

© The Author(s), under exclusive license to Springer Nature Switzerland AG 2022
H. Yin et al. (Eds.): IDEAL 2022, LNCS 13756, pp. 54–63, 2022.
https://doi.org/10.1007/978-3-031-21753-1_6

untruthful speech is to search for these contradictions, across multiple communication channels. This can be achieved using voice analysis, body language, and coherence of related actions, to accumulate evidences that the subject is lying [2,3]. The aim of this research is to develop a non-intrusive method, relying on accessible objects such as smartphones, to detect speech incoherence or lies. It relies on the use of facial recognition and artificial intelligence techniques in order to try to predict or detect deception, more specifically on the use of the Facial Action Coding System (FACS) and OpenFace. In this work, an approach/framework is presented to achieve these goals. First, some back-ground issues are presented, followed by the proposed Framework. Some conclusions and future work are then discussed.

2 State of Data

2.1 Deception Detection Techniques

The Polygraph is probably the most know device to detect deception. It relies on breathing, skin conductivity blood pressure and heart beating analysis. Other methods are also used. However, most of them are very intrusive.

The evaluation of micro-expressions is a technique that has been explored more recently. This is based on the principle that humans convey voluntary and involuntary messages through their faces. These AUs can be combined to detect expected emotions and, therefore, deception.

Ekman's research [1] showed that some facial behaviours are involuntary and can serve as an evidence that someone is untruthful about his speech. Deceptive behaviour is very subtle and different across different people. Detecting these subtle micro movements (micro facial expression), is a challenging and difficult task. With the current development of computational power and the proliferation of datasets with easy access, it becomes easier to use artificial intelligence techniques to detect such micro facial expressions that can reveal deception. The ability to train models be-came so much easier and accessible, with the proliferation of datasets, that seems the logic path to follow. In this work, the goal is to use available datasets to train some machine learning models to detect deception.

2.2 Face Recognition and Face Features Extraction

The first step to automatically detect deception using facial recognition is obviously capture and recognize face and face features. OpenCV has been for quite a while the way to go when using computer vision applications, it provides a simple and ac-curate way to process images and video [6]. Usually facial recognition evolves 3 phases: detection, analysis and recognition.

Detection consists in identifying a face in an image, normally with the use of a neural network.

Analysis is the mapping of the faces by measuring distances between nodal points.

Lastly, recognition is the attempt to confirm the identity of a person in a photo, with the last step being identifying who is on the picture.

Recognizing someone, is off course an important task. This work proposes to automatically detect deception by analysing facial expressions. Therefore, after facial recognition, some features about the identified face need to be extracted.

In this work OpenFace will be used [6]. OpenFace is capable of facial land-mark detection, head pose estimation, facial action unit recognition, and eye-gaze estimation. Furthermore, it is capable of real-time performance and is able to run from a simple webcam without any dedicated hardware. Special interest will be put in face action units recognition.

There are eight basic facial expressions: anger, contempt, disgust, fear, happiness, joy, sadness, and surprise. They are coded as combinations of Action Units (AU) of different muscles in the face according to the Facial Action Coding System (FACS) [1]. From these action units and basic facial expressions, deception detection should then be possible, as emotions are, see Table 1.

Table 1. Emotions and AUs based on the literature

Emotion	Muscle description	Associated AUs
Anger	Nostrils raised, mouth compressed, furrowed brow, eyes wide open, head erect	4, 5, 24or 38
Contempt	Lip protrusion, nose wrinkle, partial closure of eyelids, turn away eyes, upper lip raised	9, 10, 22,41, 61 or 62
Disgust	Lower lip turned down, upper lip raised, expiration, mouth open, blowing out protruding lips, lower lip, tongue protruded	10, 16, 22,25 or 26
Fear	Eyes open, mouth open, lips retracted, eyebrows raised	1, 2, 5or 20
Happiness	Eyes sparkle, skin under eyes wrinkled, mouth drawn back at corners	6, 12
Joy	Zygomatic, orbicularis, upper lip raised, nasolabial fold formed	6, 7, 12
Sadness	Corner mouth depressed, Inner corner eyebrows raised	1, 15
Surprise	Eyebrows raised, mouth open, eyes open, lips protruded	1, 2, 5,25 or 26

3 Experiment

3.1 Framework

The proposed Framework, as seen in Fig. 1 addresses all the previous discussed issues. The main goal is to use available datasets, and for each one extract AUs using OpenFace toolkit. In the next step, machine-learning models are to be created and tested, during the training phase. Information extracted by OpenFace will act as input for a classifier model to produce an answer. A recurrent neural network (RNN), due to its temporal considerations regarding each frame, will be used to make a decision. A long short-term memory is used in deep learning approach. Information processed by this network will be feed into a neural network that will produce an answer in terms of probability.

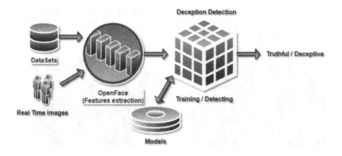

Fig. 1. Proposed framework

3.2 Dataset

In recent years, we have seen an increase in studies that utilize datasets to test deception systems, which led to the creation of different datasets. Few of them are available to use by others. This, despite being comprehensible, does not allow for sharing and testing systems across different datasets, with different techniques comparison not being made, which poses a limitation on the proposed Framework. A major problem for automatic deception systems is also the absence of real data datasets, where people are not instructed to show some predefined emotion. Table 2 summarizes some datasets that can be used as input for the Framework.

Table 2. Datasets

Dataset	Videos	Class
Real-life trial dataset [13]	121	60 truthful and 61 deceptive
Silesian deception dataset [14]	101	3 truthful and 7 deceptive each
Bag-of-lies dataset [15]	325	163 truthful and 162 deceptive
Miami University Deception detection database [16]	320	2 truthful and 2 deceptive each

3.3 Concept Proof

As a concept proof, a video from the real-life trial dataset and a video that comes with OpenFace were chosen and their features were extracted using the OpenFace tool. In order to show the tool in action, a video containing a single face and a video containing multiple faces were selected.

In both videos we can verify that the different features were recognized and posteriorly extracted, as exemplified by the different dots and the green line in Fig. 2. This will be confirmed later with the analysis of the information given by OpenFace.

(a) Single face video

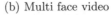
(b) Multi face video

Fig. 2. Analysed videos

In the case of the second video, we can verify that the tool can identify multiple faces at once. During this video we can also see that erratic movements make it hard for OpenFace to extract features. However, there is not a big time interval between the moment faces stop moving and the moment the features are extracted once again.

3.4 Training Details

When we apply OpenFace to a video, different attributes are extracted to a csv file. Attributes like eye gaze and head position can be useful to deception detection techniques, however, our focus is to only use the micro-expressions, and because of that only the action units are to be extracted. The extracted features are presented in Table 3.

Table 3. OpenFace attributes

Attribute	Description
Frame	Number of the frame
Face_id	The face id
Timestamp	The time of the video in seconds
Confidence	How confident the tracker is in current landmark detection
Success	If the track is successful
AU_r	The intensity of each of the 17 action units, from 0 to 5
AU_c	The presence of each of the 17 action units, either 0 or 1

Special importance will be given to attributes AU_r and AU_c, since they give us the information about each action unit. The attributes success and confidence will also be relevant, helping us filter out the data where a face is not detected.

4   Results and Discussion

4.1   Dataset Analysis

Now that we have data, a small analysis was made, giving us the data that will be fed to the model. Starting with the video from the truth set, we can confirm that the face is recognized during its full duration, since all the entries in the success column are a 1 with a corresponding confidence of 98%. Table 4 represents the results between the second 3.8 and the second 3.9.

Table 4. First video results

Timestamp	Confidence	Success	AU06_r	AU12_r	AU06_c	AU12_c
3.804	0.98	1	1.16	0.43	1	1
3.837	0.98	1	1.13	0.52	1	1
3.871	0.98	1	1.09	0.46	1	1
3.904	0.98	1	0.95	0.27	1	1

As we have seen before, action units 6 and 12 could give away a sense of happiness. We can then conclude by observing the values that between the seconds 3.8 and 3.9 happiness could be detected. However, because the intensity of action unit 6 stays around 1 and the intensity of action unit 12 barely goes over 0.50, the micro expressions only gave away a faint sense of happiness.

When it comes to the video from the lying set, because it is a video where someone is lying in trial, a search was done to see if the person expressed fear at any point.

Table 5. Second video results

Timestamp	AU01_r	AU02_r	AU20_r	AU01_c	AU02_c	AU20_c
10.744	0.89	0.67	0.54	1	1	1
10.777	0.74	0.62	0.86	1	1	1
10.811	0.60	0.44	0.96	1	1	1
10.844	0.43	0.16	0.85	1	1	1
10.878	0.50	0.34	0.54	1	1	1

As it can be seen from Table 5, it was found that, between the second 10.7 and 10.8, the presence of action units 1, 2 and 20 was found. These three action units are part of the four action units that convey fear. However, due to the intensity values, this fear was barely noticeable, since they don't go over 1, which is considered the minimum intensity.

Lastly, the video with multiple faces was analysed. During the video we can see that the subjects smile multiple times. Because of that, a search for the joy emotion was done and is presented in Table 6.

Table 6. Multiple faces video results

Face_id	Timestamp	AU06_r	AU07_r	AU20_r	AU06_c	AU07_c	AU20_c
0	23.233	2.91	3.46	2.82	1	1	1
1	23.233	0.64	0.75	1.91	0	0	1
2	23.233	1.43	0.25	1.48	1	0	1
0	23.267	3	3.51	2.8	1	1	1
1	23.267	0.68	0.37	1.97	0	0	1
2	23.267	1.35	0.59	1.52	1	0	1

As expected, a moment where the action units representative of joy, 6,7 and 12 could be found. Around the twenty-third second, the action units intensity for one of the faces is around 3, a considerable value that is between the minimum and the maximum intensities. The second face only presented the action unit 20, however, its in-tensity was of almost 2. The third face presented one more action unit than the second face, the action unit 6. However, this face had lower intensity values, being around 1.50. We can conclude that when the intensity values of the action units are high enough, the expressions become visible without any tools, since it was predicted beforehand that the joy emotion was going to be found.

For further analysis, an analysis was conducted in order to find patterns in the videos. With that in mind, the whole real-life trial dataset was analysed, divided in groups of truthful videos and deceptive videos. Figure 3 shows the number of occurrences of each action unit on the group of deceptive videos.

Looking at the graph, we can see that the action units 4 and 10 occurred more than all the other action units, possibly making them a good indicator of when someone is lying, especially action unit 4, which occurred the most by a good margin in comparison to most action units. These action units are also associated with anger, contempt and disgust, thus concluding that these emotions are prevalent in people when they lie.

On the other hand, the truthful videos were also analyzed. Graph Fig. 4 shows the same number of occurrences but this time for the truthful videos

By analyzing the graph we can see that action unit 14 is the one that was registered the most. Like in the previous set, action unit 10 was the second most found, which means we have an overlap between the two sets. Further study could be done to find out why such thing happens, however, that isn't the focus of this paper. The third most found action unit was the number 5. Because this action unit has way more occurrences in this dataset compared to the deceptive

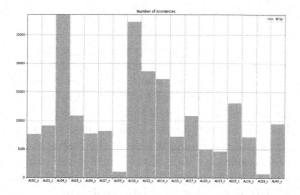

Fig. 3. Number of occurrences of each action unit in deceptive videos

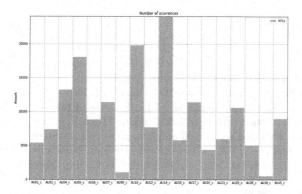

Fig. 4. Number of occurrences of each action unit in truthful videos

one, action unit 5 could be a good indicator, together with action unit 14, that someone is telling the truth.

4.2   RNN Model

A simple RNN model based on the available keras package was used for testing. The dataset was separated in train and test sets, with test being 20% of the train dataset. Then, due to the temporal nature of the data, a timeseries was generated with the train and test sets, with batch size 64. Three LSTM layers were then added to the model, followed by a single dense layer. Due to current hardware constraints, the model was tested with only 10 epochs. In the end, an accuracy of 50% was achieved. Because this is just a quick first test, the results are promising, indicating that with more work put into the model and the use of the other referenced datasets desirable numbers could be achieved.

5 Conclusion

This work proposes an approach based on camera usage, just like any ordinary smartphone, to detect deception. An overview of the existing techniques was presented, followed by the tools used nowadays for facial recognition. Machine Learning and Deep Learning are pointed out as the way to go, being the novelty in the framework the use of several datasets to better train models. The proposed framework is currently under development and testing. A summary of the collected data and the results of testing the first half of the framework was presented, accompanied by a discussion on how the obtained results can be a good indicator for future results. Following this, different models are to be trained and tested in order to achieve the best possible results with the proposed framework.

References

1. Ekman, P.: Telling lies: Clues to deceit in the marketplace, politics, and marriage (revised edition). WW Norton & Company (2009)
2. Martinez Selva, J.M.: La psicologia de la mentira (2005)
3. Vicianova, M.: Historical techniques of lie detection. Europe's J. Psychol. **11**(3), 522–534 (2015). https://doi.org/10.5964/ejop.v11i3.919
4. Hjortsjö, C.H.: Man's Face and Mimic Language, Studentlitteratur (1969)
5. Ekman, P., Rosenberg, E.L.: What the Face Reveals: Basic and Applied Studies of Spontaneous Expression Using the Facial Action Coding System (FACS). Oxford University Press (2005)
6. Baltrušaitis, T., Robinson, P., Morency, L.-P.: OpenFace: an open source facial behavior analysis toolkit. In: IEEE Winter Conference on Applications of Computer Vision (2016)
7. Carissimi, N., Beyan, C., Murino, V.: A multi-view learning approach to de-ception detection. In: 2018 13th IEEE International Conference on Automatic Face & Gesture Recognition (FG 2018), pp. 599–606. IEEE (2018)
8. Larson, J.A., Haney, G.W., Keeler, L.: Lying and its Detection: A Study of Deception and Deception Tests. University of Chicago Press, Chicago (1932)
9. Gamer, M.: Mind reading using neuroimaging: Is this the future of deception detection? European Psychologist (2014)
10. Buddharaju, P., Dowdall, J., Tsiamyrtzis, P., Shastri, D., Pavlidis, I., Frank, M.: Automatic thermal monitoring system (athemos) for deception detection. In: IEEE Computer Society Conference on Computer Vision and Pattern Recognition, CVPR 2005, vol. 2, 1179-vol. IEEE (2005)
11. Meijer, E.H., Verschuere, B.: Deception detection based on neuroimaging: better than the polygraph? J. Forens. Radiol. Imaging **8**, 17–21 (2017)
12. Lin, X., Sai, L., Yuan, Z.: Detecting concealed information with fused electroencephalography and functional near-infrared spectroscopy. Neuroscience **386**, 284–294 (2018)
13. Pérez-Rosas, V., Abouelenien, M., Mihalcea, R., Burzo, M.: Deception detection using real-life trial data. In: Proceedings of the 2015 ACM on International Conference on Multimodal Interaction (2015)

14. Radlak, K., Bozek, M., Smolka, B.: Silesian deception database: presentation and analysis. In: Proceedings of the 2015 ACM on Workshop on Multimodal Deception Detection, Seattle (2015)
15. Gupta, V., Agarwal, M., Arora, M., Chakraborty, T., Singh, R., Vatsa, M.: Bag-of-lies: a multimodal dataset for deception detection. In: Proceedings of the IEEE Conference on Computer Vision and Pattern Recognition Workshops, CA (2019)
16. Lloyd, E.P., Deska, J.C., Hugenberg, K., et al.: Miami University deception detection database. Behav. Res. **51**, 429–439 (2019). https://doi.org/10.3758/s13428-018-1061-4

Federating Unlabeled Samples: A Semi-supervised Collaborative Framework for Whole Slide Image Analysis

Laëtitia Launet[1]([✉]), Rocío del Amor[1], Adrián Colomer[1],
Andrés Mosquera-Zamudio[2], Anaïs Moscardó[2], Carlos Monteagudo[2],
Zhiming Zhao[3], and Valery Naranjo[1]

[1] Instituto de Investigación e Innovación en Bioingeniería, I3B,
Universitat Politècnica de València, Valencia, Spain
lmlaunet@i3b.upv.es
[2] Pathology Department, Hospital Clínico Universitario de Valencia,
Universidad de Valencia, Valencia, Spain
[3] Multiscale Networked Systems, University of Amsterdam, Amsterdam,
The Netherlands

Abstract. Over the last decades, deep learning-based algorithms have witnessed tremendous progress in the medical field to assist pathologists in clinical decisions and reduce their workload. For these models to reach their full potential, access to large and diverse datasets is essential, but collaborations between hospitals are highly limited by privacy-related regulations. At the same time, medical institutions do not always have specialized pathologists to diagnose biopsies and label local data. To address these limitations, federated learning gained traction to enable multi-institution model training without sharing sensitive patient data. However, this technique is still in its infancy when it comes to digital pathology applications, and does not consider institutions with unlabeled data in federations. In this paper, we introduce a novel semi-supervised federated learning approach that promotes multi-institutional training of deep learning models while integrating unlabeled collaborating data sources into the federated setup. The experimental results show a better performance for models trained under a federated setting with both labeled and unlabeled data. Optimally, this framework will also bring the promise of assisting clinical decisions in hospitals that do not have specialized pathologists.

Keywords: Federated learning · Self-training · Whole slide images · Multiple instance learning

1 Introduction

In recent years, digital pathology has witnessed a growing interest, opening the way to the use of computer-aided diagnosis systems thanks to the digitization of

© The Author(s), under exclusive license to Springer Nature Switzerland AG 2022
H. Yin et al. (Eds.): IDEAL 2022, LNCS 13756, pp. 64–72, 2022.
https://doi.org/10.1007/978-3-031-21753-1_7

glass slides into high-resolution whole slide images (WSIs). Thanks to the exponential progress of computer vision and deep learning (DL) applied to medical images, these methods have recently shown unprecedented performance for the automatic analysis of WSIs [5–7].

As histopathological image analysis remains the gold standard in terms of cancer diagnosis, artificial intelligence (AI)-based algorithms hold great potential to reduce pathologists' workload and, ultimately, maximize research progress. However, although AI experts actively work on cutting-edge models and algorithms, the scarce availability of (annotated) data samples hinders their optimal development. Indeed, gaining access to medical data is strongly restricted as it generally involves transferring data samples outside institutions' firewall. Despite data augmentation methods such as the innovating technique of generating synthetic data to cope with these limitations [15], such solutions are not optimal for DL models to reach their full potential as long as the resources at hand remain undivided. With the right resources brought together, AI experts and medical data sources such as hospitals would undeniably bring the promise of tremendous research progress in medical research. Yet sharing data across institutions comes with strong regulations in the medical field, due to the privacy concerns of the data under study.

To address data privacy limitations in this process and allow to gain access to more substantial amounts of data, federated learning (FL) [10] emerged as a learning concept that enables the collaborative training of DL models without having to exchange the data between devices or institutions. Indeed, under this paradigm, models are trained locally with each collaborative institution's own data samples before sending the weights and parameters to a central node for the creation of a consensus model. Although FL was initially developed for non-medical use [10], it has recently gained ground in the healthcare field to mitigate data sharing limitations [13]. Since then, it has also been successfully implemented for a variety of medical applications, from electronic health records to medical image analysis [4,11].

As digital pathology is a medical field that would also greatly profit from the benefits of FL [13], some studies have started to apply this concept to WSIs [2,9,12]. More precisely, Andreux et al. [2] introduced the SiloBN strategy using local-statistic batch normalization layers within DL architectures. In this work, SiloBN only shares the learned parameters of batch normalization layers across collaborators, without their statistics. In [9], Ke et al. implemented a conditional generative adversarial network for stain style transfer in a federated learning context, thus allowing discriminators to learn more than a single stain style and thus generalize better. More recently, Lu et al. [12] coupled federated learning with weakly-supervised multiple instance learning to classify WSIs using only slide-level labels. This approach demonstrated the feasibility of FL in the digital pathology field using data from multiple institutions, while also preventing patient-specific information leakage with randomized noise generation.

However, the implementations of FL in the digital pathology field are still scarce, and the previous studies do not fully leverage the datasets at hand as they only make use of labeled data for training. Integrating medical institutions

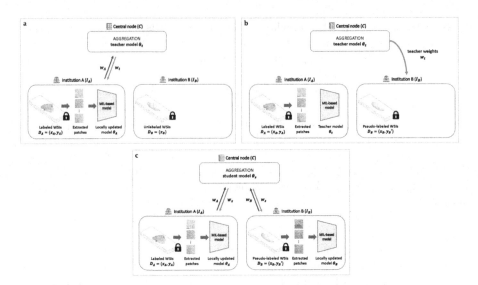

Fig. 1. Overview of the proposed framework. (a) Train a first federated teacher model θ_t using datasets from institutions with labeled samples only. (b) Infer D_B pseudo-labels using the teacher model θ_t trained with D_A. (c) Retrain a student (global) model θ_s under a federated learning setup with datasets from all collaborating institutions, including both labeled and originally unlabeled data.

with unlabeled data into federated setups would not only benefit the optimal training of DL models, but also the hospitals that are more interdisciplinary and do not count on specialized pathologists in their medical staff, by transferring clinical knowledge from one center to another. To bypass these limitations, we propose a novel framework for digital pathology combining the promise of FL collaboration with that of self-training algorithms. To the best of the authors' knowledge, no previous studies have leveraged the teacher-student approach to label the local data of nodes with unlabeled samples and retrain a federated model using both ground truth labels and pseudo-labels.

In this study, a first model is trained under a federated setting at institutions that provide labeled data only, before being used to infer the pseudo-labels of collaborators with unlabeled data. A final classification model is then retrained under a federated setup with all the participating institutions, training with both ground truth and pseudo-labels, giving less weight in the averaging aggregation to the model trained with pseudo-labels.

2   Methodology

2.1   Problem Formulation

In this work, we leverage both labeled and unlabeled datasets from different medical institutions I_A and I_B, and their corresponding local datasets D_A and

D_B. In this scenario, although I_A has specialized pathologists that could label the local datasets, I_B does not. Therefore, we have $D_A = \{x_A, y_A\}$ and $D_B = \{x_B\}$, with x representing WSIs, and y referring to their corresponding labels when available, given by expert pathologists.

2.2   Federated Model

Federated Setting. To enable the training of more robust DL models for WSI classification, we propose to use data from different institutions for training, as depicted in Fig. 1(a). Under the federated learning scenario, we aim at training a global model collaboratively without sharing the local data but only model's weights and parameters. The central node has no access to each participating institution's local data and only receives the updated weights of the model after n epochs. As not all the collaborators could label their data, in this step we train the global model at institutions that do have labeled data only.

Federated Round. During the training, each institution receives the global model's (initial) weights and then trains the model on its local data for n epochs. The resulting weights of each node are then aggregated in the central node before the updated weights are sent back to each node, thus completing a communication round, also called federated round.

Model Architecture. In this approach, we train a model under a weakly-supervised multiple instance learning assumption, using the VGG16 architecture as backbone [16], as it proved to be promising in previous studies analyzing histological images of the same lesion type [1]. More precisely, we freeze the first layers associated with low-level features and re-train the others. We also apply a trainable attention [8] patch-aggregation function, followed by a dense layer for the final prediction at the WSI level.

Federated Averaging. At the end of each federated round, the central node receives the model's weights of each node and averages them to aggregate them into a global model.

2.3   Self-trained Student Model

In this work, we combine FL and semi-supervised learning by training all models under a federated setting with a teacher-student approach [17], a paradigm that enables the use of both labeled and unlabeled data samples for training through the generation of pseudo-labels. In the teacher-student scenario, a so-called teacher model is trained using labeled data only, and then used to infer pseudo-labels of data samples lacking ground truth. A student model is then retrained using both labeled and pseudo-labeled data, what showed clear improvements compared to models trained with labeled data only [14].

In this approach, we trained all models under a federated setup, starting with the teacher model θ_t, trained at institutions that have labeled data. We then used this model to infer the pseudo-labels at the institution I_B, as depicted in Fig. 1(b). Indeed, despite having data available, I_B cannot rely on specialized pathologists to label them. The dataset for I_B therefore becomes $D_B = \{x_B, y'_B\}$, where y'_B refers to the pseudo-labels of the WSIs x_B.

A collaborative student model θ_s is then retrained, still under a federated scenario, with the data from all participating institutions as shown in Fig. 1(c). While the model trained at the institution I_A makes use of expert labels as ground truth, the one trained at I_B uses the previously inferred pseudo-labels as ground truth. For the student global model federated aggregation, we introduce a weighted average at the central node, that gives less importance to the model trained with pseudo-labels with respect to the one trained with expert ground truth.

3 Experiments and Results

3.1 Dataset

In this study, we decided to evaluate the proposed approach on WSIs of spitzoid melanocytic tumors since that type of skin lesion would, in the long run, greatly benefit from this framework. On the one hand, due to their rare occurrence in the population, obtaining large WSI datasets of such tumors is relatively limited. On the other hand, their ambiguous histological features and clinical behavior [3] make them one of the most challenging skin lesions to diagnose even for expert dermatopathologists. As of now, few studies have applied AI methods to the study of spitzoid melanocytic lesions, hence the utility of such a framework to analyze these tumors and promote the transfer of knowledge from one medical center to another.

The WSI slides of the dataset used were digitized, labeled, and annotated by expert pathologists from the University Clinic Hospital of Valencia.

3.2 Implementation Details

Data Partitioning. For the purpose of simulating a federated setting across multiple medical institutions, out of the 100 WSIs we randomly divided the dataset into 4 subsets of data. First and foremost, the validation and testing sets remain at the central node in this particular setup and consist of 18 and 22 WSIs, respectively. Regarding the training sets, the two client nodes contain subsets of different sizes that depend on the available labeled data and the samples to pseudo-label. In particular, D_A is composed of 40 WSIs labeled by expert pathologists, while D_B contains 20 unlabeled WSIs. It is important to note that all labeled datasets are balanced.

Data Preparation. Due to their vast size, the WSIs used in this study were cut down into smaller patches of 512×512 pixels without overlap ahead of training, at a $10\times$ magnification level. This process allows to further extract features from smaller images, the patches, although the final classification is performed at the WSI level by aggregating the patches' features. Patches with less than 30% of tissue were discarded.

Evaluation Metrics. To evaluate the performance of the models and allow their quantitative comparison, relevant metrics were selected including sensitivity (SN), specificity (SPC), positive predictive value (PPV), negative predictive value (NPV), F1-score (F1S), and accuracy (ACC).

Training Details. In the validation experiments, all models were trained for the same number of communication rounds and epochs per round, 10 and 5, respectively. A weights update was performed at each communication round by aggregating the weights, i.e., every 5 epochs. An Adam optimizer with a learning rate of 0.0001 was used, with a batch size of 1, that is to say, each batch corresponding to a single WSI. More precisely, as the models were developed under a multiple instance learning scenario, each batch contains one bag (WSI) of a varying number of instances (patches) depending on the histological image under study, that are aggregated by means of an attention mechanism [8].

Software and Hardware. The complete code of this framework was developed with the PyTorch 1.9.0 library under Python 3.8, and models were trained on the NVIDIA DGX A100 system.

Note that to the best of the authors' knowledge, no prior work combined self-training with federated learning to predict pseudo-labels at the nodes with unlabeled data, thus performing a comparison with the state of the art was not possible.

3.3 Validation of the Framework

In this step, we propose to optimize the aggregation of the models by varying the importance given to the institution trained with pseudo-labels, since its predictions are less reliable than that of a model trained with ground truth labels.

To do so, when performing the models' aggregation after each communication round, that is to say, averaging their weights, we give different importance to I_B weights, as it was trained with pseudo-labels. More precisely, we perform an exhaustive comparison of the performance of each student by varying the value α, given α the importance of each model in the averaging aggregation, $\alpha = 0.5$ being the scenario where its weights are as important as those of the other model when computing the average. To conduct these experiments, we thus started with a standard averaging, i.e., $\alpha = 0.5$, and decreased the importance given to the model trained with pseudo-labels until no more improvement could be noticed, i.e., $\alpha = 0.25$. The accuracy metric was chosen to compare the models' results since it is adapted for balanced datasets, as is the case in this study.

All experiments used VGG16 network architecture as a backbone, and the teacher's weights were used to initialize the training of each federated student. The teacher model was trained for 25 epochs while each of the students ran for 10 communication rounds of 5 epochs each, as for this validation step.

After performing the experiments and based on the results depicted in Fig. 2, the optimal aggregation weight for the student model was $\alpha = 0.3$, as this scenario yielded the most promising results on the validation set with an accuracy of 0.824. Therefore, we selected this value of α as a baseline to address the federated aggregation of the different institutions when involving training with pseudo-labels.

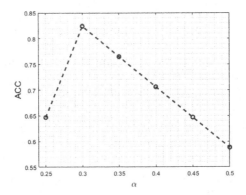

Fig. 2. Classification results on the spitzoid melanocytic lesions dataset reached during the validation set for the different values of α, which refers to the importance given to models trained with pseudo-labels when performing the weighted average at the federated aggregation step. Note that we started the experiments with $\alpha = 0.5$ and then gradually decreased the importance given to that model until the resulting accuracy stopped increasing.

3.4   Results of the Proposed Framework

To demonstrate the usefulness of the proposed method, we evaluated the teacher θ_t and student θ_s models separately on the same test set located at the central node C. In this step, the student model was trained for 20 communication rounds of 10 epochs each and using the same parameters as the teacher model and its weights to initialize training.

Table 1 shows a comparison between the teacher and student model, evaluated on the same set. Overall, classification results demonstrate a clear improvement in the predictions after retraining the model with both ground truth labels and pseudo-labels, using the best student weighted average value for aggregation as per the experiments conducted on the validation test, i.e. $\alpha = 0.3$, see Fig. 2.

Table 1. Classification results on the spitzoid melanocytic lesions dataset reached on the test set with the teacher model and best aggregation weight for the student (i.e., $\alpha = 0.3$).

	SN	SPC	PPV	NPV	F1S	ACC
θ_t	0.7500	0.8000	0.8182	**0.7273**	0.7826	0.7727
θ_s	**0.7692**	**0.8889**	**0.9091**	**0.7273**	**0.8333**	**0.8182**

Performance metrics depicted in Table 1 demonstrate the usefulness of using additional data samples in a federated setting, even when no ground truth labels are available at each institution. These results also emphasize the importance of leveraging all the data at hand even if unlabeled, especially in the case of uncommon lesions such as spitzoid tumors, where gaining access to more data is necessary to improve models' performance.

4 Conclusion

In this work, we propose an innovative framework that promotes collaboration between multiple institutions, as it not only makes use of both labeled and unlabeled data samples under a federated learning setting but also aims at promoting the transfer of clinical knowledge to clinics and hospitals that lack specialized pathologists. The obtained results show a clear improvement in the prediction of models trained with more data although initially not labeled, even for relatively small datasets.

Future research lines will include focusing on differential privacy and applying the proposed simulation to a real-world application with data coming from different medical institutions.

Acknowledgments. This work has received funding from Horizon 2020, the European Union's Framework Programme for Research and Innovation, under grant agreement No. 860627 (CLARIFY), the Spanish Ministry of Economy and Competitiveness through project PID2019-105142RB-C21 (AI4SKIN) and GVA through projects PROMETEO/2019/109 and INNEST/ 2021/321 (SAMUEL). The DGX A100 used for this work was donated by the Generalitat Valenciana (GVA), action co-financed by the European Union through the Operational Program of the European Regional Development Fund of the Comunitat Valenciana 2014–2020 (IDIFEDER/2020/030).

References

1. del Amor, R., et al.: An attention-based weakly supervised framework for spitzoid melanocytic lesion diagnosis in whole slide images. Artif. Intell. Med. (2021). https://doi.org/10.1016/j.artmed.2021.102197
2. Andreux, M., du Terrail, J.O., Beguier, C., Tramel, E.W.: Siloed federated learning for multi-centric histopathology datasets. In: Albarqouni, S., et al. (eds.) DART/DCL -2020. LNCS, vol. 12444, pp. 129–139. Springer, Cham (2020). https://doi.org/10.1007/978-3-030-60548-3_13
3. Barnhill, R.L.: The Spitzoid lesion: rethinking Spitz tumors, atypical variants, 'Spitzoid melanoma' and risk assessment. Mod. Pathol. **19**(2), S21–S33 (2006). https://doi.org/10.1038/modpathol.3800519
4. Brisimi, T.S., Chen, R., Mela, T., Olshevsky, A., Paschalidis, I.C., Shi, W.: Federated learning of predictive models from federated electronic health records. Int. J. Med. Informatics (2018). https://doi.org/10.1016/j.ijmedinf.2018.01.007
5. Campanella, G., et al.: Clinical-grade computational pathology using weakly supervised deep learning on whole slide images. Nat. Med. **25**(8), 1301–1309 (2019). https://doi.org/10.1038/s41591-019-0508-1

6. Coudray, N., et al.: Classification and mutation prediction from non-small cell lung cancer histopathology images using deep learning. Nat. Med. **24**(10), 1559–1567 (2018)

7. Hekler, A., et al.: Pathologist-level classification of histopathological melanoma images with deep neural networks. Eur. J. Cancer (2019). https://doi.org/10.1016/j.ejca.2019.04.021

8. Ilse, M., Tomczak, J.M., Welling, M.: Attention-based deep multiple instance learning. In: 35th International Conference on Machine Learning, ICML 2018 (2018)

9. Ke, J., Shen, Y., Lu, Y.: Style normalization in histology with federated learning. In: Proceedings - International Symposium on Biomedical Imaging (2021). https://doi.org/10.1109/ISBI48211.2021.9434078

10. Konecný, J., McMahan, H.B., Ramage, D., Richtárik, P.: Federated optimization: distributed machine learning for on-device intelligence. arXiv:1610.02527 (2016)

11. Li, X., Gu, Y., Dvornek, N., Staib, L.H., Ventola, P., Duncan, J.S.: Multi-site fMRI analysis using privacy-preserving federated learning and domain adaptation: ABIDE results. Med. Image Anal. (2020). https://doi.org/10.1016/j.media.2020.101765

12. Lu, M.Y., et al.: Federated learning for computational pathology on gigapixel whole slide images. Med. Image Anal. (2022). https://doi.org/10.1016/j.media.2021.102298

13. Rieke, N., et al.: The future of digital health with federated learning. NPJ Digit. Med. **3**(1), 1–7 (2020). https://doi.org/10.1038/s41746-020-00323-1

14. Shaw, S., Pajak, M., Lisowska, A., Tsaftaris, S.A., O'Neil, A.Q.: Teacher-student chain for efficient semi-supervised histology image classification. arXiv:2003.08797 (2020)

15. Shin, H.-C., et al.: Medical image synthesis for data augmentation and anonymization using generative adversarial networks. In: Gooya, A., Goksel, O., Oguz, I., Burgos, N. (eds.) SASHIMI 2018. LNCS, vol. 11037, pp. 1–11. Springer, Cham (2018). https://doi.org/10.1007/978-3-030-00536-8_1

16. Simonyan, K., Zisserman, A.: Very deep convolutional networks for large-scale image recognition. In: 3rd International Conference on Learning Representations, ICLR 2015 - Conference Track Proceedings (2015)

17. Xie, Q., Luong, M.T., Hovy, E., Le, Q.V.: Self-training with noisy student improves imagenet classification. In: Proceedings of the IEEE Computer Society Conference on Computer Vision and Pattern Recognition (2020). https://doi.org/10.1109/CVPR42600.2020.01070

Automatic Exploration of Domain Knowledge in Healthcare

Tiago Afonso and Cláudia Antunes[✉]

Instituto Superior Técnico, Universidade de Lisboa,
Av. Rovisco Pais 1, Lisboa, Portugal
{tiago.francisco.a,claudia.antunes}@tecnico.ulisboa.pt

Abstract. Throughout the years, healthcare has been one of the privileged areas to apply the information discovery process, empowering and supporting medical staff on their daily activities. One of the main reasons for its success is the availability of medical expertise, which can be incorporated in training models to reach higher levels of performance. While this has been done painfully and manually, during the preparation step, it has become hindered with the advent of AutoML. In this paper, we present the automation of data preparation and feature engineering, while exploring domain knowledge represented through extended entity-relationship (EER) diagrams. A COVID-19 case study shows that our automation outperforms existing AutoML tools, such as auto-sklearn [4], both in quality of the models and processing times.

1 Introduction

Over the centuries, physicians have cared for the ill, in both body and mind, gaining knowledge to enhance their skills. This empirical knowledge was then used to help new patients and to develop new techniques, reaching a level of complexity and scope far away from traditional medicine.

Nevertheless, research in learning algorithms has not evolved at the same pace as the remaining steps of the discovery process. In particular, data preparation continues to occupy a large part of the processing time, requiring the careful selection of a set of transformations to apply to the data. One of the biggest challenges is the need to explore the domain knowledge available, especially on choosing the most promising variables for describing the cases. While the automation of other preparation tasks has been proposed, methods for variables generation based on domain knowledge are still incipient [14].

In this paper, we describe a methodology for automating the data preparation step, including variable generation through the exploration of domain knowledge, expressed in an entity-relationship (ER) diagram. Moreover, we discuss and show how to apply it for modelling the evolution of the COVID-19 pandemic situation around the world. Experimental results show a significant increase in model performance with the generation of the new features, keeping the processing times negligible, especially when compared to an AutoML framework.

© The Author(s), under exclusive license to Springer Nature Switzerland AG 2022
H. Yin et al. (Eds.): IDEAL 2022, LNCS 13756, pp. 73–81, 2022.
https://doi.org/10.1007/978-3-031-21753-1_8

The rest of the paper is structured as follows: next in Sect. 2, we discuss how data science have been used for supporting healthcare, and the main challenge regarding the exploration of domain knowledge to enhance its results. After a description of the DANKFE automation approach in Sect. 3, we present its application in predicting the evolution of COVID-19 in diverse countries - Sect. 4. There are discussed the results achieved with the proposed approach, comparing them against results reached with another automation tool (auto-sklearn [4]). The paper concludes with a critical analysis of the difficulties faced and some guidelines for future work.

2   Background

Over the last thirty years, the application of the knowledge discovery process in the healthcare area has been a constant, having possibly benefited the most from it [7]. In some manner, the field has also contributed deeply to the advancement of the discovery process, its methodologies and methods, bringing together in a symbiotic process. The history of this cooperation is not under the scope of this work, but data science based research in cancer [11], dementia [15] and even exploring AutoML approaches [14], count with thousands of publications.

The challenges brought by healthcare to the machine learning and data science research are vast [13], with the exploration of the domain knowledge available being one of the less developed. Feature engineering is one of the most used paths to use this knowledge, usually through the manual definition of variables, which further contributes to increase the most time-consuming step of the process [14]. This can be costly, subjective and limits the process's repeatability.

To counteract this, there have been numerous works on automating feature generation, with or without the use of domain knowledge to improve induction. Data-driven approaches such as feature combination or aggregations [8], or hypothesis-based approaches using Decision Trees [10] were among the first methods. These methods do not make use of any domain knowledge, which is proven to be beneficial for data mining, even in a fragmented state [2]. Works that utilize domain knowledge extract information from known knowledge bases [6], and others from textual sources [5]. [12] uses available knowledge repositories such as ontologies to match candidate terms and extend the dataset.

The increased interest on the automation of the KDD process, led to a spread of AutoML frameworks in recent years, which goal is to return the best model for a dataset with as little human intervention as possible [9]. Auto-sklearn [4] is one of the best known among a vast number of those solutions.

3   DANKFE – DomAiN Knowledge Based Feature Engineering

Data preparation is one of the keystones in the information discovery process, occupying around 80% of its total time [14]. Furthermore, it encompasses a set of operations that require critical reasoning. Even though these operations usually do not require any creativity, they are difficult to implement without that knowledge.

The DANKFE approach provides a framework to perform such exploration, by using Extended Entity-Relationship diagrams (EER) [3] to represent domain knowledge. Given a dataset and a corresponding EER diagram, it is possible to create a new dataset by transforming and extending the original one, through the diagram exploration. This extension is accomplished through the generation of a new variable for each relationship in the EER diagram, and extending all records by computing their values for the new variable. The operations to be used to compute the new variables can be one of five distinct types: *decomposition operations* - apply an operation on a single variable from each record, extracting some component from its value (*year, month, day* are examples for dealing with dates); *algebraic operations* - apply any mathematical operation over a single record, involving one or more variables (examples are *square root* and *division*); *mapping operations* - map a variable's value into another one, possibly from different types (*weekday* is an example over dates); *aggregation operations* - are applied over a set of records that satisfy some imposed condition similar to the ones achieved with a *GROUPBY* clause in an *SQL* query (for example *sum, average, max*); *composition operations* - apply a sequence of operations similar to a mathematical composition of functions.

The preparation phase additionally requires the transformation of the data, which can be done in different ways depending on the type of operation. While the first three kinds of operations work over singular records, and are designated *record-based*, aggregation operations analyse multiple records to fill each one's value, and are called *aggregation-based*.

Extending the dataset with the addition of new *record-based* variables is straightforward and can be managed by applying the required function to each record in the dataset independently and in parallel. For efficiency reasons, each record can be extended with the values for all the new variables simultaneously. *Aggregation-based generation* becomes trickier since we need to make some computation involving more than one record before applying the generation itself. Nevertheless, if the aggregation is the same for several new variables, we can generate them simultaneously.

Furthermore, aggregation-based operations allow for the generation of more interesting variables, in particular exploring data temporality, if available. Take as an example the number of patients with COVID-19 in the last two weeks. Indeed, this can be computed by picking each day and the corresponding fourteen preceding ones, and summing the number of cases for them all. Naturally, if the data report for more than one location, we may also group records by local, and apply the same reasoning. This kind of variable is particularly useful to be used as a target variable, making possible to implement a prognosis task.

The DANKFE algorithm works as follows: relations are read from the EER diagram, and stored as a queue to be processed. If its inputs are already available, the algorithm is ready to fill the new variable for each record in the original dataset, and so the relation is processed, otherwise the relation is sent to the end of the queue. If the relationship involves an aggregation operation, then we need to collect the rows to perform the aggregation. After this is then possible to apply the list of operations creating the new values for the variable, for all the records satisfying the constraints imposed. The operations are applied as a composition of functions, beginning with the last one and sequentially applying the following ones. Whenever any row does not meet the constraints, a null value is imputed. When all rows are processed, the relation is removed from the queue.

4 Case Study: Prediction During COVID-19 Pandemic

The emergence of COVID-19 has brought a new level of demand to public health services. These institutions were not prepared to receive a significant larger number of patients than usual, and even worst, infectious ones.

The nonexistence of data in the early days was rapidly overcome with the collection of data around the world[1], which allowed the application of data science tools to understand the new levels of demand. The prediction of high risk situations was of particular interest, such as if a day would reach over 120 new cases per 100k persons in the preceding fourteen days. Being able to predict in advance if we will be in a high risk situation for a given point in time ahead is fundamental to adapt strategies and politics, such as taking healthcare administrative decisions or managing confinement measures. Other similar numbers, such as the cumulative number of deaths or new sick patients, could also be considered [1]. The data collected contains the number of newly reported cases and deaths, per country per day, and the *Eurostat 2020* data regarding country's population.

Using these as our data source, we trained models for predicting *high risk days* - days reaching a cumulative sum of new cases in the preceding fourteen days, that is over 120 per one hundred thousand persons. As seen, this target class is not explicitly provided in the data, and it was computed following the DANKFE approach proposed. Figure 1 depicts the EER diagram used, expressing simple domain knowledge to guide the process.

As usual, the diagram depicts entities through rectangles and relationships through diamonds. White rectangles represent the original variables, and blue ones the generated ones. current_date_day, current_date_month and current_date_year are decomposed from current_date, while season and current_date_weekday are generated via mapping operations. We also generate first_date as the earliest date reported per country, and nr_months as an algebraic operation between these two.

Ratio is the ratio between deaths and cases, and from the latter and population, we create cases_per_100k, both using algebraic operations.

Additionally, from cases and deaths constrained by country, several other variables are computed to extract some basic statistics, such as mean and

---

[1] https://www.ecdc.europa.eu/en/covid-19/data.

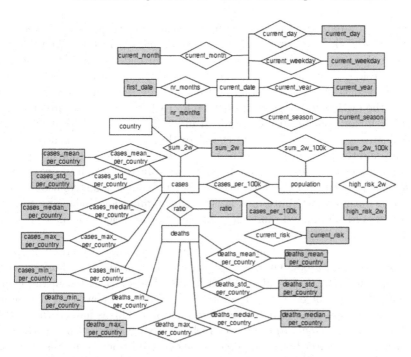

Fig. 1. EER diagram used for the datasets.

standard deviation (`cases_mean_per_country` and `deaths_std_per_country` for example). All these new variables require aggregation-based operations, constrained by a single variable (the `country`).

Even more interesting and less simple to compute, there is the `sum_2w` variable, which results by an aggregation-based operation, summing the number of `cases` again constrained by `country`, but also by the `current_date`, imposed to be in the 14 days range of the record's `current_date`. From this new variable, we then derive `sum_2w_100k` from it and `population`, just as before.

Comparing `cases_per_100k` against the reference value of 120 new cases in the previous 14 days per 100.000 inhabitants, we are able to label each day as an effective high-risk one, through the `current_risk`. This variable can then be shifted 14 days in each single country for classifying each record according to the risk 14 days ahead. This is expressed as the `high_risk_2w` variable, which label our dataset and corresponds to our target.

Now that we have an enriched and labeled dataset, we are able to enunciate our analysis goal as: given a current date, country and its corresponding number of cases and deaths, predict if the country will be in high risk in fourteen days.

4.1 Experimental Results

The validation of the DANKFE approach is done through its application to the data and EER described above, addressing the goal established. For this matter, we compared the quality of the classification models trained over the original data against the ones learnt from its preparation: first, just using scaling and balancing transformations, and then, through the feature generation process. Beside the performance of the models learnt, we studied the impact on the time spent training and predicting those models, as well as the importance given by the models to the generated variables. Additionally, we compared those results with the resulting from training a model for the same problem through auto-sklearn, while also evaluating the algorithm's scalability. (Datasets, EER diagrams and the python code, will be made publicly available upon acceptance).

As a manner to have more sound results, we split the data into five datasets, one for each continent, all of them reaching unbalanced distributions. When comparing variable distribution, it is clear that scaling and balancing transformations should be applied, and the same for feature selection for some of the continents.

The set of experiments consisted on three steps: first, by training models over the original datasets, without applying any data preparation beside missing value imputation by mean value, which is identified as *Baseline* in the charts. The second one involved just the application of usual preparation techniques, scaling through z-score and balancing by replication. This is named *Base+Prep*. Third, we have the *Generation+Prep* which covered the train of models over the extended datasets after being prepared as before. Additionally, auto-sklearn was run over the original dataset (*auto-sklearn*).

For each dataset, several training technique (Naive Bayes, KNN, Decision Trees, Random Forests and Gradient Boosting, all implemented with the sklearn package) were used to find the 10 best models trained over an equal number of random data partitions, and exploring a fixed grid of hyperparameters. While running auto-sklearn, only one model was found. AUC was chosen to assess models quality, due to the unbalanced nature of all datasets.

Results in Fig. 2 (left) show that on average, all datasets benefit both from preparation and feature generation guided by domain knowledge, increasing their performance. While data preparation, such as scaling and balancing can improve results in the baseline, feature generation increases AUC by 5 to 10% points (pp) for the generality of methods used. It is interesting to see that auto-sklearn presents the same performance as the best of techniques over the baseline (GradientBoosting), and that Naive Bayes, the simplest technique, is the one that benefits most, gaining more than 30pp.

Figure 2 (right) shows the average time per record spent by each training technique (including the time spent running DANKFE-II for extending the dataset). In this figure, we can see the time spent by the new approach increases significantly when compared against mining the original dataset. This increase stays somewhat constant independent of the model used. However, while feature generation takes longer, it significantly increases model performance as seen above.

When compared to auto-sklearn, our results are much better. This AutoML tool spends a fixed time of one hour to find the best model, which in average corresponds to around 300 milliseconds per record on the datasets used. Using

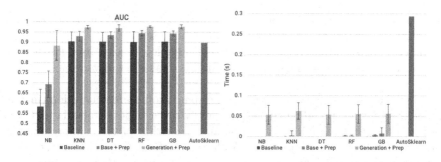

Fig. 2. Models quality (left) and processing times running different algorithms (right).

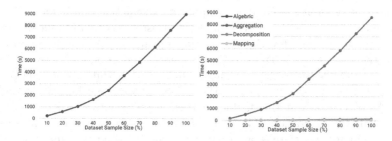

Fig. 3. Scalability study: total time on variable generation (left) per types of variables generated (right).

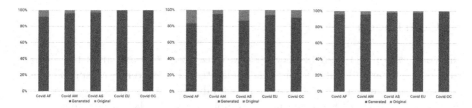

Fig. 4. Average feature importance for original and generated variables for Decision Trees (left), Random Forests (middle) and Gradient Boosting (right).

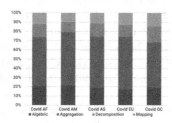

Fig. 5. Time comparison per types of variable generated.

our approach, on average we spent between 10% and 20% of the time spent by auto-sklearn, depending on the training technique used.

In terms of scalability, we took different samples of a dataset (with 135k records) with progressive sizes, and generated the same set of variables for each one.

Figure 3 shows the total amount of time spent on feature generation (including the time spent on reading the original and writing the extended datasets) on the left. From this chart it is clear that our algorithm presents approximately a linear growth on the dataset size. Additionally, from the chart on the right, that shows the time spent per type of operation, we see that the time spent per non-aggregation operations is residual, and almost all the time is spent on generating aggregation-based variables.

At last, the variables generated also proved extremely useful in increasing the performance of the models, as seen in Fig. 4. Most datasets and models greatly benefited with generated variables, making up almost all feature importance used for training Decision Trees, Random Forests and Gradient Boosting.

The type of variable generated also influences the time spent (Fig. 5). On average, aggregation operations are the ones that take the longest to run (at least 50% of the overall time), due to their need to check other records for the GROUP BY operation. The other 3 types of operations (decomposition, algebraic and mapping) take approximately similar times to run, since these operations do not have any row dependence when computing the value to fill each record.

5   Conclusion

The cooperation between healthcare and data science researchers has been a fruitful endeavor, allowing for mutual benefits. The existence of specific challenges, in particular the need to explore available domain knowledge, makes the healthcare field one of the best suited to motivate and validate new methodologies to do it.

In this paper, we propose to use the DANKFE approach, which defines a framework for performing data preparation, in particular feature generation, through the exploration of an EER diagram. Experimental results show an improvement on the quality of models between 5 and 10% points when compared to the original data, and better results in the same range when compared with auto-sklearn. In terms of processing times, the results are even better, with the DANKFE approach spending about 20% of auto-sklearn's processing time, and a linear growth.

Beside the types of operations proposed, new ones shall be added, and simultaneously the DANKFE-II algorithm shall be optimized to generate more than one variable per time. Additionally, the specification of operations shall be simplified, following the work done for axiom interpretation in the context of ontologies and knowledge representation.

Acknowledgments. This work was supported by national funds by Fundação para a Ciência e Tecnologia (FCT) through project VizBig (PTDC/CCI-CIF/28939/2017).

References

1. Adam, D.: A guide to R - the pandemic's misunderstood metric. Nature **583**(7816), 346–348 (2020)
2. Donoho, S., Rendell, L.: Feature construction using fragmentary knowledge. In: Liu, H., Motoda, H. (eds.) Feature Extraction, Construction and Selection. SECS, vol. 453, pp. 273–288. Springer, Boston (1998). https://doi.org/10.1007/978-1-4615-5725-8_17
3. Elmasri, R., Navathe, S.B.: The Enhanced Entity-Relationship (EER) Model, pp. 107–135. Addison-Wesley, Boston (2000)
4. Feurer, M., Eggensperger, K., Falkner, S., Lindauer, M., Hutter, F.: Auto-sklearn 2.0: hands-free autoML via meta-learning. arXiv preprint arXiv:2007.04074 (2020)
5. Gabrilovich, E., Markovitch, S.: Wikipedia-based semantic interpretation for natural language processing. J. Artif. Intell. Res. **34**, 443–498 (2009)
6. Galhotra, S., Khurana, U., Hassanzadeh, O., Srinivas, K., Samulowitz, H., Qi, M.: Automated feature enhancement for predictive modeling using external knowledge. In: 2019 International Conference Data Mining Workshops, pp. 1094–1097. IEEE (2019)
7. Garg, A., Mago, V.: Role of machine learning in medical research: a survey. Comput. Sci. Rev. **40**, 100370 (2021)
8. Hu, Y.J., Kibler, D.: Generation of attributes for learning algorithms. In: AAAI/IAAI, vol. 1, pp. 806–811 (1996)
9. Hutter, F., Kotthoff, L., Vanschoren, J.: Automated Machine Learning: Methods, Systems, Challenges. Springer, Cham (2019). https://doi.org/10.1007/978-3-030-05318-5
10. Markovitch, S., Rosenstein, D.: Feature generation using general constructor functions. Mach. Learn. **49**, 59–98 (2004)
11. Momeni, Z., Hassanzadeh, E., Abadeh, M.S., Bellazzi, R.: A survey on single and multi omics data mining methods in cancer data classification. J. Biomed. Inform. **107**, 103466 (2020)
12. Salguero, A.G., Medina, J., Delatorre, P., Espinilla, M.: Methodology for improving classification accuracy using ontologies: application in the recognition of activities of daily living. J. Ambient Intell. Hum. Comput. **10**(6), 2125–2142 (2019)
13. Shilo, S., Rossman, H., Segal, E.: Axes of a revolution: challenges and promises of big data in healthcare. Nat. Med. **26**(1), 29–38 (2020)
14. Waring, J., Lindvall, C., Umeton, R.: Automated machine learning: review of the state-of-the-art and opportunities for healthcare. Artif. Intell. Med. **104**, 101822 (2020)
15. Yang, H., Bath, P.A.: The use of data mining methods for the prediction of dementia: evidence from the English longitudinal study of aging. IEEE J. Biomed. Health Inform. **24**(2), 345–353 (2020)

On Studying the Effect of Data Quality
on Classification Performances

Roxane Jouseau[✉], Sébastien Salva, and Chafik Samir

University of Clermont Auvergne, Clermont-Ferrand, France
roxane.jouseau@doctorant.uca.fr, {sebastien.salva,chafik.samir}@uca.fr

Abstract. During the last decade, data have played a key role for learning and decision making models. Unfortunately, the quality of data has been ignored or partially investigated as a pre-processing step. Motivated by applications in various fields, we propose to study data quality and its impact on the performance of several learning models. In this work, we first study the difficulty of repairing errors by introducing a list of elementary repairing tasks ranging from easy to complex with an increasing level. Then, we form categories from the state-of-the-art cleaning and repairing methods. We also investigate if it is always efficient to repair data. By including standard classifications models and public dataset, our work enables their use in different contexts and can be extended to other machine learning applications.

Keywords: Data quality · Data engineering · Data cleaning · Data repairing · Classification · Machine learning

1 Introduction

The field of data cleaning and data repairing is very active. The literature is rich in methods and tools to tackle data cleaning tasks, e.g.: HoloClean [15], CleanML [11], ZeroER [16], BoostClean [10]. Confronted with this plethora of approaches, data scientists have to choose which repairing method to use. This choice is still very complex. Indeed, they have different computational times and effectiveness. Moreover, choosing a method solely according to those criteria overshadows another question: what does specific methods requires? Some approaches require complex metadata that should be accounted for since the production of this metadata takes part in the repairing process, if it was not previously available. Data scientists are not always aware of the difficulties of using data repairing tools or if they perform better enough to justify the extra work that is necessary to produce the required metadata. The amount of errors in the data also has an impact, as repairing a dataset with only a few erroneous entries would be approached differently than repairing a severely degraded dataset. In some cases, data are too deteriorated to be exploited, and we only discover this information after spending time and effort repairing them.

© The Author(s), under exclusive license to Springer Nature Switzerland AG 2022
H. Yin et al. (Eds.): IDEAL 2022, LNCS 13756, pp. 82–93, 2022.
https://doi.org/10.1007/978-3-031-21753-1_9

All these factors lead us to the question: Is it always better to repair data? This paper aims to partially answer this question, by focusing on the case of unstructured and structured numeric dataset in the context of classification. We tackle the question through five criteria:

- C1: The perceived difficulty of using a method according to experts. This is a piece of crucial information when choosing a repairing method as it can lead to spending a huge time working on metadata.
- C2: The impact of the data degradation. This investigates how classification models perform with untreated errors in comparison to repaired data. This comparison will be helpful in identifying how repairing methods improve the accuracy and f1 score of classification tasks. Moreover, it studies how different levels of degradation affect the results of classification
- C3: The effectiveness of the repairing tool. This aims to observe the repairing effectiveness of repairing tools. Moreover, it compares the difference in effectiveness between repairing methods that are simple to use and complex ones.
- C4: The impact of the type of error present in data. This allows us to identify differences in accuracy and f1 scores of classification methods with different types of error in training data. This information is valuable for the decision-making process of repairing data.
- C5: The impact of the classification model. This aims to verify whether the accuracy and f1 score of classification models are affected similarly by data errors.

Other works related to this problematic are given in: CleanML [11]. It investigates the impact of data cleaning on classification, in [6] which proposes a taxonomy of the data cleaning literature, and in [1] which investigates whether data cleaning tools are robust enough to capture most real world errors.

In this paper, we restrict our study to 5 types of errors from literature: missing values [11,15], exact and partial duplicates [8,11,16], domain value violations [10] and outliers [8,11]. To study these errors, we create deteriorated datasets by injecting data errors in clean datasets. Artificially injecting errors allow us to control the quantity of errors in data, their type, and to have access to a reference version for evaluation.

Next, we selected repairing methods based on the type of data they target, the error types to repair, and the availability of the metadata needed to use them on the datasets we chose. For missing values, the choice of the algorithm is very dependent on the repairing approach we want to take as the detection is trivial. A simple approach would be to impute the values with statistical data such as the median (R_med) or mean value (R_mean) of the attribute or by taking into account the correlation between dataset attributes (R_correl) [11]. For the repairing duplicated data, we consider exact or partial copies. Dataset with keys or not, if dataset have primary keys, using a key-collision (R_key) method is a simple solution [11]. Otherwise data need to be compared for equality. For partial duplicates, it's more complex as we first need to define a measure of when two records are considered partial duplicates before trying to repair them. The definition of a threshold for partial duplicates is case-specific, but tools like ZeroER (R_ZER) [16], which we use in this study, can detect them. For outliers, we considered 4 methods based on: the standard deviation (R_std), the interquartile range (R_quart), the interquantile range (R_quant) and the data linter (R_linter) [8,11]. For domain value violations, a combination of regular expressions and types checks is usually used to detect this type of error (R_check) [10].

Since this study focuses on classification, we have selected several well known classification models through scikit-learn [13]: Logistic regression (Cl_LR), K-Nearest Neighbors (Cl_KNN), Decision tree (Cl_tree), Random forest (Cl_rdForest), Ada boost (Cl_AdaB), Naive Bayes (Cl_NB), XGboost (Cl_XGB), Support vector classification (Cl_SVC), Gaussian process (Cl_GP), Multi-layer perceptron (Cl_MLP), Stochastic gradient descent (Cl_SGD), and Gradient boosting (Cl_GB). We chose these models in order to cover multiple classification models.

To answer the question "Is it always better to repair data?", we first present the context and scope of our work. Then, we study C1 and propose a method to evaluate how difficult is a repairing method to use. Secondly, the criteria C2 to C5 are studied through experiments that allow the observation of the impact of data cleaning and repairing on classification tasks and analyze results. These criteria allowed us to identify two categories of error type: low impact and high impact on accuracy and f1 score. Moreover we also observed that repairing methods perform similarly at very low and very high levels of errors (10% and 80%) but differ in between. This experiment was conducted on datasets adapted to classification that have originally very low percentages of errors. They also cover very different domains and have different dimensions and sizes. Last but not least, they are free to use, and are also used in many papers and hence increase the reproductibility of our results.

The paper is organized as follow: Sect. 2 investigates C1 and proposes an evaluation of the difficulty of using a repairing method. Section 3 presents the experiment we designed with its empirical setup and the analysis of the results corresponding to C2, C3, C4, and C5. Finally, we discuss the possible threats to validity and our results in Sect. 4 and conclude in Sect. 5.

2   C1: The Perceived Difficulty of Using a Method According to Experts

Repairing methods are traditionally evaluated and compared using effectiveness (their ability to produce the desired outcome), accuracy (how close they are to the ideal outcome), and performance (how fast they are). However, this gives a fair comparison. For instance, some tools require more complex metadata inputs than others ([7,15] etc.). The time and efforts put into creating this metadata are often complicated to quantify and therefore disregarded. That is one of the reasons why data scientists rarely use them in the industry.

To deal with this, we propose an evaluation process which breaks down the repairing methods discussed in the introduction into steps and sub-steps, until

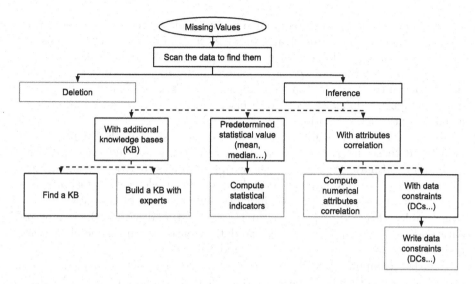

Fig. 1. Elementary tree decomposition for missing values.

we obtain elementary tasks (i.e. small actions difficult to split) describing the actions needed to apply these methods. Including creating the metadata. Given an error type e, and a repairing method R_{ei}, we build a tree expressing the steps of R_{ei}. For any other repairing method R_{ej} we complete the same tree with new

Table 1. Difficulty ratings of the elementary tasks of data repairing.

Elementary tasks	Estimated difficulty (Easy, Medium, Medium +, Hard)
Compute statistical indicators	1.89
Delete data	1
Mining regexp	2
Mining data constraints	2.38
Write data conversion scripts	1.63
Compute attributes correlation	1.75
Write data harmonisation scripts	2.44
Define similarity metrics	2.88
Data scientist check the data (for miscoding)	2.13
Write data format rules (regexp...)	2.44
Write data constraints (DCs...)	2.71
Build a knowledge base with experts	3.33
Set a threshold (for partial duplicates detection)	2.67
Write a probabilistic model	3
Define a metric (for outliers detection)	3.56

sub-steps and branches when required. The final nodes of the tree are elementary tasks. For example, Fig. 1, illustrates the tree achieved for the error type missing values with regards to the repairing methods [5,7,11,15]. These elementary tasks (nodes in red) are then evaluated by experts in terms of difficulty to complete them independently of the complete approaches. These evaluations allow us to finally compute a difficulty score for every repairing method as a summation of its elementary task evaluations.

To quantify the difficulty of each elementary task, we asked a panel of 8 industry data scientists to rank them on a four values scale: easy, medium, medium+, and hard. We chose a scale with four values to avoid having answers in the middle. We compute the difficulty score of an elementary task dt_t as a weighted arithmetic mean over the difficulty rankings given by the data scientists panel Table 1. $dt_t = \sum_{v=1}^{n} \frac{\delta_t(v)}{n}$, with $1 \leq dt_t \leq 4$, and with n the number of data scientists, and $\delta_{task}(v) = 1$, if the data scientist v ranked the elementary task t easy, 2 if medium, 3 if medium+ and 4 if hard.

We then compute the difficulty score of a repairing method dm_k using the difficulty scores of the elementary tasks included in the method. $dm_k = \frac{\sum_{i \in M} dt_i}{\sum_{j \in T} dt_j}$, with $0 \leq dm_k \leq 1$, and with T the set of all elementary tasks, and M the set of all elementary tasks used by the repairing method k. Table 2 illustrates the scores obtained for the repairing methods introduced in the introduction.

Table 2. Difficulty scores of repairing methods.

Repairing method	Elementary tasks included	Total estimated difficulty score
R_med, R_mean	Compute statistical indicators	0.053
R_correl	Compute statistical indicators, compute attributes correlation	0.102
R_key	Delete data	0.028
R_ZER	Set a threshold for partial duplicates detection	0.075
R_std, R_quart, R_quant	Compute statistical indicators, define a metric for outlier detection, delete data	0.18
R_linter	Compute statistical indicators, delete data	0.081
R_check	Write data conversion scripts	0.046

Table 1 shows that there are clear disparities in the perceived difficulties of using repairing methods. For instance R_correl is more complex than R_mean or R_med according to experts. We are now able to use these evaluations to compare two repairing methods before using them. This is especially useful in the case where we expect the two repairing methods to have similar effectiveness. We develop more on this subject in the following section.

3 How Good Is a Repairing (Study of C2 to C5)

Below, we study the criteria C2 to C5 presented in the introduction through experiments on deteriorating and repairing numerical data. Due to space limitation we will only present a brief summary of our conclusions for C4 and C5. A technical report with additional figures is available [9].

3.1 Empirical Setup

All dataset used in these experiments are adapted to classification. They have very low percentages of errors. They are all numerical with different properties to cover a large panel of applications. Some are structured, some are not structured, some are from *real-world* data, whereas others are curated. They also come in various sizes. Dataset are: Mnist, Fashion-Mnist, Olivetti, Iris, Adult, Breast cancer, and Wine [3,13,17]. We decided to limit the global computing time to under a week for each dataset. For this reason, we do not use the complete datasets for Fashion-mnist and Adult, but reduced versions of them (700 entries for Fashion-mnist and 2000 for Adult).

In this experiment, we start by splitting datasets to training and test. Then we apply a panel of modifications: We first inject the training dataset with one type of error e at a percentage p varying from 0 to 95% with increments of 5%. We apply each repairing method R_{ei} to different copies of the deteriorated dataset to obtain repaired datasets. We then use repaired dataset to train several classification models Cl_X. Finally, we compute the accuracies and f1-scores by means of the testing sets. We executed the complete process 30 times to reduce the bias for each percentage p. We summarize all the steps of the experiment in Fig. 2.

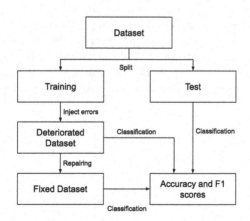

Fig. 2. Protocol of the experiment.

We remind that we use the errors, repairing methods and classification models presented in the introduction. Our experiments aim to study C2 to C5: Impact of the data degradation, effectiveness of the repairing tool, impact of the type of error, and impact of the classification model. The following sections show the experimental results related to the accuracy. For space reasons but the experimental results for the f1 scores follow those of the accuracies.

3.2   C2: Impact of the Degradation of the Data on Repairing Effectiveness

In order to observe the impact of the degradation of data on the effectiveness of the repairing methods for each type of error, we randomly injected increasing percentages of errors in the datasets from 0 to 95% over the total amount of data. We trained the classification models on these data before and after repairing. In Fig. 3, we respectively depict the mean accuracies of the classification models before and after repair, on all the datasets by error type as a function of the percentage of errors injected in the data. From Fig. 3 (left), we identify two distinct categories of error types: 1. the data degradation level has little to no impact on the accuracy, and 2. the data degradation level seems to have a big impact. The error types: domain value violations, exact duplicates, and partial duplicates belong to the first category, while missing values and outliers belong to the second category. Therefore, the impact of repairing is more interesting to observe for the second category. We also note a slight improvement of the accuracy for partial duplicates. This improvement actually is only observed for two datasets: wine and breast cancer [13]. It is caused by the fact that in this scenario creating partial duplicates act as enriching the data. For the other datasets, data is denser this is why adding partial duplicates has no effect on them. Domain value violations belonging to the first category can be explained by the fact that we need to repair domain value violations since we can't use an attribute with mixed data types for learning. From Fig. 3 (right) we observe the same categories as we did before repairing the data. However the repairing of outliers seems to be less effective than the repairing of missing values. Since outliers are defined relatively to the rest of the data the repairing methods R_std, R_quart, R_quant and R_linter were defined statistically dependant to the data. They therefore become very quickly undetectable since having a lot of outliers

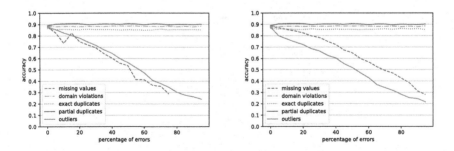

Fig. 3. Accuracy score before (left) and after (right) repairing.

populate the extremes of data values and they stop being distant from the data. The notion of a large amount of outliers is still relatively small for example, if 10% of data values are outliers, this is a significant number and cannot be considered distant from the data anymore.

We can clearly see that the impact of the degradation on data is very different depending on which category of error we observe. For the first category, the degradation of the data have very little impact on accuracy and so does repairing errors. However in the second category the degradation of the data has a strong impact on the accuracy. For example, outliers very quickly become too many to be able to detect them with simple statistical indicators. Repairing the missing values works better for higher percentages of degradation but we can see that we fall under a mean accuracy of 0.8 after 30% of missing values present in the data.

3.3   C3: Effectiveness of the Repairing Tools

To study the effectiveness of the repairing tools considered in the paper, we randomly injected errors at increasing percentages from 0 to 95% in our training sets. We then repaired them with different repairing methods, trained the classification models on those datasets and computed their respective accuracies and f1-scores. We only show the repairing of missing values and outliers in Figs. 4 and 5 as their degradation has the most impact on accuracy as we saw in Sect. 3.2. Figure 4 shows the accuracies for 10%, 25%, 50%, and 75% of missing values after repairing with the methods R_med, R_mean, and R_correl. Figure 5 depicts

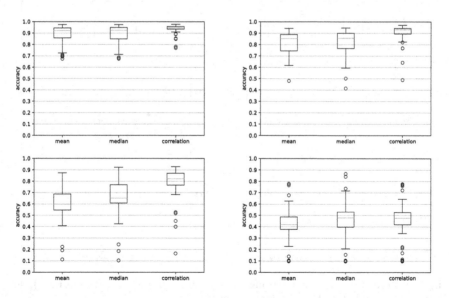

Fig. 4. Effectiveness of some repairing tools for 10% (top left), 25% (top right), 50% (bottom left), and 75% (bottom right) of missing values.

the accuracies for 10%, 25%, 40%, and 50% of outliers after repairing with the methods R_std, R_quart, R_quant, and R_linter.

For missing values, with Fig. 4 we observe that the repairing method using attributes' correlation is more efficient than imputation with the mean or median before 75% of missing values but performs similarly afterward. Repairing with R_mean or R_med seems to give very similar results before 50% of missing values but starting from 50% R_med gives slightly better results.

The choice of the repairing method, in this case, cannot be based solely on effectiveness as the methods perform more and more similarly as the percentage of missing values increases. Assessing an approximate percentage of errors in the dataset first and choosing a repairing tool afterward seems to be a good strategy in terms of efficiency and effectiveness (use of the simplest tool for the best outcome).

For outliers (Fig. 5) the effectiveness of the different repairing tools seems to be less distinct than for missing values. However, the inter-quartiles and quantiles approaches seem to perform similarly, which is not surprising as they have similar concepts. Around 50% of degraded cells and up, none of the repairing methods we studied seem to perform well in terms of accuracy. This makes sense as outliers are, by definition, entries distant from the rest of the data. If there is a significant amount of them, they are not statistically distant anymore and thus undetectable with statistical indicators such as the standard deviation, quantiles, or quartiles. With high levels of outliers, repairing tools become less and less effective and their interquartile ranges seems to grow.

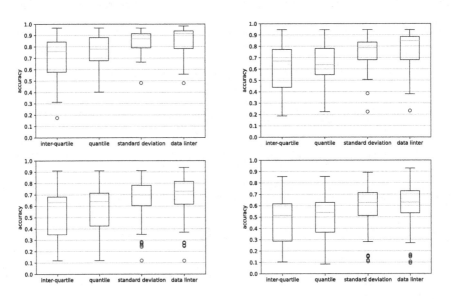

Fig. 5. Effectiveness of some repairing tools for 10% (top left), 25% (top right), 40% (bottom left), and 50% (bottom right) of outliers.

Overall, our experiment shows that the repairing methods does have an impact on the accuracy, some methods do perform better than other on average, but at high levels of degradation the effectiveness of the different repairing methods seems to be leveled out. For very low percentages of degradation of the data (10% and under) most methods perform well. But at higher levels of degradation (75% and up for missing values and 50% and up for outliers) most methods perform equivalently poorly. So depending on the accuracy we aim to reach, a relatively simple method could possibly give the desired results.

3.4   C4 and C5: Impact of the Type of Error and Impact of the Classification Model

In this section, we present a brief overview of our conclusions for C4 and C5, we do not detail these criteria for the sake of space.

The study of C4 shows that the type of error has a strong impact on the accuracies and f1 scores. We identified again the two categories of errors highlighted in Sect. 3.2. C4 also shows that outliers are the most complicated type of error to repair as they quickly become statistically significant and are thus harder to detect. Moreover partial duplicates should not always be repaired as we risk deleting non-duplicated data, they can even be considered as data enrichment in some cases.

The study of C5 shows that even for low percentages, the impacts of outliers on the classification models' are not equivalent for all models both before and after repairing. This difference grows as the percentage of outliers increases. But only up to a certain point as around 80% of degradation and up the models all tend to offer similar accuracies. We can therefor deduce that the choice of the classification model has an impact even after repairing, especially in the presence of outliers and in the presence of a high proportion of missing values.

4   Discussion

4.1   Is It Always Better to Repair Data?

Our work studied the impact of repairing methods on classification tasks to answer this question. We investigated it through five criteria. From them, we observed that it is interesting to repair data when the errors detected in the data are missing values and outliers (C2 and C4). Besides, C3 showed that for 10% of deterioration or less of the data, simple repairing methods (evaluated with C1) gave similar results to the one that can be obtained with more complex repairing methods. Hence, repairing data in this case tends to not bring any difficulty and is strongly recommended. C5 showed that it is interesting to repair outliers when these classification models Cl_AdaB, Cl_NB, Cl_SGD, Cl_XGB, and Cl_GP are used. On the contrary, data should not be repaired when partial duplicates are similar to data enrichment and hence artificially improve accuracies and f1 scores (C2 and C4). C2 and C3 also showed that at high degradation levels (starting from 75% for missing values and 50% for outliers), the repairing methods did not allow to improve accuracies and f1 scores. Hence using them at these degradation levels seems to be pointless. For the other situations, the decision of repairing data depends on several factors and has to be evaluated by data scientists. These factors correspond to the data scientist skills, the time they have to perform this task, if a complex repairing method using metadata is required, the percentage of errors present in the data, and the classification model used. We believe that our criteria may help them in the decision of repairing data.

4.2   Threats to Validity

In this section, we address the internal and external threats to the validity of our work. We identified four internal threats: 1. The number of experts who answered the survey to rank the elementary tasks (Sect. 2), 2. The implementation of the classification models, 3. The hyper-parametrization of the classification models, 4. The parameters we chose for the repairing tools. To limit the impact of these internal threats the criterion C1 was studied by 8 data scientists working in the industry and experts in data repairing and machine learning. We implemented the classification models using scikit-learn [13], a widely used library. We hyper-parameterized the classification models on the original datasets using a grid search in order to avoid poor performance that can be caused by bad parametrization. Finally, for the repairing tools we chose widely used parameters such as the standard deviation, minimum and maximum or quantiles.

We also identified four external threats: 1. The choice of datasets, 2. The choice of the classification models, 3. The evaluation of the difficulty to do elementary tasks, and 4. The generation of errors. We also tried to limit these threats by choosing datasets that cover a variety of applications. We also included widely used datasets such as Mnist and Fashion-mnist, which increases the reproductibility of our results. The classification models we chose represent a wide selection of different classification approaches. By surveying experts and using the weighted mean of their evaluations, we limit the external threat of the evaluation of the difficulty to do elementary tasks. Moreover, the elementary tasks were obtained by decomposing repairing methods from multiple papers [1,2,4,5,7,8,11,12,14–16,18]. For the generation of errors, we generated them randomly by means of a uniform distribution in datasets and repeated the process 30 times to reduce bias.

5   Conclusion

We investigated the question: Is it always better to repair data? And studied five criteria C1, C2, C3, C4, and C5. There is not a single common answer for all situations to our research question but we were able to answer it for specific situations and give elements to help answer it in other situations such as the difficulty to use a repairing method.

Extensions of this work to other applications is a line of future work. Additional research could also include more data types than numeric, especially more complex data types such as time series, which would imply more possible types of errors.

References

1. Abedjan, Z., et al.: Detecting data errors: where are we and what needs to be done? Proc. VLDB Endowment **9**(12), 993–1004 (2016)
2. Ataeyan, M., Daneshpour, N.: A novel data repairing approach based on constraints and ensemble learning. Expert Syst. Appl. **159**, 113511 (2020)
3. Blake, C., Merz, C.: UCI repository of machine learning databases (1998)
4. Chu, Xu, I.F.I., Papotti, P.: Holistic data cleaning: putting violations into context. In: IEEE 29th International Conference on Data Engineering (ICDE) (2013)
5. Chu, X., Ilyas, I.F.: Qualitative data cleaning. Proc. VLDB Endowment **9**(13), 1605–1608 (2016)
6. Chu, X., Ilyas, I.F., Krishnan, S., Wang, J.: Data cleaning: Overview and emerging challenges. In: SIGMOD Proceedings of the 2016 International Conference on Management of Data, pp. 2201–2206 (2016)
7. Chu, X., et al.: Katara: a data cleaning system powered by knowledge bases and crowdsourcing. In: Proceedings of the 2015 ACM SIGMOD International Conference on Management of Data, pp. 1247–1261 (2015)
8. Hynes, N., Sculley, D., Terry, M.: The Data Linter: Lightweight, Automated Sanity Checking for ML Data Sets (2017)
9. Jouseau, R., Salva, S., Samir, C.: Technical report for: on studying the effect of data quality on classification performance. Technical report, University of Clermont Auvergne, September 2022. https://hal.archives-ouvertes.fr/hal-03782760
10. Krishnan, S., Franklin, M.J., Goldberg, K., Wu, E.: Boostclean: automated error detection and repair for machine learning. arXiv preprint arXiv:1711.01299 (2017)
11. Li, P., Rao, X., Blase, J., Zhang, Y., Chu, X., Zhang, C.: CleanML: a study for evaluating the impact of data cleaning on ml classification tasks. In: 36th IEEE International Conference on Data Engineering (ICDE 2020) (virtual) (2021)
12. Li, Y., Vasconcelos, N.: Repair: Removing representation bias by dataset resampling. Proceedings of the IEEE/CVF Conference on Computer Vision and Pattern Recognition, pp. 9572–9581 (2019)
13. Pedregosa, F., et al.: Scikit-learn: machine learning in Python. J. Mach. Learn. Res. **12**, 2825–2830 (2011)
14. Qahtan, A., Tang, N., Ouzzani, M., Cao, Y., Stonebraker, M.: Pattern functional dependencies for data cleaning. Proc. VLDB Endowment **13**(5), 684–697 (2020)
15. Rekatsinas, T., Chu, X., Ilyas, I.F., Ré, C.: HoloClean: holistic data repairs with probabilistic inference. Proc. VLDB Endowment **10**(11), 1190–1201 (2017)
16. Wu, R., Chaba, S., Sawlani, S., Chu, X., Thirumuruganathan, S.: Zeroer: entity resolution using zero labeled examples. In: Proceedings of the 2020 ACM SIGMOD International Conference on Management of Data, pp. 1149–1164 (2020)
17. Xiao, H., Rasul, K., Vollgraf, R.: Fashion-mnist: a novel image dataset for benchmarking machine learning algorithms (2017)
18. Zhang, X., Ji, Y., Nguyen, C., Wang, T.: DeepClean: data cleaning via question asking. In: 2018 IEEE 5th International Conference on Data Science and Advanced Analytics (DSAA), pp. 283–292 (2018)

A Binary Water Flow Optimizer Applied to Feature Selection

Fagner José de Matos Macêdo[(✉)][iD] and Ajalmar Rêgo da Rocha Neto[iD]

Computer Science Department, Federal Institute of Ceará, IFCE, Fortaleza, Brazil
fagnerjmatosifce@gmail.com, ajalmar@ifce.edu.br

Abstract. In this work, a binary version of the Water Flow Optimizer (WFO) algorithm, called Binary Water Flow Optimizer (BWFO), is introduced addressing the feature selection problem. WFO is an evolutionary algorithm inspired by the way water flows in nature. In this new approach, the BWFO uses the laminar flow and turbulent flow operators in a binary version, using the Optimum-Path Forest (OPF) classifier as a fitness function. The proposed approach is evaluated through a comparative analysis made with classical methods of dimensionality reduction, more specifically with the Principal Component Analysis (PCA) and Linear Discriminant Analysis (LDA) and with the metaheuristics Binary Water Wave Optimization (BWWO), Binary Bat Algorithm (BBA) and Binary Cuckoo Search (BCS). The computational results demonstrate that the approach is a valid and effective alternative to the feature selection problem.

Keywords: Feature selection · Water flow optimizer · Dimensionality reduction

1 Introduction

Machine learning is becoming increasingly important in computer science research, but it is also playing an increasing role in our everyday lives, whether in text and voice recognition, e-mail spam filters or diagnosing medical diseases [7]. However, even with advances, the high dimensionality of datasets can affect the performance or efficiency of the most accurate approaches [6]. Thus, dimensionality reduction arises to alleviate problems related to noise, as well as the high dimensionality of datasets that can affect the predictor. According to [8], dimensionality reduction addresses the process of converting a high-dimensional dataset to a lower-dimensional dataset, so that the information is kept in a clear and summarized way. Dimensionality reduction can be performed in two different ways: via feature selection, in which a subset of the most relevant attributes is selected, and via feature extraction, in which information from the attribute set is derived to build a new attribute subspace.

Among the feature extraction methods, classical methods such as Principal Component Analysis (PCA) and Linear Discriminant Analysis (LDA) are well

© The Author(s), under exclusive license to Springer Nature Switzerland AG 2022
H. Yin et al. (Eds.): IDEAL 2022, LNCS 13756, pp. 94–103, 2022.
https://doi.org/10.1007/978-3-031-21753-1_10

known and widely used for the task of dimensionality reduction [1,10,11]. Among the feature selection methods, several works have addressed the feature selection task by modeling the problem as an optimization task, through evolutionary metaheuristics.

The authors [3], proposed a binary version for the Water Wave Optimization (WWO) metaheuristic using the Optimum-Path Forest (OPF) classifier as a fitness function, to evaluate each solution during the research phase of the algorithm. In the same context, [4] presented a binary version of the Bat Algorithm (BA). In this work, the search is driven by Binary BA, and candidate solutions are evaluated with the OPF classifier. The authors [9], still in the same context, proposed a binary version of the Cuckoo Search (CS) metaheuristic, which is inspired by the behavior of cuckoo birds. Thus, it can be seen that the use of evolutionary metaheuristics has stood out efficiently in the feature selection task.

Recently, a new metaheuristic known as Water Flow Optimizer (WFO) was proposed, having as inspiration two types of water flows existing in nature, they are: laminar flow and turbulent flow. WFO is a recent algorithm that shows significant results. Therefore, the present work proposes a binary version of the WFO metaheuristic, called Binary Water Flow Optimizer (BWFO), applied to feature selection, so that the new binary algorithm is an efficient and effective alternative in the dimensionality reduction task.

The remaining of this paper is organized as follows: Sect. 2 presents the WFO metaheuristic, its structure and operators which are necessary to understand our proposal presented in Sect. 3. Then, in Sect. 4 the results obtained are presented and discussed. Finally, the paper is concluded in Sect. 5.

2   Water Flow Optimizer

The Water Flow Optimizer (WFO) algorithm is an evolutionary method based on water flows for solving global optimization problems [2]. In WFO, each solution is a water particle, and its position is analogous to the value of the solution, and the potential energy of the particle is similar to its fitness value. Laminar flow and turbulent flow present in fluid mechanics serve as inspiration for WFO operators, which were designed to perform regular and irregular stochastic searches in optimization.

In laminar flow, all particles move in parallel straight lines. In turbulent flow there is an irregular movement of water particles, which causes the straight flow to be interrupted, due to the rapid flow of water, the particles randomly collide and move over adjacent layers [2].

2.1   Laminar Operator

The regular motion denoted by laminar flow was modeled by the following equation:

$$y_i(t) = x_i(t) + s * \vec{d} \quad \forall i \in \{1, 2, ..., n\} \tag{1}$$

$$\vec{d} = x_b(t) - x_k(t), \quad (b \neq k, \quad f(x_b(t)) \leq f(x_k(t))), \tag{2}$$

where t is the iteration number, i is the index of the particle in the population of solutions and n is the size of the population of water particles employed, $x_i(t)$ is the position of the i-th particle in the t-th iteration, $y_i(t)$ denotes its possible position of motion, the random number $s \in U([0, 1])$ represents the shifting coefficient of a water particle and the vector $\vec{d}$ is determined by two different particles selected in the iteration, where $x_b(t)$ is the best solution found, and $x_k(t)$, is a randomly selected particle, such that, $f(x_b(t))$ and $f(x_k(t))$, is the potential energy of $x_b(t)$ and $x_k(t)$, respectively [2].

2.2 Turbulent Operator

The irregular motion in turbulent flow is simulated through a mathematical transformation in a dimension of the problem, which in turn is considered in the operator as a layer [2]. The particle motion is produced by the oscillation of a randomly selected dimension of each particle, where the operation to be performed on the dimension is determined by the control parameter called eddying probability p_e, which is the threshold for the application of turbulence operations, in that the eddying transformation ψ occurs with probability p_e and the general move over-layer operation φ occurs with probability $1 - p_e$.

The eddying function in the turbulent operator is constructed by a similar equation to a special Archimedean spiral, for its resemblance to a eddy

$$\psi(x_{t,j_1}(t), x_{k,j_1}(t)) = x_{i,j_1}(t) + \rho * \theta * cos(\theta) \tag{3}$$

$$\rho = |x_{i,j_1}(t) - x_{k,j_1}(t)|, \tag{4}$$

where θ is a random number within the range $[-\pi, \pi]$ and ρ is dynamically determined by a self-adaptive rule. The φ function aims to simulate the general behavior of particles over layers, defined by the following equation:

$$\varphi(x_{i,j_1}(t), x_{k,j_2}(t)) = (ub_{j_1} - lb_{j_1}) * \frac{x_{k,j_2}(t) - lb_{j_2}}{ub_{j_2} - lb_{j_2}} + lb_{j_1}, \tag{5}$$

where lb and ub, respectively, represent the lower and upper bounds. The linear transformation can inherit the scale of a randomly selected particle to boundaries in a selected dimension.

2.3 Algorithm

The WFO algorithm is presented in Algorithm 1. The evolution of water particles in the WFO is achieved by executing the laminar (lines 5–7) and turbulent flow (lines 9–18) operators in each iteration, where the laminar operator is executed with probability p_l, and the turbulent operator is executed with probability $1 - p_l$. At each iteration, the algorithm evaluates (lines 19–23) its particles, in which particles performed a movement that increases their potential energy are selected for the next iteration, otherwise, the particle proceeds to the next iteration with its previous state.

Algorithm 1. Algorithm WFO

1: Initialize particles (solutions) in the population with random position vectors.
2: Calculate the potential energy of each particle.
3: **while** a given stopping criterion is not satisfied **do**
4:    **if** $rand() < pl$ **then**
5:       Determine the laminar direction, $d \leftarrow x_b - x_k$
6:       **for** each particle $i \in \{1, 2, ..., n\}$ **do**
7:          $s \leftarrow rand(), y_i \leftarrow x_i + s * \overrightarrow{d}$
8:    **else**
9:       **for** each particle $i \in \{1, 2, ..., n\}$ **do**
10:         Initialize a trial position, $y_i \leftarrow x_i$.
11:         Select a random particle, $k \in \{1, 2, ..., n\}$ and $k \neq i$.
12:         Select a random dimension, $j_1 \in \{1, 2, ..., m\}$.
13:         **if** $rand() < p_e$ **then**
14:            $\rho \leftarrow |x_{i,j_1} - x_{k,j_1}|, \theta \leftarrow 2 * \pi * rand() - \pi$
15:            $y_{i,j_1} \leftarrow x_{i,j_1} + \rho * \theta * cos(\theta)$
16:         **else**
17:            Select a random dimension, $j_2 \in \{1, 2, ..., m\}$ and $j_2 \neq j_1$.
18:            $y_{i,j_1} \leftarrow (ub_{j_1} - lb_{j_1}) * \frac{x_{k,j_2} - lb_{j_2}}{ub_{j_2} - lb_{j_2}} + lb_{j_1}$
19:    **for** each particle $i \in 1, 2, ..., n$ **do**
20:       **if** $f(y_i) < f_i$ **then**
21:         $f_i \leftarrow f(y_i), x_i \leftarrow y_i$
22:       **if** $f_i < f_b$ **then**
23:         $f_b \leftarrow f_i, x_b \leftarrow x_i, b \leftarrow i$

---

3 Proposal: Binary Water Flow Optimizer (BWFO)

In this work, a new binary version of the WFO algorithm is proposed for the feature selection task, called Binary Water Flow Optimizer (BWFO). In this version, the WFO operators were modified so that their outputs return binary values, where the value 1 determines if the characteristic will remain for the construction of the new dataset and the value 0 defines if the characteristic will be excluded.

3.1 Binary Laminar Flow Operator

The application of the new approach to the laminar flow operator is employed on each water particle of the current iteration, in turn each particle may or may not have its layers changed randomly, decided by a random number generated for each particle layer in the iteration, modeled by the following equation:

$$y_{i,j}(t) = \begin{cases} x_{b,j}(t), & \text{se } s_j < rand() \\ x_{i,j}(t), & \text{otherwise.} \end{cases} \tag{6}$$

where $s \in [0,1]$ is a random number that acts as a threshold for mutation, j is the j-th layer of the particle, i is the index of the i-th particle in the iteration t and b the index of the best particle in the iteration. Thus, each layer chosen for replacement is replaced by the layer of the current best particle in the iteration. The binary laminar flow operator makes the water particles get layers of the best particle of the current generation, performing a new movement towards the best particle.

3.2   Binary Turbulent Flow

The new turbulent flow operator uses the eddying probability parameter p_e to determine whether the turbulence operation will apply to the layers. When the turbulence operator is applied, a water particle x_k is randomly selected from the population, the layer of the particle to be modified receives the value of the particle layer x_k if both are different, otherwise the layer of the current particle receives randomly a value $v \in \{0,1\}$, thus randomly assuming the value 0 or 1 as follows:

$$y_{i,j}(t) = \begin{cases} x_{k,j}(t), & \text{se } x_{k,j}(t) \neq x_{i,j}(t) \\ v \in \{0,1\}, & \text{otherwise.} \end{cases} \tag{7}$$

The binary turbulent flow operator aims to adhere to the characteristics of the other particles or to undergo a kind of mutation when the selected layer of the particle x_k has the same value as the particle x_i, thus bringing more diversity to the research within space of search.

3.3   Framework BWFO

The Algorithm 2, shows the general idea for the BWFO algorithm. The laminar probability parameter p_l, continues as a decision threshold for the application of laminar or turbulent flow. In turbulent flow, the eddying probability parameter p_e is also used as a threshold for applying the turbulence operation.

To evaluate the solutions in the feature selection problem, the Optimum-Path Forest (OPF) classifier was used as a fitness function in the optimization problem, with the fitness value being the accuracy returned by the OPF. The OPF was chosen because it is a fast pattern recognition technique [5].

Therefore, like the original algorithm, the BWFO is also a simple algorithm to implement, being an algorithm with few parameters, such as the laminar probability parameter and the eddying probability parameter.

Algorithm 2. Algorithm BWFO

Require: Training set Z_1 and validation Z_2 λ-labeled and a number T of iterations.
Ensure: most accurate particle.
1: Randomly initialize a population of n solutions (particles)
2: Calculate the potential energy of each particle (accuracy)
3: **while** a given stopping criterion is not satisfied **do**
4:   **if** $rand() < p_l$ **then**
5:     Select the index of the best current particle b
6:     **for** each particle $i \in \{1, 2, ..., n\}$ **do**
7:       **for** each dimension $j \in \{1, 2, ..., m\}$ **do**
8:         $s \leftarrow rand()$
9:         **if** $s < rand()$ **then**
10:           $y_{i,j} \leftarrow x_{b,j}$
11:   **else**
12:     **for** each particle $i \in \{1, 2, ..., n\}$ **do**
13:       **for** each dimension $j \in \{1, 2, ..., m\}$ **do**
14:         Randomly select the index of a particle k
15:         **if** $rand() < p_e$ **then**
16:           **if** $y_{i,j} \neq x_{k',j}$ **then**
17:             $y_{i,j} \leftarrow x_{k,j}$
18:           **else**
19:             $v \leftarrow randint(0, 1), y_{i,j} \leftarrow v$
20:   **for** each particle $i \in 1, 2, ..., n$ **do**
21:     **if** $f(y_i) > f_i$ **then**
22:       $f_i \leftarrow f(y_i), x_i \leftarrow y_i$
23:       **if** $f_i > f_b$ **then**
24:         $f_b \leftarrow f_i, x_b \leftarrow x_i, b \leftarrow i$

---

4   Simulations and Discussions

The experiments were carried out using the metaheuristics BWWO [3], BBA [4] and the BCS [9] as methods of comparison to the BWFO, that use the same methodology as the BWFO to carry out the experiments. Furthermore, classical dimensionality reduction methods such as PCA and LDA were also used.

Initially, the datasets were normalized, shuffled and partitioned into three subsets: training set (Z_1), validation (Z_2) and test (Z_3), where 30% of the data was destined for the training set, 30% for Z_2 set and the remaining 40% for Z_3. During the execution of the BWFO, new subsets Z_1', Z_2' and Z_3' are generated according to the binary values present in each generated solution. Then, the OPF algorithm is trained with Z_1' and classify Z_2' to assess the fitness of each candidate solution. This process is repeated for 30 iterations or when the best fitness solution has not evolved for five iterations. After satisfying the stopping criteria, the BWFO returns its best solution to create the new reduced dataset in order to evaluate it with the subset Z_3'.

The simulations were conducted using the datasets[1] described in Table 1, obtained from the UCI Machine Learning Repository[2].

Table 1. List of datasets evaluated in this work.

Dataset	Abbreviation	#Patterns	#Features	#Labels
Australian Credit Approval	ACA	690	14	2
Breast Cancer Wisconsin	BCW	699	10	2
Dermatology	DM	366	33	6
Vertebral Column	VC	310	6	3
Sonar	SNR	208	60	2
Pima Indians Diabetes	PID	768	8	2
Divorce Predictors	DIV	170	54	2
Parkinson's Disease	PDS	756	753	2
Turkish Music Emotion	TKM	400	50	4
Urban Land Cover	UBL	168	147	9
Arrhythmia	ARR	452	279	16

In order to make a fair comparison between the methods, the experiments were conducted using the same data standards for each approach, and the same classifier, the OPF. Therefore, each method takes as a parameter exactly the same subsets Z'_1, Z'_2 and Z'_3 used in the BWFO. Furthermore, for PCA and LDA, the number of components and discriminants are chosen based on the number of features selected by the BWFO in the round.

Since PCA and LDA compared to metaheuristics do not need a validation set, only the subsets Z'_1 and Z'_3 were used by them for their transformations, each one having a number of components (dimensions) equal to the number of characteristics returned by the BWFO. Table 2 summarizes the parameter settings of each approach used for the experiments.

Table 2. Parameter settings for each algorithm.

Algorithm	Parameter
BWFO	$p_l = 0.3$ e $p_e = 0.7$
BWWO	$kMax = min(12, dim/2)$, $hMax = 3$ e $\alpha = 0.01$
BBA	$\alpha = 0.9$ e $\gamma = 0.9$
BCS	$\alpha = 0.1$ e $P_\alpha = 0.9$

---

[1] The Arrhythmia dataset was modified to consider only two classes in the problem, considering the normality and the existence of arrhythmia in the heartbeats.

[2] https://archive.ics.uci.edu/ml/.

Table 3 presents the results obtained by the methods used in this paper. These results refer to the accuracy (ACC) on the set Z_3', the standard deviation of the ACC and the average pruning rate (PR) on the number of features of the datasets, measured during 20 independent runs, with population size equal to 100 for the metaheuristics.

Note that in the table only the values of the number of characteristics of the BWFO, BCS, BWWO and BBA can be viewed, since the number of components and discriminants used in the PCA and LDA, respectively, are equal to the number of characteristics selected by the BWFO, and OPF keeps the number of original attributes in the dataset.

Table 3. Results obtained in this work (ACC and PR are expressed in percentage).

Dataset	Metric	BWFO	BCS	BWWO	BBA	OPF	PCA	LDA
PID	Acc(%)	63.66	63.80	63.74	63.48	**64.07**	62.81	63.01
	Std	2.55	2.38	2.54	2.73	2.84	3.41	2.70
	PR(%)	68.57	56.25	43.75	43.75			
VC	Acc(%)	**78.63**	78.22	**78.63**	**78.63**	77.82	74.56	74.44
	Std	3.04	3.31	3.04	3.04	2.64	5.16	6.03
	PR(%)	43.33	44.17	43.33	43.33			
ACA	Acc(%)	81.57	**83.22**	81.61	81.99	78.57	75.84	78.57
	Std	2.51	3.43	3.21	2.39	2.09	2.01	2.54
	PR(%)	55.36	74.64	50.00	60.71			
BCW	Acc(%)	93.77	94.76	94.38	94.36	**94.89**	94.80	94.25
	Std	2.53	1.66	1.99	1.97	1.70	1.39	1.47
	PR(%)	46.50	51.50	43.00	47.00			
DM	Acc(%)	95.35	95.14	95.33	95.15	95.27	**95.43**	91.96
	Std	1.64	2.18	1.93	1.70	1.18	1.39	2.84
	PR(%)	50.00	59.56	46.76	49.71			
SNR	Acc(%)	76.70	75.44	77.17	77.09	**77.24**	76.69	60.65
	Std	4.86	3.79	5.18	5.28	4.73	5.76	7.36
	PR(%)	58.67	66.00	52.83	47.83			
DIV	Acc(%)	97.36	97.21	97.14	97.53	**97.58**	97.50	84.21
	Std	1.31	1.82	1.67	1.68	1.46	1.61	3.63
	PR(%)	59.81	72.22	50.93	52.04			
PDS	Acc(%)	**76.51**	75.39	75.31	76.09	75.65	75.13	59.91
	Std	2.83	2.65	2.94	3.71	3.71	3.77	3.67
	PR(%)	57.18	71.40	50.39	66.21			
TKM	Acc(%)	**75.27**	72.44	74.40	74.39	74.59	73.72	67.12
	Std	1.88	1.90	2.62	3.11	3.06	2.50	3.55
	PR(%)	54.10	65.40	45.80	49.50			
UBL	Acc(%)	**83.27**	83.07	83.26	82.94	82.31	82.15	72.74
	Std	1.95	2.13	1.89	1.87	2.00	2.09	5.92
	PR(%)	56.29	72.76	51.12	62.99			
ARR	Acc(%)	61.23	**61.78**	60.86	60.03	58.99	59.12	58.46
	Std	4.92	3.45	3.93	3.37	4.58	4.41	4.41
	PR(%)	57.01	71.85	51.02	68.30			

It can be seen in Table 3 that the BWFO obtained good results in terms of accuracy. In the tests performed, the proposed method surpassed classical dimensionality reduction techniques such as PCA and LDA in almost all experiments, having been surpassed by PCA in the BCW, DM and DIV datasets and by LDA in the BCW dataset. On the other hand, against the other metaheuristics, the BWFO surpassed the BCS in seven opportunities, being them in the VC, DM, SNR, DIV, PDS, TKM and UBL datasets. Against the BWWO, the proposed method was ahead in the DM, DIV, PDS, TKM, UBL and ARR datasets. In relation to the BBA, the BWFO had better results in terms of accuracy in 6 opportunities, being them in the PID, DM, PDS, TKM, UBL and ARR datasets.

The results demonstrate that, in terms of accuracy the BWFO is a good alternative for the task of selecting features, since in the experiments the proposed method surpassed the BCS, BWWO and BBA in good opportunities. However, the BWFO does not significantly reduce the number of attributes of the datasets when compared to the other methods. The BCS greatly reduces the size of the set, despite having a greater impact on accuracy. Taking into account datasets with a larger dimension, the BWFO stands out in relation to the other methods.

5   Conclusion

In this work, a new feature selection technique based on the WFO algorithm was proposed, called Binary Water Flow Optimizer (BWFO). The proposed BWFO brings a binary version for the laminar flow and turbulent flow operators keeping the basic structure of the algorithm. To evaluate the quality of solutions in the feature selection problem, the OPF classifier is used as a fitness function.

The experiments were carried out by evaluating the BWFO compared with classical techniques of dimensionality reduction, such as PCA and LDA, and with the evolutionary methods also used in the feature selection task, the BCS, BWWO and BBA. The methods were applied to eleven datasets. All algorithms were investigated according to two criteria: accuracy and feature pruning rate. Based on the experiments, the BWFO surpassed in most tests the classical techniques of dimensionality reduction, PCA and LDA, in addition to surpassing in some good opportunities the BCS, BWFO and BBA algorithms. Therefore, it can be said that the proposed method is a valid alternative for feature selection problems.

References

1. Balakrishnama, S., Ganapathiraju, A., Picone, J.: Linear discriminant analysis for signal processing problems. In: Proceedings IEEE Southeastcon'99. Technology on the Brink of 2000 (Cat. No.99CH36300), pp. 78–81 (1999)
2. Luo, K.: Water flow optimizer: a nature-inspired evolutionary algorithm for global optimization. IEEE Trans. Cybern. **52**, 7753–7764 (2021)
3. Macêdo, F., Barbosa, G., Neto, A.R.: A binary water wave optimization algorithm applied to feature selection. In: Anais do XVI Encontro Nacional de Inteligência Artificial e Computacional, pp. 448–459. SBC, Porto Alegre, RS, Brasil (2019)

4. Nakamura, R.Y.M., Pereira, L.A.M., Costa, K.A., Rodrigues, D., Papa, J.P., Yang, X.S.: BBA: a binary bat algorithm for feature selection. In: 2012 25th SIBGRAPI Conference on Graphics, Patterns and Images, pp. 291–297 (2012)
5. Papa, J., et al.: Feature selection through gravitational search algorithm, pp. 2052–2055, May 2011
6. Papa, J.P., Rosa, G.H., Papa, L.P.: A binary-constrained geometric semantic genetic programming for feature selection purposes. Pattern Recogn. Lett. **100**, 59–66 (2017)
7. Raschka, S.: Python Machine Learning. Packt Publishing (2015)
8. Ray, S., Analytics, B.: Beginners guide to learn dimension reduction techniques. https://www.analyticsvidhya.com/blog/2015/07/dimension-reduction-methods/. Accessed 30 Mar 2019
9. Rodrigues, D., et al.: BCS: a binary cuckoo search algorithm for feature selection. In: 2013 IEEE International Symposium on Circuits and Systems (ISCAS 2013), pp. 465–468, May 2013
10. Su, H., Wang, X.: Principal component analysis in linear discriminant analysis space for face recognition. In: 2014 5th International Conference on Digital Home, pp. 30–34 (2014)
11. Zhao, W., Chellappa, R., Krishnaswamy, A.: Discriminant analysis of principal components for face recognition. In: Proceedings Third IEEE International Conference on Automatic Face and Gesture Recognition, pp. 336–341 (1998)

Benchmarking Data Augmentation Techniques for Tabular Data

Pedro Machado$^{(\boxtimes)}$, Bruno Fernandes , and Paulo Novais

ALGORITMI Center, University of Minho, Braga, Portugal
pedrofcmachado26@gmail.com, bruno.fernandes@algoritmi.uminho.pt,
pjon@di.uminho.pt

Abstract. Imbalanced learning and small-sized datasets are usual in machine learning problems, even with the increased data availability provided by recent developments. The performance of learning algorithms in the presence of unbalanced data and significant class distribution skews is known as the "imbalanced learning problem". The models' performance on such problems can drastically decrease for certain classes with an uneven distribution because the models do not learn the distributive features of the data and present accuracy too favorable for a specific set of classes of data. As an example, this can have negative consequences when talking about cancer detection since the model may poorly identify unhealthy patients. Hence, data augmentation techniques are usually conceived to evaluate how models would behave in non-data-scarce environments, generating synthetic data that mimics the characteristics of real data. By applying those techniques, the amount of available data can be increased, balancing the class distributions. However, there are no standardized data augmentation processes that can be applied to every domain of tabular data. Therefore, this study aims to identify which characteristics of a dataset provide a better performance when synthesizing samples by a data augmentation technique in a tabular data environment.

Keywords: Data augmentation · Imbalanced data · Machine learning

1 Introduction

In recent years, the imbalanced learning problem has become a highly frequent topic among academia, industry, and government funding agencies. The fundamental issue with the imbalanced learning problem is the ability of imbalanced data to significantly decrease the performance of machine learning algorithms [4]. These algorithms, when faced with imbalanced data, do not learn the distributive features of the data and present accuracies too favorable to a specific set of classes of data, in this case, the majority classes, compromising the performance of the other classes (the minority classes) because of that bias. In fairness, a dataset is considered imbalanced when it exhibits an unequal distribution between its classes. Nevertheless, the community usually considers that

© The Author(s), under exclusive license to Springer Nature Switzerland AG 2022
H. Yin et al. (Eds.): IDEAL 2022, LNCS 13756, pp. 104–112, 2022.
https://doi.org/10.1007/978-3-031-21753-1_11

imbalanced data corresponds to a large unequal distribution and, in some cases, extremes.

In an imbalanced data problem, the real problem can't be solved with data treatment and/or model changes since the limitation is in the data itself. The same goes for a small-sized dataset, since the model cannot learn enough features to classify the problem in a real-time situation. Therefore, data augmentation appears as a way to surpass that limitation [3].

The present article, by identifying more favorable properties in a dataset when synthesizing samples, aims to conceive and benchmark several candidate models to overcome the imbalanced learning problem as well as increase the amount of available data without a loss in quality. With this in mind, we used classical techniques, such as SMOTE, a very uncommon clustering technique like Gaussian Mixture Model (GMM), and deep learning ones, such as Variational Autoencoder (VAE). Finally, this manuscript is structured as follows: the next section describes the literature review on the addressed domains; the third section presents the conducted experiments as well as the achieved results for this benchmark; the last section summarizes the obtained conclusions and outlines future work.

2   State of Art

Data Augmentation refers to methods for constructing iterative optimization or sampling algorithms via the introduction of unobserved data or latent variables [2]. With these techniques, we can increase the amount of data, thus balancing the target variable in an imbalanced dataset. Data augmentation can be applied to images or tabular data, but this article will focus on the latter. When used alongside images, techniques tend to apply transformations to samples of datasets like geometric transformations, flipping, color modification, cropping, etc. One other way is to introduce new synthetic images created by machine learning algorithms, for example, VAEs. Instead, if we are dealing with tabular data, we cannot apply simple transformations to samples, but instead synthesize samples (new or duplicated) based on the class distributions and features.

In [11], Data Augmentation is utilized to surpass the data limitations of the minority class, in this case, fraudulent transactions. The classification performance improved considerably and overfitting was alleviated, demonstrating the benefits of using a these techniques. These techniques can also be applied to automated skin lesion analysis by applying traditional color and geometric transformations, and more unusual augmentations such as elastic transformations, random erasing, and a novel augmentation that mixes different lesions, as stated in [8]. They prove the importance of data augmentation techniques in both training and testing, leading to more performance gains than simply obtaining new images.

In this study, we focused on some of the most popular data augmentation techniques, namely, SMOTE, GMM, and VAE. These techniques are present in multiple data augmentation studies and are going to be developed and evaluated in order to augment the used datasets.

SMOTE. One of the classic data augmentation techniques is SMOTE. It over-samples the minority class by creating synthetic samples [1]. One way to solve the imbalance problem is to duplicate minority samples. However, this does not provide any new information to the machine learning algorithm training on the data. Therefore, instead of duplicating minority samples, SMOTE synthesizes new examples from that class.

This technique synthesizes the minority class by operating in the feature space. It selects examples that are close in the feature space and introduces synthetic samples along the line drawn from these examples. SMOTE is effective because the new synthetic samples from the minority class are somewhat close in feature space to real samples from that same class. This makes the created samples plausible.

Gaussian Mixture Model. The GMM's generative nature provides an opportunity to explore its performance as a data augmentation technique, contrarily to other clustering algorithms, such as K-means. The use of a simple radial distance metric by k-means to assign cluster membership results in poor performance and a typical circular form for the clusters. This algorithm has no built-in way of accounting for non-circular clusters (oblong or elliptical), which do not represent the true shape of the data points sometimes. Moreover, this algorithm does not have a probabilistic nature when forming clusters.

Therefore, GMMs are an extension of the ideas behind k-means. This algorithm aims to model the data as a combination of multiple multi-dimensional Gaussian probability distributions and it works on the basis of the Expectation-Maximization algorithm. Because of this, the EM algorithm finds the maximum likelihood, i.e., finds a set of parameters that results in the best fit for the joint probability of the data sample [7]. Due to the generative nature of GMM, it can generate synthetic data close to the distribution of the fitted data [10]. After the algorithm fits the data and learns its distribution, it can generate an arbitrary number of samples from the learned distribution.

Variational Autoencoder. Nowadays, deep learning has gained a lot of interest and has made some amazing improvements regarding its performance. From the deep learning models, the family of generative models has also increased in popularity, showing a magnificent ability to produce highly realistic samples of various kinds, such as images, text, and sounds. These families of models, like all deep learning models, rely on huge amounts of data, well-structured architectures, and smart training techniques. One of these popular deep learning generative models is the Variational Autoencoder. In short, a VAE is an autoencoder whose encoding distribution is regularized during the training in order to ensure that its latent space[1] has good properties, allowing us to generate some new data [9].

---

[1] Latent space is a representation of compressed data in which similar data points are closer together in space. It is useful to learn the features.

A VAE consists of an encoder and a decoder, just like an autoencoder, but the loss term and the encoded layers of the autoencoder are altered in order for the model to be used as a generative model [5]. Its training is adjusted to avoid overfitting, making sure that the latent space has good properties that enable the generative process. On the other hand, an autoencoder is trained to encode and decode with as few losses as possible, making no difference how the latent space is organized. The main distinction between the two encoding layer algorithms is that they encode an input as a distribution throughout the latent space rather than a single point [9]. With this in mind, the VAE avoids having some points in the latent space that would provide meaningless information once decoded.

3   Experiments

The data generated was used in two different ways in order to evaluate the data augmentation techniques. First, it was added to the original training data that trained the classifiers and then evaluated based on the test data. The second approach is to train the classifiers only with synthetic data and evaluate them with the test data. Note that all the test data is real.

3.1   Data

In this experiment, multiple datasets were chosen to perform a good comparison of these data augmentation techniques in generating new data. Moreover, these datasets are inserted into different domains, such as health and fraud detection, and are imbalanced. Furthermore, these datasets were also chosen due to being mainly composed by continuous or categorical features. The chosen datasets were the following:

- **Adult**. The adult dataset was extracted from the census bureau and has information about multiple adults. This dataset serves as a binary classification, predicting if a certain adult has an income superior to fifty thousand in a year. The target class is clearly imbalanced, as the majority class (income superior to fifty thousand) is three times more frequent than the minority class. The dataset has over $30K$ instances.
- **Breast Cancer**. Another health domain analyzed in this experiment was breast cancer prediction. This dataset was obtained from the University of Wisconsin Hospitals, Madison by Dr. William H. Wolberg and contains samples of clinical cases gathered periodically. The dataset contains a target class imbalanced with 66% of the instances belonging to benign cases and the rest being malignant. This dataset, beyond being imbalanced, is also small in size, since it only has 570 instances.
- **Credit Card Fraud**. Fraud detection is also a recurrent domain where imbalanced data is present. Therefore, in this dataset, transactions made by credit cards in September 2013 by European cardholders are analyzed. This dataset originally had more than $280K$ instances but was reduced (while maintaining the target class ratio) to $85K$ due to computational reasons.

3.2 Assessment Metrics

In order to compare the performances of all the data augmentation techniques, it is required to define how their performances can be compared. Therefore, we need to define which metrics fit better into an imbalanced data problem. Traditionally, the most often used metrics are *accuracy* and *error rate*. Although accuracy provides an easy way to describe the model's performance, it can mislead in certain situations. Therefore, accuracy and error rate do not provide enough information about a classifier's functionality in terms of the sort of classification required.

As a means to provide comprehensive assessments of imbalanced learning problems, the research community adopted other evaluation metrics, such as *precision, recall, F-measure*[2], and *G-mean*.

First, precision is a metric that measures how many correct positive predictions the model makes (a measure of exactness)[3]. Therefore, precision calculates the accuracy of the positive class and is sensitive to data distribution. Second, recall is a metric that measures how many correct positive predictions were produced out of all possible positive predictions. Unlike precision, which only gives information on the correct positive predictions of all positive predictions, recall indicates the missed positive predictions and it is not sensitive to data distributions. Moreover, recall is also known as sensitivity. When used correctly, recall and precision can evaluate an imbalanced learning problem adequately. Nevertheless, the F-measure metric combines the two previous metrics as a weighted focus on either recall or precision. Finally, the G-mean (Geometric mean) metric evaluates the balance of classification between the majority and minority classes. Even if the negative cases are accurately identified, a low G-Mean suggests poor performance in the classification of positive cases.

3.3 Experimental Results

Regarding the synthetic data generated by all data augmentation techniques, the experiments have produced a variety of findings. These analyses consist, mainly, of:

1. Comparing each feature's distribution throughout statistical methods;
2. Training the classifiers only with synthetic data;
3. Training the classifiers with real and synthetic data.

In all cases, all the techniques implemented generated the same amount of synthetic data. In this case, this amount is the size of the training dataset (i.e., seventy percent of the entire dataset). As a result, all the synthetic data generated can be compared to one another and to the original data.

In order to compare the synthetic data of each data augmentation technique, we first compared how each feature of the real and synthetic datasets behaves,

---

[2] It is also known as *F-score*.

[3] In this case, the positive class is considered the minority.

i.e., if they possess the same distribution. Ideally, a synthetic dataset should have properties very similar to the original one. Therefore, we implemented some statistical methods to compare each feature distribution on the real and synthetic datasets. However, due to the very different behavior of continuous and categorical features, it was necessary to apply different statistical tests. The categorical feature distributions were analyzed by the chi-square test and the continuous features by the Kolmogorov-Smirnov test.

In the adult dataset, most of the features are categorical, with only one continuous feature. The statistical tests showed that the data augmentation techniques had almost no difficulty representing the original categorical features in the synthetic data. However, the techniques couldn't represent the continuous feature distributions since all of the techniques failed the test. In regards to the Breast Cancer dataset, all features are continuous since they are medical measures. SMOTE showed as the best technique to represent the feature distributions as the other techniques couldn't. Finally, the credit card fraud dataset had similar properties to the previous dataset, containing only continuous features. However, all the data augmentation techniques had difficulties representing similar continuous feature distributions.

The results of the data augmentation techniques throughout all datasets indicated the increased difficulty in representing continuous feature distributions, with SMOTE being the technique with the best representations. However, the categorical feature distributions were much easier to represent, with GMM and VAE being the ones with better results.

Moreover, one important factor noticed during the training of more complex and computationally resource-demanding techniques, such as VAE, was the fact that continuous features should be normalized in order to reduce computational cost and avoid crashes during training. These kinds of crashes are detectable when the loss is *NaN* during training.

In regards to the data distribution of one of the datasets, as we can see in Fig. 1, most data augmentation techniques can represent the entirety of the data distribution. Also, VAE seems to be the technique with the most difficulty in separating what seems to be the two target classes, but it doesn't appear to affect the classifier's performance as we will observe.

We can still perform two additional crucial analyses to determine how reliable the generated data is after the analysis of the synthetic data properties. First, we are going to train the machine learning classifiers with real data and then compare the results with training with only synthetic data. Note that there were multiple classification models, but we only represented the best model for each case.

During the experiments and implementation of the data augmentation techniques, there were some obstacles with regard to the performance of some techniques, mainly the Variational Autoencoder. The implemented VAE suffered from the phenomenon called *posterior collapse* [6]. As a result, the minority class was unable to be synthesized, and the solution was to add a weight decay on the loss function as well as change the latent space dimension.

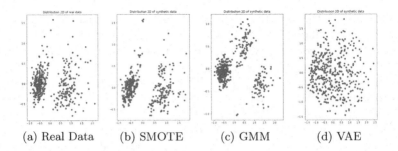

(a) Real Data   (b) SMOTE   (c) GMM   (d) VAE

Fig. 1. Comparison of two dimension data throughout all data augmentation techniques on the Breast Cancer Dataset. (a) Real Data. (b)SMOTE. (c) GMM. (d) VAE.

As observed in Table 1, synthetic data can achieve similar training scores in comparison with training with real data. SMOTE and VAE demonstrated better performance in generating samples on the three datasets. In this experiment, the VAE showed better performance for datasets with fewer categorical features (in this experiment, the Breast Cancer and Credit Card fraud datasets), while GMM indicated difficulties in generating good synthetic data. These experiments demonstrated that data augmentation techniques such as SMOTE or VAE can synthesize data in order to replace the real data in an efficient way. This could be very interesting in datasets where some data is sensitive and privacy matters.

Table 1. Best classifier performance on different kinds of training data.

Dataset	DA Technique	Minority class			Majority class			G-Mean
		Precision	Recall	F1Score	Precision	Recall	F1 Score	
Adult	—	0.6957	0.6327	0.6626	0.8868	0.9123	0.8993	0.7597
	SMOTE	0.6088	0.6173	**0.6130**	0.8781	0.8742	**0.8762**	**0.7643**
	GMM	0.2352	0.6033	0.3385	0.7504	0.3780	0.5028	0.4776
	VAE	0.4565	0.2742	0.3427	0.7958	0.8965	0.8431	0.4958
Breast	—	1.0	0.9524	0.9756	0.9737	1.0	0.9867	0.9759
Cancer	SMOTE	1.0	0.8095	0.8947	0.9024	1.0	0.9487	0.8997
	GMM	0.6111	0.5238	0.5641	0.7500	0.8108	0.7792	0.6517
	VAE	0.9090	0.9524	**0.9302**	0.9722	0.9459	**0.9589**	**0.9492**
Credit	—	1.0	0.8667	0.9286	0.9998	1.0	0.9999	0.9309
Card	SMOTE	0.9286	0.8667	0.8966	0.9998	0.9998	0.9998	0.9309
Fraud	GMM	0.0027	0.8667	0.0054	0.9995	0.4411	0.6121	0.6005
	VAE	0.9286	0.8667	**0.8967**	0.9998	0.9999	**0.9999**	**0.9309**

With those results in mind, the following analyses focus on training the classifiers with an increased amount of data (in this case, twice the original data size). At Table 2, we get to see that the data augmentation techniques that achieved better results in Table 1 got the best results with the addition of real data into the classifier's training. We can also observe that these techniques increase or maintain the classifier's performance in both major and minor classes. One example of that is the dataset Breast Cancer, where the application of a Variational Autoencoder made the classifier's performance go up in both classes' f1-score and g-mean metrics. This implies that the data augmentation can increase not only the minority class's performance but all classes' performances as well. Another interesting finding was the performance gained by the GMM technique when joining its synthetic and real data. This may be explained by the variety of generated samples that, in this case, benefited the classifier training.

Table 2. Best classifier performance while training with real and synthetic data.

Dataset	DATechnique	Minority Class			Majority Class			G-Mean
		Precision	Recall	F1Score	Precision	Recall	F1 Score	
Adult	SMOTE	0.6935	0.6263	0.6583	0.8850	0.9123	0.8884	0.7559
	GMM	0.7149	0.6301	**0.6698**	0.8870	0.9203	**0.9034**	**0.7615**
	VAE	0.7032	0.6199	0.6589	0.8839	0.9171	0.9002	0.7540
Breast Cancer	SMOTE	1.0	0.9048	0.9500	0.9487	1.0	0.9737	0.9512
	GMM	1.0	0.9048	0.9500	0.9487	1.0	0.9734	0.9512
	VAE	0.9545	1.0	**0.9767**	1.0	0.9730	**0.9863**	**0.9864**
Credit Card Fraud	SMOTE	1.0	0.8000	0.8889	0.9996	1.0	0.9998	0.8944
	GMM	1.0	0.7333	0.8462	0.9996	1.0	0.9998	0.8563
	VAE	1.0	0.8667	**0.9286**	0.9998	1.0	**0.9999**	**0.9308**

4   Conclusion

In this study, we went through a benchmark of different data augmentation techniques in multiple datasets of various domains. We observed that classical techniques such as SMOTE are competitive with more recent and powerful techniques like VAE. Also, the introduction of a not so frequent technique like GMM gave a new look to cluster models as a possibility to generate samples. Even though the Variational Autoencoder is more complex and susceptible to training problems such as the *posterior collapse*, it is a very powerful technique.

Furthermore, VAE was shown to be a better solution for a dataset with more continuous features. On the contrary, SMOTE had a better performance in a dataset with more categorical features. One other important factor to take into account is the normalization of continuous features during the preprocessing of the data. This avoids higher losses that may stall the training process.

Regarding the obtained results, the data augmentation techniques showed a great capability to create almost identical datasets to the real ones and have very similar scores. Moreover, these techniques can combat the imbalanced data problem by increasing the performance of the minority class, and they can also increase the size of a dataset without a classifier's performance loss. Future work will focus on further benchmarking techniques and new analyses of the classifier's performances.

Acknowledgment. This work has been financed by *FCT - Fundação para a Ciência e Tecnologia* within the scope of project DSAIPA/AI/0099/2019.

References

1. Chawla, N.V., Bowyer, K.W., Hall, L.O., Kegelmeyer, W.P.: Smote: synthetic minority over-sampling technique. J. Artif. Intell. Res. **16**, 321–357 (2002)
2. van Dyk, D.A., Meng, X.L.: The art of data augmentation. J. Comput. Graph. Stat. **10**, 1–50 (2001)
3. Fernandes, B., Silva, F., Alaiz-Moretón, H., Novais, P., Analide, C., Neves, J.: Traffic flow forecasting on data-scarce environments using ARIMA and LSTM networks. In: Rocha, Á., Adeli, H., Reis, L.P., Costanzo, S. (eds.) WorldCIST 2019. AISC, vol. 930, pp. 273–282. Springer, Cham (2019). https://doi.org/10.1007/978-3-030-16181-1_26
4. He, H., Garcia, E.A.: Learning from imbalanced data. IEEE Trans. Knowl. Data Eng. **21**(9), 1263–1284 (2009). https://doi.org/10.1109/TKDE.2008.239
5. Islam, Z., Abdel-Aty, M., Cai, Q., Yuan, J.: Crash data augmentation using variational autoencoder. Accident Anal. Prev. **151**, 105950 (2021). https://doi.org/10.1016/j.aap.2020.105950
6. Lucas, J., Tucker, G., Grosse, R.B., Norouzi, M.: Understanding posterior collapse in generative latent variable models. In: DGS@ICLR (2019)
7. McLachlan, G.J., Krishnan, T.: The EM Algorithm and Extensions, vol. 382. Wiley, New York (2007)
8. Perez, F., Vasconcelos, C., Avila, S., Valle, E.: Data augmentation for skin lesion analysis. In: Stoyanov, D., et al. (eds.) CARE/CLIP/OR 2.0/ISIC -2018. LNCS, vol. 11041, pp. 303–311. Springer, Cham (2018). https://doi.org/10.1007/978-3-030-01201-4_33
9. Rocca, J.: Understanding variational autoencoders (vaes), March 2021. https://towardsdatascience.com/understanding-variational-autoencoders-vaes-f70510919f73. Accessed 11 Jul 2022
10. Sarkar, T.: How to use a clustering technique for synthetic data generation, September 2019. https://towardsdatascience.com/how-to-use-a-clustering-technique-for-synthetic-data-generation-7c84b6b678ea. Accessed 10 Jul 2022
11. Shao, M., Gu, N., Zhang, X.: Credit card transactions data adversarial augmentation in the frequency domain. In: 2020 5th IEEE International Conference on Big Data Analytics (ICBDA), pp. 238–245 (2020). https://doi.org/10.1109/ICBDA49040.2020.9101344

Deep Learning Based Predictive Analytics for Decentralized Content Caching in Hierarchical Edge Networks

Dhruba Chakraborty$^{(\boxtimes)}$, Mahima Rabbi , Maisha Hossain ,
Saraf Noor Khaled , Maria Khanom Oishi , and Md. Golam Rabiul Alam

Brac University, 66 Mohakhali, Dhaka 1212, Bangladesh
mahima.rabbi.mohi@gmail.com

Abstract. The content-centric network is a state-of-the-art networking architecture for content distribution and content caching. However, it is inefficient to cache every content in each network device. The modern edge computing technology opens the door for content caching on the edge of the network. However, still, we have to decide which content we should cache and which content we should replace from the cache. Deep learning-based predictive analytics can play an important role in selecting content for caching purposes. In this research, we will use LSTM-based Recurrent Neural Network(RNN) for decentralized content caching at the hierarchical edge of the network.

Keywords: Predictive analytics · Content · Caching · Edge networking · Deep learning · Recurrent Neural Network(RNN) · Long short-term memory(LSTM) · Decentralized · Hierarchical

1 Introduction

In the very first era of networking, it was just a connection between computers for sharing mostly research data or important files. Only some of the sophisticated researchers and high-level people got to have the benefit of networking. But, in modern times, there are thousands of fields in networking. People from every stage in society get help from networking in their day to day life. In this context, content has become the most powerful weapon in the networking field. Starting from media streaming sites, social networking sites, online news portals and many others are spreading digital well-being to human beings through content.

Content centric networks are getting richer day by day with the help of thousands of content providing sites and its users. However, efficiently caching the contents is more important. In efficient content caching, files get fetched from the closest server. As a result, lots of time gets saved.

However, there is a significant issue when deciding which contents to cache and which contents to replace from the cache. Because of the limited cache size, we need to cache contents that are more important to the users. But, it is harder to decide which content is more important to the user. Again, another

© The Author(s), under exclusive license to Springer Nature Switzerland AG 2022
H. Yin et al. (Eds.): IDEAL 2022, LNCS 13756, pp. 113–121, 2022.
https://doi.org/10.1007/978-3-031-21753-1_12

issue might be, whether to cache in the regional cache or the central cache. To make the purpose easier, we used deep learning based predictive analytics which help us to decide which file to cache and which file to replace from the cache depending on its importance. Moreover, predictive analytics helps us to cache the contents in a decentralized way which helps us to save a lot of computational power usage and time. Our proposed model improves the conventional caching in the following ways:

– Using our proposed model, the system can cache the movies much faster than conventional models.
– Moreover, our proposed system does not need to make any prediction in the central cache server. As a result, a major amount of computational power and a huge amount of time gets saved.

2   Literature Review

Edge networking is a distributed computing paradigm that brings computation and data storage as close to the point of request as possible in order to deliver low latency and save bandwidth. However, according to [2] edge computing is a modern technology on data center and cloud computing architectures to help create efficiencies. According to [4], popular content and objects can be stored and served from edge locations, which are closer to the end users. According to [3] 'The Emergence of Computing', this placement at the edge helps to increase operational efficiency and contributes many advantages to the system. According to [5], using the cloud as a centralized server simply increases the frequency of communication between user devices which limits applications that require real-time response. Hence, there has been a need for looking 'beyond the clouds' towards the edge of the network, referred to as edge computing. However, according to [11] Recurrent Neural Networks work just fine when we are dealing with short-term dependencies. According to [6] Long Short- Term Memory (LSTM) networks are a type of recurrent neural network capable of learning order dependence in sequence prediction problems. According to [5], LSTM models are quite popular due to their special design property related to carefully avoiding vanishing and exploding gradient problems when building deep layer neural network models.

3   Related Works

This part aims to critically review previous relevant works in the field of Predictive Analytics in the context of Efficient Content Caching at edge networks. Observing different techniques used in different relevant research works, we found many challenges in efficiently caching the contents through prediction.

Content caching on the edge of the network is so important because if not cached, the data will be accessed by the user directly from the main server through the cloud. Which will increase the latency. According to [4], popular content and objects can be stored and served from edge locations, which are closer to the end users. This operation is also beneficial from the end user perspective since edge caching can dramatically reduce the overall latency to access the content and increase the sense of overall user experience.

Again, edge computing is another factor in terms of content caching. According to [5], using the cloud as a centralized server simply increases the frequency of communication between user devices, such as smartphones, tablets, wearable and gadgets, referred to as edge devices, and geographically distant cloud data centers. This is limiting for applications that require real-time response. Hence, there has been a need for looking 'beyond the clouds' towards the edge of the network, referred to as edge computing. Computing on edge nodes closer to application users could be exploited as a platform for application providers to improve their service. Although, the cache memory at the edge of the network is limited. So, we have to make a decision about what content to cache and what content to replace from the cache. That's where deep learning based predictive analytics comes in useful.

Recurrent Neural Network (RNN) is significantly useful for solving the efficient content caching prediction problem because it not only utilizes the current state but also uses the previous state data using sequence. According to [4], Unlike the hidden neuron in FNN, the output of RNN depends on both the current output of the previous layer and the last hidden state. However, using RNN might not be appropriate in some cases as there might be data vanishing gradient problems. Recurrent Neural Networks work just fine when we are dealing with short-term dependencies [11]. To solve that issue, LSTM (Long Short-Term Memory) comes handy. Long Short- Term Memory (LSTM) networks are a type of recurrent neural network capable of learning order dependence in sequence prediction problems [6]. According to [5], LSTM models are quite popular due to their special design property related to carefully avoiding vanishing and exploding gradient problems when building deep layer neural network models. With LSTMs, the information flows through a mechanism known as cell states. This way, LSTMs can selectively remember or forget things [11].

Hierarchical LSTM that considers both check-in time and event taxonomy structure from check-in sequences to provide accurate predictions on a user's future check-in location category. Each category is also projected into an embedding via hierarchical LSTM, resulting in new representations with greater semantic implications. The efficiency of the suggested Hierarchical LSTM is set to be demonstrated by experimental results that Hierarchical LSTM increases Accuracy by 4.22% on average, and Hierarchical LSTM learns a superior taxonomic representation for clustering categories, culminating in a 1.5X increase in Silhouette Coefficient [10].

Several cloud-based apps use a data center as a centralized computer to analyze data from edge devices like smartphones, tablets, and wearable. This strategy puts ever-increasing requirements on communication and computing resources, thereby lowering Quality of Service and Experience. Edge Computing is based on the idea of transferring part of this computing burden to the network's edge in order to take use of computational capabilities that are presently underutilized in edge nodes such base stations, routers, and switches. This position paper examines the difficulties and possibilities that come as a result of this new computing path [3].

4 Methodology

The aim of using predictive analytics for decentralized content caching at the hierarchical edge network is to cache the most popular contents at the edge of the network and thus decrease the latency. With a view to doing so, the model systematically processes the input data and outputs different results.

4.1 System Architecture

For an efficient content centric caching, we needed better prediction which we achieved using a multi-layered system which can be seen in the figure above. In our proposed system, we have 2 layers that are interconnected and highly optimised for efficient content caching. The first layer or the Regional Servers have the lowest hop count from the end users. Which means that the regional servers have the lowest latency with respect to the end users. In the second and the final layer or the central cache server, is the closest server to the cloud. In the very beginning of the caching timeframe, the most demanding movies are stored in the regional cache servers based on the highest predicted hit counts. After filling up all the cache servers with their maximum capacity, they send back the remaining popular movies to the central server. The most popular movies get stored in the central cache server with the maximum capacity and the rest are dropped. In the next time frame, when a new movie comes, the system checks if the movie is available in the regional cache server or not. If the movie is available, then it simply streams from the cache server at a faster speed. If the movie is not available in the regional cache, it looks for the movie into the central cache server. If the movie is available on the central server, it will stream from there. If the movie is not available in the central cache server too, then it will cache the movie to the regional cache server or the central cache server based on the predicted hit count value of the movie. Thus, the regional cache server and the central cache server makes the streaming job much faster.

4.2 Dataset Preprocessing

We have chosen MovieLens dataset which consists of 27753444 ratings and 1108997 tag applications across 58098 movies. We have categorized genres and extracted daily from timestamp. Then we have joined the data sets (movies, rating) that are coming from our previous step, data set cleanup. Then we have divided the data set into three parts for building prediction models on three regional cache servers. Then we have sorted our joined data sets in ascending order based on times- tamp, userId, movieId. After sorting the datasets, we have created label from tstamp_day and movieId. After the preprocessing, we are only keeping the movieId, tstamp day and label in our final data sets because the other entities are not useful for our use case.

4.3 Model Specification

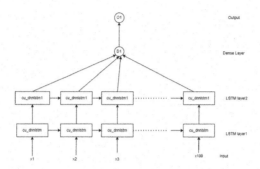

Fig. 1. Model architecture

Model Architecture. We have built our LSTM model (Fig. 1) to predict hit counts of movies in a time series manner. The LSTM model is a many-to-one RNN model because our model takes movieID and timestamp as input and gives us hit count as predicted output.

Model Flowchart. We are using LSTM based Recursive Neural Network (RNN) for solving our problem. We could use other Deep Learning based models for this work as per Fig. 2. But, unlike many other algorithms, LSTM based RNN remembers the previous sequence by keeping them in memory. As a result, the output gets more and more accurate day by day.

5   Implementation

5.1   Constructing the Model

In the very beginning we are importing the three datasets that were produced in the dataset preprocessing phase. After importing the datasets, we are dropping the duplicate values as the duplicate values will cause an under fitting problem in our models. Then we are splitting the datasets into train and test datasets. We are taking 70% for the train and 30% for the test dataset. After splitting the datasets, we are reshaping the train and test datasets because we will need to transform our datasets in a shape so that our model grants them. In the next step we are creating our LSTM model. Here we have a 64 cell LSTM layer, a 32 cell LSTM layer, a dropout layer and finally a 1 cell dense layer where we used

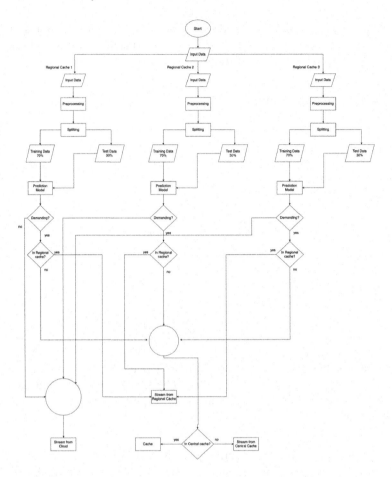

Fig. 2. The flow chart of the predictive analytics model

rectified linear activation function or ReLU as activation function. To compile the model we have used adam optimizer and mean squared error as our loss function. In the dense layer we have only used one cell because our LSTM model is many to one and we want a single output in the output layer. Thus the model takes movieId and timestamp day as input and outputs a single output which is hit count. Then we are saving our model for future caching (Fig. 3).

5.2 Content Caching and Replacing

In the very beginning of the caching, we loaded 3 models that were saved in the implementation stage. For real life scenarios, in spite of using the test data of the existing dataset, we have created a synthetic dataset made with 60 movies and corresponding movie size for those movies. Then we have trained the 3 models individually with the same dataset which refer to 3 regional cache servers. We

Model: "sequential_2"		
Layer (type)	Output Shape	Param #
cu_dnnlstm_2 (CuDNNLSTM)	(None, 2, 64)	17152
cu_dnnlstm_3 (CuDNNLSTM)	(None, 32)	12544
dropout_1 (Dropout)	(None, 32)	0
dense_1 (Dense)	(None, 1)	33
Total params: 29,729 Trainable params: 29,729 Non-trainable params: 0		

Model: "sequential"		
Layer (type)	Output Shape	Param #
cu_dnnlstm_1 (CuDNNLSTM)	(None, 2, 64)	17152
cu_dnnlstm_1 (CuDNNLSTM)	(None, 32)	12544
dropout (Dropout)	(None, 32)	0
dense (Dense)	(None, 1)	33
Total params: 29,729 Trainable params: 29,729 Non-trainable params: 0		

Model: "sequential_1"		
Layer (type)	Output Shape	Param #
cu_dnnlstm_2 (CuDNNLSTM)	(None, 2, 64)	17152
cu_dnnlstm_3 (CuDNNLSTM)	(None, 32)	12544
dropout_1 (Dropout)	(None, 32)	0
dense_1 (Dense)	(None, 1)	33
Total params: 29,729 Trainable params: 29,729 Non-trainable params: 0		

Fig. 3. Summary of the models

have 15000 MB space for each regional cache server and 25000 MB space for the central cache server. Then we ran knapsack on all the regional servers using predicted hit count as value and movie size as weights. The items returned by the knapsack algorithm are sent to the edges respectively.

After filling up the regional caches with their maximum capacity, the rest of the movies from all the regional caches that could not be cached will be sent to the central cache server. However, there is also a storage scarcity. So, we need to choose which movie to cache. To ensure the most popular movies on the central server, we ran knapsack with the regional cache excluded movies and their predicted hit counts.

In the second time slot, when there will be a new movie request by a user in a particular region, the system will search for the movie in the regional cache server. If it finds the movie in the regional cache, it will simply load the movie from the regional cache server much faster compared to the cloud.

If the movie is not in the regional cache, it will search in the central cache server. If the movie is available in the central cache, it will stream the movie from there. It will be slower than regional cache but still much faster than cloud. However, if the movie is not in the central cache too, then the model will run with movieId and corresponding timestamp and it will output a predicted hit count. Then it will look into the central cache. If the predicted hit count is greater than any of the movies in the central cache, then the movie will be replaced by the new movie. However, if the hit count is less than all the movies, then the movies won't be cached.

6   Result Analysis

The main goal of our prediction model was to predict the hit count of the movies so that we can accurately predict the popularity of the movies in an autonomous way. Predicting the hit count accurately helps us efficiently cache most popular movies on the regional and central cache servers. And, our prediction result here is highly satisfactory. We ran our models with Adam optimizer and mean squared error loss function individually on all three regional cache servers respectively. After completing 100 epochs on each of the regional servers, the prediction accuracy was 99.79%, 99.70% and 99.49% on the three servers respectively. By evaluating the three LSTM models we can also see that the loss values are pretty acceptable (Fig. 4).

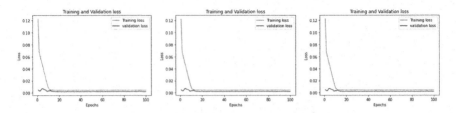

Fig. 4. loss vs val_loss of the models

It can also be seen in Fig. 5 that the time required to stream a content from the caches is much lower compared to cloud. Moreover, the accuracy is considerably higher compared to conventional caching methods (Fig. 6).

Fig. 5. Time required to stream a content (in ms)

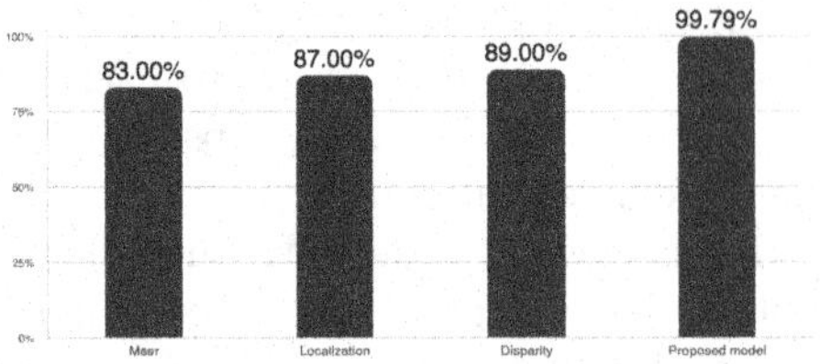

Fig. 6. Comparison with conventional caching methods

Finally, it is visible that using the central cache server in a decentralized way uses a lot less power compared in Fig. 7 to not using in a decentralized way.

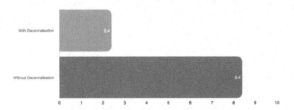

Fig. 7. Power usage comparison (in kWh per day)

7   Conclusion

The aim of our research was to build a complete system model which accurately predicts contents by determining efficiently what to cache and what to replace from the cache. Our system model will reduce the total hop count and thus decrease the latency to access the contents. Our work also helps to decrease computational power used in conventional ways. However, there are a lot of fields that we can improve in future. Among them, increasing the regional cache server count, the overall caching on the central server would be more efficient and accurate and considering the base stations under the regional servers, there will be three layers for caching contents and the overall caching would be more efficient and contents would be more accessible.

References

1. Jithin: What Is Content Caching? Interserver Tips, 11 November 2016. https://www.interserver.net/tips/kb/what-is-content-caching/
2. Archive | Design Automation Conference (dac.com). http://www2.dac.com/events/videoarchive.aspx?confid=170
3. Satyanarayanan, M.: The emergence of edge computing. Computer **50**(1), 30–39 (2017). IEEE Xplore. https://doi.org/10.1109/MC.2017.9
4. '(PDF) Challenges and Opportunities in Edge Computing'. ResearchGate. www.researchgate.net, https://doi.org/10.1109/SmartCloud.2016.18
5. Narayanan, Arvind, et al. 'DeepCache: A Deep Learning Based Framework For Content Caching'. Proceedings of the 2018 Workshop on Network Meets AI & ML, Association for Computing Machinery, 2018, pp. 48–53. ACM Digital Library, https://doi.org/10.1145/3229543.3229555
6. Brownlee, J.: A gentle introduction to long short-term memory networks by the experts. Machine Learning Mastery, 23 May 2017. https://machinelearningmastery.com/gentle-introduction-long-short-term-memory-networks-experts/
7. Long Short Term Memory (LSTM). OpenGenus IQ: Computing Expertise & Legacy, 16 Febuary 2019. https://iq.opengenus.org/long-short-term-memory-lstm/
8. Understanding LSTM Networks - Colah's Blog, 27 August 2015. https://colah.github.io/posts/2015-08-Understanding-LSTMs/
9. Phi, M.: Illustrated Guide to LSTM's and GRU's: A Step by Step Explanation'. Medium, 28 June 2020. https://towardsdatascience.com/illustrated-guide-to-lstms-and-gru-s-a-step-by-step-explanation-44e9eb85bf21
10. Liu, C.-H., et al.: Hierarchical LSTM: modeling temporal dynamics and taxonomy in location-based mobile check-ins. In: Yang, Q., Zhou, Z.-H., Gong, Z., Zhang, M.-L., Huang, S.-J. (eds.) PAKDD 2019. LNCS (LNAI), vol. 11440, pp. 217–228. Springer, Cham (2019). https://doi.org/10.1007/978-3-030-16145-3_17
11. Long Short Term Memory | Architecture Of LSTM. Analytics Vidhya, 10 December 2017. https://www.analyticsvidhya.com/blog/2017/12/fundamentals-of-deep-learning-introduction-to-lstm/
12. Loss Function. DeepAI, 17 May 2019. https://deepai.org/machine-learning-glossary-and-terms/loss-function
13. Introduction to Loss Functions. https://algorithmia.com/blog/introduction-to-loss-functions/#whats-a-loss-function

Explanations of Performance Differences in Segment Lining for Tunnel Boring Machines

Hans Aoyang Zhou[1]([✉])[iD], Aymen Gannouni[1][iD], Tala Bazazo[1][iD], Johannes Tröndle[2], Anas Abdelrazeq[1][iD], and Frank Hees[1][iD]

[1] Cybernetics Lab, RWTH Aachen University, Aachen, Germany
{hans.zhou,aymen.gannouni,tala.bazazo,anas.abdelrazeq,
frank.hees}@ima.rwth-aachen.de
[2] Herrenknecht AG, Schwanau, Germany
troendle.johannes@herrenknecht.de
https://cybernetics-lab.de/

Abstract. The tunnel lining process with segments is a labour-intensive task, for which the expertise of Tunnel Boring Machines' operators is crucial. For this task, human expertise can be evaluated based on the average time of building a tunnel ring. Data-driven identification of the different levels of operators' expertise can help to understand the causes of possible discrepancies. Consequently, bridging possibly existing gaps in expertise can be achieved through more training offered to less experienced operators or through support from user-assistance systems. In order to make the expertise more tangible, we trained deep learning models to classify expertise profiles of erector operators based on time series data accrued during the process. Afterwards, we investigate these with explainable artificial intelligence techniques to identify features with the highest influence on the performance prediction and derive regions of interest in ring-building sequences leading to specific performance classifications. Finally, we discuss how the observations from our study can contribute to designing assistance systems that support operators toward a more efficient ring-building process.

Keywords: Tunnel Boring Machines · Time series classification · XAI

1 Introduction

Tunnel Boring Machines (TBM) are specialized machines for excavating and building tunnels in a safe and efficient manner. They operate in an alternating fashion between *advance* and *ring building* phases. During the advance phase, the machine's cutting wheel excavates soil from the tunnel face, through a drilling motion. During the ring building phase, human construction operators place and fixate multiple segments into a ring-shaped arrangement with an erector. Due to the alternating work process, both phases influence equivalently the efficiency of the overall performance of TBM.

© The Author(s), under exclusive license to Springer Nature Switzerland AG 2022
H. Yin et al. (Eds.): IDEAL 2022, LNCS 13756, pp. 122–133, 2022.
https://doi.org/10.1007/978-3-031-21753-1_13

With the increasing digitization of construction sites, data-driven solutions have become a popular research topic for optimizing TBM performance. Although both work phases influence TBM efficiency, current research on data-driven solutions focuses mainly on the advance phase. Most commonly proposed solutions predict either the machine's surrounding ground conditions [2] or excavation performance [3]. This leaves the ring-building phase nearly unexplored for data analysis.

Similarly to the advance phase, the ring building phase is a complex process, that requires precision in controlling heavy segments. Thus, the required time to complete a ring fluctuates strongly and depends on the experience of the erector operator. Usually, this form of expertise is hidden in the way how the erector is controlled and limited to rules that only become applicable when encountered during the ring building process. A formalization of this form of expertise in an explicit way would both significantly contribute to improving the training of erector operators, as well as enable the development of more user-centred assistance systems [22].

One possibility to automatically learn complex patterns from available data is via deep learning. These models learn abstract representations from the presented data that allow an approximation of any functional relationship between input and output variables. Despite the promising abilities of these models, the learned representations are often not human-understandable, which results in the necessity of methods for explainable artificial intelligence (XAI). These methods aim to uncover the decision process of deep learning models to improve the trustworthiness of their prediction.

In this work, we propose to use the combination of deep learning with XAI to extract representations that allow differentiating between performance classes of ring-building sequences. First, we train and compare different deep learning time-series models to predict performance classes of ring-building sequences. Afterwards, we use the XAI method `IntegratedGradients` [16] to derive global and local explanations for the best-performing models.

The remaining sections of this paper are structured as follows. In Sect. 2 we review current applications of machine learning in the domain of TBM. Section 3 describes our proposed approach of extracting representations with deep learning (see Sect. 3.1) and XAI (see Sect. 3.2). Afterwards, we present our results in Sect. 4 which compares, on one hand, the performance of different state of the art time-series models (see Sect. 4.1) and presents on the other hand the extracted representations that were learned by the models (see Sect. 4.2). In Sect. 5 we review our approach and how the knowledge gained is beneficial from a domain perspective. Finally we conclude our contributions in Sect. 6.

2   Related Work

With the general advancement of sensor technology and the resulting abundance of data, new applications of data-driven solutions for TBM have been investigated. As mentioned previously, these solutions usually use historical operational data to improve the advance phase. We categorize them regarding their target variable into the following:

1. Prediction of performance indicators like penetration rate or advance rate [1,8,19], cutter head torque, thrust or pressure [12,15,17], as well as a combination of multiple indicators simultaneously [21].
2. Prediction of geological conditions like rock mass classification [5,9].

Undoubtedly, the proposed models show promising prediction results, but only in [1,17,19,21] the authors further evaluate during post-processing with feature importance methods the plausibility of their models. To the best of our knowledge, neither data-driven solutions were presented for the ring-building phase of TBM nor XAI methods were applied to derive new insights from those deep learning models.

3   Methods

We propose a two-step approach for extracting hidden representations from time-series data to gain more insights into the ring-building process. First, we compare different time-series models regarding their accuracy in predicting performance classes. Afterwards, we apply XAI methods to identify which regions of the ring building sequence resulted in the prediction of the best-performing models. Thereby, we derive which features and their respective regions within the ring-building sequence have the highest impact on ring-building performance.

3.1   Performance Classification

The task of training a model for performance prediction can be formalized as a multivariate time series classification problem. For that, we segment a 3-month-long data sample into ring building sequences $X = [X_1, X_2, \ldots, X_N]$, as proposed in [22]. Each building sequence consists of values from 10 sensor measurements capturing the movement of the erector and are described in Table 1. Because the ring building process varies from ring to ring, each individual building sequence X_i has a different length T_i. With the overarching optimization goal of tunneling efficiency in mind and T_i given, we categorize X_i into three performance classes $K_i \in [\text{fast}, \text{average}, \text{slow}]$ according to Eq. (1). This results in a dataset $D = \{(X_1, K_1), (X_2, K_2), \ldots, (X_N, K_N)\}$ with $N = 637$ pairs of build sequence and performance class.

$$K_i = \begin{cases} \text{fast} & \text{if } T_i < 30\,\text{min} \\ \text{average} & \text{if } 20\,\text{min} \leq T_i < 40\,\text{min} \\ \text{slow} & \text{if } 40\,\text{min} \leq T_i \end{cases} \tag{1}$$

Predicting the ring building sequence length based on the building sequence induces potential bias in the training process. More precisely, the class label can easily be determined by counting the number of time steps of the building sequence. Therefore, we propose two approaches to negate the effect of bias during model training:

Table 1. Description of erector variables with respective units.

Erector variable	Unit	Description
position	°	Rotation along tunneling axis
travel_right_stroke	mm	Translation along tunneling axis
travel_left_stroke	mm	Translation along tunneling axis
telescope_blue_stroke	mm	Translation radial to tunneling axis
telescope_red_stroke	mm	Translation radial to tunneling axis
head_tilt_stroke	mm	Tilt of erector head
head_rotate_stroke	mm	Rotation of erector head
vacuum_normal_segment_pressure	bar	Erector head suction plate pressure
vacuum_key_segment_pressure	bar	Erector head suction plate pressure
vacuum_tank_pressure	bar	Suction tank pressure

1. Fixed Window Size. The idea of fixed window size is to limit the amount of available input information to a fixed sequence length. This method is similar to the image data augmentation method of random cropping. Instead of showing the model the whole building sequence, we randomly sample a fixed subsequence T_S from the whole building sequence.
2. Linear Interpolation. The idea of interpolation is to artificially increase the build sequence length to the longest build sequence of the dataset. This method is comparable to rescaling images to bigger shapes. Here we first insert evenly empty time stamps into shorter build sequences and afterwards use linear interpolation to fill in the thereby created empty data points. By rescaling the building sequence length, we preserve the structure of the ring building sequence and fix the length of the build sequences to a fixed length T_S.

During experimentation, we compare the effect of both data pre-processing approaches on classification performance. Here, we distinguish between models that were trained with and without interpolation. Additionally, we test different T_S that are determined via a window size factor $\omega \in [0.25, 0.5, 1.0]$. In the case of interpolation, ω describes the fraction of the input sequence length relative to the longest build sequence. Whereas, in the case of no interpolation, ω describes the fraction of the input sequence length relative to the shortest build sequence. This distinction is necessary, due to the fact that in the case of no interpolation the shortest ring building sequence is the longest possible sequence size that can be used without padding for training.

After introducing our preprocessing steps, we have reviewed state-of-the-art deep learning architectures used for time series classification from [6]. The choice has been made with the aim to cover different types of neural architectures and ensure their trainability, testability, and especially explainability. Based on this, we selected the following deep neural models:

- Standard 1D fully convolutional network (FCN) [18].
- Hybrid network architecture consisting of residual and convolutional layers (ResCNN) [23].
- Convolution-based architecture consisting of multiple inception modules (InceptionTime) [7].
- Combination of a Long Short Term Memory with a fully convolutional neural network for multivariate time series classification (MLSTM-FCN) [10].

For each model, we tune all hyperparameters using the framework *Tune* [14]. More specifically, we use the Asynchronous Successive Halving Algorithm (ASHA) [13], a scheduling strategy that tries multiple training runs in parallel and stops low performing runs early. We combine ASHA with a Tree-structured Parzan estimator approach (TPE) [4] that learns a probability distribution of the expected improvement of the model performance over the hyperparameter space. This distribution is used for sampling hyperparameter configurations of the next run. For each model architecture, we run 100 runs with different architecture-specific hyperparameters (e.g. number of layers, kernel sizes, etc.) and training-specific hyperparameters (e.g. batch size, learning rate, etc.).

3.2   Model Evaluation

With the identification of the best-performing models, our aim is to use them to derive differentiating factors of ring building performance. We describe in the following our selection process on how we picked from multiple XAI methods the most suitable one for further model evaluation. For our selection process we used the python library `captum` [11] to apply different pre-implemented XAI methods to our time-series deep learning models. All the implemented methods calculate attribution values for each input variable at a given time step. The supported algorithms in `captum` are categorized into three families: *primary attribution*, *neural attribution*, and *layer attribution* [11]. As we use different types of deep neural network architectures, we aimed at explaining their predictions in the most possible model-agnostic way to avoid any influences due to their architectures. Therefore, primary attribution methods were selected as both neural and layer attributions might depend on the neural architectures of the models.

From the remaining list of all available primary attribution methods, our model selection process is guided by applicability and robustness. First, we tested whether the method can be applied ad-hoc to the best-performing models. Afterwards, we compared the robustness of each remaining method by comparing the standard deviation of the resulting global feature importance values for each variable and model architecture. The global feature importance is derived by averaging over all attribution values for every instance of the test set. It indicates how much each variable contributes on average towards the decision of the model. The standard deviation of the global feature importance indicates how strong the feature importance fluctuates based on given input and thus indicates the robustness of the method. Our comparison resulted in us selecting `IntegratedGradients` [16] as the XAI method for further model evaluation.

With `IntegratedGradients` as our selected model evaluation method, we calculate for each class (fast, average, slow) the global feature importance of every variable and measure its average impact on the respective class prediction. Furthermore, we use `IntegratedGradients` to visualize local attribution values to emphasize within a build sequence the most important regions of interest that contributed to the prediction of the model for each class (fast, average, slow) separately. This enables a thorough investigation of the decision process for each model.

4   Results

4.1   Model Performance Comparison

In our experiments, we evaluate for each model architecture the influence of interpolation and window size factor. We trained 10 models with different random seeds for each model architecture and report their average accuracy with standard deviation on a separate test set in Table 2.

Our results show that overall interpolation has a positive effect on model performance. On the contrary, a bigger window size factor only contributes positively to the model performance for ring building sequences without interpolation. This indicates, that due to building sequences without interpolation being on average shorter than those with interpolation, a successful performance classification of building sequences requires a necessary minimum amount of information.

Table 2. Accuracy comparison of time series models for performance classification.

Int	ω (T_S)	FCN	ResCNN	InceptionTime	MLSTM-FCN
False	1.0 (1230)	0.66 ± 0.17	0.70 ± 0.07	0.70 ± 0.03	$\mathbf{0.74 \pm 0.05}$
	0.5 (615)	0.63 ± 0.03	0.64 ± 0.02	0.65 ± 0.03	0.64 ± 0.02
	0.25 (307)	0.56 ± 0.04	0.58 ± 0.06	0.60 ± 0.03	0.59 ± 0.04
True	1.0 (4958)	0.72 ± 0.06	0.70 ± 0.07	0.90 ± 0.04	0.54 ± 0.05
	0.5 (2479)	0.75 ± 0.08	0.86 ± 0.04	$\mathbf{0.92 \pm 0.03}$	0.83 ± 0.04
	0.25 (1239)	0.81 ± 0.09	0.84 ± 0.05	0.89 ± 0.05	0.80 ± 0.06

4.2   Feature Representation Extraction

Global Feature Importance. Global feature importance values quantify how much each feature contributes on average towards a correct model prediction. In Fig. 1 we compare feature importance values for the best performing model architectures MLSTM-FCN and InceptionTime and derive the following observations. First, the global feature importance values change significantly with the target class and the model architecture. The MLSTM-FCN network focuses primarily on the telescope variables and neglects mostly the vacuum and head stroke variables. Whereas the InceptionTime model switches its focus between the telescope stroke variables and the travel stroke variables dependent on the target class. Both models neglect almost completely the head stroke variables. In other words, both models focus on translational movements towards the radial direction of the tunnel ring but neglect movement patterns during fine adjustment of the segment. The neglection of erector variables responsible for fine adjustment is surprising because the required time for fine adjusting the segment distinguishes experienced erector operators from inexperienced ones.

By comparing the importance values of the vacuum pressure variables between both architectures, the MLSTM-FCN architecture does not take them into consideration, whereas the InceptionTime model considers them for especially slow ring building sequences. From that, we conclude that the MLSTM-FCN in contrast to the InceptionTime model does not take vacuum pressure variables into consideration, which is expected, as they should have no impact on ring building performance. In order to further evaluate model decision behaviour, we present in the following local attribution values for exemplary ring-building sequences.

Local Feature Attribution. In order to better understand the local impact of each erector variable, we investigate their attribution values within the building sequence. For that, we plotted in Fig. 2 a ring-building sequence of each class and overlaid only positive attribution values of our applied XAI method `IntegratedGradients`. Overall it can be observed, that both models highlight areas that correspond to an extension and thus movement of the erector machine. On contrary, attribution values of the MLSTM-FCN architecture are more evenly distributed compared to the InceptionTime architecture which focuses on specific areas.

From our results we conclude, that our trained models were not able to learn complex human behaviour patterns that enables differentiation between human expertise, but rather specific clues in the building sequence that allow an estimation of the ring building time. The MLSTM-FCN architecture uses an accumulation of multiple segment placements to derive the performance classes. Whereas on the contrary, InceptionTime uses distinct areas within the building sequence to derive more precisely than MLSTM-FCN the correct performance class. This suggests that due to the interpolation of the multivariate time series, some form of bias was introduced presumably within the vacuum pressure variables that highly simplified the decision process of the InceptionTime model.

Although the MLSTM-FCN does not use the head stroke variables for differentiating between performance classes, an equal distribution of attribution values along the extended telescope indicates that the duration at which the telescope is extended has the highest impact on predicting the performance class. Precisely, this timespan where the telescope is extended corresponds to the actual fine adjustment of the segment. This indirectly confirms the assumption that the time spent adjusting and precisely fitting the segment to its final position has the highest impact on ring building performance.

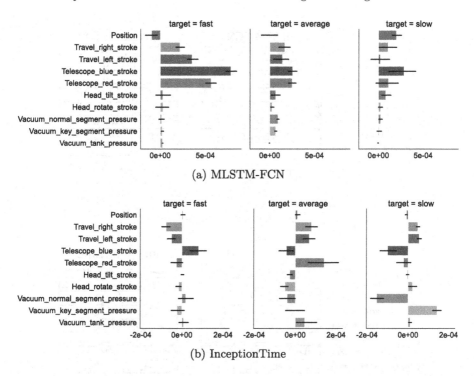

Fig. 1. Global feature importance values for both models (a) MLSTM-FCN and (b) InceptionTime. The global feature importance values are displayed with their respective standard deviation and divided based on their respective target class. A positive value indicates a positive impact on prediction performance and vice versa.

5   Discussion

Before drawing conclusions from our work, we want to review our proposed approach of extracting hidden insights from deep learning models trained with erector data of TBM. To begin with, labelling ring sequences purely by sequence length leads to two major issues. First, this labelling approach may induce bias, without proper data preprocessing, as demonstrated by using linear interpolation. Second, the segregation threshold of 20 and 30 min between classes, although approved by domain experts, is arbitrary, where the number of classes could be further increased until it becomes a regression task. Our decision to train the models for this specific classification problem is of pragmatical nature. This setup simplifies the model evaluation process because on the one hand performance metrics of classification tasks are well established and on the other hand, transferring the concept of attribution values from an image-based to a time-series-based classification problem is intuitively easy to perform. An alternative to labelling building sequences purely by time is through either more expressive quality metrics (e.g. placement accuracy) or the individual who controlled the

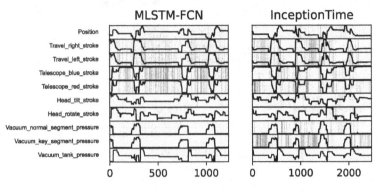

(a) Slow ring building sequence

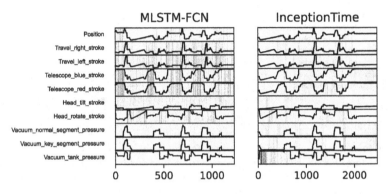

(b) Average ring building sequence

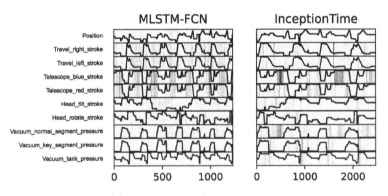

(c) Fast ring building sequence

Fig. 2. Local attribution values of MLSTM-FCN (left) and InceptionTime (right) for example ring building sequences of performance classes (a) slow, (b) average and (c) fast. For each performance class and erector variable the attribution values are overlaid (green) on top of the variable sequences (black). (Color figure online)

segment erector during the sequence. These approaches were infeasible because the required labels are either not captured within the data or protected due to the data privacy of the individual operators. This labelling approach consequently lead to a model behaviour where the performance class was estimated by counting the amount of time where the erector telescope was extended.

Besides the model training, our selection process of the most suitable XAI method is also not free from author bias. To tackle this issue, two quantitative metrics *sensitivity* and *infidelity* introduced by [20] were tested, but no method significantly outperformed the other methods. This demonstrates that a suitable out-of-the-box evaluation method for selecting a suitable XAI method for deep learning models for time-series data is still an open problem.

Although no detailed movement patterns of the segment lining process were extracted by our approach, we were able to confirm the assumption that the fine adjustment process is most likely responsible for the performance of the ring building process. Based on this confirmation, TBM developers should focus on improving the efficiency of the fine adjustment process.

6 Conclusion

With our contribution, we demonstrate how to leverage process data from segment erectors of TBM to gain insights into the segment lining process. We compared different model architectures regarding their ability to differentiate between performance classes of segment lining sequences and evaluated the effect of common preprocessing steps of time-series data. By applying `IntegratedGradients`, a thorough model evaluation was executed on a global and local level, which confirmed that fine adjustments during the segment placement have the highest impact on ring building time. Additionally, our evaluation revealed potential bias induced by linear interpolation of ring sequences. Through our findings, we show that although deep learning models show promising performance, they are highly dependent on the task and the data preprocessing. Nonetheless, with current state-of-the-art DL and XAI methods, it was possible to verify to a certain degree hidden assumptions about the segment lining process from TBM, that potentially advance research and development of more efficient machines. With our work, we show that similar to humans, teaching real domain expertise in a supervised way, requires a more sophisticated labelling approach than key performance indicators like ring building time.

Acknowledgment. The research project (no. 21250 N) is funded by the German Federal Ministry for Economic Affairs and Energy (BMWi) via the German Federation of Industrial Research Associations (AiF) with the Mechanical Engineering Research Federation (FKM) as the responsible AiF association. The funding is part of the Industrial Collective Research (IGF) program and based on a resolution of the German Bundestag. Simulations were performed with computing resources granted by RWTH Aachen University under project rwth0817.

References

1. Abolhosseini, H., Hashemi, M., Ajalloeian, R.: Evaluation of geotechnical parameters affecting the penetration rate of TBM using neural network (case study). Arab. J. Geosci. **13**(4), 1–11 (2020). https://doi.org/10.1007/s12517-020-5183-5, https://link.springer.com/article/10.1007%2Fs12517-020-5183-5

2. Ayawah, P.E., et al.: A review and case study of artificial intelligence and machine learning methods used for machines. Tunn. Undergr. Space Technol. **125**, 104497 (2022). https://doi.org/10.1016/j.tust.2022.104497

3. Baghbani, A., Choudhury, T., Costa, S., Reiner, J.: Application of artificial intelligence in geotechnical engineering: a state-of-the-art review. Earth Sci. Rev. **228**, 103991 (2022). https://doi.org/10.1016/j.earscirev.2022.103991

4. Bergstra, J., Bardenet, R., Bengio, Y., Kégl, B.: Algorithms for hyper-parameter optimization. In: Advances in Neural Information Processing Systems, vol. 24 (2011)

5. Erharter, G.H., Marcher, T., Reinhold, C.: Application of artificial neural networks for underground construction - chances and challenges - insights from the BBT exploratory tunnel Ahrental Pfons. Geomech. Tunnel. **12**(5), 472–477 (2019). https://doi.org/10.1002/GEOT.201900027

6. Fawaz, H.I., Forestier, G., Weber, J., Idoumghar, L., Muller, P.A.: Deep learning for time series classification: a review. https://doi.org/10.1007/s10618-019-00619-1, https://arxiv.org/pdf/1809.04356

7. Ismail Fawaz, H., et al.: InceptionTime: finding alexnet for time series classification. https://doi.org/10.1007/s10618-020-00710-y, https://arxiv.org/pdf/1909.04939

8. Jing, L.J., Li, J.B., Zhang, N., Chen, S., Yang, C., Cao, H.B.: A TBM advance rate prediction method considering the effects of operating factors. Tunnel. Undergr. Space Technol. **107**, 103620 (2021). https://doi.org/10.1016/j.tust.2020.103620

9. Jung, J.H., Chung, H., Kwon, Y.S., Lee, I.M.: An ANN to predict ground condition ahead of tunnel face using TBM operational data. KSCE J. Civil Eng. **23**(7), 3200–3206 (2019). https://doi.org/10.1007/s12205-019-1460-9, https://link.springer.com/article/10.1007%2Fs12205-019-1460-9

10. Karim, F., Majumdar, S., Darabi, H., Harford, S.: Multivariate lstm-fcns for time series classification. Neural networks: the official journal of the International Neural Network Society **116**, 237–245 (2019). https://doi.org/10.1016/j.neunet.2019.04.014

11. Kokhlikyan, N., et al.: Captum: a unified and generic model interpretability library for Pytorch. http://arxiv.org/pdf/2009.07896v1

12. Li, J., Li, P., Guo, D., Li, X., Chen, Z.: Advanced prediction of tunnel boring machine performance based on big data. Geosci. Front. **12**(1), 331–338 (2021). https://doi.org/10.1016/j.gsf.2020.02.011

13. Li, L., et al.: A system for massively parallel hyperparameter tuning. In: Conference on Machine Learning and Systems (2020). https://doi.org/10.48550/arXiv.1810.05934, https://arxiv.org/pdf/1810.05934

14. Liaw, R., Liang, E., Nishihara, R., Moritz, P., Gonzalez, J.E., Stoica, I.: Tune: a research platform for distributed model selection and training. https://arxiv.org/pdf/1807.05118

15. Qin, C., et al.: Precise cutterhead torque prediction for shield tunneling machines using a novel hybrid deep neural network. Mech. Syst. Sig. Process. **151**, 107386 (2021). https://doi.org/10.1016/j.ymssp.2020.107386

16. Sundararajan, M., Taly, A., Yan, Q.: Axiomatic attribution for deep networks. http://arxiv.org/pdf/1703.01365v2

17. Wang, Q., Xie, X., Yu, H., Mooney, M.A.: Predicting slurry pressure balance with a long short-term memory recurrent neural network in difficult ground condition. Comput. Intell. Neurosci. **2021**, 6678355 (2021). https://doi.org/10.1155/2021/6678355

18. Wang, Z., Yan, W., Oates, T.: Time series classification from scratch with deep neural networks: a strong baseline. https://arxiv.org/pdf/1611.06455

19. Xu, H., Zhou, J., Asteris, P.G., Jahed Armaghani, D., Tahir, M.M.: Supervised machine learning techniques to the prediction of tunnel boring machine penetration rate. Appl. Sci. **9**(18), 3715 (2019). https://doi.org/10.3390/app9183715

20. Yeh, C.K., Hsieh, C.Y., Suggala, A.S., Inouye, D.I., Ravikumar, P.: On the (in)fidelity and sensitivity for explanations. http://arxiv.org/pdf/1901.09392v4

21. Zhang, Q., Yang, K., Wang, L., Zhou, S.: Geological type recognition by machine learning on in-situ data of EPB tunnel boring machines. Math. Probl. Eng. **2020**, 1–10 (2020). https://doi.org/10.1155/2020/3057893

22. Zhou, H.A., et al.: Towards a data-driven assistance system for operating segment erectors in tunnel boring machines. In: 2021 14th International Symposium on Computational Intelligence and Design (ISCID), pp. 263–267. IEEE (2021). https://doi.org/10.1109/ISCID52796.2021.00068

23. Zou, X., Wang, Z., Li, Q., Sheng, W.: Integration of residual network and convolutional neural network along with various activation functions and global pooling for time series classification. Neurocomputing **367**, 39–45 (2019). https://doi.org/10.1016/j.neucom.2019.08.023

On Autonomous Drone Navigation Using Deep Learning and an Intelligent Rainbow DQN Agent

Andreas Karatzas[1], Aristeidis Karras[1], Christos Karras[1(✉)], Konstantinos C. Giotopoulos[2], Konstantinos Oikonomou[3], and Spyros Sioutas[1]

[1] Computer Engineering and Informatics Department,
University of Patras, Patras, Greece
{akaratzas,akarras,c.karras,sioutas}@ceid.upatras.gr
[2] Department of Management Science and Technology,
University of Patras, Patras, Greece
kgiotop@upatras.gr
[3] Department of Informatics, Ionian University, Corfu, Greece
okon@ionio.gr

Abstract. Drones are intelligent devices that offer solutions for a continuously expanding variety of applications. Therefore, there would be a significant improvement if these systems could explore space automatically and without human-supervision. This work integrates cutting-edge artificial intelligence techniques that allow drones to travel independently. Following an overview of reinforcement learning methods built for discrete action space settings, a multilayer Perceptron model is constructed for feature extraction along with Hybrid neural networks. The agent employed in the experiments is a Rainbow DQN agent trained on the AirSim simulator. The experimental results are encouraging as the agent was tested for 16 missions and the accuracy was higher than 93%. In particular, the success for action selection was 97% and 93 for mission success. Finally, future work related to the navigation of autonomous drones is discussed including current concepts and methods of integration with more sophisticated algorithmic approaches.

Keywords: Drones · Autonomous drone navigation · Deep learning · Reinforcement learning · Intelligent agents · Swarm intelligence · UAV

1 Introduction

The area of autonomous robotics is rapidly growing due to the range of tasks that robots can perform [7]. Examining a number of missions done by such robots, it is discovered that there is an increase in risk. In many of these instances, the integrity of the human workers sent at the point of dispatch is at danger. As the operator is removed from the hazardous shipping point, robots arrive to handle

© The Author(s), under exclusive license to Springer Nature Switzerland AG 2022
H. Yin et al. (Eds.): IDEAL 2022, LNCS 13756, pp. 134–145, 2022.
https://doi.org/10.1007/978-3-031-21753-1_14

the situation. Because of their usefulness, robots have been mass-produced. In addition to ground robots, there are also flying robots which keep the scientific community busy. Researchers are concerned about the autonomy of such systems [6]. This specific difficulty is intriguing, as groups of unmanned aerial vehicles (UAVs) now take off frequently. Coordinating various systems with a shared goal might provide a solution in a variety of circumstances, such as forest patrols, identification of suspicious behaviour in stalking instances, etc.

When studying designs of autonomous computer systems for unmanned aerial vehicles, the subject of accessible information arises, i.e., what information is available that may be utilised by the system to solve the autonomy problem? The standard sensors of an unmanned aerial vehicle are: i) Camera(s) ii) Radar(s) iii) a Time-of-flight sensor and iv) a LiDar Sensor [8].

Other than the first sensor, the others contribute to obstacle detection in an unknown areas. The camera of the UAV gathers all required information. Using image processing techniques mixed with new approaches of artificial intelligence, there are presently systems that can do remarkably well in such tasks given the right circumstances [2]. Typically, the sole stipulation is that there must be sufficient light for the camera to capture the required information in fine detail. Smaller resolution might entail higher uncertainty about the solution the system software will compute at its output, and hence there might be an increase in risk.

2   Preliminaries

A challenge in reinforcement learning (RL) may be characterised as a Markov process decision consisting of four sets [1] $\langle S, A, R, T \rangle$. S is a limited collection of training circumstances experienced by the agent. The information provided by the environment at the input of each training cycle for the agent configures the current state of the agent. The A is a limited collection of responses to a circumstance from which an agent may pick. $R(s, a)$ computes the reward offered to a state s after an action a. $HT(s, a, \hat{s}) \rightarrow [0, 1]$ is a potential transfer function. It specifies the likelihood of transitioning from state s to state $\hat{s}$ after action a. The functions $R(s, a)$ and $T(s, a, \hat{s})$ solely rely on the current values of s, a, and $\hat{s}$. The probability density of a Markov decision process is defined as follows:

$$\Pr\{s_{t+1} = \hat{s}, r_{t+1} = r \mid s_t, a_t, r_t, s_{t-1}, a_{t-1}, \ldots, r_1, s_0, a_0\} \qquad (1)$$

Equation (1) is a specific example of the more general situation in which the present state also relies on earlier states and actions (2).

$$\Pr\{s_{t+1} = \hat{s}, r_{t+1} = r \mid s_t, a_t\} \qquad (2)$$

When Eq. (1) is equivalent to Eq. (2) for all $\hat{s}, r$ and the sequence of form $s_t, a_t, r_t, s_1, a_1, r_1, s_0, a_0, r_0$ then the case holds the Markov property.

When investigating RL algorithms, the value of the Markovian property is substantial. However, in the majority of cases, agents do not have access to all of the data hence, there is concealed state data, preserving the Markov condition. Such data are described as Markov decision processes that are partly observable.

2.1   Value Function

Each training cycle is comprised of the agent observing a state and executing one action. The mapping between state and action is known as policy π. Specifically, the chance of selecting an action a in a given state s under policy is specified by the probability function V^π as $\pi(s, a) \to [0, 1]$ as defined by Eq. (3).

$$V^\pi(s) = E_\pi \left\{ R_t \mid s_t = s \right\} = E_\pi \left\{ \sum_{k=0}^{\infty} \gamma^k r_{t+k+1} \mid s_t = s \right\} \tag{3}$$

In Eq. (3), E_π is the expected value received by agent if follow policy π for a state s_t. Equation (3) can be rewritten, as shown in Eq. (4), as the sum of the rewards presented in the training cycles performed by the agent.

$$V_\pi(s_t) \equiv r_t + \gamma r_{t+1} + \gamma^2 r_{t+2} + \cdots \implies V_\pi(s_t) \equiv \sum_{k=0}^{\infty} \gamma^k r_{t+i} \tag{4}$$

In Eq. (4), the factor $\gamma \in [0, 1]$ is a degradation value of this sum so that the most recent rewards gain more weight. An equally important concept is the function Q^π, also known as the energy function and payoff for policy π. Q^π is the expected payoff having chosen an action a in a state s and following policy π. The function $Q^\pi(s, a)$ is given by Eq. (5).

$$Q^\pi(s, a) = E_\pi \left\{ R_t \mid s_t = s, a_t = a \right\} = E_\pi \left\{ \sum_{k=0}^{\infty} \gamma^k r_{t+k+1} \mid s_t = s, a_t = a \right\} \tag{5}$$

The goal when training a RL agent is to approximate the policy that maximizes the sum of rewards over time. The optimal policy is denoted by π^* and the value function which returns the maximum accumulated reward to the agent in a state s following the optimal policy π^* is denoted by $V^*(s)$ and is called the reward function. $Q^*(s, a)$ is the optimal function of reward, and means the expected reward of choosing an action a in a state s following the optimal policy.

2.2   Multilayer Perceptron Neural Networks

Perceptron multilayer neural networks are composed of many layers of perceptron-type neurons. Perceptron neurons receive input data through the forward propagation technique and are trained using the reverse propagation algorithm. Input vector is received by the first layer of Perceptron neurons. Subsequent layers change the input vector using two parameters: the numerical weight of synapses between neurons at the various levels and the non-linear activation function that follows the output of each neuron.

A multi-layer Perceptron neural network employs a collection of numerical samples created by certain controlling features for each class in the set of classes throughout its training. The format of the produced sample set is X matrix, as shown in Eq. (6).

$$X = \begin{bmatrix} | & | & \cdots & | \\ x^{(1)} & x^{(2)} & \cdots & x^{(m)} \\ | & | & \cdots & | \end{bmatrix} \tag{6}$$

In Eq. (6), each vector $x^{(i)}$ has dimension n, where n is the number of the attributes used to code the classes. Therefore, it holds that $X \in \mathbb{R}^{n_x, m}$, where m is the number of training samples. Same as the registry X, there is also the Y register in which the desired output of the after model is stored from the transformation of the input data, which is shown in Eq. (7). Each element $y^{(i)}$ of register Y has a number representing the class it belongs to the sample i. Therefore, it holds that $Y \in \mathbb{R}^{i,m}$.

$$Y = \begin{bmatrix} y^{(1)} & y^{(2)} & \cdots & y^{(m)} \end{bmatrix} \tag{7}$$

From the forward propagation of the X matrix in the model, intermediates arise transformations of the matrix X. The matrix A in each hidden level j of the neural network is defined as in Eq. (8).

$$A^{[j]} = \begin{bmatrix} | & | & \cdots & | \\ a^{(1)[j]} & a^{(2)[j]} & \cdots & x^{(m)[j]} \\ | & | & \cdots & | \end{bmatrix} \tag{8}$$

The synapse weights of a multi-layer Perceptron neural network for one layer j are described in Eq. (9).

$$W^{[j]} = \begin{bmatrix} w_{1,1}^{[j]} & w_{1,2}^{[j]} & \cdots & w_{1,m}^{[j]} \\ w_{2,1}^{[j]} & w_{2,2}^{[j]} & \cdots & w_{2,m}^{[j]} \\ \vdots & \vdots & \ddots & \vdots \\ w_{n,1}^{[j]} & w_{n,2}^{[j]} & \cdots & w_{n,m}^{[j]} \end{bmatrix} \tag{9}$$

In Eq. (9), the subscript n refers to the number of neurons in layer j-1, while the index m refers to the number of neurons in level j. At each level j, there is also a special neuron called a threshold. Threshold has no synapses with the previous level. A constant is stored in the threshold which is propagated only at the respective level. Considering b_j the threshold at a level j, then Eq. (10) describes the forward propagation result algorithm from a level j-1 to a level j.

$$Z^{[j]} = \left(W^{[j]}\right)^T \times A^{[j-1]} \tag{10}$$

Threshold $b^{[j]}$ is added to each element of register $Z^{[j]}$. Then, it takes effect in the register a non-linear function called the activation function. With the activation function, the model can categorize the samples of the set education. In case there is no trigger function, then the model performs a simple linear transformation of the input data and therefore will not learn complex patterns.

If $g(\cdot)$ is the activation function, then at the end of a cycle forward propagation of the neural network, the result shown in Eq. (11).

$$A^{[j]} = g\left(Z^{[j]} + b^{[j]}\right) \tag{11}$$

3   Methodology

3.1   Deep Q Networks

Deep Q-nets were developed as a result of the necessity for an algorithm capable of tackling a broad variety of issues. It integrates reinforcement learning concepts with artificial neural networks. The development of complex topologies for artificial neural networks, which use additional layers of neurons for more effective generalisation and pattern recognition, made it feasible for machines to learn categories from raw data. Deep Q networks primarily take tensors as input; hence, deep convolutional networks are used for the identification of needed standards. At the model's input, the tensor shells are pictures of a single channel.

Deep Q networks solve challenges involving agent-environment interaction. In particular, it posits that the agent may observe a circumstance, choose an action, and get a reward. The agent's objective is to choose the sequence of behaviours that accumulates the greatest potential amount of rewards over time. Specifically, a neural network with deep convolution is employed to approximate the ideal energy-reward function shown in Eq. (12).

$$Q^*(s, a) = \max_{\pi} \mathbb{E}\left[r_t + \gamma r_{t+1} + \gamma^2 r_{t+2} + \ldots \mid s_t = s, a_t = a, \pi\right] \tag{12}$$

Equation (12) expresses the maximum sum of rewards r_t reduced by a factor γ in each training cycle t. This sum can be achieved following the behavior defined by the policy function $\pi = P(a \mid s)$ in a state s performing action a.

Neural networks like other non-linear methods until the discovery of deep Q networks were considered unsuitable for approximating the function energy-reward (also known as Q function). The difficulties are enough, such as for example small changes in the Q function that can produce large ones changes in agent policy. The concept of deep Q networks uses experience repetition mechanism that solves the problem of correlating similar data in a time sequence by reducing the rate of change of the data distribution. The second element of deep Q networks is the iterative update of the Q function in order to more accurately approximate the optimal energy-action function Q^*.

A tuple is stored to implement the iterative mechanism $e_t = (s_t, a_t, r_t, s_{t+1})$ at each training cycle t forming a data set $D_t = e_1, \ldots, e_t$. During learning, refreshes are applied following the method of learning Q on samples (chunks) of experiences $(s, a, r, \hat{s}) \sim U(D)$, which are chosen in a random fashion following a uniform distribution from the set D_t. For implementation of the iterative renewal mechanism introduces a new neural network itself architecture with the model used to train the agent.

The second model $\hat{Q}$ aims to estimate the optimal energy. Every t training cycles, $\hat{Q}$ is synchronized with model Q. This change stabilizes $\hat{Q}$ algorithm as it deals with small but often unwanted changes in agent policy that would occur at the end of each training cycle. Thus, based on [5], the refresh learning following the learning method Q in an iteration i is configured from the cost function described in Eq. (13).

$$L_i\left(\theta_i\right) = \mathbb{E}_{(s,a,r,\hat{s})\sim U(D)}\left[\left(r + \gamma \max_{\hat{a}} Q\left(\hat{s},\hat{a};\theta_i^-\right) - Q\left(s,a;\theta_i\right)\right)^2\right] \quad (13)$$

In Eq. (13), the θ_i are the parameters of the neural network Q during iteration i and the θ_i^- are the parameters of the neural network $\hat{Q}$ used for the estimation of the best possible energy during repetition i. Traffic information of the agent in the environment is formed by combining 4 consecutive single frames channel. To avoid overtraining, except for the conversion done in each frame from color to grayscale, the frame is also cropped from 210×160 to 84×84. Thus, variable s, which is the convolutional neural network input is a $4 \times 84 \times 84$ tensor size.

3.2 Double Deep Q Networks

One of the new concepts that appeared in deep Q networks was the usage second neural network to estimate the optimal policy, in parallel with one neural network operating during agent training. In this way, the agent is not affected by small changes in policy and the training process converges as only after a number t training cycles does synchronization occur of the Q model with the $\hat{Q}$ model. However, this can also lead to overestimation of the objective as the same maximization operator is used, as shown in Eq. (13).

With the algorithm of double deep Q networks a separate operator is introduced maximization for the $\hat{Q}$ model. Therefore, there is a separate selection operator energy and a separate energy evaluation operator. Also, the policy that is modeled by the ϵ-greedy method is evaluated by the Q model, but the reward estimated by the model $\hat{Q}$. Based on [9], the policy update for the double deep algorithm Q of networks can be described as in Eq. (14).

$$Y_t^{\text{Double DQN}} \equiv R_{t+1} + \gamma Q\left(S_{t+1}, \operatorname*{argmax}_a Q\left(S_{t+1}, a; \theta_t\right), \theta^-\right) \quad (14)$$

3.3 Learning with Multiple Training Cycles

The method adopted for solving problems with RL methods uses the model of Markovian decision processes. According In this decision model, an agent interacts in an environment for a series training cycles t. In each training cycle, the agent receives information for the environment state $S_t \in \mathcal{S}$, where S is the set of all possible situations. The agent uses this information to choose an action A_t from the set of all possible actions $\mathcal{A}$. Based on the agent's behavior in state $S_t \in \mathcal{S}$, a payoff $R_{t+1} \in \mathbb{R}$ is calculated and the agent moves to next state

$S_{t+1} \in S$ with a state transfer probability $p(\hat{s} \mid s, a) = \hat{s} \mid S_t = s, A_t = a)$, for $a \in \mathcal{A}$ and $S, \hat{s} \in S$. The behavior of the agent is determined by policy $\pi(a \mid s)$ follows, which is a probability distribution over the set $S \times \mathcal{A}$.

During the training of the agent, the optimal policy π^* is formulated that maximizes the estimated reduced total reward, as described in Eq. (15).

$$G_t = R_{t+1} + \gamma R_{t+2} + \gamma^2 R_{t+3} + \cdots = \sum_{k=0}^{T-t-1} \gamma^k R_{t+1+k} \tag{15}$$

Algorithms following the time difference method aim to maximizing the amount of G_t. The state-value function describes the estimated performance when the agent is in a state s and follows policy π, as shown in Eq. (16).

$$u_\pi = \mathbb{E}\left[G_t \mid S_t = s\right] \tag{16}$$

At the center of the agent's training is also the energy function-value, which is the estimated payoff when the agent chooses an action a over a state s following policy π, as shown in Eq. (17).

$$q_\pi = \mathbb{E}_\pi\left[G_t \mid S_t = s, A_t = a\right] \tag{17}$$

Equation (17) can be calculated iteratively by observing new rewards based on previous estimates of q_π and using the renewal rule that shown in Eq. (18).

$$Q\left(S_t, A_t\right) \leftarrow Q\left(S_t, A_t\right) + \alpha\left[R_{t+1} + \gamma Q\left(S_{t+1}, A_{t+1}\right) - Q\left(S_t, A_t\right)\right] \tag{18}$$

In Eq. (18), the constant $\alpha \in (0, 1]$ is the cycle length parameter education. The time difference method can be extended to more cycles education. By carefully choosing a parameter $n > 1$ improved results can be obtained results when training an agent. The result is Eq. (19), which shows the update rule used for learning with multiple cycles of training [3].

$$Q_{t+n}\left(S_t, A_t\right) \leftarrow Q_{t+n-1}\left(S_t, A_t\right) + \alpha\rho_{t+1}^{t+n}\left[G_{t:t+n} - Q_{t+n-1}\left(S_t, A_t\right)\right] \tag{19}$$

In Eq. (19), the estimated return $G_{t:t+n}$ for an agent using learning n training cycles is given by Eq. (20).

$$G_{t:t+n} = \sum_{k=0}^{n-1} \gamma^k R_{t+k+1} + \gamma^n Q_{t+n-1}\left(S_{t+n}, A_{t+n}\right) \tag{20}$$

In Eq. (19) and Eq. (20), the quantity Q_{t+n-1} is the estimate of the function q_π at time $t + n - 1$ and the subscript $t : t + n$ denotes the renewal duration. Also, in Eq. (19), the term ρ_{t+1}^{t+n} sets the sampling mode so that they are selected proportionally the most important samples and is described by Eq. (21).

$$\rho_t^{t+n} \prod_{k=t}^{\tau} \frac{\pi\left(A_k \mid S_k\right)}{\mu\left(A_k \mid S_k\right)} \tag{21}$$

In Eq. (21), the variable $\tau = \min(t + n - 1, T - 1)$ is the training cycle until the end of the refresh step.

3.4   Rainbow Agent

A Rainbow agent is a combination of all the previous methods (Sects. 3, 3.2, and 3.3). First, the learning duration is increased, as described in Sect. 3.3 using n training cycles for learning. The multiple cycle learning method training is also combined with the dual deep Q network method using the action obtained by following the ϵ-greedy method in the state S_{t+n}. Finally, the architecture of the model is conflicting, and on the pathways used to assess benefit and value noise is introduced from a NoisyNet model. Rainbow agent hyperparameters are given in Table 1, as they were optimized. The Rainbow agent [4] successfully combines all previous enhancement learning techniques, as shown in Fig. 1.

Table 1. Hyperparameter table of a Rainbow agent.

Hyperparameter	Value
Minimum training cycles before learning	80.000
Adam optimizer Training rate	0.0000625
Parameter ϵ	0.0
Parameter σ_0 for NoisyNet models	0.5
Synchronization interval between $\hat{Q}$ and Q model	32.000
Numerical stability parameter ϵ for the Adam optimizer	0.00015
Precedence exponent ω	0.5
Importance parameter β for priority sampling	$0.4 \longrightarrow 1.0$
Parameter n for learning with multiple training cycles	3
Sets of groups in the N value distribution	51
V_{MIN}, V_{MAX}	$[-10, 10]$

3.5   Problem Formulation

On the complex problem of autonomy in navigation for unmanned aerial vehicles a Rainbow agent was created [4]. Two different ones were created architectures at the Q model level. In the first case, the kernel that uses the agent to extract the necessary patterns from the environment is a multilayer Perceptron neural network, while the second kernel was hybrid. As a core characterizes the model that processes the input data before passing through the advantage and value paths of the overall Q model. The tasks of the agent are as follows:

1. The agent is asked to move from a reference point $\langle x_0\ y_0\ z_0 \rangle$ to a target point $\langle x_1\ y_1\ z_1 \rangle$, with $x_0 \neq x_1, y_0 \neq y_1$ and $z_0 \neq z_1$.
2. The agent shall not collide with any object during its transition in the target.
3. The agent should try to reach the shortest known[1] route.

---

[1] The agent is not aware of the map, i.e. the layout of the obstacles. However, he knows the direction where the target is located. Therefore, it can estimate the direction to shift it with a larger one pace to target.

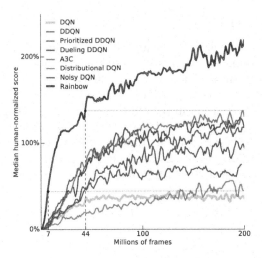

Fig. 1. Comparison between the rainbow agent and reinforcement learning methods.

To satisfy all 3 missions, the agent needs more information from the image given by the depth camera. Therefore the classical deep architecture Q network should be transformed according to the needs of the problem. As result, instead of convolutional neural network, tests were done using 2 different cores: using a) Multilayer Perceptron Neural Network and b) Hybrid neural network which consists of: Convolutional Neural Network Core, Core Perceptron Multilayer Neural Network, Fully cohesive layer for combining features from cores.

The idea behind the multi-layer Perceptron neural network is that it can process the information it receives from the depth camera along with the necessary information about its location in the environment and the location of the target. The state of S_t agent at time t is a vector of: i) The preprocessing image of depth D_t, ii) The position vector of the agent $P_t = \langle\, x_t\ y_t\ z_t\,\rangle$, iii) The position vector of the agent $P_{t-1} = \langle\, x_{t-1}\ y_{t-1}\ z_{t-1}\,\rangle$ in the previous time, iv) The position vector of the target $T = \langle x_T\quad y_T\quad z_T\rangle$ and v) A scalar l_t floating point that informed the agent how many actions he has done up to time t. The agent could do up to 4 times more actions than the optimal estimated number of actions. The optimal estimated number of actions was obtained by summing the number of steps that the agent pays if he always chooses a reducing action among its distance from the target.

Preprocessing on the depth image consists of a basic image transformation in gray scale followed by a transformation to smaller dimensions and finally a crop of the result to its center so that it has the same dimension the length by the width. Thus, the depth image is transformed from $3 \times 210 \times 160$ to 84×84. Finally, the depth image is normalized to the range $[0, 1]$ by dividing it by the maximum possible value which is 255. The result is the matrix D_t.

The matrix D_t is transformed into a column vector D_t^{fl} and then joins with remaining input parameters P_t, P_{t-1}, T and l_t shaping its state agent S_t at time

t. The available actions are 6. In the experiments of the work, the displacement step was 1 m. The payoff at each step was given by Eq. (22).

$$R_t = \begin{cases} -100 & \text{in the event of a collision with an obstacle,} \\ 100 & \text{upon completion of the mission,} \\ -10 & \text{if } l_t <= 0, \\ -10 & \text{in case of early landing} \\ \mathcal{D}_{t-1} - \mathcal{D}_t & \text{otherwise} \end{cases} \qquad (22)$$

In Eq. (22), the quantity $\mathcal{D}_t$ is the distance of the agent from the target in time moment t. The same quantity $\mathcal{D}_{t-1}$ is the distance of the agent from the target in time $t - 1$. The distance is calculated according to Eq. (23).

$$\mathcal{D}_t = P_t - \mathcal{T} \qquad (23)$$

4   Experimental Results

During the experiments the performance of the agent is evaluated across 16 different targets. To create a simulation interface, the software AirSim was used[2]. The results for the Multilayer Perceptron Neural Network are shown in Table 2. Moreover, the training evaluation is shown in Fig. 2. The results for the Hybrid Neural Network are shown in Fig. 3.

Table 2. Percentage of success/failure for action selection and mission completion.

Evaluation	Result	Percentage
Percentage of action selection as per target	Success	97.3%
Percentage of action selection as per target	Failure	2.7%
Percentage of success across 16 missions	Success	93.8%
Percentage of success across 16 missions	Failure	6.2%

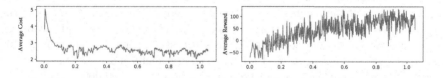

Fig. 2. Results of the rainbow agent as per average cost and average reward.

The Perceptron multi-layer neural network core agent has been shown to work very satisfactory. However, it showed inability to make a decision in cases where was faced with an obstacle of great dimensions. To solve the problem, a hybrid

---

[2] https://microsoft.github.io/AirSim/.

architecture was tried. The entry of the new Q model consists of the matrix D_t and the vector V_t which is the concatenation of P_t, P_{t-1}, T and l_t. The model Q consists of a deep convolutional neural network for efficient extraction features from the D_t matrix and from a deep multi-layer neural network Perceptron for feature extraction from vector V_t. The characteristics of 2 cores are joined at a common plane, called the union plane. The level union is an input layer for a deep multilayer Perceptron neural network, the association model.

The computational resources required to train the agent were very high more than available. Notable was the maximum possible memory size of repeating experiences according to the system memory used for the experiments. The maximum experience memory size was 50,000 samples. Also, for for speed reasons, prioritized replay memory was implemented using segment tree[3]. The partition tree belongs to the family of binary trees and therefore the agent instead of 50,000 items in memory could only load $2^{\lfloor \log_2 50.000 \rfloor} = 2^{15} = 32.768$ experiences. As a result, model training was not able to be completed fully.

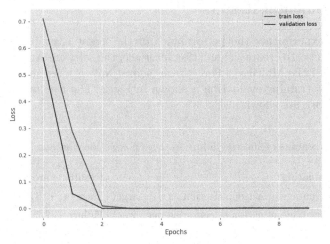

Fig. 3. Training of multi-layer neural network as per feature extraction from V_t.

After training the 2 kernels, it remained to train the union model. Keeping the parameters of the 2 cores fixed, during training they were adjusted only the union model parameters and the parameters in the value paths and advantage respectively. As in the previous experiments, it is stated that available computing resources were extremely limited for his needs problem and how the indicative number of training cycles, as can be seen in Fig. 1, is 44,000,000 cycles instead of the 450,000 training cycles run by model.

5   Conclusions and Future Work

Autonomous UAV navigation is an emerging concept in robotics where most environments are unknown. Therefore, techniques for automated navigation are

---

[3] https://github.com/segment_tree.py.

increasing including modern AI systems that can analyse huge volumes of data to uncover patterns leading to successful solutions. Using reinforcement learning, the software learns a function that maximises the reward within an environment. Rainbow-type agents were chosen because of their exceptional performance in limited-action situations. As noted, an agent with high performance and optimum policy was built by using preprocessed image as input for pattern recognition from a tensor allowing a deep convolutional neural network to train using more data. Moreover, a better strategy may be established than the one utilised here to build the agent using a multi-layer neural network core like Perceptron as the current computing resources are insufficient to perform such tasks.

Future directions of this work include the construction of a model that utilizes a set of sequential images at its input, which contain all the information the agent needs in terms of the device, in terms of surroundings and in terms of the goal. Perhaps a convolutional model will emerge which can be designed to detect many different classes within an image to perform better than the convolutional core used for the experimental evaluation. Ultimately, fine-tuning optimizations include the use of more sophisticated algorithmic choices along with this work.

References

1. Karras, A., Karras, C., Giotopoulos, K.C., Giannoukou, I., Tsolis, D., Sioutas, S.: Download speed optimization in P2P networks using decision making and adaptive learning. In: Daimi, K., Al Sadoon, A. (eds.) Proceedings of the ICR'22 International Conference on Innovations in Computing Research. ICR 2022. Advances in Intelligent Systems and Computing, vol. 1431, pp. 225–238. Springer, Cham (2022). https://doi.org/10.1007/978-3-031-14054-9_22
2. Amer, K., Samy, M., Shaker, M., ElHelw, M.: Deep convolutional neural network based autonomous drone navigation. In: Thirteenth International Conference on Machine Vision, vol. 11605, pp. 16–24. SPIE (2021)
3. De Asis, K., Hernandez-Garcia, J., Holland, G., Sutton, R.: Multi-step reinforcement learning: a unifying algorithm. In: Proceedings of the AAAI Conference on Artificial Intelligence, vol. 32 (2018)
4. Hessel, M., et al.: Rainbow: combining improvements in deep reinforcement learning. In: Thirty-Second AAAI Conference on Artificial Intelligence (2018)
5. Mnih, V., et al.: Human-level control through deep reinforcement learning. Nature **518**(7540), 529–533 (2015)
6. Roghair, J., Niaraki, A., Ko, K., Jannesari, A.: A vision based deep reinforcement learning algorithm for UAV obstacle avoidance. In: Arai, K. (ed.) IntelliSys 2021. LNNS, vol. 294, pp. 115–128. Springer, Cham (2022). https://doi.org/10.1007/978-3-030-82193-7_8
7. Siciliano, B., Khatib, O., Kröger, T.: Springer Handbook of Robotics, vol. 200. Springer, Heidelberg (2008). https://doi.org/10.1007/978-3-540-30301-5
8. Um, J.S.: Drones as Cyber-Physical Systems. Springer, Singapore (2019). https://doi.org/10.1007/978-981-13-3741-3
9. Van Hasselt, H., Guez, A., Silver, D.: Deep reinforcement learning with double q-learning. In: Proceedings of the AAAI Conference on Artificial Intelligence, vol. 30 (2016)

An Intelligent Decision Support System for Road Freight Transport

Hugo Silva Carvalho[1], André Pilastri[1], Arthur Matta[1], Luís Miguel Matos[3], Rui Novais[2], and Paulo Cortez[3(✉)]

[1] EPMQ, CCG ZGDV Institute, Guimarães, Portugal
{hugo.carvalho,andre.pilastri,arthur.matta}@ccg.pt
[2] ABMN - Business Solutions, Azurém, Guimarães, Portugal Guimarães, Portugal
rui.novais@abmn.pt
[3] ALGORITMI Research Centre/LASI, Department of Information Systems, University of Minho, Guimarães, Portugal
{luis.matos,pcortez}@dsi.uminho.pt

Abstract. This paper presents an Intelligent Decision Support System (IDSS) to optimize transport and logistics activities in a set of Portuguese companies currently operating in the freight transport sector. This IDSS comprises three main modules that can be used individually or chained together, dedicated to: a geographic clustering detection of transport services; a transport driver suggestion; and a route and truckload optimization. The IDSS was entirely designed and developed to support real-time data and it consists of an end-to-end solution (E2ES), given that it covers all the main transport and logistics processes since the registration in the database to the optimized transport plan. The entire set of functionalities inserted in the IDSS was designed and validated by freight transport sector experts from the different companies that will use the proposed system.

Keywords: Intelligent decision support system · Transport and logistics · Geospatial clustering · Resource optimization · Route optimization · TAM 3

1 Introduction

Due to the global market competition, current world market dynamics (e.g., effect of the COVID-19 pandemic) and other issues (e.g., sustainability), there is an increased pressure to improve the road freight transport sector. Until recently, the planning of transport activities was mainly focused on cost minimization [7]. However, the transport sector represents a substantial source of greenhouse gas (GHG) emissions. As a result, one of the main goals of supply chain management programs is to develop more sustainable transport solutions [9]. Under this context, Intelligent Decision Support Systems (IDSS) [3] represent a potential valuable solution, since they can be used to extract valuable insights from the

© The Author(s), under exclusive license to Springer Nature Switzerland AG 2022
H. Yin et al. (Eds.): IDEAL 2022, LNCS 13756, pp. 146–156, 2022.
https://doi.org/10.1007/978-3-031-21753-1_15

freight transport data, allowing to address several objectives (e.g., reduce costs and gas emissions).

In this work, we propose an end-to-end solution (E2ES) that consists of an IDSS that targets the main processes and tasks involved in planning freight transport. The IDSS prototype was designed in conjunction with a consultant company that is responsible for the transport management of several Portuguese freight transport companies. Moreover, the business knowledge and data obtained proved crucial throughout the development, testing, and validation of the IDSS. In particular, the IDSS includes: an adaptive distance-constrained geospatial clustering algorithm, for the detection of possible clusters of loading or unloading locations; a tool for ranking drivers in the allocation to a transport service by considering their distance to a loading location; and an overall routing optimization process that considers the rules stipulated by the European Commission (EC) in Regulation No 561/2006 regarding driving time and rest periods, and that includes the truckload and sequencing operations. In order to evaluate the proposed IDSS, the inexistence of a system or set of methods and procedures dedicated to the optimization of the sector in the freight transport companies led us to assemble a questionnaire using the Technology Acceptance Model (TAM) 3 model [13] and perform open interviews.

2   Related Work

Within the freight transport sector, there are several studies and proposed frameworks to enhance the optimization of transport planning. However, most approaches address specific and isolated topics within the transport sector, and within our knowledge there is no complete solution that includes the whole set of tasks and activities required to optimize a full transport plan.

Regarding the use of unsupervised Machine Learning (ML) techniques, clustering has been applied to freight transport and logistics issues to understand the geographic distribution of demand [6] and streamline logistics operations, enabling the use of GPS data to analyze the repetitiveness of commercial vehicle travel [11].

As for the driver management module, the rules stipulated by the European Commission (EC), their strictness, and respective oversight and monitoring by law enforcement agencies, are contributing to the evolving scientific research on this subject. These related works fall mainly under the field of mixed integer linear programming for driver activities scheduling, considering rest periods and breaks [4].

Over the last sixty years, load optimization, sequencing, and vehicle routing issues have been a subject of an intensive and growing research, in which the analyzed task is commonly known as a Vehicle Routing Problem (VRP). Due to the wide range of operating rules and restrictions found across the various industry sectors, diverse variants of the VRP have emerged. Over time, complexities such as travel times, time windows for delivery and collection of goods, and legislation establishing the driver working hours have been assimilated into the problem [5]. For instance, [14] explores current research trends, recent achievements, and new challenges in the topic while providing a short overview of current and emerging VRP variants.

As a result, the existing range of solutions and methodologies in the freight transport sector focus essentially on specific topics without taking into account the complexity and vast scope of the theme and the interdependence of all the variables encountered in the process of planning a transport service from start to finish. Also of note is the lack of adaptation of these solutions to real-world applications and preparation of their deployment in the industry with a continuous integration and continuous deployment (CICD) philosophy. In this paper, in contrast to the related works, an end-to-end solution (E2S2) is proposed, comprising a set of components capable of carrying an adaptive optimized transport plan throughout the whole process.

3   Materials and Methods

3.1   Problem Formulation

The set of Portuguese companies covered in the development of this IDSS is currently operating in the freight transport sector and is primarily composed of the following components: the main logistics center, a fleet of vehicles, and drivers. The main logistics center represents the company headquarters and it is also used as a warehouse and distribution center. The primary role is operationalizing logistics activities and freight transport. We highlight that the transport planning and resource allocation are performed using a software system in which the proposed IDSS will be incorporated. This system works as the cornerstone of the entire transport and logistics business.

Regarding the drivers, these represent the human resources that perform transport activities. In addition to being essential to the execution of the transport services, these resources require a particular organization and management to comply with the requirements set forth by the European Commission (EC) Regulation No 561/2006, which establishes driving and rest periods.

Each business that holds a fleet of vehicles can diverge in their types, models, sizes, capacities, and purposes. The digital transition philosophy in these Portuguese freight transport companies has led to data storage of the mentioned tasks and components. However, data analysis, pattern recognition, and automation are not optimized or even implemented. The range of transport planning activities is managed and arranged based on intuition and human experience of the transport software system users. Often, this manual approach is non optimal. The lack of interoperability and divergent (often ad-hoc) human logistics decisions lead to a misuse of resources and, consequently, to an unsustainable development.

3.2   Proposed IDSS

In this work, we propose an IDSS that follows a three-tier architecture that assumes three logical and physical computing elements, as presented in Fig. 1. The Data Layer is accountable for extracting and processing data from the different databases regarding the surrounding factors involved in freight transport planning, including data on freight services to be planned, data associated with the various drivers (relevant to comply with European Union (EU) rules) and data on the physical properties of trailers. In addition, there is a middleware with PTV [8], a mobility software prepared to provide distances and routes that consider the truck dimensions and legal restrictions. Its outputs are sent to the Processing Layer.

The Processing Layer of the IDSS is composed of three modules: Groupage Detection Module (GpDM), Resource Optimization Module (RsOM), and Route Optimization Module (RtOM).

Currently, the company plans routes based on specific Portuguese districts (e.g., Viseu, Coimbra, Castelo Branco), creating long and more expensive routes, thus spending more resources. To tackle this issue, the GpDM assumes an unsupervised ML algorithm in charge of planning clusters of route points to travel based on a maximum distance, thus producing better route planning and cost management (e.g., time and fuel). The developed distance-constrained geospatial clustering (DcGC) algorithm can group geographic locations based on their geographical positional similarities and a maximum adjustable distance between transports inserted in the same cluster. This way, it is possible to improve the current approach implemented with an adaptive method to create dynamic zones according to the reality of each moment of the component execution. This functionality was designed to address the inefficiency of the current grouping of transports that, despite belonging to the same zone, are highly dispersed in the vertical or horizontal geographical spectrum. It is also important to note the particularity of its performance in cross-border areas where the problem of geographic coordinates separation from very close locations is addressed. The DcGC pseudocode is presented in Algorithm 1.

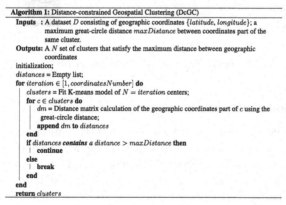

Algorithm 1: Distance-constrained Geospatial Clustering (DcGC)

Inputs : A dataset D consisting of geographic coordinates $\{latitude, longitude\}$; a maximum great-circle distance $maxDistance$ between coordinates part of the same cluster.

Outputs: A N set of clusters that satisfy the maximum distance between geographic coordinates

initialization;
$distances$ = Empty list;
for $iteration \in [1, coordinatesNumber]$ **do**
    $clusters$ = Fit K-means model of $N = iteration$ centers;
    for $c \in clusters$ **do**
        dm = Distance matrix calculation of the geographic coordinates part of c using the great-circle distance;
        append dm **to** $distances$
    end
    if $distances$ **contains** a distance $> maxDistance$ **then**
        continue
    else
        break
    end
end
return $clusters$

Afterwards, a Time Difference (TD) leaderboard of possible drivers is calculated and provided by the Resource Optimization Module (RsOM) module. The TD leaderboard is calculated using the following equation:

$$TD = |PID - AD_d| \tag{1}$$

where PID represents the expected start date by the company, and AD_d represents the driver availability date. This ranking is then presented to the driver manager, who will choose the most suitable driver to travel the route provided by the GpDM module.

The last module is the Route Optimization Module (RtOM), which is responsible for creating and optimizing the route to be covered based on two objectives: sequencing the road to be covered in the shortest possible distance; maximizing the load carried by the various vehicles on these routes. Furthermore, this module provides a better selection of vehicles depending on each route created. One of the main advantages lies within the ability of each module to be executed individually, and if requested, the outputs will be returned to the IDSS user.

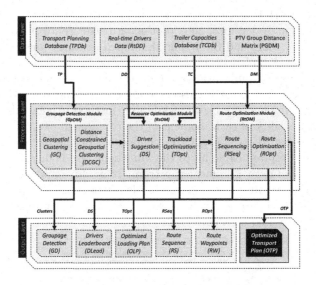

Fig. 1. Proposed IDSS architecture.

3.3 Evaluation Methodology

The proposed IDSS was developed by a research team comprising Artificial Intelligence (AI), Software Development (SD), and freight transport experts. Several individual components of the proposed IDSS (e.g., GpDM module) are evaluated by considering specific logistics scenarios. To globally validate the full IDSS, we adopted the Technology Acceptance Model (TAM) 3, which allowed us to develop a questionnaire with ten questions to assess the system acceptance and evaluation. TAM 3 is composed of six constructs: Perceived Usefulness (PU); Perceived Ease of Use (PEOU); Perception of External Control (PEC); Job Relevance (REL); Output Quality (OUT); and Behavioral Intention (BI). Each question included a 5-point Likert scale [10] choice with a range of 1 (strongly disagree) to 5 (strongly agree) for each response. Direct feedback from the consultant company responsible for transport management was also captured. Moreover, we also map the IDSS tool functionalities, comparing them with current logistics practices and procedures (termed here as the AS-IS approach).

4 Results

4.1 Developed IDSS Prototype

The designed IDSS was exclusively developed using the Python language. Prototyping and deployment were accomplished using `Flask`. Finally, for the set of critical and proposed features in this paper, namely ML and optimization techniques, `scikit-learn`[1], `ortools`, and `PTV Developer`[2], were the tools used. It is also important to note that, at a preliminary stage of development, to overcome the lack of data derived from General Data Protection Regulation (GDPR) bureaucratic issues, the `RanCoord`[3] package was used. The package offers an easy procedure to generate random coordinates within a set of geographic boundaries [12].

Groupage Detection Module (GpDM) is responsible for detecting possible load groupages using zone grouping and clustering techniques. Figure 2 (*plot a* and *plot b*) depicts the transition from static to adaptive and dynamic zones, elevating the compactness of the clusters. This illustrates the grouping of 71 transport load zones to be carried out, spread over four districts within the Portuguese territory. The AS-IS approach considers the district borders, thus obtaining four grouping zones. These zones are divided into 26, 23, 22, and 4 loading locations respectively.

Coincidentally, the proposed approach using the DcGC algorithm, executed with a standard maximum distance used by the different transport companies of 55 km, also clustered the data into four clustering zones, consisting of 23,14, 16, and 22 locations. The equal number of clusters makes it possible to compare the efficiency and primacy of DcGC. The enhancement of transport groupage can be verified through each approach's distance matrix of the different clusters. The AS-IS method obtained an average distance between geographic points of the same cluster of 19.45 km, composed of 23.14, 19.04, 20.21, and 15.44. On the other hand, the DcGC algorithm allowed a remarkable reduction of this distance to 15.42 km, composed of averages of 13.98, 18.76, 15.71, and 13.23 km. The contribution of the algorithm extends to balancing the number of locations grouped in the same cluster. The standard deviation of the number of sites to be grouped in each cluster was reduced from 8.60 to 3.80.

The Resource Optimization Module (RsOM) includes the driver suggestion (DS) and the optimization of truckload allocation (TOpt). Table 2 illustrates how this component behaves, including the set of input attributes in the Transport and Drivers sections. The Leaderboard section represents the result of this micro-framework, assembling qualified drivers to execute a transport arranged

---

[1] https://scikit-learn.org/stable/index.html.

[2] https://developer.myptv.com/.

[3] Publicly available https://github.com/hugodscarvalho/rancoord.

by TD. The resulting leaderboard (DLead) relevance lies within the decision-making support throughout the driver selection process, enhancing its efficiency. The information regarding driving hours will allow the manager to comply with European standards. The second component of the module relates to truckload optimization (TOpt) by formulating the problem as a packaging problem. The goal of issues of this nature is to find the best combination to pack a set of items of given sizes or weights into containers with fixed capacities. Using the SCIP for mixed integer programming (MIP) solver [1,2], we developed an algorithm capable of optimizing the allocation of loads in different vehicles to maximize the occupancy of the vehicle and minimize the number of vehicles needed to pack and transport them, turning it into a multi-objective optimization. Regarding a set of 11 loads stored in a specific warehouse of one of the freight transport companies, it was possible to find the best combination of loads with a minimized solution of 4 vehicles and maximized occupation of the different vehicles of 99.60%, 100.00%, 77.10%, and 98.80% with a capacity established at 25 tons, the standard legal net load value.

Table 1. Route Optimization Module (RtOM) execution.

Vehicle ID	Load	Occupancy	Sequence
1	19	95.45%	LC - HM4 - HM8 - LC
2	25	100.00%	LC - HM3 - HM9 - HM13 - HM12 - LC
3	23	92.00%	LC - HM7 - HM6 - LC
4	24	96.00%	LC - HM2 - HM10 - HM11 - LC
5	23	92.00%	LC - HM1 -HM5 - LC

The last module of the IDSS focuses on the Route Optimization Module (RtOM). As exemplified in Table 1 and Fig. 2 (*plot c*), the component takes as input parameters a set of loads to be transported to 13 geographically separated hypermarkets (HM) from the Central Region of Portugal from a logistics center and, as a second input, a set of 7 available vehicles and their respective physical profiles. This component of the IDSS generates a load allocation to the set of vehicles and the sequencing of hypermarkets to be carried out by each one. The solution presented in Table 1 and illustrated in Fig. 2 (*plot c*) is a multi-objective optimization that takes into account: (i) minimizing the distance traveled by each vehicle; (ii) maximizing the occupancy of each vehicle. Objective (ii) maximizes the occupancy of each vehicle, allowing the usage of only 5 vehicles of the 7 available. Finally, the second component of the module only increases the value of the solution provided by the previous one. Rather than just returning the sequence of locations to cross, it returns the complete route to be taken by each vehicle through a set of waypoints depicted in Fig. 2 (*plot c*). The user of the IDSS can specify whether to avoid tolls or not.

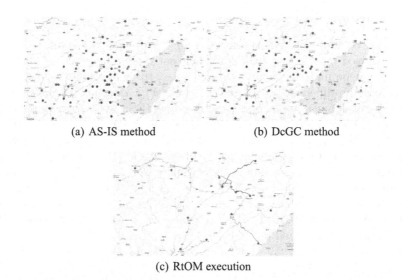

(a) AS-IS method   (b) DcGC method

(c) RtOM execution

Fig. 2. Grouped geographic coordinates of transport loading locations using the AS-IS method (left plot) versus DcGC proposed method (right plot) and grouped geographic coordinates of transport loading locations (bottom plot).

Table 2. Driver Suggestion (DS) module attributes and output

Context	Attribute	Description
Transport	Transport ID	Transport identification
	Expected start datetime	Expected start date
	Start Latitude	Initial Transport latitude point
	Start Longitude	Initial Transport longitude point
	Expected driving time	Estimated driving hours
Driver	Driver ID	Driver identification
	Availability datetime	Availability date and time
	Day driving hours	Expected daily driving time
	Week driving hours	Expected weekly driving time
	Fortnight driving hours	Expected Fortnight driving time
	Availability latitude	Latitude geographical point
	Availability longitude	Longitude geographical point
Leaderboard	Driver ID	Driver identification
	Time window	Time duration to reach transport location
	Distance	Distance to transport location
	Day driving hours	Expected daily driving time
	Week driving hours	Expected weekly driving time
	Fortnight driving hours	Expected Fortnight driving time

4.2   Evaluation

The results columns include the average of five freight transport managers' responses based on a Likert scale and their corresponding standard deviation (SD). It must be underlined that, to avoid bias, these managers correspond to people from the different transport companies who were not in the research, design, or development process of the proposed IDSS. The average responses range from 3.60 to 4.60, and the standard deviation (SD) is between ±0.16 and ±0.40, revealing a positive evaluation and acceptance of the proposed IDSS and compliance among the various people involved in the questionnaire. The most positive responses are inserted in Job Relevance (REL1 and REL2) and Behavioral Intention (BI1), with an average rating of 4.60. To increase the value of this evaluation, the feedback given during all technical sessions of design, presentation, and enrichment of the IDSS, was also aggregated. The designed TAM 3 questionnaire is presented in Table 3.

Table 3. The adopted TAM 3 questionnaire and results (5 responses)

Construct	Items	Question	Results	
			Mean	SD
Perceived Usefulness (PU)	PU1	Using the IDSS improves my performance in my job	4.00	± 0.40
	PU2	The IDSS is (potentially) useful in my job	4.40	±0.24
Perceived Ease of Use (PEOU)	PEOU1	I find the IDSS to be very easy to use	3.60	±0.24
	PEOU2	It's easy to get the information that I want from the IDSS	3.60	±0.24
Perceptions of External Control (PEC)	PEC1	I have the knowledge to use de IDSS	3.80	±0.16
Job Relevance (REL)	REL1	In my job, the usage of the IDSS is important	4.60	±0.24
	REL2	The use of the IDSS is pertinent to my various job-related task	4.60	±0.24
Output Quality (OUT)	OUT1	The quality of the output I get from the IDSS is high	4.40	±0.24
	OUT2	I have no difficulty telling others about the results of the IDSS	4.00	±0.40
Behavioral Intention (BI)	BI1	Assuming I had access to the IDSS, I intend to use it	4.60	±0.24

Regarding the GpDM, the feedback was unanimously positive. These managers pointed out a remarkable optimization of the clustering process of transport services, evidenced by the examples of comparison between the AS-IS process and the developed DcGC algorithm. Table 4 summarizes the main features introduced by the proposed IDSS, which substantially enhance the current processes and procedures (AS-IS).

Table 4. Comparison between the current functionalities (AS-IS) and proposed IDSS tasks and procedures.

Functionalities	AS-IS	IDSS	IDSS module
Group transport services by zone	✓	✓	GC
Cluster transport services by geographic radius		✓	DcGC
Driver suggestion		✓	DS
Optimized truckload allocation		✓	TOpt
Optimized route sequencing		✓	RSeq
Route based on vehicle profile	✓	✓	ROpt
Route optimization		✓	ROpt

Concerning the RsOM module, the truckload optimization component was particularly emphasized because of the transition between intuition and human experience to an optimal combinatorial truckload optimization algorithm. On the other hand, the driver suggestion component, despite its value recognition to the decision-making process and its clear relevance, the suggestion of its improvement regarding increased automation and inclusion of the driving rules and rest times stipulated by the European Union was given by four of the five managers. Lastly, RtOM was the best verbally rated module with a clear satisfaction of the complexity and benefits provided by its sequencing and detailed routing of an optimized transport plan for each required vehicle. Overall, all respondents concluded that the proposed IDSS (including its three modules) is valuable for optimizing transport and logistics activities and represents a possible positive digital disruption for the companies.

5 Conclusions

This paper presents an Intelligent Decision Support System (IDSS) to optimize transport and logistics activities as an end-to-end solution (E2ES) in a set of Portuguese companies currently operating in the freight transport sector. The proposed IDSS follows a three-tier architecture: Data Layer – responsible for extracting and processing data from different data sources accordingly to the respective component to be addressed; Processing Layer – which represents the core of the proposed IDSS, consisting of the set of freight transport crucial procedures and tasks, namely: (a) Groupage Detection Module (GpDM); (b) Resource Optimization Module (RsOM); (c) Route Optimization Module (RtOM); and Output Layer – which includes the results and outputs from the middle layer.

The prototype was evaluated by five managers who were not in the research, design, or development process of the proposed IDSS to avoid bias. The adopted evaluation methodology was TAM 3, comprising questionnaires and feedback gathered throughout technical sessions of the design, presentation, and enrichment of the IDSS. Overall, a very positive evaluation and acceptance were obtained. In future work, we intend to add new modules and components to the IDSS articulated with the suggestions and the expected evolution of the freight transport sector.

Acknowledgment. The authors would like to express the most significant recognition to the project on which this IDSS has arisen, "aDyTrans - Dynamic Transportations Platform" reference NORTE-01-0247-FEDER-045174, supported by Norte Portugal Regional Operational Programme (NORTE 2020), under the PORTUGAL 2020 Partnership Agreement, through the European Regional Development Fund (ERDF).

References

1. Achterberg, T.: SCIP: solving constraint integer programs. Math. Program. Comput. **1**(1), 1–41 (2009)
2. Achterberg, T., Berthold, T., Koch, T., Wolter, K.: Constraint integer programming: a new approach to integrate CP and MIP. In: Perron, L., Trick, M.A. (eds.) CPAIOR 2008. LNCS, vol. 5015, pp. 6–20. Springer, Heidelberg (2008). https://doi.org/10.1007/978-3-540-68155-7_4
3. Arnott, D., Pervan, G.: A Critical Analysis of Decision Support Systems Research, pp. 127–168. Palgrave Macmillan UK, London (2015)
4. Bernhardt, A., Melo, T., Bousonville, T., Kopfer, H.: Scheduling of driver activities with multiple soft time windows considering European regulations on rest periods and breaks. Schriftenreihe Logistik der Fakultät für Wirtschaftswissenschaften der htw saar 12, econstor, Saarbrücken (2016)
5. Braekers, K., Ramaekers, K., Van Nieuwenhuyse, I.: The vehicle routing problem: state of the art classification and review. Comput. Ind. Eng. **99**, 300–313 (2016)
6. Cao, B., Glover, F.: Creating balanced and connected clusters to improve service delivery routes in logistics planning. J. Syst. Sci. Syst. Eng. **19**(4), 19:453–19:480 (2010)
7. Demir, E., Bektaş, T., Laporte, G.: A review of recent research on green road freight transportation. Eur. J. Oper. Res. **237**(3), 775–793 (2014)
8. PTV Group: Mobility and Transportation Software. PTV Route Optimiser
9. Pan, S., Ballot, E., Fontane, F.: The reduction of greenhouse gas emissions from freight transport by pooling supply chains. Int. J. Prod. Econ. **143**(1), 86–94 (2013)
10. Robinson, J.: Likert scale. In: Michalos, A.C. (ed.) Encyclopedia of Quality of Life and Well-Being Research, pp. 3620–3621. Springer, Dordrecht (2014). https://doi.org/10.1007/978-94-007-0753-5_1654
11. Sharman, B.W., Roorda, M.J.: Analysis of freight global positioning system data: clustering approach for identifying trip destinations. Transp. Res. Rec. **2246**(1), 83–91 (2011)
12. Carvalho, H.S., Pilastri, A., Cortez, P.: Rancoord (2022)
13. Venkatesh, V., Bala, H.: Technology acceptance model 3 and a research agenda on interventions. Decis. Sci. **39**(2), 273–315 (2008)
14. Vidal, T., Laporte, G., Matl, P.: A concise guide to existing and emerging vehicle routing problem variants. Eur. J. Oper. Res. **286**(2), 401–416 (2020)

Endowing Intelligent Vehicles with the Ability to Learn User's Habits and Preferences with Machine Learning Methods

Paulo Barbosa[1], Flora Ferreira[1,2], Carlos Fernandes[1], Wolfram Erlhagen[2], Pedro Guimarães[1,2], Weronika Wojtak[1,2], Sérgio Monteiro[1], and Estela Bicho[1(✉)]

[1] Research Center Algoritmi, University of Minho, 4800-058 Guimarães, Portugal
{w.wojtak,sergio.monteiro,estela.bicho}@dei.uminho.pt
[2] Research Center of Mathematics, University of Minho, 4800-058 Guimarães, Portugal
{fjferreira,wolfram.erlhagen}@math.uminho.pt

Abstract. A private vehicle frequently carries the same passengers who routinely take specific objects with them, have their vehicle comfort preferences, and visit the same places at relatively the same time of a given day or day of the week. Thus, developing intelligent vehicles that are able to reduce the cognitive workload of the drivers by learning and adapting to their occupants' routines is of the highest interest. In this paper, we present two independent models based on machine learning methods, including artificial neural networks and linear and ridge regressions, to learn the habits and preferences of the vehicle's users. The first model is responsible for predicting the next vehicle trip state, i.e., the departure location and time, and the driver, passenger, and object states. The second model anticipates the comfort setting inside the cockpit - temperature, cockpit mirror, and driver seat poses. The developed models were trained, evaluated, and validated with different datasets in the Portuguese city of Braga. The results prove that the vehicle efficiently learns the routines of several users with varying complexities. Prediction errors happen in cases of an exceptional, one-time deviation from routine behavior.

Keywords: Learning driver habits · Human mobility patterns · Time and space prediction · Deep learning · Intelligent vehicles

---

*The work received financial support from European Structural and Investment Funds in the FEDER component, through the Operational Competitiveness and Internationalization Programme (COMPETE 2020) and national funds (Project "Easy Ride: Experience is everything", ref POCI-01-0247-FEDER-039334), and R&D Units Project Scope: UIDB/00319/2020 and UIDB/00013/2020.

© The Author(s), under exclusive license to Springer Nature Switzerland AG 2022
H. Yin et al. (Eds.): IDEAL 2022, LNCS 13756, pp. 157–169, 2022.
https://doi.org/10.1007/978-3-031-21753-1_16

1   Introduction

In our day-to-day lives, we regularly follow sequences of actions in repetitive tasks. These routine behaviors influence nearly every aspect of our lives. Consequently, the ability of intelligent technology to make predictions regarding a person's routine could help improve inefficient behavior. In the specific context of driver routine prediction, a cognitive vehicle endowed with the ability to learn its occupants' routines could improve user punctuality by triggering an alarm to depart from a specific location. Moreover, it could assist a driver by anticipating certain events, items, or occupants. While human mobility patterns are typically shaped by many factors, several studies have reported that individuals' daily routines are generally consistent in time and space [3,6,16]. To put it simply, one tends to visit the same places at relatively the same time of the day. For instance, for many drivers, weekdays consist of leaving home in the mornings, dropping the kids at school, and heading to the office; later, they perform the same trips in reverse order by departing from work and picking up the kids from school and returning home. Thus, predictions about the future driver intent can be learned from past spatio-temporal sequences [11]. Different approaches - most of them statistical - to tackle this issue have been proposed. Traditional Markov Models, for instance, perform well for a set of behaviors but require defining destinations a priori [15]. More recently, dynamical systems models based on Deep Neural Networks and Dynamic Neural Fields [4,17] addressed the learning problem by taking inspiration from processing mechanisms in the brain. However, the focus was on learning and recall of spatio-temporal trip information, ignoring passenger, object and comfort status.

In this paper, we present two independent approaches based on machine learning methods, including artificial neural networks and linear and ridge regressions, to learn the habits and preferences of the vehicle's users. The first concerns each trip state. It comprises a hierarchical density-based clustering algorithm followed by a deep neural network and answers the following questions: Where to go? When to go? How long to stay there? Who the next driver(s)/passenger(s) is(are)? Which objects come in(out)? The second approach applies either regression or a feed-forward neural network (depending on which performs better) to predict the trip's desired temperature, cockpit mirror, and driver seat poses. The learning occurs implicitly, i.e., the users are not questioned about their destinations and preferences in advance. The developed models were trained, evaluated, and validated with different datasets in the Portuguese city of Braga, each of which contained the routine of a vehicle collected in eleven weeks. The present work extends our preliminary results in [4].

2   Overview of Applied Techniques

Before detailing the implemented models and their respective results, we give a brief overview of the clustering methods used for points of interest extraction and the regression or Neural Network techniques applied for learning.

2.1   Clustering Approaches for Point of Interest (POI) Extraction

Clustering techniques constitute unsupervised learning algorithms to discover natural grouping in data, proving useful in POI extraction. Generally, the more frequent a destination is, the easier it is to detect. As a result, these algorithms consistently identify a person's home or workplace but might fail to recognize scarcely visited locations such as the cinema or a restaurant. Several approaches, such as neighborhood-based, hierarchical, and density-based, are commonly employed to tackle POI extraction [10]. We will, however, only focus on the latter, namely on Density-Based Spatial Clustering of Applications with Noise (DBSCAN), for its ability to deal with varying shapes and sizes for clusters [14]. However, DBSCAN has known limitations. Firstly, its performance relies heavily on the radius and the minimum number of points required to mark a location as a cluster or, in our application, a POI (defined by the number of neighbors parameter). Moreover, driver routines are not densely consistent, i.e., density varies from location to location. Thus, employing DBSCAN with a fixed set of values might lead to poor performance.

Hierarchical clustering algorithms provide an intuitive method for clustering, as urban areas and human mobility are also hierachically organized [9,12]. Hierarchical density-based clustering (HDBSCAN) combines hierarchical and density-based spatial clustering techniques by looking for data regions that are denser than their surroundings, thereby assuming the presence of noise. The main advantage is that, contrarily to DBSCAN, it handles data with multiple (varying) densities. One has to prespecify only the minimum number of points required to define a cluster.

2.2   Artificial Neural Networks

Artificial Neural Networks (ANN) represent a modeling framework which is inspired by neurocomputational mechanisms in the brain. ANNs have proven capable of computing large amounts of data, extracting meaningful information from it, learning input-output mappings, and making accurate predictions based on that knowledge. As a result, they have found applications in various industries, such as Anomaly Detection [1], Natural Language Processing [5], and Robotics [18], to name a few. One particular kind of ANN is the Recurrent Neural Network (RNN). They have proven especially useful in learning time series and dealing with temporal data in general. However, when processing large datasets, traditional RNN architectures experience the problem of exploding gradients, i.e., cumulative gradient errors result in very large neural weight updates during training. Several network architectures have been proposed to tackle this problem [13], but Long Short-Term Memory Networks (LSTM) are arguably the most successful. These networks comprise a set of recurrently connected cells, each with an internal memory state and three additional gates - input, output, and forget - which control the memory flux and enable the internal cell state to store and access information over long periods. On the other hand, Feed-Forward Neural Network (FFNN) architectures perform well in simple classification and regression tasks. Here, the information processing does not involve cycles or loops but the information processing moves in only one direction, from the input to the output layer.

2.3 Regressions

Regressions are instances of supervised learning algorithms, i.e., these receive labeled data as input. They aim to calculate the relationship between variables - assuming a direct correlation amongst them - by plotting either a line, a plane, or a hyperplane depending on the problem's dimensionality. Regression algorithms differ in the way they compute their coefficients. Linear regression, for instance, utilizes the Ordinary Least Squares iterative method as a loss function. However, this strategy is prone to overfitting the data since it does not differentiate between the importance of different coefficients. The ridge regression technique modifies the loss function to tackle this issue by introducing a bias (for more details, see [2]).

3 Methodology

As previously mentioned, we developed two independent models, each aiming to tackle a specific problem.

3.1 Predicting the Next Vehicle Trip State

The developed framework relies on the supposition that driving is mostly a routine and, by observing previous spatiotemporal sequences, one can make predictions regarding the future driver intent. However, any routine behavior varies, i.e., it is not deterministic, but presents variations to some extent from trial to trial, requiring a probabilistic model framework.

Following a data cleansing process, we employ POI extraction as detailed previously and filter the relevant attributes. It is worth noting that the clustering algorithm labels some POIs as unknown locations when the frequency of visits is below a certain threshold value. Data processing consists of identifying categorical features followed by a scaling of the results between 0 and 1. For a schematic representation of this process, see Fig. 1.

Fig. 1. Schematic representation of the data processing stage for the models responsible for next vehicle trip state prediction.

We created several deep learning models differing only in the produced output - departure time, day of the week, next destination, next driver, passenger, and object states inside the vehicle. The first two networks output a scalar, while the remaining networks produce a probability vector (Figs. 2 and 3, respectively). The implemented architectures consist of a LSTM followed by a FFNN. The former aims to uncover high-level patterns from the data, while the latter is responsible for processing the data and making predictions.

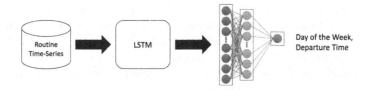

Fig. 2. Schematic representation of the developed architecture to predict the day of the week and departure time.

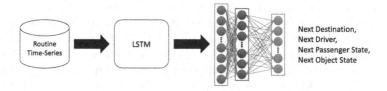

Fig. 3. Schematic representation of the developed architecture to predict next destination, next driver, passenger, and object states inside the vehicle.

All models utilize cross-entropy as loss function, except the one concerning the departure time, for which mean squared error proved a better fit. The hyperparameters of the model were tuned by performing a random search. This method tests arbitrary values within a set specified domain and keeps the combination that produced the least amount of error as the best performance. The chosen evaluation metrics for the classification tasks (such as predicting the next destination) are Top-n accuracy and F1-score. Alternatively, for regression tasks (i.e., prediction of the departure time), we chose the mean squared error and the coefficient of determination R^2. The latter measures the percentage of variance in the dependent variable that the model explains, and its best possible score is 1.0. In the results section, however, we multiplied the score by 100 to display it as a percentage value.

3.2   Predicting the Comfort Setting

The models built to tackle the comfort setting prediction problem are independent from each other. Despite being very similar in structure, - all processing steps from the data processing to the tuning are the same - their output format varies as follows:

- The Temperature Model predicts a vector with one element containing the value of the temperature inside the cockpit in degrees Celsius.
- The Cockpit Mirror Pose Model predicts a vector with two elements corresponding to the values of the pan and tilt angles, respectively.
- The Driver Seat Pose Model predicts a vector with three elements corresponding to the values of the driver seat's position, height, and angle, respectively.

The data processing stage is relatively straightforward and consists solely of encoding the data in the driver column, splitting it into training, testing, and validation sets (the latter is only necessary when the selected model is a neural network), and filtering the relevant columns. For instance, in the case of the Cockpit Mirror Pose Model, only the encoded driver information and the cockpit mirror pan and tilt angle columns are kept.

Afterward, the processed data serves as input to three different architectures: two based on regression techniques and one on artificial neural networks. Regarding model tuning and optimization, there are a few points worth mentioning. Linear regressions, due to their mathematical simplicity, do not integrate this step. Conversely, we tested different solvers to compute the coefficients on ridge regressions, namely single value decomposition, and stochastic average gradient descent. Moreover, the tolerance value (i.e., the precision of the solution) was also tuned (ranging from 0.0001 to 0.1). It is well known that a feed-forward neural network with only a single hidden layer approximates any function [8]. Consequently, the defined models contain only three layers - input, hidden, and output. While the first and the last layers have a fixed number of neurons according to the number of drivers and the number of features they predict, the number of neurons in the hidden layer can fluctuate. Hence, we employed a brute-force technique to find its optimal value. With values ranging between 2 and 128, we trained and evaluated the resulting networks and kept the one that produced the best results. Additionally, we included early stopping and model checkpoints. The number of epochs varied between 100 and 1000 by increments of 100, and the batch size value was 32. Mean Absolute Error (MAE) was chosen as the evaluation metric for all developed models since it is less sensitive to outliers when compared to Mean Squared Error (MSE), and the error value is easily interpreted.

4   Results

The current section presents the results of the developed models for ten datasets with varying routine deviations.

4.1   Datasets

Each collected dataset contains a vehicle's routine, with either one or two drivers, covering eleven weeks. Essentially, each row corresponds to a trip and includes information regarding its respective location, time, day, driver, passenger, object, and comfort settings inside the vehicle. The datasets can hold up to 4 passengers - besides the driver - and multiple instances of the same item. For example, it allows for the presence of different handbags. To collect the data, we used a simple mobile application [7].

Figure 4 presents the first five trips from one of the collected datasets representing a Monday routine. Initially, the dataset solely contains the columns with blue headers. However, before feeding it to the models, we extract the different POIs with HBSCAN - introduced in Sect. 2.1 - and add these as a new column (Fig. 4 - red header). To obtain an adequate grasp to which extent a user - or a set of users - of a vehicle maintains a routine, we must analyze their destinations. Figure 5 presents the extracted POIs using the HDBSCAN clustering algorithm for the tenth dataset. It is relevant to mention that we set its m_{pts} parameter to 4, which means we only memorize a destination if its visit frequency is greater or equal to 4. Otherwise the algorithm classifies it as an unknown destination. A POI is labeled with an asterisk and an unknown destination with a question mark. In this example, Vehicle 10 has 10 POIs and 3 unknown locations.

Time of the Day	Event Type	GPS	Location	Driver	Passengers	Objects	Cockpit Mirror Pose	Driver Seat Pose	Temperature
08:21	Departure	41.562, -8.422	POI1	D1	P1, P2	BackPack (2), Briefcase (1)	14.34, -12.66	18.13, 3.53, 14.26	19
08:43	Arrival	41.542, -8.413	POI2	D1	-	Briefcase (1)	14.34, -12.66	18.13, 3.53, 14.26	19
08:45	Departure	41.542, -8.413	POI2	D1	-	Briefcase (1)	14.34, -12.66	18.13, 3.53, 14.26	19
08:59	Arrival	41.559, -8.398	POI3	D1	-	-	14.34, -12.66	18.13, 3.53, 14.26	19
17:04	Departure	41.559, -8.398	POI3	D1	-	Briefcase (1)	14.34, -12.66	18.13, 3.53, 14.26	19

Fig. 4. First five events (departure or arrival) representing a Monday routine of a specific dataset. (Color figure online)

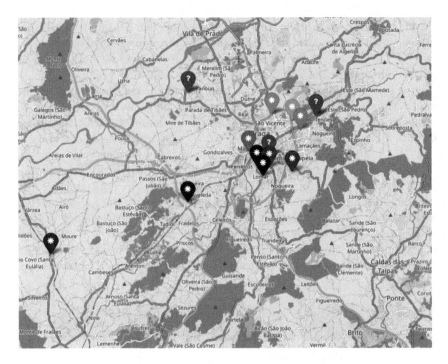

Fig. 5. Representation of the extracted points of interest for Vehicle 10. Unknown locations are illustrated with a question mark.

4.2 Next Trip State of a Vehicle

For brevity, instead of displaying the respective performance for each dataset, we show tables concerning the "best"- and "worst"-performing datasets - 4 and 10, respectively - and the mean achieved score. It is relevant to note that, as mentioned above, we built several models to tackle the problem of predicting the next trip state of a vehicle. For each model, the "best"- and "worst"-performing datasets vary. Consequently, the selection is based on the average score for all models. The factors that determine how well the models perform are the number of users and the extent to which they keep to their routine. We split the presentation of results into two parts: the first concerns the models employed to tackle spatio-temporal predictions (Table 1) and the second concerns the prediction about the passenger and object state (Table 2). For the deep learning-based models, we used the first seven weeks for training, the eighth and ninth for validation, and the remaining for testing.

Table 1. Next destination, time, and day prediction.

Metric	Training			Validation			Testing		
–	Veh. 4	Veh. 10	Mean	Veh. 4	Veh. 10	Mean	Veh. 4	Veh. 10	Mean
Next destination prediction									
Top-1 Acc.	98.13%	46.20%	92.33%	96.15%	34.43%	83.52%	94.89%	40.21%	83.43%
Top-2 Acc.	100.0%	75.44%	97.45%	100.0%	59.02%	92.48%	99.27%	64.95%	92.94%
Top-3 Acc.	100.0%	85.96%	98.60%	100.0%	80.33%	95.67%	100.0%	77.32%	95.42%
Top-4 Acc.	100.0%	92.40%	99.24%	100.0%	85.25%	97.33%	100.0%	85.57%	97.12%
Top-5 Acc.	100.0%	98.25%	99.83%	100.0%	91.80%	98.27%	100.0%	93.81%	98.51%
F-1 Score	98.16%	36.39%	**91.28%**	96.20%	25.36%	**82.33%**	95.19%	31.37%	**81.98%**
Time prediction									
R^2 Score	88.10%	81.30%	86.33%	62.30%	74.40%	74.33%	70.10%	64.30%	71.26%
MAE	0.87	0.69	0.79	1.58	1.02	1.15	1.45	1.16	1.18
Day prediction									
Accuracy	98.74%	96.69%	97.02%	94.71%	95.90%	93.81%	94.71%	95.90%	93.81%
F-1 score	98.73%	96.68%	97.00%	94.70%	96.01%	93.82%	94.70%	96.01%	93.82%

Table 2. Passenger and object state prediction.

Metric	Training			Validation			Testing		
–	Veh. 4	Veh. 10	Mean	Veh. 4	Veh. 10	Mean	Veh. 4	Veh. 10	Mean
Passenger state prediction									
Top-1 Acc.	96.75%	98.90%	97.45%	90.87%	97.54%	92.41%	93.80%	96.91%	92.09%
Top-2 Acc.	99.82%	100.0%	99.89%	97.12%	100.0%	98.19%	98.54%	100.0%	98.81%
Top-3 Acc.	99.82%	100.0%	99.98%	99.04%	100.0%	99.47%	99.27%	100.0%	99.62%
Top-4 Acc.	100.0%	100.0%	100.0%	100.0%	100.0%	99.72%	100.0%	100.0%	99.86%
Top-5 Acc.	100.0%	100.0%	100.0%	100.0%	100.0%	99.92%	100.0%	100.0%	99.95%
F-1 score	96.82%	98.91%	97.46%	90.89%	97.58%	92.72%	94.00%	97.19%	92.18%
Object state prediction									
Top-1 Acc.	99.28%	83.70%	95.78%	97.12%	61.48%	89.95%	98.54%	73.71%	91.62%
Top-2 Acc.	100.0%	99.72%	99.77%	100.0%	94.26%	98.00%	100.0%	99.48%	98.57%
Top-3 Acc.	100.0%	100.0%	100.0%	100.0%	100.0%	99.19%	100.0%	100.0%	99.51%
Top-4 Acc.	100.0%	100.0%	100.0%	100.0%	100.0%	99.55%	100.0%	100.0%	99.56%
Top-5 Acc.	100.0%	100.0%	100.0%	100.0%	100.0%	99.79%	100.0%	100.0%	99.62%
F-1 score	99.28%	83.17%	95.74%	96.96%	59.21%	89.97%	98.50%	74.00%	91.61%

By examining the results of Tables 1 and 2, one might argue that the system slightly overfits the training data as the evaluated performance is better overall. However, people's routines tend to change with time, and, therefore, it is expectable for the system's performance to drop as time goes on. Independent of this issue, the developed models generally perform well on all datasets. Vehicle 4 contains the routine with the smallest number of deviations from typical behavior. On the other hand, Vehicle 10 displays the highest level of variability. Nevertheless, regarding the problem of next destination prediction, the system has a mean F-1 Score - over all datasets - of 91.28%, 82.33%, and 81.98% for the training, validation, and testing data, respectively (see entries in bold in

Table 1). However, by observing the remaining tables, one cannot draw the same conclusions. As mentioned above, we based the decision about the "best"- and "worst"-performing datasets on the mean performance for all models. Consequently, when looking at the Passenger State Prediction columns, we see Vehicle 10 performs better than Vehicle 4 and also better than the mean for all routines. This pattern can be understood since although Vehicle 10 presents a less stable routine, the majority of trips are performed by the same driver, rendering the driver prediction problem trivial.

4.3   Next Trip's Comfort Setting

Table 3. Comfort setting results.

Feature	–	Linear regression		Ridge regression		Neural networks	
		Train	Test	Train	Test	Train	Test
Temp. (°C)	V08	.001	**.001**	.016	.016	.008	.008
	V06	.052	.125	.060	.131	.047	**.118**
	Mean	.044	**.038**	.053	.048	.045	**.038**
Pan (°)	V09	.001	**.001**	.001	**.001**	.002	.002
	V10	.340	.643	.401	.644	.352	**.626**
	Mean	.289	.348	.295	**.347**	.305	.388
Tilt (°)	V08	.001	**.001**	.001	**.001**	.001	**.001**
	V03	.374	**.560**	.374	**.560**	.373	.593
	Mean	.266	**.359**	.266	.360	.302	.403
Position (cm)	V09	.001	**.001**	.001	**.001**	.001	**.001**
	V05	.399	.387	.399	**.382**	.870	.750
	Mean	.275	.353	.276	**.352**	.554	.642
Height (cm)	V08	.001	**.001**	.001	**.001**	.001	**.001**
	V07	.514	.447	.514	.430	.555	**.083**
	Mean	.284	.342	.283	.340	.301	**.313**
Angle (°)	V08	.001	**.001**	.001	**.001**	.001	**.001**
	V05	.380	**.378**	.395	.412	.660	.650
	Mean	.277	**.307**	.279	.310	.540	.576

Table 3 presents the results of the employed models to tackle comfort setting predictions for the collected datasets, each divided by a solid line, dashed lines, on the other hand, separate different features within each model - temperature and cockpit mirror, and driver seat positions, respectively. A feature, in its turn, consists of three rows containing the Vehicle that performed best and worst across all models and the mean obtained by all datasets. The evaluation metric for all entries is Mean Absolute Error (MAE). Vehicle 5 often performs worse than the other datasets because it represents various frequently-alternating drivers with different comfort preferences. Conversely, Vehicles 8 and 9 produce the smallest amount of error as frequent driver changes are not common in their routines.

One point to stress is that there is no model that consistently performs better than the others. Considering the driver seat pose, the lowest MAE values for the position, height, and angle resulted from ridge regression, neural networks, and linear regressions, respectively. In general, the three models perform well with MAE $< 0.131\,°$C for Temperature, MAE < 0.644 ° for Tilt and Pan, MAE $< 0.750\,$cm for Position and Height, and MAE < 0.650 ° for Angle, in the test set. It is important to stress that a perfect prediction score is impossible as one never adjusts the Vehicle's seat or cockpit mirror precisely the same way.

5   Conclusion

In this paper, we presented a DL-based approach capable of learning and memorizing both the ordinal and temporal properties of a Vehicle, including all its different driver, passenger, and object routines. Moreover, we also built models to predict and adjust the comfort setting inside the cockpit, namely the temperature, the cockpit mirror, and the driver seat poses. Experiments with several Vehicles' routine data validate the developed architecture, which performs well for multiple drivers, passengers, and objects. Regarding Vehicle trips, prediction errors happen in cases of an exceptional, one-time deviation from routine behavior. It is important to stress however that an observed larger temporal mismatch between the real event and its prediction in some cases is the result of a highly irregular timing pattern for a specific location. Moreover, for the comfort setting prediction problem, the MAE values indicate that the prediction error is consistently below one degree or centimeter, depending on the feature.

In the future, we intend to increase the sample size for a more robust validation of the developed models. Moreover, a limitation for real-world applications is that in the current experiments the model were fed with pre-processed data, collected a priori. Consequently, future work will focus on augmenting the system with the capacity to identify who and what enters/exits a Vehicle. Integrating such a system in the current modeling framework would represent an embedded-ready solution. Additionally, integrating other factors such as the current traffic conditions into the long-term memory could produce more effective route selection/recommendations without requiring input from the driver. For instance, considering that, on a particular day, the traffic is slower than usual, the system could alert the driver to depart earlier than usual from a stop location so that he/she arrives still in time at the next destination. Currently, the predicted temperature value depends solely on the user's preference. External conditions such as, for instance, the temperature outside the Vehicle are not considered. Future work will integrate such factors, increasing the robustness of the system.

References

1. Andropov, S., Guirik, A., Budko, M., Budko, M.: Network anomaly detection using artificial neural networks. In: 2017 20th Conference of Open Innovations Association (FRUCT), pp. 26–31. IEEE (2017)
2. Belkin, M., Hsu, D., Ma, S., Mandal, S.: Reconciling modern machine-learning practice and the classical bias-variance trade-off. Proc. National Acad. Sci. **116**(32), 15849–15854 (2019)
3. Eagle, N., Pentland, A.S.: Eigenbehaviors: identifying structure in routine. Behav. Ecol. Sociobiol. **63**(7), 1057–1066 (2009)
4. Fernandes, C., Ferreira, F., Erlhagen, W., Monteiro, S., Bicho, E.: A deep learning approach for intelligent cockpits: learning drivers routines. In: International Conference on Intelligent Data Engineering and Automated Learning, pp. 173–183. Springer (2020). https://doi.org/10.1007/978-3-030-62365-4_17
5. Goldberg, Y.: Neural network methods for natural language processing. Synth. Lect. Hum. Lang. Technol. **10**(1), 1–309 (2017)
6. Gonzalez, M.C., Hidalgo, C.A., Barabasi, A.L.: Understanding individual human mobility patterns. Nature **453**(7196), 779–782 (2008)
7. Guimarães, P., Ferreira, F., Silva, A.C., Erlhagen, W., Monteiro, S., Bicho, E.: A data recording mobile application to create datasets of vehicle users' routines. In: 2022 IEEE International Conference on Autonomous Robot Systems and Competitions (ICARSC), pp. 167–172. IEEE (2022)
8. Guliyev, N.J., Ismailov, V.E.: On the approximation by single hidden layer feedforward neural networks with fixed weights. Neural Netw. **98**, 296–304 (2018)
9. Ibrahim, R., Shafiq, M.O.: On predicting taxi movements modes in Porto city using classification and periodic pattern mining. In: 2019 IEEE 21st International Conference on High Performance Computing and Communications; IEEE 17th International Conference on Smart City; IEEE 5th International Conference on Data Science and Systems (HPCC/SmartCity/DSS), pp. 1197–1204. IEEE (2019)
10. Jia, R., Khadka, A., Kim, I.: Traffic crash analysis with point-of-interest spatial clustering. Accid. Anal. Prev. **121**, 223–230 (2018)
11. Park, S.H., Kim, B., Kang, C.M., Chung, C.C., Choi, J.W.: Sequence-to-sequence prediction of vehicle trajectory via LSTM encoder-decoder architecture. In: 2018 IEEE Intelligent Vehicles Symposium (IV), pp. 1672–1678. IEEE (2018)
12. Roth, C., Kang, S.M., Batty, M., Barthélemy, M.: Structure of urban movements: polycentric activity and entangled hierarchical flows. PloS One **6**(1), e15923 (2011)
13. Schmidhuber, J.: Deep learning in neural networks: an overview. Neural Netw. **61**, 85–117 (2015)
14. Schubert, E., Sander, J., Ester, M., Kriegel, H.P., Xu, X.: DBSCAN revisited, revisited: why and how you should (still) use DBSCAN. ACM Trans. Database Syst. (TODS) **42**(3), 1–21 (2017)
15. Simmons, R., Browning, B., Zhang, Y., Sadekar, V.: Learning to predict driver route and destination intent. In: 2006 IEEE Intelligent Transportation Systems Conference, pp. 127–132. IEEE (2006)
16. Song, C., Qu, Z., Blumm, N., Barabási, A.L.: Limits of predictability in human mobility. Science **327**(5968), 1018–1021 (2010)

17. Wojtak, W., et al.: Towards endowing intelligent cars with the ability to learn the routines of multiple drivers: a dynamic neural field model. In: International Conference on Computational Science and Its Applications, pp. 337–349. Springer (2021). https://doi.org/10.1007/978-3-030-86973-1_24
18. Yang, S.X., Meng, M.: An efficient neural network approach to dynamic robot motion planning. Neural Netw. **13**(2), 143–148 (2000)

Duplication Scheduling with Bottom-Up Top-Down Recursive Neural Network

Vahab Samandi$^{(\boxtimes)}$, Peter Tiňo , and Rami Bahsoon

School of Computer Science, University of Birmingham,
Birmingham B15 2TT, UK
{vxs899,P.Tino,r.bahsoon}@cs.bham.ac.uk

Abstract. Scientific workflows can be represented as directed acyclic graphs (DAGs) with nodes corresponding to individual tasks and directed edges between the nodes signifying the order of task execution. The nodes contain informative attributes related to task-specific data transfer/storage and scheduling length. Given an available amount of (cloud) computational resources, the overall workflow scheduling length can sometimes be reduced by making certain "critical tasks" run in parallel (task duplication) on multiple resources. In this way, a carefully designed spread of computational effort across the resources can result in a more efficient computational structure and hence a shorter scheduling length. However, task duplication algorithms deciding which tasks to duplicate in a given workflow can themselves be computationally expensive. Here we propose a novel Bottom-Up Top-Down Recursive Neural Network (BUTD RecNN) model that is able to learn from historical duplication decisions on workflows (represented as DAGs) to efficiently produce duplication recommendations for new unseen workflows. The approach is tested on collections of Montage workflows.

Keywords: Graph classification · Recursive neural network · Workflow scheduling · Duplication-based scheduling algorithm

1 Introduction

Today's applications consist of an immense number of interactive tasks. These applications include scientific workflow and big data that forms large scale datasets in a vast range of scientific domains, such as climate science, bioinformatics, astronomy, and others [6,7,10]. These tasks generally require enormous processing power that is beyond the ability of a single machine. With the emergence of distributed computing, the computing power needed for processing these large datasets is provided. Workflows are applications or computational tasks logically connected by data- and control-flow dependencies in the form of direct acyclic graphs (DAGs). Each node represents a task with a specific amount of computation workload, and each edge represents a precedence constraint (data dependency) between two tasks. The precedence constraints among tasks and

© The Author(s), under exclusive license to Springer Nature Switzerland AG 2022
H. Yin et al. (Eds.): IDEAL 2022, LNCS 13756, pp. 170–178, 2022.
https://doi.org/10.1007/978-3-031-21753-1_17

the structural complexity of the directed acyclic graphs highly affects the performance of the workflow scheduling algorithms. This is due to the fact that these workflows are parallel program graphs with data- and control-dependency among their tasks.

Workflow scheduling is a particular form of task scheduling in which tasks are mapped into the distributed resources for execution such that the task-precedence and their execution priority conditions are satisfied. The most important performance measures of workflow scheduling algorithms is the makespan (scheduling length) of a workflow and the scheduling algorithm time complexity. Among different types of workflow scheduling, duplication-based scheduling algorithms (DSA) [3,5,9,11,12,18,19] aim to reduce the scheduling length (makespan) of the workflow by duplicating some of the workflow tasks on the available resources. The aim is to avoid the excessive data transmission time from one resource to another one in distributed computing (e.g., cloud). However, the main issue for designing an efficient task duplication scheduling is identifying the target tasks to be duplicated, which causes an increase in the time complexity of the algorithm. To design a DSA with low algorithm time complexity, we propose to use a specialized neural network model that can learn from historical workflow and duplication data to identify suitable tasks for duplication. The workflow data is in the form DAGs. Initial work on classification of structured data (logical terms) in tress and DAGs was presented in [11]. Recursive neural network [13] is a generalized form of recurrent neural network, structured as a deep graph, rather than an RNN chain-like structure. Recursive neural network potential applications were discussed in [2]. Frasconi et al. [8], applied recursive network on structural data. Recursive neural network has also been used in natural language processing and computer vision applications by Socher et al. [15,16]. Sperduti et al. [17] have introduced a generalized recursive neuron that can be used to all the supervised networks developed for the classification to be generalized to structures.

The novel contribution of this paper is a bottom-up top-down recursive neural network model (BUTD RecNN) that extends the regular recursive neural network. This model process input graphs (DAGs) upward and downward recursively. We apply this model to a set of Montage workflows [1] to predict the tasks suitable to be duplicated to reduce the scheduling length of workflows. The architecture diagram of this model is illustrated in Fig. 1. The rest of this paper is organized as follows: Sect. 2 defines some preliminary concepts on graphs and briefly explains the regular recursive neural network. We then introduce the bottom-up top-down recursive neural network model. In Sect. 3, we discuss and report the experiment and training the learning model. In Sect. 4, we discuss the results and performance comparison, and in Sect. 5, a conclusion is drawn.

2 Preliminaries

Graphs : a directed graph G is specified by a set of vertices V_G, a set of directed edges E_G: $(v_i, v_j) \in E_G$ signifies a connection from vertex $v_i \in V_G$ to $v_j \in V_G$.

A graph G' is called a subgraph of a graph G, if $V_{G'} \subseteq V_G$, and $E_{G'} \subseteq E_G$. For a vertex $v \in V_G$ of graph G, the sets of its parents and children, $Pa(v)$ and $Ch(v)$, are defined as $Pa(v) = \{w \in V_G | (w,v) \in E_G\}$ and $Ch(v) = \{w \in V_G | (v,w) \in E_G\}$, respectively. The number of input and output edges of v is then $in\_deg(v) = |Pa(v)|$ and $out\_deg(v) = |Ch(v)|$, respectively.

We assume a consistent ordering on children and parents of vertices. Let $Ch(v)_i \in V_G$ and and $Pa(v)_j \in V_G$ denote the ith child and j^{th} parent of v, respectively. The *max_degree* of a graph G is $max_{v \in V_G}\{out\_deg(v), in\_deg(v)\}$. $|V| = n$ is the number of nodes in a graph.

In this work we consider directed acyclic graphs (DAGs) that are finite vertex graphs. In a domain D, the maximum out-degree among graphs is defined as *max_degree* of domain.

2.1 Recursive Neural Network

Recursive neural network is an extension of recurrent neural network from linearly order data to graph structures. For a vertex $v \in V_G$ the output $y_j(v)$ of the j-th state unit is computed as

$$y_j(v) = f\left(\sum_{i=1}^{N_L} w_{ij}\, l_i(v) + \sum_{k=1}^{out\_deg(v)} \sum_{q} v_{kjq}\, y_q(Ch(v)_k)\right) \quad (1)$$

where $f()$ is e.g. a sigmoidal activation function. The output of the neuron for a vertex v is computed recursively on the output computed for all the vertices pointed by it[1].

2.2 Bottom-Up Top-Down Recursive Neural Network

We introduce a bottom-up top-down recursive neural network model (BUTD RecNN) by extending the regular recursive neural network. The proposed model can be trained using available input information from both directions (Bottom-up and Top-down) of the DAG-structured pattern. The model includes two networks for graph encoding in the two opposite directions (see Fig. 1).

---

[1] The number of recursive connections of a recursive neuron should be equal to the *max_degree* of the domain D, even if not all of them will be used for computing the output of a vertex v with $out\_deg(v) < max\_degree$.

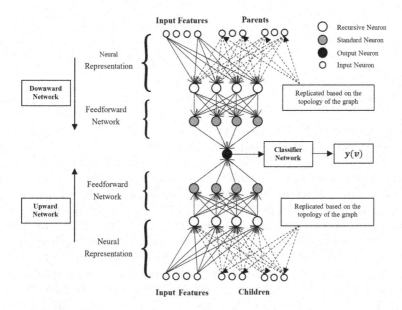

Fig. 1. Bottom-up top-down recursive neuron network architecture

Outputs of the j-th recursive neuron for the upward and downward networks, $y_j^\uparrow(v)$ and $y_j^\downarrow(v)$, respectively, for vertex v of graph G are defined as

$$
\begin{cases}
y_j^\uparrow(v) = f\left(\sum_{i=1}^{N_L} w_{ij}^\uparrow l_i(v) + \sum_{k=1}^{out\_deg(v)}\sum_q v_{kjq}^\uparrow\, y_q^\uparrow(Ch(v)_k) + b_j^\uparrow\right) & \text{Upward Network} \quad (2)\\[3mm]
y_j^\downarrow(v) = f\left(\sum_{i=1}^{N_L} w_{ij}^\downarrow l_i(v) + \sum_{k=1}^{in\_deg(v)}\sum_q v_{kjq}^\downarrow\, y_q^\downarrow(Pa(v)_k) + b_j^\downarrow\right) & \text{Downward Network} \quad (3)
\end{cases}
$$

where $f()$ is an activation function, $w_{ij}^\uparrow$ and $w_{ij}^\downarrow$ are the upward and downward weights, respectively, N_L is the number of units encoding the label attached to the current node v. $v_{kjq}^\uparrow$ and $v_{kjq}^\downarrow$ are the weights on the recursive connections for the upward and downward networks, respectively, with the corresponding bias terms $b_j^\uparrow$ and $b_j^\downarrow$.

We compute the final output $y(v)$ for a vertex v of a graph G as

$$
y(v) = f\left(\sum_{i=1}^{N_L} w_i\, l_i(v) + \sum_j a_j^\uparrow\, y_j^\uparrow(v) + \sum_k b_k^\downarrow\, y_k^\downarrow(v) + b\right) \tag{4}
$$

where $a_j^\uparrow$ and $b_k^\downarrow$ are connection weights and b is the output bias.

The BUTD RecNN structure is trained by repeatedly scanning each workflow DAG upward (2) and downward (3), evaluating the output for each node (duplicate vs. do not duplicate) (4) and confronting with the desired targets. The weight updates are driven by the gradient flow through (2–4). In particular, to each node a logistic regression classifier (4) is applied to predict a given target t. We aim to minimize the cross-entropy error between the target value

$t(v) \in \{0, 1\}$ at node v and the predicted value $y(v) \in (0, 1)$ at that node. During training, error backpropagates through to the recursively used weights and updates the weights using Stochastic Gradient Descent.

3   Experiments

To train the BUTD RecNN model, we used a set of 20,000 labeled Montage scientific workflows. Montage [1] is a portable software toolkit developed by NASA that allows for making a mosaic of astronomical images. The input files are from astronomical sources in the Flexible Image Transport System (FITS) format. We first executed a set of 20,000 Montage workflows on the CloudSim [4] by implementing critical Path first duplication (CPFD) algorithm. We then profile the workflows, collecting information such as average CPU utilization, memory usage, task runtime, input data size, output data size, and duplication status (duplicated/not-duplicated) of each task as the input features to the learning model. The set of 20,000 labeled workflows creates the training dataset, which is used to train the BUTD RecNN model. The data set is split into 80% training set, 10% validation set, and 10% test set. The model is trained to predict nodes/tasks of a workflow that are suitable to be duplicated. The model classifies nodes into one of the two classes as 0 (do-not-duplicate) or 1 (duplicate). We finally applied our learning model to a set of 1000 unlabeled Montage workflows. The model predicted class 1 (nodes to be duplicated) with around 62% accuracy. Our duplication-based scheduling algorithm BUTD RecNN uses the predicted model output to make a decision for duplicating a tasks.

Our algorithm BUTD RecNN was compared with two commonly used DSAs and a list scheduling algorithm for the performance measurement. These algorithms include critical path first duplication (CPFD) [9], fast and scalable scheduling (FSS) [5] algorithm and heavy node first (HNF) [14] algorithm. We applied these algorithms to schedule the workflows on the CloudSim resources. The CPFD algorithm classifies the tasks into three classes: Critical path nodes (CPN), In-Branch nodes (IBN), and Out-Branch nodes (OBN). A CPN is a node on the critical path. A graph's critical path (CP) is defined as a set of vertices and edges from an entry node to an exit node, establishing a path with the maximum computation and communication cost. An IBN is a node that is not a CPN node and from which there is a path to a CPN. An OBN is a node that is neither a CPN nor an IBN. The CPFD examines these nodes based on their relative importance. CPNs are scheduled first, then IBNs, and finally OBNs, without violating the precedence constraints. In the task assignment process of the FSS algorithm, only essential tasks of the critical path required to establish a path from a node to the entry node are duplicated. While the other two duplication algorithms (CPFD and FSS) have their strategies to duplicate a task, Our algorithm duplicates tasks based on the output of the BUTD RecNN prediction model. Our BUTD RecNN algorithm duplicates a task only if it is predicted by the learning model to be duplicated and also based on the following conditions: for a task that is predicted to be duplicated, our algorithm duplicates it if the node is a fork node. It duplicates the critical path fork nodes first and then the non-critical path fork nodes. A *fork* node is a node with out-degree grater than 1, and a *join* node is a node with In-degree grater than 1. A node can be both a join and fork node as these are not exclusive terms. Our algorithm ignores duplication of join nodes with an out-degree equal to one even if the learning model has predicted it to be duplicated. Our implementation extends some of the existing classes of the CloudSim. A simulation environment for all experiments consists of one Datacenter, one Host and five VMs. Each Processing unit has a speed of 1,000 millions instructions per second (MIPS), 4 GB RAM, 10 Gbps bandwidth, and a storage capacity of 1 TB. To produce consistent results, we applied the same CloudSim configurations for all scheduling algorithms we used in this experiment.

4   Results and Performance Comparison

Generally, the performance of the scheduling algorithms are compared according to the following metrics:

Schedule Length (makespan) – the overall schedule length of a workflow. A workflow scheduling algorithm that generates the lowest makespan is the best. In our experiment, we used average makespan values over the tested workflows.

Time Complexity – the algorithm's running time to obtain the workflow schedule. An algorithm with the lowest time complexity is the most scalable algorithm. However, there exists a trade-off between the makespan of a workflow and the time complexity of an algorithm, i.e., algorithm which generates a shorter scheduling length for a given workflow possibly has a higher time complexity.

Speedup – speedup relative to the no-duplication scenario.

Table 1. Comparison of scheduling algorithms

Algorithms	Categories	Time complexity	Average makespan	Average speedup
CPFD	DSA	$O(V^4)$	83973.4	3.59
BUTD RecNN	DSA	$O(V^2)$	114931	2.62
FSS	DSA	$O(V^2)$	124447.31	2.42
HNF	List Scheduling	$O(V(logV))$	154156.23	1.95

Figure 2(a) shows the makespans obtained by executing 1000 Montage workflows on the CloudSim using CPFD, BUTD RecNN, FSS, and HNF scheduling algorithms. The makespans obtained by using these algorithms are sorted in ascending order. We can see that the computationally expensive CPFD algorithm has the shortest makespans among the algorithms and BUTD RecNN slightly outperforms FSS. The comparison results show that the BUTD RecNN algorithm can further minimize the makespan of a workflow while having the same time complexity $(O(V^2))$ as the FSS algorithm. We could achieve this by employing a neural network model that helps in identifying target nodes to be duplicated. All three duplication-based algorithms (CPFD, BUTD RecNN, FSS) outperform HNF list-scheduling algorithm with respect to the average scheduling length (makespan).

Table 1 summarizes the performance comparison of the four scheduling algorithms. Generally, list-scheduling algorithms have minimum time complexity among different scheduling algorithms. HNF has the least time complexity as $O(V(logV))$. BUTD RecNN and FSS have $O(V^2)$ time complexity, and CPFD has the worst time complexity as $O(V^4)$. The average speedup ranking of the algorithms is as follows: CPFD → BUTD RecNN → FSS → HNF. An algorithm with a higher average speedup performs better in the parallel processing of resources. The result shows CPFD has the highest speedup ratio, and BUTD RecNN outperforms FSS and HNF.

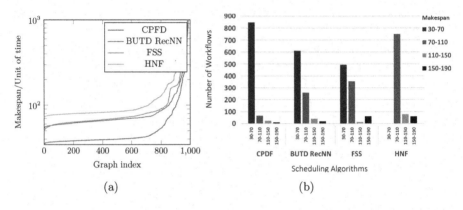

Fig. 2. Performance comparison of CPFD, BUTD RecNN, FSS, HNF algorithms according to the makespan.

Figure 2(b) shows the performance comparison with respect to the scheduling length (makespan) for the set of 1000 unseen Montage workflows. The range of makespan for the entire set of workflows is 30–4500/unit of time. Around 92% of the whole set of makespans is in the range of 30–190/Unit of Time. The remaining 8% workflows are the heavy Communication/Computation cost workflows in the range of 190–4500/Unit of Time. CPFD has the maximum percentage of about 85% of the shortest makespans in the range of 30–70/Unit of Time. This indicates that CPFD has the best performance in reducing the makespan of workflows by duplicating tasks. However, as mentioned earlier, it suffers from the high algorithm time complexity. Our algorithm, BUTD RecNN, with a much lower time complexity than CPFD, has 59% of workflow makespan in the range of 30–70/Unit of Time which outperforms FSS and HNF. FSS has a better performance than HNF, having 49% of workflow makespan in the range of 30–70/Unit of Time.

5   Conclusion

Machine learning-based workflow scheduling algorithms can assist in intelligent allocation of tasks on the resources. This work proposes a neural network based duplication scheduling algorithm through supervised learning modeling and data-driven profiling. The algorithm aims to minimize makespans by avoiding high communication costs among tasks. We introduce a bottom-up top-down recursive neural network model by extending the regular recursive neural network. The model can be trained using available input information from both the directions (bottom-up and top-down) of workflows. With the help of the learnt model, we could predict the suitable nodes for duplication in our duplication-based scheduling algorithm. This work provides a significant step towards the ultimate goal - a high-performance duplication scheduling algorithm having a low time complexity.

References

1. Berriman, G.B., et al.: Montage: a grid enabled image mosaic service for the national virtual observatory. Astronom. Data Analysis Softw. Systs. **314**, 593 (2004)
2. Bottou, L.: From machine learning to machine reasoning. Technical report. arXiv. 1102.1808. pp. 394–396 (2011). https://doi.org/10.48550/arXiv.1102.1808
3. Bozdag, D., Ozguner, F., Ekici, E., Catalyurek, U.A.: task duplication based scheduling algorithm using partial schedules, In: International Conference on Parallel Processing, pp. 630–637 (2005). https://doi.org/10.1109/ICPP.2005.15
4. Calheiros, R. N., Ranjan, R., Beloglazov, A., De Rose, C. A. F., Buyya, R., CloudSim: a toolkit for modeling and simulation of cloud computing environments and evaluation of resource provisioning algorithms. Softw. Pract. Exper. Vol. 41, 1, pp. 23–50. (2011). https://doi.org/10.1002/spe.995
5. Darbha, S., Agrawal, D.P.: A fast and scalable scheduling algorithm for distributed memory systems. In: Proceedings Seventh IEEE Symposium on Parallel and Distributed Processing, pp. 60–63 (1995). https://doi.org/10.1109/SPDP.1995.530665
6. Deelman, E.: The future of scientific workflows. Int. J. High Performance Comput. Appl. **32**(1), 159–175 (2018)
7. Ferreira da Silva, R., Filgueira, R., Pietri, I., Jiang, M., Sakellariou, R., Deelman, E.: A Characterization of Work-flow Management Systems for Extreme-Scale Applications. Future Gen. Comput. Syst. **5**, 228–238 (2017)
8. Frasconi, P., Gori, M., Sperduti, A.: On the efficient classification of data structures by neural networks. In: Proceedings of the Fifteenth International Joint Conference on Artifical Intelligence - Volume 2, Morgan Kaufmann Publishers Inc., San Francisco, CA, USA, pp. 1066–1071 (1997)
9. Ishfaq, A., Yu-Kwong, K.: On exploiting task duplication in parallel program schedulingIEEE . Trans. Parallel Distrib. Syst. **9**(9), 872–892 (1998). https://doi.org/10.1109/71.722221
10. Juve, G., Chervenak, A., Deelman, E., Bharathi, S., Mehta, G., Vahi, K.: Characterizing and profiling scientific workflows. Future Gen. Comput. Syst. 29(3), 682–692 (2013). https://doi.org/10.1016/j.future.2012.08.015

11. Kruatrachue, B., Lewis, T.: Grain size determination for parallel processing. IEEE Softw. **5**(1), 23–32 (1988). https://doi.org/10.1109/52.1991
12. Li, G., Chen, D., Wang, D., Zhang, D.: Task clustering and scheduling to multiprocessors with duplication, In: Proceedings International Parallel and Distributed Processing Symposium, pp. 8 (2003). https://doi.org/10.1109/IPDPS.2003.1213079
13. Pollack, J.B.: Recursive distributed representations. Artif. Intell. **46**(1–2), 77–105 (1990). https://doi.org/10.1016/0004-3702(90)90005-K
14. Shirazi, B., Wang, M., Pathak, G.: Analysis and evaluation of heuristic methods for static task scheduling. J. Parallel Distrib. Comput. **10**(3), 222–232 (1990). https://doi.org/10.1016/0743-7315(90)90014-G
15. Socher, R., Huang, E.H., Pennington, J., Ng, A.Y., Manning, C.D.: Dynamic pooling and unfolding recursive autoencoders for paraphrase detection, In: Proceedings of the 24th International Conference on Neural Information Processing Systems, Curran Associates Inc., Red Hook, NY, USA, pp. 801–809 (2011)
16. Socher, R., Perelygin, A., Wu, J., Chuang, J., Manning, C. D., Ng, A., Potts, C.: Recursive Deep Models for Semantic Compositionality Over a Sentiment Treebank, In: Proceedings of the 2013 Conference on Empirical Methods in Natural Language Processing, Association for Computational Linguistics, Seattle, Washington, USA, pp. 1631–1642 (2013)
17. Sperduti, A., Starita, A.: Supervised neural networks for the classification of structures. IEEE Trans. Neural Netw. **8**(3), 714–735 (1997). https://doi.org/10.1109/72.572108
18. Yao, F., Pu, C., Zhang, Z.: Task Duplication-Based Scheduling Algorithm for Budget-Constrained Workflows in Cloud Computing. IEEE Access **9**, 37262–37272 (2021). https://doi.org/10.1109/ACCESS.2021.3063456
19. Yeh-Ching C., Ranka, S.: Applications and performance analysis of a compile-time optimization approach for list scheduling algorithms on distributed memory multiprocessors, In: Proceedings of the ACM/IEEE Conference on Supercomputing, pp. 512–521 (1992). https://doi.org/10.1109/SUPERC.1992.236653

Towards a Low-Cost Companion Robot for Helping Elderly Well-Being

J. A. Rincon[1(✉)] , C. Marco-Detchart[1] , V. Julian[1] , C. Carrascosa[1] , and P. Novais[2]

[1] Valencian Research Institute for Artificial Intelligence (VRAIN), Universitat Politècnica de València, Camino de Vera s/n, 46022 Valencia, Spain
{jrincon,carrasco}@dsic.upv.es, {cedmarde,vjulian}@upv.es
[2] ALGORITMI Centre, Universidade do Minho, Braga, Portugal
pjon@di.uminho.pt

Abstract. The use of robot assistants and companion robots has been shown to improve the well-being of the elderly, enabling them to improve their quality of life. Nowadays, many older people live alone and do not have the resources to acquire specialized systems to monitor and accompany them. But at the same time, they are under the supervision of specialized personnel that make them feel safe. Loneliness can be a crucial element that affects many people's lives; not having someone to talk to or hang out with is a current concern. This work explores this issue by developing a small companion robot and a heart signal monitoring system (ECG). The presented approach is inexpensive, easy to maintain, and integrates machine learning (ML) models and the Alexa voice assistant.

Keywords: Edge-AI · Robotics · Assistant · Companion robot · Deep learnig · TinyML · IoT · IoMT · AWS · AWS-IoT

1 Introduction

Life expectancy has increased considerably in recent decades, which has led to an ageing population every year. This increase has prompted the European Union and many countries to propose new policies for care services focused on this group of people. Currently, programmes are being promoted to care for the elderly in their homes, thus avoiding investing resources in residences and medical staff. One of the effects on the human body during the ageing process is the systemic deterioration of many of its systems. Consequently, this is reflected in reduced motor, functional and cognitive ability. The use of drugs is a widely used practice to keep blood pressure levels, depression and other problems constant. These drugs can trigger side effects, being a significant concern in older people and increasing the risk of cardiovascular events [11] and mortality [8]. Steptoe et al. [10] proposes that well-being in older people is essential and that evidence suggests that health systems should focus their efforts not only on

© The Author(s), under exclusive license to Springer Nature Switzerland AG 2022
H. Yin et al. (Eds.): IDEAL 2022, LNCS 13756, pp. 179–187, 2022.
https://doi.org/10.1007/978-3-031-21753-1_18

illness and disability. Instead, they should also support and positively enhance the psychological states of older people.

The last pandemic (COVID-19) has also led to staff shortages in health and social care and a reduced level of support provided [5]. Assistive robots can play an essential role in this era by providing emotional support, companionship and automated home care for people with dementia.

In terms of this new trend, new solutions try to take advantage of the continuous advances in computer science, Artificial Intelligence (AI) and robotics. In addition to providing companionship to the elderly, companion robots may have other functionalities. They can be related to supporting independent living, such as basic activities like eating, bathing, toileting and dressing. They can also monitor people who need continuous care, helping maintain physical and psychological safety. The latter is perhaps one of the most relevant tasks that companion robots could have in the future.

However, trying to improve psychological states is not an easy task since specialised persons are needed to accompany the elderly daily. This daily accompaniment represents a high economic cost for families and homes that care for these people. One way to give a solution to this problem is the use of companion robots. Various researchers have addressed the role of companion robots in improving child or older people's daily lives. Pransky [9] introduces an interesting perspective on the different profiles that this type of robot could adopt. In one of his perspectives, Pransky raises the possibility of a "nanny robot". This robot would perform tasks such as playing with the child, feeding the child, etc., but, on the other hand, it would make the child have no human interaction and see interaction with the robot as the "norm". A companion robot [4] can be defined as a particular type of robot whose main characteristic is to accompany older people, children, single people or people living alone. This type of robot can interact naturally with the person, usually through speech recognition [1].

The latter has become particularly important in recent years, with the emergence of voice assistants such as Alexa, Siri, and Google Home. They are allowing people to directly talk to these devices for playing their favourite songs or telling them a joke. Not to be left behind, these devices can do even more, such as controlling other appliances (lights, ovens, heating, etc.). This allows companies and developers to design their applications. In the case of Alexa, Amazon offers tools that enable users and companies to build their applications or Skills, which their assistants will then use. It also allows interconnection with its entire ecosystem of services, giving users the scalability that other tools or systems do not have. This scalability enables rapid and secure prototyping of individual applications.

Numerous research studies highlight the benefits of voice assistants for older people and people with dementia. The research presented by S. Jiménez [6], Clemente [3], Beh [2] are the first approximations of the use of these voice assistants in the elderly. Although there are errors that are associated with the interaction capabilities of the assistants, it is still a possible useful tool to be used in this population.

Taking this into account, this paper explores the use of robot assistants by developing a small companion robot which includes a heart signal monitoring system. In this sense, the proposed approach tries to be inexpensive, easy to maintain, and also, integrates machine learning models and the Alexa voice assistant in order to offer an easy interaction with the potential users.

2   System Description

This section describes the system proposed in this research. The system presented below is a companion robot which integrates the Alexa voice assistant. This companion robot, called "E-Bot", is intended to provide a solution for the care of elderly people and/or people living alone. E-Bot allows older people to interact with it, providing the possibility of a continuous contact with caregivers, family members, or doctors. At the same time, E-Bot allows connection to wearable devices, capable of measuring physiological variables such as heart rate (ECG), skin resistance and oxygen saturation (SPO2) and an AI-enabled camera, all connected through Amazon AWS. This connection allows doctors and caregivers to observe the physiological signals over extended periods, analyzing the signals stored in the database.

E-Bot has been designed to be simple and with a friendly look to reduce possible rejection by the user. Its relatively small size makes it easy to transport, and at the same time, E-Bot is economical and easy to maintain, upgrade and/or repair. Thanks to the integration of a voice assistant like Alexa, E-Bot can be customised through Skills, which can be updated and improved to help the people. These skills help to create exercise routines, enhancing the device's ability to interact, connect with other devices such as blood pressure monitors, or even connect with other robots.

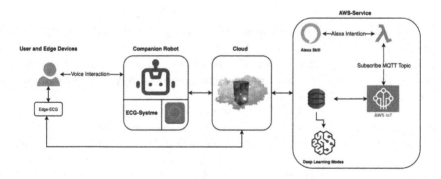

Fig. 1. Main view of the proposed architecture

E-Bot is divided into different parts; the first part integrates the Alexa voice assistant. This is embedded within the M5Stack-AWS, where the user can interact with the E-Bot Skill and all the applications that Alexa offers to its users.

The following part is in charge of capturing ECG signals, using an M5Stick-C Plus and a BMD101 cardio chip. These elements were attached to a chest strap, which integrates two electrodes that serve as a signal acquisition interface between the cardio chip and the skin. The ECG signal is captured and sent to the AWS via the AWS-IoT. These two elements are independent, allowing caregivers or family members to activate the heart signal capture from their Alexas.

2.1 Hardware Description

E-Bot was built using an M5Stack Core2 AWS (Fig. 2), which is a development system for AWS IoT EduKit. This system allows IoT applications to be built using AWS services. The M5Stack Core2 AWS incorporates a Microchip ATECC608 Trust&GO encryption chip, in addition to the existing features of the standard M5Stack Core2. In terms of processing, the Core2 for AWS comes with an ESP32-D0WDQ6-V3. It has two 32-bit Xtensa LX6 cores and the main frequency of up to 240Mhz, with Bluetooth and 2.4GHz Wi-Fi.

Fig. 2. M5Stack Core2 AWS

This development system was explicitly designed to be used with AWS-IoT and Alexa services. The M5Stack-AWS is a compact system which integrates a microphone and a speaker. This makes this device ideal for E-Bot construction. The robot's body is ergonomic and can be placed on any surface, making the E-Bot an easy to transport tool (Fig. 3).

As mentioned above, E-Bot can communicate with devices capable of measuring physiological variables. One of these variables is the heart signal (ECG). To capture this signal conventionally, a series of differential amplifiers of instrumentation is necessary. As incorporating these devices would make the capture system too large, we have used a cardio chip. This cardio chip (BMD101) is a system that allows the acquisition of ECG signals easily and quickly; only two electrodes in contact with the skin and a micro-controller with serial communication protocol are provided. This cardio chip is currently used in wearable devices such as smartwatches (Fig. 4).

As mentioned above, one of the main objectives of E-Bot and the set of connected wearable devices is monitoring signals to detect possible problems.

(a) E-Bot Virtual Design   (b) E-Bot Real Image

Fig. 3. E-Bot robot.

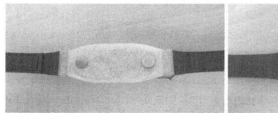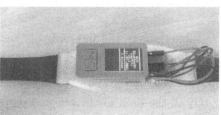

(a) Chest strap back view   (b) Chest strap front view

Fig. 4. ECG chest strap.

This detection is done at two levels; the first is an analysis of the heart signal using models trained to be used within the M5Stick-C Plus, and the second one is to employ more robust models in the Cloud part. In this paper, we have focused on detecting these anomalies in the Edge, i.e. embedding the model within the M5Stick-C Plus and determining whether the ECG signal is normal or abnormal. However, regardless of the pre-analysis result, all signals with their respective time stamps are sent to the AWS for further analysis and storage. The description of how the training was conducted in the Edge will be described in detail in the following subsection.

2.2 Software Description

To classify whether a signal is normal or abnormal, we have used the dataset located in Kaggle[1] [7]. This dataset comprises two signal datasets from two famous datasets in ECG signal classification. The first is the MIT-BIH arrhythmia dataset and the second is the PTB diagnostic ECG dataset. The number of data stored in the two datasets is large enough to train a neural network. However, in our experiments, we selected the PTB diagnostic ECG dataset[2]. This selected dataset has been used to explore the classification of heart signals

---

[1] https://www.kaggle.com/datasets/shayanfazeli/heartbeat.
[2] https://www.physionet.org/content/ptbdb/1.0.0/.

using a deep neural network architecture to obtain a sufficiently stable and small model which can fit within an embedded device.

Table 1. Model parameters.

Layer (type)	Output shape	Param #
dense_38 (Dense)	(None, 15)	2820
dense_39 (Dense)	(None, 10)	160
dense_40 (Dense)	(None, 5)	55
dense_41 (Dense)	(None, 1)	6

Table 1 shows the internal configuration of the model, which is composed of the following hyperparameters: *Optimizer: Adma, Learning Rate: 0.001, Loss: binary crossentropy, metrics: accuracy, epochs: 150, shuffle: True* and *batch size: 32*.

Of all the data of the selected dataset, 80% was used for training, 10% for testing and the last 10% for validation. Figure 5 shows two ECG signals, Fig. 5(a) shows a normal ECG signal, and Fig. 5(b) shows an abnormal ECG signal. Figure 5(a) shows that it is a normal heart signal, as its characteristic shape can be seen. The QRS complex which represents ventricular depolarization can be observed, however, Fig. 5(b) does not show this particular characteristic of a heart signal. Therefore, the system will determine that this signal is an abnormal signal.

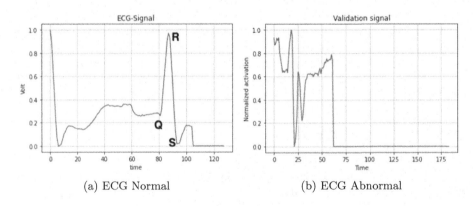

(a) ECG Normal   (b) ECG Abnormal

Fig. 5. ECG signals extracted from the dataset.

A confusion matrix was obtained from the training process, which is shown in Fig. 6. As can be seen in the confusion matrix, the results obtained with the portion used for validation are quite good. These are around 90% classification, which means that our proposed strap can determine whether an ECG signal is normal or abnormal.

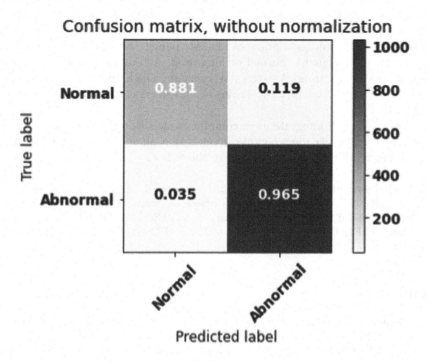

Fig. 6. Confusion matrix

Once the ECG signal has been classified, the system sends the data to Amazon AWS, which is stored in a DynamoDB database. This database stores all the information coming from the chest strap, along with a timestamp that shows the day, month and year, as well as the hour, minutes and seconds. The stored data can be downloaded in CSV format or analysed using the AWS lambda function. The Fig. 7 shows a series of signals obtained during the laboratory experiments carried out during the construction of the E-Bot. Moreover, the result of the analysis is communicated by the robot to the patient, if considered necessary. In addition, the robot can interact with the patient to transmit possible messages or advice from the caregivers if necessary.

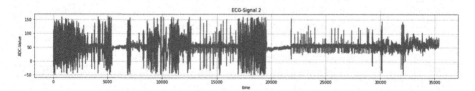

Fig. 7. Signal extracted from the AWS database.

3   Conclusions and Future Works

This work focuses on constructing a companion robot for older people or people living alone. The E-Bot companion robot is a low-cost and easy-to-repair robot. E-Bot integrates the Alexa voice assistant with a heart signal monitoring system (ECG). With the approach presented in this paper, the system can determine whether an ECG signal is normal or abnormal. At the same time, the system can communicate with an Amazon AWS service, which allows storing the data captured by the chest strap. Physicians or caregivers can then view this stored data for analysis.

Future work will integrate more complex models using Amazon Deep Learning services and create more complex models that allow in-depth analysis of the signals sent by the systems connected to the E-Bot.

Acknowledgements. This work was partly supported by the Spanish Government (RTI2018-095390-B-C31),Universitat Politecnica de Valencia Research Grant PAID-10-19, Consellería d'Innovació, Universitats, Ciencia i Societat Digital from Comunitat Valenciana (CIPROM/2021/077, BEST 2022 and APOSTD/2021/227) and the European Social Fund (Investing In Your Future).

References

1. Alnuaim, A.A., et al.: Human-computer interaction for recognizing speech emotions using multilayer perceptron classifier. J. Healthc. Eng. 2022 (2022)
2. Beh, J., Pedell, S., De Kruiff, A., Reilly, A.: Alexa, what day is it again?: Virtual assistants empowering people living with dementia at home. In: Design for People Living with Dementia, pp. 108–120. Routledge (2022)
3. Clemente, C., Greco, E., Sciarretta, E., Altieri, L., et al.: Alexa, how do i feel today? Smart speakers for healthcare and wellbeing: an analysis about uses and challenges. Sociol. Soc. Work Rev. **6**(1), 6–24 (2022)
4. Dautenhahn, K., Woods, S., Kaouri, C., Walters, M.L., Koay, K.L., Werry, I.: What is a robot companion-friend, assistant or butler? In: 2005 IEEE/RSJ International Conference on Intelligent Robots and Systems, pp. 1192–1197. IEEE (2005)
5. Greenberg, N.E., Wallick, A., Brown, L.M.: Impact of COVID-19 pandemic restrictions on community-dwelling caregivers and persons with dementia. Psychol. Trauma: Theory Res. Pract. Pol. **12**(S1), S220 (2020)
6. Jiménez, S., Favela, J., Quezada, Á., Alanis, A., Castillo, E., Villegas, E.: Alexa to support patients with dementia and family caregivers in challenging behaviors. In: World Conference on Information Systems and Technologies, pp. 336–345. Springer (2022). https://doi.org/10.1007/978-3-031-04826-5_33
7. Kachuee, M., Fazeli, S., Sarrafzadeh, M.: ECG heartbeat classification: a deep transferable representation. In: 2018 IEEE International Conference on Healthcare Informatics (ICHI), pp. 443–444. IEEE (2018)
8. Maust, D.T., et al.: Antipsychotics, other psychotropics, and the risk of death in patients with dementia: number needed to harm. JAMA Psychiatr. **72**(5), 438–445 (2015)

9. Pransky, J.: Social adjustments to a robotic future. Wolf and Mallett, pp. 137–59 (2004)
10. Steptoe, A., Deaton, A., Stone, A.A.: Subjective wellbeing, health, and ageing. Lancet **385**(9968), 640–648 (2015)
11. Stoner, S.C.: Management of serious cardiac adverse effects of antipsychotic medications. Mental Health Clin. **7**(6), 246–254 (2017)

Zero-Shot Knowledge Graph Completion for Recommendation System

Zhiyuan Wang[1] , Cheng Chen[2] , and Ke Tang[1(✉)]

[1] Guangdong Key Laboratory of Brain -Inspired Intelligent Computation,
Department of Computer Science and Engineering, Southern University of Science
and Technology, Shenzhen 518055, China
`wangzy2020@mail.sustech.edu.cn`, `tangk3@sustech.edu.cn`
[2] Research Institute of Trustworthy Autonomous Systems, Southern University
of Science and Technology, Shenzhen 518055, China
`chenc3@sustech.edu.cn`

Abstract. Knowledge graphs are structured representations of actual
entities and relations. They are widely used to improve the performance
of downstream tasks such as recommendation systems and semantic
searching. Knowledge graph completion (KGC) is a technology for discovering the missing relations between the entities in a knowledge graph
(KG). Existing methods leverage known relations on a KG to build a
model to predict missing relations. Such methods implicitly require a
substantial number of relations to be known in advance, which might not
be available in practice. To cope with the cold-start scenario for KGC,
i.e., no relation is known in advance, we propose a zero-shot approach in
this paper. Our approach converts the KGC process to an optimization
problem. It uses the Evolutionary Strategy (ES) algorithm to optimize
a model used to complete the KG according to the performance of the
recommendation system constructed based on the completed KG. Experiments on a movie dataset demonstrate that our approach can complete
the KG in the cold-start scenario and improve the performance of the
recommendation system built based on the completed KG.

Keywords: Knowledge graph completion · Zero-shot ·
Recommendation system · Evolutionary strategy

1 Introduction

Recently, knowledge graphs (KGs) have been widely applied in the domains such
as recommendation systems and semantic searching. A KG can contain a wealth
of real-world facts, supplying much prior knowledge for downstream tasks to
improve performance [5]. The KG expresses the relations between entities in
triple form (e_h, r, e_t), with e_h, e_t, and r representing the subject, object and the
type of this relation. In KG studies, the subject and object in a relation are also
called head entity and tail entity.

© The Author(s), under exclusive license to Springer Nature Switzerland AG 2022
H. Yin et al. (Eds.): IDEAL 2022, LNCS 13756, pp. 188–198, 2022.
https://doi.org/10.1007/978-3-031-21753-1_19

KGs are usually of large scale, making manually constructing a KG expensive. Hence, it is crucial to build the KG automatically. The principal automatic construction method extracts a KG from external text data by named entity recognition and relation extraction tasks. The quality of external text data limits the quality of the extracted KG. To enhance the quality of the extracted KG, knowledge graph completion (KGC) is also used widely in the automatic construction of KG.

The existing KGC methods focus on learning a model to make link predictions on the known KG. These methods always train a model by supervised learning, where the training data is derived from the relation triples in the known KG [5]. Therefore, the performance of existing KGC methods presents a high dependence on the quality of the known KG. This makes existing methods limited in cold-start scenarios where known knowledge does not exist. This paper addresses this limitation by exploring the zero-shot KGC with the recommendation system as the downstream task.

The rest of this paper is organized as follows. The existing studies about the KGC task are presented in Sect. 2. Then in Sect. 3, the problem description of the zero-shot KGC and the proposed approach are introduced. Section 4 analyzes experimental studies. And finally, conclusions are given at the end of this paper in Sect. 5.

2   Related Work

The KGC task means using graph data mining technology to discover the missing relations for a KG. Current studies about the KGC task mainly focus on how to make link predictions on an existing KG. Given the head entity and relation type, the algorithm predicts the most likely tail entity from a list of candidate tail entities. The candidate tail entities may consist of a subset or the whole of the entities in the KG. To evaluate the performance of a KGC method, current studies usually use a test set consisting of multiple relation triples. For each relation triple (e_h, r, e_t) in the test set, suppose the algorithm has known its head entity e_h and relation type r and sorts the candidate tail entities in descending order by the probability predicted by a trainable model. The evaluation metric is the rank of the ground truth tail entity e_t of the relation triple in the descending sorted list of candidate tail entities [5].

Previous KGC studies can be classified into embedding-based methods and path-based methods. Path-based methods take the path between entities in the KG as the feature. Then use the machine learning model to predict whether a relation triple exists or not according to the relation type and path between the head entity and tail entity. PRA is one of the earlier studies using the random walk to calculate the feature value [7]. Following that, some studies use RNN to model the path between entities, such as Path-RNN [9]. Reinforcement learning is also used in path-based KGC methods such as DeepPath [15] and MINERVA [3], and these methods train an agent to explore the path on the KG. CogKR combines multiple paths to a subgraph and uses the subgraphs as the feature

[4]. Embedding-based methods find suitable embedding representation for entities and relation types in KGs by the representation learning approach. Then embedding-based methods use a calculation model to predict whether a relation triple is true or not according to the embedding representations of its head entity, tail entity, and relation type. The main difference among the embedding-based methods lies in the design of the calculation models. For example, the displacement model is represented by TransE [2]; the tensor decomposition model is represented by RESCAL [11] and TuckER [1]; the convolutional neural network is represented by ConvKB [10]. Moreover, some methods apply the graph neural network to the KGC tasks [8,12]. These methods use graph neural networks to import the topological structure feature of the KG to the embedding representation of entities and relation types.

Most previous KGC studies use the existing KG as the supervisory signal and optimize the model to improve the link prediction ability on the existing KG. It also means these KGC methods require a high-quality initial KG. While the external data is insufficient, it is impossible to get an initial KG automatically, so the previous KGC methods cannot be applied. This paper proposes a KGC approach to solve this problem. Our approach converts the KGC process to an optimization problem and uses the performance of the downstream recommendation system constructed based on the completed KG as the supervisory signal. This makes our approach unlimited from the quality of the initial KG. It can optimize the KGC model even if there is no relation triple in the initial KG.

3   Our Approach

3.1   Framework

In the cold-start scenario, there is no relation triple in the KG, and the existing KGC methods cannot learn a model that is used to make link predictions without training data. For this scenario, we propose a zero-shot KGC method called KGCDT (Knowledge Graph Completion based on Downstream Task). Our zero-shot approach essentially uses the performance of the recommendation system constructed based on the completed KG as the feedback signal to optimize the model that is used to make link predictions. From this fundamental idea, KGCDT can be divided into two parts: 1) **Optimizer**: Generate the models used to complete the KG and update the models according to their performance from the Evaluator; 2) **Evaluator**: Evaluate the performance of the models generated by the Optimizer. It uses the model to complete the KG with a known entity set and relation type set, then constructs a recommendation system based on the completed KG. The performance of the recommendation system represents the model's performance. The framework of our approach is shown in Fig. 1.

3.2   Problem Formulation

The zero-shot KGC problem is formulated as follows. In a cold-start scenario for KGC, there is an initial KG $\mathcal{G} = (V, \emptyset)$ with no relations. Let $V = \{v_1, v_2, \cdots\}$

and $R = \{r_1, r_2, \cdots\}$ be the sets of entities and relation types in the KG $\mathcal{G}$, respectively. In addition to the entities and relation types of the KG, we have a user-item interaction data Y which is used to construct the downstream recommendation system to evaluate the completed KG. Let $U = \{u_1, u_2, \cdots\}$ and $M = \{m_1, m_2, \cdots\}$ denote the sets of users and items, and the $y_{um} \in Y$ represents the user u's feedback to the item m.

Given the initial KG $\mathcal{G}$ as well as the interaction data Y, this paper uses a model $\mathcal{F}$ to complete the initial KG $\mathcal{G}$ and obtain a completed KG $\mathcal{G}'$. Based on the completed KG $\mathcal{G}'$, we can construct a recommendation system $\mathcal{Z}$ using the interaction data Y. The performance of $\mathcal{Z}$ can represent the advancement of downstream tasks by the completed KG $\mathcal{G}'$, which is also can be seen as the effectiveness of the KGC process. The goal of our approach is to optimize the parameter Θ of the model $\mathcal{F}$, which is used to complete the initial KG $\mathcal{G}$, from the feedback signal of the downstream recommendation system $\mathcal{Z}$, to maximize the AUC (Area Under Curve, the Curve specifically refers to the Receiver Operating Characteristic curve) performance of the recommendation system $\mathcal{Z}$.

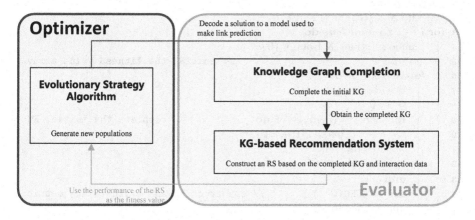

Fig. 1. The framework of the KGCDT.

3.3   Zero-Shot KGC

The zero-shot KGC approach called KGCDT, as the framework in Sect. 3.1, can be transformed into an optimization problem. In this optimization problem, the independent variable is the parameter Θ of the model $\mathcal{F}$ that is used to complete the initial KG $\mathcal{G}$ and obtain a completed KG $\mathcal{G}'$. The optimization objective of the problem is to maximize the performance of the recommendation system $\mathcal{Z}$, which is the downstream task of the completed KG $\mathcal{G}'$. The algorithm of KGCDT is presented in Algorithm 1.

In the optimization problem transformed from KGCDT, the optimization objective is not differentiable to the independent variable, because the ranking operation during the process of completing the initial KG $\mathcal{G}$ is not differentiable.

Therefore, the gradient-based methods commonly used in existing KGC methods cannot be applied to this optimization objective. To address this limitation, we use the evolutionary strategy (ES) algorithm as the optimization method, which is corresponding to the Optimizer part of the KGCDT. The ES algorithm is a population-based method, it will generate a solution set and update it according to the fitness value of the solutions in the set. The fitness value is calculated by the Evaluator part. The solution of the ES algorithm is a d_Θ dimension real value vector with the d_Θ as the dimension of Θ. It can be decoded into the parameter Θ of the model $\mathcal{F}$ by filling the values of each dimension in the solution to Θ in order.

Algorithm 1: KGCDT

Data: The relation type set R; the candidate head entity set H_r and candidate tail entity set T_r for each relation type $r \in R$; the interaction data Y

Result: The solution $\mathbf{x}^\star$ with the best fitness value has been found and the completed KG $\mathcal{G}'$ corresponds to $\mathbf{x}^\star$

1 initialze μ, σ ;
2 generate $S = \{(h, r) | r \in R, h \in H_r\}$;
3 **for** $i \leftarrow 1$ **to** $numGens$ **do**
4 $\quad$ sample solutions $\mathbf{X}$ from $\mathcal{N}\left(0, \sigma^2\right)$;
5 $\quad$ $Q \leftarrow \emptyset$;   // initial the fitness value array
6 $\quad$ **foreach** $\mathbf{x} \in \mathbf{X}$ **do**
7 $\quad\quad$ $\Theta \leftarrow$ decode $\mathbf{x}$ directly;
8 $\quad\quad$ $\mathcal{G}' \leftarrow \emptyset$;
9 $\quad\quad$ **foreach** $s = (h, r), s \in S$ **do**   // complete the initial KG
10 $\quad\quad\quad$ $t \leftarrow \arg\max_{t'} \mathcal{F}\left((h, r, t'), \Theta\right)$;
11 $\quad\quad\quad$ $e \leftarrow (h, r, t)$;
12 $\quad\quad\quad$ append e to $\mathcal{G}'$;
13 $\quad\quad$ **end**
14 $\quad\quad$ auc $\leftarrow$ CFKG$(\mathcal{G}', Y)$;   // construct RS and evaluate auc metric
15 $\quad\quad$ append auc to Q;
16 $\quad$ **end**
17 $\quad$ adjust μ and σ according to $\mathbf{X}$ and Q;
18 **end**
19 **return** *the solution $\mathbf{x}^\star$ with the best fitness value and the corresponding $\mathcal{G}'$*

To evaluate a solution, the Evaluator completes the initial KG by the model $\mathcal{F}$ that has the parameter Θ decoded from the solution and obtains the completed KG $\mathcal{G}'$. The performance of the recommendation system $\mathcal{Z}$ that is constructed based on the completed KG $\mathcal{G}'$ and interaction data Y is the fitness value of this solution. The fitness function can be written as:

$$fitness = \mathrm{AUC}\left(\mathcal{Z}\left(\mathcal{G}', Y\right)\right) \tag{1}$$

Specifically, the solution evaluation process has two steps, knowledge graph completion and KG-based recommendation system building. The details of these two steps are described below.

In the first step, our approach completes the initial KG $\mathcal{G}$ by the model $\mathcal{F}$ with the parameter Θ decoded from the solution which needs to be evaluated. In a cold-start KGC scenario, entity set V and relation type R are known, but the relation triple does not exist in the initial KG. By analyzing the semantic information of the relation types, our approach can obtain the candidate head entity set H_r and candidate tail entity set T_r for each relation type $r \in R$. Therefore, it can generate a set S consisting of the (head_ent, rel_type) pairs:

$$S = \{(h, r) | r \in R, h \in H_r\} \tag{2}$$

For each (head_ent, rel_type) pair in S, one or more tail entities will be predicted by the model $\mathcal{F}$, thus obtaining the completed KG $\mathcal{G}'$. In this paper, we choose TransE [2] as the model $\mathcal{F}$ used to predict the tail entity for the (head_ent, rel_type) pair. The TransE model uses a score function to predict the tail entity, and the triple with a higher score more possibly be true. The score function of the TransE model is:

$$\text{score}(e) = -\|\mathbf{h} + \mathbf{r} - \mathbf{t}\| \tag{3}$$

where $e = (h, r, t)$ is a relation triple, and the $\mathbf{h}$, $\mathbf{t}$ and $\mathbf{r}$ are the embedding vector of the head, tail entity and the relation type. For each (head_ent, rel_type) pair in S, we can obtain a tail entity that can form the triple with the pair and get the highest score under the TransE model. The triples formed by the pairs and the tail entities will make up the completed KG.

The second step is building a recommendation system based on the completed KG $\mathcal{G}'$. Recommendation system is one of the most widely used downstream tasks of the KG. The aim of incorporating KGs into the construction process of recommendation systems is to improve the performance of recommendation systems. Therefore, the performance of the recommendation system based on the completed KG $\mathcal{G}'$ can reflect the quality of the KGC. Our approach uses the CFKG algorithm [16] to construct a recommendation system $\mathcal{Z}$ using the completed KG $\mathcal{G}'$ and the interaction data Y. The AUC metric of the recommendation system $\mathcal{Z}$ is the fitness value of the solution decoded into the parameter Θ.

4   Experiments

To investigate the effectiveness of our approach, we evaluate it on a movie recommendation dataset and KG in this section. In the research scenario of this study, there is no existing KGC method that can be applied. Therefore, we compare the performances of the recommendation systems constructed based on different KGs in the experiments to clarify the effectiveness of KGCDT.

4.1   Dataset

Our experiment's interaction data is a movie dataset sampled from MovieLens-1M, which is widely used in the recommendation system benchmark. The KG data comes from a recent KG-based recommendation system study, and it is extracted from Microsoft Satori [14]. MovieLens-1M consists of approximately 1 million explicit ratings (ranging from 1 to 5) on the MovieLens website. In this study, we transform its explicit feedback data from rating value to implicit feedback data. Each interaction record with a value larger than or equal to 4 is marked with 1, indicating that the user is interested in this item. Furthermore, the interaction record that its item is not in the entity set of the KG will be removed.

Table 1. Statistic for the dataset.

#Users	#Items	#Records	#Entities	#Relation types	#Relations
6,036	2,347	37,902	7,008	7	20,195

Even after converting the explicit feedback to the implicit feedback, there are still too many interaction records in the dataset. The high record density means we can build an excellent enough recommendation system without the KG. The experiments cannot distinguish different KGs' performance gain to the recommendation system. Therefore, we sample the interaction data again to reduce the records density. The statistics of the interaction data and KG data are shown in Table 1.

Table 2. Completion configuration for each relation type.

Relation type	#Candidate head entities	#Candidate rail entities	#Tail entities need to be completed
film.star	2445	1251	2
film.genre	2445	23	2
film.writer	2445	2441	1
film.direct	2445	1567	1
film.rating	2445	5	1
film.language	2445	73	1
film.country	2445	11	1

4.2   Data Pre-processing

To mimic the cold start scenario, we ignore the relation triples in the original KG, but keep the entity list and relation type list. Based on prior knowledge, we set the candidate head and tail entities for each relation type and generate a set S consisting of all the (head_ent, rel_type) pairs that need to complete the tail entity. For each relation type, we set the number of the tail entities that need to be completed for each (head_ent, rel_type) pair in set S. The configuration of the completion process is shown in Table 2.

4.3   Experimental Setup

We evaluate our approach in recommendation scenarios in the top-k recommendation task. In this task, the recommendation system will recommend k items for each user. The evaluation metric is recall@k, precision@k, and nDCG@k [6], $k \in \{1, 5, 10\}$.

In the experiments, we construct the recommendation systems based on three KGs and evaluate their performance. The first KG, named **KGCDT**, is the completion result of the model $\mathcal{F}$ with a parameter Θ that is optimized by the KGCDT approach. The second KG, named **Initial**, is the completion result of the model $\mathcal{F}$ with a parameter Θ that is initialized randomly. The last KG, named **Empty**, is a KG with no relation triple.

While constructing the recommendation systems based on the KGs and evaluating their performance, we apply 3-Fold cross-validation to the interaction data Y and get three pairs of train sets and test sets. The ratio of the train set to test set is 2:1. For each KG, we learn a recommendation system based on the train set data and the KG, then evaluate the performance on the test set. For each pair of train set and test set, we repeat the learning and evaluation processes 10 times, i.e., we repeat training and testing 30 times for each KG. The performance of the recommendation system is the average value of 30 evaluation results. It is worth noting that when we optimize the KGC model based on the performance of the recommendation system, we can use only the train set, and the test set is sealed.

In the optimization process, we use the PGPE method [13] to sample solutions from a probability distribution and adjust the probability distribution parameters according to the fitness values of the solutions. The hyperparameters of both the ES algorithm and the solution evaluation process are given as follows. For the ES algorithm, the population size is set to 41; the maximum generation is set to 300; the σ is initialized to 0.1 with decay rate and lower bound are 10^{-3} and 0.01; and the learning rate of the mean value μ is initialized to 10^{-3} with decay rate and lower bound are both 10^{-4}. For the solution evaluation process, the embedding dimension of the TransE model F is set to 64; the embedding dimension of the CFKG is set to 64; the optimizer of the CFKG is the Adam optimizer with batch size and learning rate are 4096 and 0.01; the epoch number of CFKG is searched in [1, 1200] by applying 3-Fold cross-validation on the train set.

4.4 Experiments Result and Comparisons

The average performance and the standard deviation of the 30 repeated testing scores of the recommendation systems based on the three KGs are shown in Table 3, where the best averaged performance is marked in bold. The ‡ flag and the last row show the statistical analysis of the Wilcoxon signed-rank test at a 0.05 significance level. It can be seen that our approach can complete the KG efficaciously in a cold-start scenario and improve the performance of the downstream recommendation system. In the top-k recommendation task, the recommendation system based on **KGCDT** has better averaged performance on recall@k, precision@k, and nDCG@k metrics than the recommendation systems based on **Initial** and **Empty**. And the advantage of all the metrics except recall@10 and precision@10 of **Initial** are statistically significant.

Table 3. The mean value and standard deviation of the performance in the top-k recommendation task. The best averaged metrics are in boldface. The ‡ flag indicates the entry is not significantly different from the **KGCDT** KG. The results of the significance check come from Wilcoxon signed-rank test. And the last row shows the statistical analysis, where the three values indicate the number of metrics that the averaged performance of the recommendation system based on **KGCDT** is statistically better, the same, or worse than that of the corresponding KG.

Metric	KGCDT	Initial	Empty
Precision@1	**0.9605** ± 0.0020	0.9565 ± 0.0027	0.8817 ± 0.0172
Precision@5	**0.3842** ± 0.0021	0.3807 ± 0.0056	0.3024 ± 0.0130
Precision@10	**0.2017** ± 0.0010	0.2009 ± 0.0018‡	0.1632 ± 0.0066
Recall@1	**0.3144** ± 0.0007	0.3130 ± 0.0009	0.2884 ± 0.0056
Recall@5	**0.6168** ± 0.0035	0.6111 ± 0.0094	0.4828 ± 0.0206
Recall@10	**0.6430** ± 0.0028	0.6404 ± 0.0060‡	0.5168 ± 0.0205
nDCG@1	**0.9605** ± 0.0020	0.9565 ± 0.0027	0.8817 ± 0.0172
nDCG@5	**0.7006** ± 0.0037	0.6936 ± 0.0090	0.5726 ± 0.0207
nDCG@10	**0.7093** ± 0.0032	0.7038 ± 0.0074	0.5851 ± 0.0206
Statistics analysis	–	7 − 2 − 0	9 − 0 − 0

5 Conclusion and Future Work

This paper proposes a zero-shot KGC approach in a cold-start scenario. Our approach sets the performance of the downstream recommendation system as the optimization objective and applies the ES algorithm to optimize the model used to predict the missing relations according to the feedback signal from the downstream recommendation system. Our experiments show that the recommendation system based on the KG completed by our KGCDT approach has an excellent performance in the top-k recommendation tasks.

For the future work, we plan to investigate the KGC that is based on different downstream tasks. In addition, the relationship between KG quality and downstream task performance is also worth exploring.

References

1. Balazevic, I., Allen, C., Hospedales, T.M.: TuckER: tensor factorization for knowledge graph completion. In: Proceedings of the 2019 Conference on Empirical Methods in Natural Language Processing (EMNLP), pp. 5184–5193. ACL (2019). https://doi.org/10.18653/v1/D19-1522
2. Bordes, A., Usunier, N., García-Durán, A., Weston, J., Yakhnenko, O.: Translating embeddings for modeling multi-relational data. In: Advances in Neural Information Processing Systems (NeurIPS), pp. 2787–2795 (2013). http://proceedings.neurips.cc/paper/2013/hash/1cecc7a77928ca8133fa24680a88d2f9-Abstract.html
3. Das, R., et al.: Go for a walk and arrive at the answer: reasoning over paths in knowledge bases using reinforcement learning. In: International Conference on Learning Representations (ICLR). OpenReview.net (2018). http://openreview.net/forum?id=Syg-YfWCW
4. Du, Z., et al.: CogKR: cognitive graph for multi-hop knowledge reasoning. IEEE Trans. Knowl. Data Eng. 1–1 (2021). https://ieeexplore.ieee.org/document/9512424
5. Ji, S., Pan, S., Cambria, E., Marttinen, P., Yu, P.S.: A Survey on knowledge graphs: representation, acquisition, and applications. IEEE Trans. Neural Networks Learn. Syst. **33**(2), 494–514 (2021). https://ieeexplore.ieee.org/document/9416312
6. Järvelin, K., Kekäläinen, J.: Cumulated gain-based evaluation of IR techniques. ACM Trans. Inf. Syst. **20**(4), 422–446 (2002). https://doi.org/10.1145/582415.582418
7. Lao, N., Mitchell, T., Cohen, W.W.: Random walk inference and learning in a large scale knowledge base. In: Proceedings of the 2011 Conference on Empirical Methods in Natural Language Processing (EMNLP), pp. 529–539. ACL, Edinburgh, Scotland, UK. (2011). http://aclanthology.org/D11-1049
8. Li, Z., Zhao, Y., Zhang, Y., Zhang, Z.: Multi-relational graph attention networks for knowledge graph completion. Knowl. Based Syst. **251**, 109262 (2022). https://doi.org/10.1016/j.knosys.2022.109262
9. Neelakantan, A., Roth, B., McCallum, A.: Compositional vector space models for knowledge base completion. In: Proceedings of the 53rd Annual Meeting of the Association for Computational Linguistics (ACL), pp. 156–166. ACL, Beijing, China (2015). http://aclanthology.org/P15-1016
10. Nguyen, D.Q., Nguyen, T.D., Nguyen, D.Q., Phung, D.Q.: A novel embedding model for knowledge base completion based on convolutional neural network. In: Proceedings of the 2018 conference of the north american chapter of the association for computational linguistics: Human language technologies (NAACL-HLT), pp. 327–333. ACL (2018). https://doi.org/10.18653/v1/n18-2053
11. Nickel, M., Tresp, V., Kriegel, H.P.: A Three-way model for collective learning on multi-relational data. In: Proceedings of the 28th International Conference on Machine Learning (ICML), pp. 809–816. Omnipress (2011). http://icml.cc/2011/papers/438_icmlpaper.pdf

12. Schlichtkrull, M., Kipf, T.N., Bloem, P., van den Berg, R., Titov, I., Welling, M.: Modeling relational data with graph convolutional networks. In: Gangemi, A., et al. (eds.) ESWC 2018. LNCS, vol. 10843, pp. 593–607. Springer, Cham (2018). https://doi.org/10.1007/978-3-319-93417-4_38

13. Sehnke, F., et al.: Parameter-exploring policy gradients. Neural Netw. **23**(4), 551–559 (2010). https://doi.org/10.1016/j.neunet.2009.12.004

14. Wang, H., Zhang, F., Zhao, M., Li, W., Xie, X., Guo, M.: Multi-task feature learning for knowledge graph enhanced recommendation. In: The World Wide Web Conference (WWW), pp. 2000–2010. ACM (2019). https://doi.org/10.1145/3308558.3313411

15. Xiong, W., Hoang, T., Wang, W.Y.: DeepPath: a reinforcement learning method for knowledge graph reasoning. In: Proceedings of the 2017 Conference on Empirical Methods in Natural Language Processing (EMNLP), pp. 564–573. ACL (2017). https://doi.org/10.18653/v1/d17-1060

16. Zhang, Y., Ai, Q., Chen, X., Wang, P.: Learning over knowledge-base embeddings for recommendation. CoRR abs/1803.06540 (2018). http://arxiv.org/abs/1803.06540

The Covid-19 Influence on the Desire to Stay at Home: A Big Data Architecture

Regina Sousa, Daniela Oliveira, Ana Carneiro, Luis Pinto, Ana Pereira, Ana Peixoto, Hugo Peixoto, and José Machado[(✉)]

ALGORITMI Research Centre/LASI, University of Minho, Braga, Portugal
{regina.sousa,daniela.oliveira}@algoritmi.uminho.pt,
{pg46983,pg47428,pg46978,pg46988}@alunos.uminho.pt,
{hpeixoto,jmac}@di.uminho.pt

Abstract. The COVID-19 pandemic has had an impact on many aspects of society in recent years. The ever-increasing number of daily cases and deaths makes people apprehensive about leaving their homes without a mask or going to crowded places for fear of becoming infected, especially when vaccination was not available. People were expected to respect confinement rules and have their public events cancelled as more restrictions were imposed. As a result of the pandemic's insecurity and instability, people became more at ease at home, increasing their desire to stay at home. The present research focuses on studying the impact of the COVID-19 pandemic on the desire to stay at home and which metrics have a greater influence on this topic, using Big Data tools. It was possible to understand how the number of new cases and deaths influenced the desire to stay at home, as well as how the increase in vaccinations influenced it. Moreover, investigated how gatherings and confinement restrictions affected people's desire to stay at home.

Keywords: Big data · Business intelligence · COVID-19 · Stay at home · Behavior analysis

1 Introduction

Since 2020, the COVID-19 pandemic has had a significant impact on all socioeconomic factors. One of the most fluctuating factors was the population's obligation to stay at home, depending on the implementation of restrictions that limited mobility and public gatherings. Working remotely has become a way of life, and people have begun to value their homes even more, preferring homes with open and green spaces. However, it is often impossible to perform work activities from home in several jobs, which leads to unemployment or the search for alternative job. The pandemic has affected people's mental health in terms of

© The Author(s), under exclusive license to Springer Nature Switzerland AG 2022
H. Yin et al. (Eds.): IDEAL 2022, LNCS 13756, pp. 199–210, 2022.
https://doi.org/10.1007/978-3-031-21753-1_20

depression, anxiety, stress, sleep problems, and psychological distress as a result of the uncertainties and quarantine periods [6, 7]. Preferences to stay at home and confinement measures are felt and met differently depending on the culture of each country. Taking this into account, by studying the factors and impacts of the Covid-19 pandemic, this work intends to serve as a base model for governments to better understand different metrics of evaluations, more specifically people's adherence to stay at home and its impact in the economy. The analysis of this metrics may help implement more effective legislation for each country in order to reduce the magnitude of future pandemics. Public datasets, as well as several big data tools, were used and analyzed, and the architecture was created. Following that, the main findings will be discussed, followed by a critical analysis and some final thoughts in the conclusion.

2   State of the Art

The work hereby presented falls within the current multidisciplinary area that combines topics of the most recent worldwide pandemic, COVID-19, and its influence on various socio-cultural factors. Therefore, this section will focus not on the technologies used for the development of the work, but rather on an overview of what works have already been published in this specific area. Considering the fact that this is a fairly recent subject, three scientific articles were initially studied. Having said this, some of the references of these first three articles were also read and studied.

In this first article [1] the authors implemented a discrete choice experiment survey to quantify respondents' willingness to stay at home depending on five attributes: the length of a new stay-at-home order, state-level increases in the number of newly confirmed cases, increases in the number of unemployment insurance claims, the probability of schools opening, and whether a mask wearing mandate is implemented. Implementing this survey, that targets three states in the U.S. (California, Georgia, and Illinois), allowed them to construct hypothetical scenarios by varying different attributes, delineating the trade-offs that are vital to individuals' decision-making and providing flexibility to measure respondents' willingness to stay at home. In the second research [2], authors seek to test the effect of attitude towards the behavior of following the stay-at-home policy during COVID-19 pandemic by collecting data through an online survey with 148 respondents in the Greater Area of Jakarta, Indonesia. In this article it shows that attitude towards the behavior, subjective norm and perceived behavioral control positively and significantly affect intention to follow the stay-at-home orders during the COVID-19 pandemic. However, perceived susceptibility and perceived severity of COVID-19 do not significantly influence the intention to follow stay-at-home orders. Finally, in this third article [3] the authors examined the intention to stay home due to COVID-19 during a second wave of the pandemic by an online survey conducted from citizens of Pakistan. The findings of the study highlight a positive and significant effect of fear of COVID-19, attitudes to stay at home behavior, knowledge about COVID-19 and health consciousness on the intention to stay at home.

3   Materials and Methods

To draw some pertinent conclusions about the amount of data from different indicators about this topic, a four-part architecture was designed. Initially, datasets were chosen from various sources, and one component of the architecture is the merging of these datasets, which will then serve as the foundation for all subsequent work. The data storage is represented in the second component.

Having said that, the core component of the work emerges, the ETL process. In this step the data is firstly extracted from the data storage, subsequently processed and treated, and finally loaded into a visualization place. Finally, using PowerBI, this architecture culminates in the data visualization through the creation of dashboards. Figure 1 meticulously depicts all of these components.

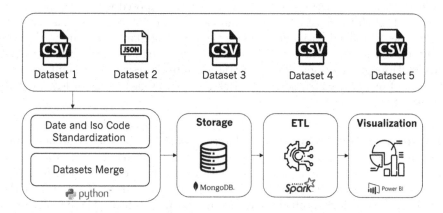

Fig. 1. Phases and tools of the implemented pipeline. Adapted from [14].

3.1   Sources and Data Collection

To correctly answer the use cases listed above, 5 datasets were chosen, one of which was provided by the teachers, and the rest were found in the websites listed in the references below. All datasets are in CSV format, except the vaccination dataset which is in JSON format. Despite the observations in each dataset being carried out in different periods of time, countries do not always have information for each day within that period.

Dataset 1 - WHO COVID: The first dataset represents global information associated with the COVID-19 pandemic in each of the 236 countries from January 3, 2020 to March 1, 2022. This dataset serves as the foundation for the subsequent datasets. This dataset includes the number of new COVID-19 cases per day, the number of accumulated COVID-19 cases, and the number of new

COVID-19 deaths per day, in addition to the attributes that represent the country, country code, World Health Organization (WHO) region, and date as well as the total number of COVID-19 deaths [4].

Dataset 2 - Vaccination: This dataset depicts the evolution of the vaccination process in 235 countries from January 22, 2021 to March 27, 2022. Some metrics and attributes were used to represent this vaccination process by country; however, it should be noted that not all observations have values for all of these parameters. There are 14 attributes in this dataset that represent the vaccination process [5].

Dataset 3 - Restrictions on Public Gatherings: The third dataset associates the public gatherings restrictions that were imposed with each day (from January 1, 2020 to March 21, 2022) and each of the 186 countries. As a result, the dataset is divided into four columns: the first represents the country, the second the country code, the third the day, and the fourth the restrictions applied (**restriction_gatherings**). These restrictions are represented by numbers scale of 0 to 4 from "No restrictions" to "Restrictions on gatherings of less than 10 people" [8].

Dataset 4 - Stay at Home: This dataset associates the stay-at-home recommendation with each day (from January 1, 2020 to March 21, 2022) and each of the 186 countries. As a result, the dataset has four columns: the first represents the country, the second the country code, the third the day, and the fourth the restrictions applied (**stay_home_requirements**). These restrictions are represented by values ranging from 0 to 3, with 0 representing "No measures," 1 representing "Recommended not leaving the house," 2 representing "Require not leaving the house with exceptions for daily exercise, grocery shopping, and 'essential' trips," and 3 representing "Require not leaving the house with minimal exceptions" [9].

Dataset 5 - Residential Mobility: This dataset associates each day (from February 17, 2020 to January 31, 2022) and each of the 129 countries, with a metric that evaluates people's willingness to stay at home (**Increase in Residential Stay**). This metric measures the increase in the percentage of people who prefer to stay at home during the COVID-19 pandemic, compared to before the pandemic [10].

Final Dataset Acquisition: To merge the five datasets, the implemented architecture makes use of the Pandas - Python library. This library was chosen because it is appropriate for heterogeneous datasets made up of tables containing various types of data, allowing for better use of these objects. Each dataset is organized by country and registration date; these columns were chosen to perform the merge operation across all datasets. To be consistent with the other datasets, each 3-digit country code in the first dataset had to be converted into a 2-digit country code. The PyCountry library was used for this purpose. Furthermore, in order to maintain consistency across all datasets and perform the merge operation, the dates had to be standardized to yyyy-mm-dd format. Finally, in order to correctly merge all of the datasets into a single final dataset, it was necessary to ensure that each data column used in the merge operation had the same names.

3.2 Storage

This architecture component is in charge of storing the data obtained at the end of the previously described phase. The original file was stored with missing values and was not handled in case the original dataset was required without the synthetic data. Furthermore, by storing this file in its raw form, ghost data, generated in future phases, can be easily identified, The PyMongo [11] library was used to accomplish this, keeping the pipeline consistent with the use of Python and the integration of its phases. This library allows MongoDB and Python to communicate. Furthermore, MongoDB is a well-known tool for securely and flexibly storing JSON documents. This database includes a query-language that enables you to extract information from the original document in an quick and efficient way.

3.3 ETL Process

The third component of the *pipeline* is the **ETL** process. In first place, the data stored in MongoDB (*Extract*) is obtained. Then the data suffers transformations, using PySpark (*Transform*). Finally, the resulting data is stored in an excel file *xlsx* (*Load*), to later be transferred to a data visualization platform. For the transformation in the ETL process, it was used the Python library, PySpark [12]. This library offers users an Apache Spark API that applies the methods and features of this tool in a Python environment, thus maintaining consistency with the Python libraries that were also used in the previous phases. This tool was chosen because is effective in the implementation of complex and iterative methods. The data processing part was divided in three big tasks, described below:

Renaming Countries with Unrecognized Characters. In order to facilitate the visualization of the results, the name of the countries that had unrecognized characters were changed. In this case, only the country Ivory Coast (in English) had this problem, and its name was changed from "Côte d'Ivoire" to "Cote de Ivoire".

Removing Redundant Columns. After a thoughtful analysis, it was concluded that in the *dataset* there were some redundant columns that had no relevance for the use case, and so they could be removed.

Handling Null Values. To handle the null values of the dataset, it was necessary to analyze different cases and handle them accordingly, as explained in the following sections.

Registers Before the Beginning of Vaccination. To deal with the null values, each country's records prior to the start of vaccination were examined. COVID-19 vaccination was made available to the public in late 2020 and early 2021. (depending on the country). As a result, records prior to 2021 for vaccination fields in the merged dataset had null values that should be replaced with zero. As a result, there was a need for each country to check the date of the first vaccination record and set these vaccination fields to zero for earlier dates.

Removal of Countries with a Great Amount of Null Values. Following the completion of the previous step, it is possible to identify the countries that do not have enough values to perform a cohesive generation of synthetic values. These countries have a significant number of null values, which would not lead to a confidence analysis or generate synthetic values after the correct distribution of values.

Countries that do not reliably ensure data consistency, defined as having at least one column with at least 60% of null values, were eliminated. This elimination criterion was chosen because the remaining countries in the data set have at least 50% of non-null values, allowing for more accurate synthetic value generation.

Generating Synthetic Values. Finally, synthetic values were generated for the fields that remained null after the previous steps. This transformation is split into two parts. Initially, the fields were filled with the weekly average. The values were then filled in for the null values that are encompassed in a week with no value filled in using the previous and/or following weeks in that country.

The *toPandas* method was used to pass the *Spark dataframe* to the *Pandas dataframe*. The *transform* method was also applied to the previous method's result to obtain the doctor of the values.

This approach made it possible to obtain realistic values, within the same distribution as the next and/or previous week's values.

3.4 Data Visualization

Data visualization is the last step of our data pipeline, and it includes graphical representation of data and dashboard creation in order to analyze the COVID-19 pandemic influence in the stay at home desire.

The tool chosen to perform data visualization was PowerBI, which allows the creation of representative graphics well-fitted to the problem [13]. This tool was chosen because it is intuitive and easy to use and as the data that will be used for visualization is not obtained in real-time, there is no need to use more complex tools.

After storing the processed and treated EXCEL file, it will be imported into PowerBi where it will be possible to create a graphical representation of the use case.

4   Results and Discussion

Some dashboards have been created in order to have an intuitive data interpretation. A temporal analysis was performed for each country and WHO region to assess the impact of the number of COVID-19 deaths and infections on public gatherings and stay-at-home restrictions. The effect of the vaccination process on the preference to residential stay was also investigated. As a result, it was possible to analyze and compare the countries with unusual and exceptional values in order to obtain a more general and comprehensive result. At the worldwide level, the number of new cases and deaths per country is presented in Fig. 2. This panel allows a simplified view of pandemic impact in the world and how it relates to the number of cases and deaths, and not necessarily how these metrics change with time.

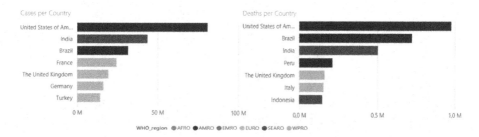

Fig. 2. COVID-19 new cases and deaths per country. Adapted from [14].

In general, it can be seen that there is a balance between the number of new cases and deaths, as countries with more cases also have more deaths. Therefore, the presented work is focused on the analysis of the three countries with the highest number of new cases and deaths: **United States of America, India and Brazil.**

4.1   Use Case 1: United States of America

Firstly, through the Fig. 3 it can be seen that, the USA is negatively decadent as it is the country with the highest number of new cases and, consequently, deaths. It can also be seen that the total number of cases during this analysis was 81 million and about 1 million deaths, i.e. about 1% of the total number of COVID-19 cases. Furthermore, there is a strong correlation between the restrictions implemented and the population's desire to stay at home. As it can be observed, in the first line graphic: "Residential Stay Per Day", the desire to stay at home peaked at the beginning of the pandemic, which means that people preferred to stay at home during this period. Furthermore, by looking at the restriction graphs, it can be concluded that they all followed the same trend in the first half of 2020. As time went on, the restrictions changed, which affected people's desire to stay at home. Lastly, it can be seen that the restrictions on level 3 stays were never implemented, which could possibly lead to more infections and consequently more deaths among the population. Moreover, Fig. 3, also shows that USA is not one of the top countries when it comes to vaccination rate, with only about 76% of people vaccinated during the analysis period. Despite being a significant number, many people were vaccinated throughout the year 2021, so that peak was only reached in 2022. This may contribute to the number of cases and deaths seen above. In addition, the graph shows that vaccination in the US started in January 2021 which immediately contributed to the easing of stay-at-home restrictions. However, restrictions on public gatherings varied throughout the year reflected in periods of more and fewer restrictions.

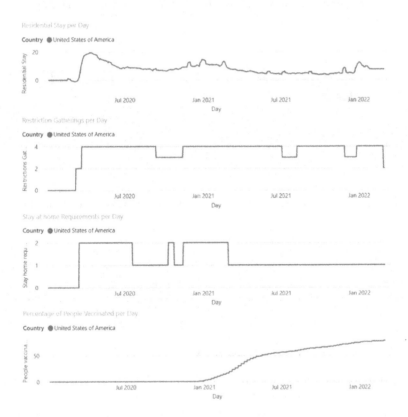

Fig. 3. Evolution of the level of contingency measures over time in USA. Adapted from [14].

4.2   Use Case 2: India

Turning to **India**, it can be seen that the total number of new cases reached 44 million, and the number of deaths was about 500,000 (1% of cases). In Fig. 4, it can be seen that India experienced a strong variation in restrictions over time. It can also be seen that at the beginning of the imposition of pandemic restrictions, people's desire to stay home grew proportionally with the increase in restrictions. However, the restrictions on staying at home increased slowly, reaching and maintaining their peak for a short period of time in 2020, which may have resulted in more infections among people. When analyzing a country like India, it is important to keep in mind that the population is incredibly high and how this may affect the results. Consequently, this statement may be why India has the most recent cases and deaths.

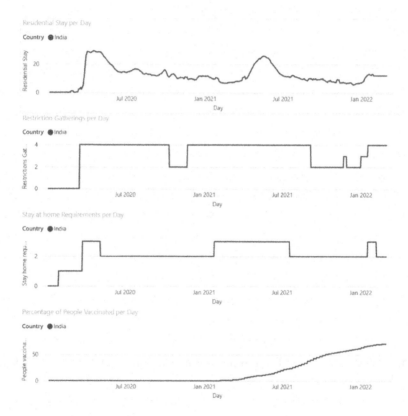

Fig. 4. Evolution of the level of contingency measures over time in India. Adapted from [14].

The last line graphic in Fig. 4 represents the number of people vaccinated, around 69% in India. This is also not very significant considering the number of COVID-19 cases. In this country, vaccination began in January 2021, but the number of vaccinated people grew slowly compared to the USA. This delay and the percentage of people vaccinated in one year can indicate cases and deaths in this country. When vaccinations peaked, restrictions on public gatherings and staying home were eased, and people were less willing to stay home.

4.3   Use Case 3: Brazil

Lastly, **Brazil**, which reached 32 million cases and about 700 thousand deaths, corresponding to 2% of COVID-19 cases. It can be observed in Fig. 5 that restrictions implemented had a strong increase at the beginning of the pandemic.

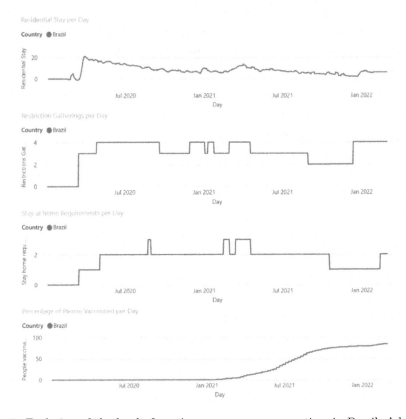

Fig. 5. Evolution of the level of contingency measures over time in Brazil. Adapted from [14].

As time went by, confinement rules varied and only during a short period of time maintained at their peak, which may have caused such a large number of new cases and deaths.

As it concerns of vaccination in Brazil also started in January 2021, and reached 82% of people vaccinated during the analysis period. In this country, there is a gradual growth in the number of people vaccinated, which is also reflected in the easing of restrictions on public gatherings and stay-at-home. This decrease after the peak also led to a decrease in people's willingness to stay at home. Thus, there is a strong correlation and mutual influence between these metrics.

5 Conclusions

The above analysis reflects differences in access to healthcare and the credibility of each country's population towards the vaccine emergency. The three countries chosen showed equivalent trends in the peak of preference to stay home in the face of increased restrictions, followed by their slowing. This easing of restrictions was proportional to the increase in the number of people vaccinated over time. The potential US world was able to reach its vaccination peak earlier than India and Brazil, as expected. As for the preference for staying at home, while this decreased over time for the US and Brazil, India did not maintain this trend.

Through this analysis we can say that, for futures pandemics, governments should not only apply a higher level of stay-at-home restrictions but also apply them earlier when the population does not have access to vaccines. Applying these restrictions immediately could be an intricate process due to several limiting socioeconomic factors, however by doing so countries could lower the number of new cases and deaths specially in the begin of the pandemic, when the level of contamination is the highest. Moreover, as shown in most countries in the COVID-19 pandemic, we can recommend, for futures pandemics, governments to maintain a higher level of restriction gatherings and thus helping lowering the number of cases and deaths. Finally, both restrictions have a direct influence on the desire to stay at home by making the population feel safer at home during the increase of deaths and new cases.

The architecture designed and implemented allowed us to develop a business intelligence tool on one of the most important topics today, also affirming the need for increasingly open data science and data sharing with the community for common purposes such as a pandemic.

Acknowledgements. This work has been supported by FCT-Fundação para a Ciência e Tecnologia within the R&D Units Project Scope: UIDB/00319/2020.

References

1. Li, L., Long, D., Rouhi Rad, M., Sloggy, M.R.: Stay-at-home orders and the willingness to stay home during the COVID-19 pandemic: a stated-preference discrete choice experiment. PLoS ONE **16**(7), e0253910 (2021)
2. Sumaedi, S., et al.: Factors influencing intention to follow the "stay at home" policy during the COVID-19 pandemic. Int. J. Health Gov. (2020)
3. Soomro, B.A., Shah, N.: Examining the intention to stay home due to COVID-19: a pandemic's second wave outlook. Health Educ. **121**(4), 420–435 (2021)
4. WHO Coronavirus (COVID-19). http://covid19.who.int/data. Accessed 10 May 2022
5. Our World in Data: COVID-19 Dataset by Our World in Data, 16 January 2021. http://github.com/owid/COVID-19-data/blob/master/public/data/vaccinations/vaccinations.json. Accessed 10 May 2022
6. Lakhan, R., Agrawal, A., Sharma, M.: Prevalence of depression, anxiety, and stress during COVID-19 pandemic. J. Neurosci. Rural Pract. **11**(04), 519–525 (2020)

7. Coughenour, C., Gakh, M., Pharr, J.R., Bungum, T., Jalene, S.: Changes in depression and physical activity among college students on a diverse campus after a COVID-19 stay-at-home order. J. Community Health **46**(4), 758–766 (2021). https://doi.org/10.1007/s10900-020-00918-5

8. Our World in Data: Restrictions on public gatherings in the COVID-19 pandemic, 1 January 2020. http://ourworldindata.org/grapher/public-gathering-rules-COVID. Accessed 8 May 2022

9. Our World in Data: Stay-at-home requirements during the COVID-19 pandemic, 1 January 2020. http://ourworldindata.org/grapher/stay-at-home-COVID. Accessed 8 May 2022

10. Kumar, A.: Worldwide Residential Mobility in COVID-19. Kaggle, 2 February 2022. www.kaggle.com/datasets/aestheteaman01/people-staying-in-home-during-COVID19. Accessed 12 May 2022

11. MongoDB: PyMongo 4.1.1 documentation. Pymongo Documentation (2008). http://pymongo.readthedocs.io/en/stable/tutorial.html. Accessed 2022

12. Apache Spark: PySpark Documentation. PySpark Documentation (2022). http://spark.apache.org/docs/latest/api/python/index.html. Accessed 2022

13. Microsoft: Tutorial: Introdução ao serviço de criação no Power BI. Microsoft Documentation, 1 July 2022. http://docs.microsoft.com/pt-pt/power-bi/fundamentals/service-get-started. Accessed 16 May 2022

14. Filipa, P., Luis, P., Luisa, C., Rita, P.: Impacto da Pandemia COVID-19 na Preferência Stay at home. Unpublished Internal Document, University of Minho (2022)

Distance-Based Delays in Echo State Networks

Stefan Iacob[(✉)] [iD], Matthias Freiberger[iD], and Joni Dambre[iD]

IDLab-AIRO, Ghent University-imec, 9052 Ghent, Belgium
{stefanteodor.iacob,matthias.freiberger,joni.dambre}@ugent.be

Abstract. Physical reservoir computing, a paradigm bearing the promise of energy-efficient high-performance computing, has raised much attention in recent years. We argue though, that the effect of signal propagation delay on reservoir task performance, one of the most central aspects of physical reservoirs, is still insufficiently understood in a more general learning context. Such physically imposed delay has been found to play a crucial role in some specific physical realizations, such as integrated photonic reservoirs. While delays at the readout layer and input of Echo State Networks (ESNs) have been successfully exploited before to improve performance, to our knowledge this feature has not been studied in a more general setting. We introduce inter-node delays, based on physical distances, into ESNs as model systems for physical reservoir computing. We propose a novel ESN design that includes variable signal delays along the connections between neurons, comparable to varying axon lengths in biological neural networks or varying length delay lines in physical systems. We study the impact of the resulting variable inter-node delays in this setup in comparison with conventional ESNs and find that incorporating variable delays significantly improves reservoir performance on the NARMA-10 benchmark task.

Keywords: Echo state networks · Bio-inspired computing · Evolutionary algorithms · Variable delays

1 Introduction

Echo State Networks (ESNs) [12] offer a promising low-energy alternative for error backpropagation that has gained attention in recent years. Multiple layers of reservoirs can be combined in deep ESNs and have proven successful at several practical applications [6,7,18]. Due to ESNs essentially being recurrent neural networks (RNNs) with fixed-weight input and hidden layer, they can be approximated in a time-continuous way by many physical systems [20] which is commonly referred to as physical reservoir computing. In physical reservoirs,

This project has received funding from the European Union's Horizon 2020 research and innovation programme under the Marie Skłodowska-Curie grant agreement No 860949.

© The Author(s), under exclusive license to Springer Nature Switzerland AG 2022
H. Yin et al. (Eds.): IDEAL 2022, LNCS 13756, pp. 211–222, 2022.
https://doi.org/10.1007/978-3-031-21753-1_21

delays between nodes tend to vary [5] due to, amongst other reasons, design constraints and imperfections in the manufacturing process. We argue that these variations in inter-node delays, which are often imposed by the spatial layout of the reservoir, i.e. the physical distance of reservoir nodes between each other, are not merely to be tolerated as they are intrinsic to the underlying physical processes, but can be embraced as beneficial to reservoir performance.

Similarly, despite artificial neural networks being in general far removed from biological systems, one can also draw inspiration from a simple and often overlooked aspect of biological networks: neurons have a spatial location, which means that there is physical distance between neurons. Animal axon lengths vary in the order of millimeters, and signal delays can vary in terms of milliseconds [3]. This variation in delay has been shown to have a qualitative effect on plasticity [19]. In contrast, rate-based ANNs and ESNs in particular have no such feature, meaning that all inputs to a neuron are processed simultaneously, disregarding possible variation in timing.

ESNs and physical reservoirs bear promise as a low-energy solution for time-series tasks. Therefore, the understanding of temporal processing in reservoirs is essential for improving task performance. The use of strong non-linearities in information-processing dynamical systems decreases the memory capacity of the system, whereas less non-linearity limits computational power. This observation is commonly referred to as the memory-nonlinearity-tradeoff [4]. We argue that the use of variable-length connection delays allows for a simple way to introduce linear memory at various timescales in the reservoir, mitigating the issue of too rapidily fading memory in highly nonlinear systems. In more concrete terms, delay lines of different lengths meeting at a single node allow that node to combine information from two different time points without any decay in memory.

The novel contribution presented in this work is a systematic exploration of the impact of varying propagation delays on task performance in a substrate-agnostic setting. We use inter-node distance-dependent delays in ESNs, with nodes modeled as points in physical space, and distances computed as a euclidean norm of the difference of their coordinates. This begs the question of how to optimize their spatial locations. We present a novel, spatially represented ESN implementation which we refer to as distance-based delay network (DDN), where neuron locations are sampled from a Gaussian mixture distribution. We optimize the spatial structure of this network by tuning the parameters of the location distribution with a genetic algorithm. We show a strong and significant improvement in performance on the NARMA-10 benchmark task compared to conventional ESNs. Our baseline performance is in accordance with other conventional ESN implementations, whereas our best models perform better than some deep ESNs. However, showing an absolute improvement of the state of the art for time series prediction is beyond the scope of this project. We simply show in a well controlled experimental setting that variations in connection delay are exploitable by optimizing neuron location distributions in a physical space, all else being equal.

Although the use of time delays in rate-based neural networks is limited, timing is inherently considered to some extent in spiking neural network implementations due to the temporal nature of spike encoding. Jeanson et al. explored the use of simulating axonal delays for robot control [16]. However, the added computational complexity of SNNs is an added challenge for practical applications. Therefore we study the effect of delays in a more general sense. The use of delay lines in ESNs has been previously explored in for both readout connections [11] and input connections [15]. The former work introduced learnable readout delays, which was shown to improve network performance. The latter work showed an improvement in performance of untuned reservoirs by using input delays To our knowledge, no previous work explores the use of interneuron signal delays in a rate-based echo state network by freely varying neuron positions.

2 Methods

In this section we describe our approach to the implementation of an ESN that includes varying signal delays in its connections (both input and recurrent). This is opposed to conventional ESNs, where the network activation at time t is instantaneously processed to produce the recurrent input at time $t + 1$. These conventional ESNs can be represented by the following equation [13]

$$x(n + 1) = (1 - a)x(n) + a \cdot f(\mathbf{W}_{\mathrm{res}}x(n) + \mathbf{b}_{\mathrm{res}} + \mathbf{W}_{\mathrm{in}}v(n)) \tag{1}$$

where $\mathbf{x}(n)$, a, $\mathbf{W}_{res}$, $\mathbf{W}_{in}$, $\mathbf{b}$, f, and $\mathbf{v}(n)$ refer to the network state at timestep n, the decay rate, the (recurrent) reservoir weight matrix, the input weight matrix, the bias weights of the reservoir, the activation function, and the input at timestep n respectively. This equation needs to be augmented for the proposed distance based delay networks (DDNs), as we do not directly take the current network state $\mathbf{x}$ as input for our activation function, but rather the sum of incoming delayed input signals to each neuron.

With DDNs, we propose an ESN whose delays are dependent on neuron locations. Any signal travelling between two neurons is delayed by an amount of time steps proportional to the physical distance between them. Given a set of 2D neuron coordinates, we can define a N by N Euclidean distance matrix $\mathbf{D}$, with N indicating the number of neurons and element $D_{i,j}$ corresponding to the distance between neuron i and neuron j. Since the simulation approach that we use requires a finite number of distance values, the elements of $\mathbf{D}$ are discretized over the set of $\{1, 2, .., D_{\mathrm{max}}\}$, where D_{max} corresponds to the maximum number of possible delay steps. This means that the longest connection implements a delay of D_{max} simulation steps, and the shortest connections applied only one simulation step delay. We refer to $\mathbf{D}$ as a delay matrix. In order to extend $\mathbf{W}_{\mathrm{res}}$ from Eq. 1 to incorporate varying delays, instead of a single weight matrix, we define a set of "masked" weight matrices $\mathbf{W}_{D=d}$ where for each element

$$W_{i,j,D=d} = \delta_{d,D_{i,j}} \cdot W_{i,j} \tag{2}$$

where $d \in [1, D_{max}]$ and δ is the Kronecker delta operator. As such, the weights in $\mathbf{W}_{D=d}$ corresponding to connections with a delay different from d are set to 0. Using the same notational conventions as in Eq. 1, we can formalize the state update mechanism of a DDN as

$$\mathbf{x}(n+1) = (1-a)\mathbf{x}(n) + a\mathbf{y}(n) \tag{3}$$

$$\mathbf{y}(n) = f\left(\sum_{d=0}^{D_{max}} \left(\mathbf{W}_{D=d}^{res}\mathbf{x}(n-d) + \mathbf{W}_{D=d}^{in}\mathbf{v}(n-d)\right) + \mathbf{b}_{res}\right) \tag{4}$$

In concrete terms, this means that, in order to compute the current reservoir activation, we consider the historical activation from up to D_{max} earlier timesteps. We multiply the activation of each of these previous timesteps with their corresponding masked weight matrix, such that we obtain the corresponding delayed neuron input.

Notably, we are now left with the challenge of selecting optimal neuron coordinates. The coordinates could be treated as hyperparameters to be optimized. However, this is inconvenient due to the large amount of neuron coordinates ($2N$). Moreover, for many physical implementations, it is likely unrealistic that location can be assigned precisely. Therefore, we sample the neuron locations from a 2D Gaussian Mixture Model (GMM) distribution, with a fixed number of Gaussians, described by

$$p(\theta) = \sum_{i=1}^{K} \phi_i \mathcal{N}(\mu_i, \Sigma_i) \tag{5}$$

These Gaussians represent K clusters of neurons. Instead of optimizing each neuron location, we optimize the parameters of this distribution. These consist of the mixture parameters ϕ, means μ, correlations, and variances (each existing for the x and y coordinates), which are used to compute the covariance matrices Σ. As such, with K being the number of Gaussians in our GMM and N the number of neurons in our reservoir, we have a drastic reduction from $2N$ location parameters to K values for the mixture parameters, $2K$ values for the means, K values for the correlations and $2K$ values for the variances, resulting in $6K$ parameters. Note that in ESNs, N is usually in the order of hundreds, whereas in our experiments we use $K \leq 4$.

Additionally, network architecture is dependent on reservoir connectivity, which is the fraction of non-zero weights. Using the cluster representation of neurons, connectivity can be defined within and between each cluster rather than a single connectivity parameter. We define a $K+1$ by $K+1$ connectivity matrix. Each element (i, j) in this matrix indicates the fraction of non-zero weights (with respect to full connectivity) from cluster i to cluster j (the last row and column correspond to the input neuron). As such, the diagonal elements indicate the recurrent connectivity of each cluster. We performed experiments with single-cluster, as well as four-cluster networks, in both DDNs and standard ESNs (i.e. $K = 1$ and $K = 4$). Note that even though standard ESNs do not take into account neuron locations, the presence of multiple clusters still influences the size of the connectivity matrix. This allows us to study the effect of optimizing a multi-cluster connectivity matrix and optimizing neuron locations as two independent factors.

Additionally, like other ESN implementations, optimization of several other hyperparameters is necessary. Specifically, we optimize input weight scaling, reservoir weight scaling, bias scaling, the decay parameter, the fraction of inhibitory neurons, next to the previously mentioned between- and within-cluster connectivity, and location distribution parameters. The scaling parameters are scalars ranging between 0 and 1 that are multiplied with the corresponding subset of weights. The decay parameter, referred to in Eq. 3 as a, also ranges between 0 and 1.

We make use of a hyperbolic tangent activation function. The values of W_{res}, W_{in} and bias weights $\mathbf{b}_{res}$ are uniformly sampled between -0.5 and 0.5. We use 300 neurons for our reservoirs. For our readout layer we use ridge regression [10] with 5-fold cross-validation.

To optimize the hyperparameters we make use of the Covariance Matrix Adaptation Evolution Strategy (CMA-ES) [8,9]. At each generation, a population of 20 parameter sets (i.e., individuals) is generated. From each parameter set, five networks are generated and evaluated. A fitness score is determined for each parameter set based on the average performance of these five networks, which is fed back to the evolutionary algorithm. In turn, a new population of parameters is returned based on the best performing individual. We run CMA-ES for 99 generations.

In the remainder of this paper we refer to the different parameter sets in a population from one generation as different *candidates*. We refer to the generation of networks based on a set of parameters as *sampling networks from a candidate*, due to the fact that most network aspects are random.

We optimize the hyperparameters of both standard ESNs and DDNs with CMA-ES in order to compare performance. In case of a standard ESN with one cluster (our baseline models), location parameters are not used and the connectivity parameter is just one scalar defining the percentage of non-zero connections in the whole network.

To validate the benefit of variable delays in ESNs, we use a 10th-order Non-linear Auto-Regressive Moving Average (NARMA) task [2]. This system is described by the following equation.

$$y(t+1) = 0.3y(t) + 0.05y(t) \sum_{i=0}^{9} y(t-i) + 1.5u(t-9)u(t) + 0.1 \qquad (6)$$

Here $y(t)$ and $u(t)$ are respectively the output and input at time t. We generate a training sequence of 8000 samples, a validation sequence of 4000 samples and a test sequence of 10000 samples. We only use the test sequence after evolution, to evaluate our best parameter settings.

NARMA-10 is a commonly used benchmark for ESNs, hence validating our model on this task allows us to understand how DDNs compare to state-of-the-art

ESNs. We generate our training, validation and test data with an input sequence that is uniformly distributed between 0 and 0.5 as input. These inputs are sequentially fed to the NARMA-10 system to generate the task labels. To train the model, the same input sequence is used as input for the reservoir. A linear regression is performed with the reservoir activity as independent variables and the task labels as dependent variables. However, the first 400 samples are discarded before regression, to account for initialization of the reservoir (see [14]).

In the case of conventional ESNs, the input at time t corresponds with the network activity at time t and the label at time t. However, this only makes sense if there are no connection delays. Due to the introduction of delays in the proposed DDN reservoir, it takes an unknown amount of additional simulation steps until the information that is relevant for a particular label is sufficiently present in the reservoir activity. Therefore, it is necessary to shift the input values in time with relation to the labels. For this purpose, we introduce a new lag parameter l (indicating the lag of relevant information), which controls how many time steps the labels are shifted. We optimize l by doing a grid search for each network evaluation, i.e., instead of evaluating a network once, we evaluate it multiple times using different positive integer values for l and pick the one that performs best. We measure performance using normalized root mean squared error (NRMSE) [5], considering the lag parameter by pairing every network activation vector $\mathbf{x}(n)$ with label $y(n - l)$.

3   Results

The purpose of our experiments is to establish if the introduction of variable connection delays improves ESN performance. Therefore, we compare the test performance on a fixed dataset of a baseline network (i.e. a network without delays), and a variable delay DDN. Both are tuned using CMA-ES as described in Sect. 2. Our variable delay model has delays that can range between 1 and 20 simulation steps. The actual range of delays will depend on the sampled neuron positions, which in turn depend on location distribution parameters selected by the evolutionary strategy.

We present the results of tuning single cluster and four-cluster networks based on the baseline ESNs and variable delay DDNs with CMA-ES in Fig. 1a and Fig. 1b respectively. We show the average achieved validation score for the best found set of hyperparameters in a generation as well as the average validation scores achieved by the whole population for 99 generations. The validation performance of a single set of hyperparameters was obtained by training 5 networks previously generated using the set in question. Subsequently, we have evaluated the performance of each network on the validation set, and computed the average of all resulting errors.

We can see that in both experiments, the variable delay models achieve a lower validation NRMSE than the baselines. This difference is especially pronounced in the single cluster experiment. However note that the best candidates

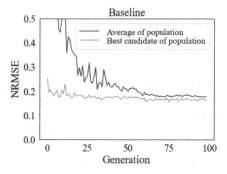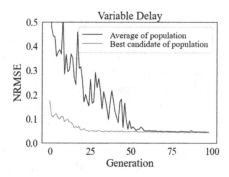

(a) Validation performance per generation for single-cluster experiment. We notice that both conditions in this experiment converged to a stable NRMSE. The best performance and average performance are close, and the variance within a population is low (not reported here). We see that the average baseline and variable delay validation NRMSE converge around 0.18 and 0.05 respectively.

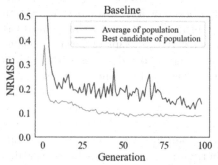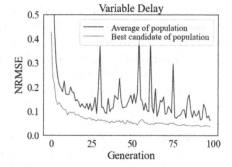

(b) Validation performance per generation for four-cluster experiment. Although it appears that the best baseline model has converged to a stable NRMSE, we see that the average NRMSE of both conditions have not yet stabilized. Moreover, in the variable delay condition, both average and best performances are still improving during the final generations. **The average validation scores over the last 10 generations for the** baseline and variable delay models are 0.14 and 0.09.

Fig. 1. NRMSE on validation set throughout the CMA-ES optimization. We show the average performance per generation averaged over all candidate solutions, as well as the average performance of the best candidate within each generation. The best performing candidate solution of the single-cluster baseline networks achieves an average NRMSE of 0.1579 in generation 81 when evaluated on the NARMA-10 validation set. For the best performing single-cluster variable delay network, we get an average NRMSE of 0.04234 at generation 86. Analogously, the best four-cluster baseline candidate has a validation NRMSE of 0.08319 in generation 53 and the best four-cluster variable delay candidate a validation NRMSE of 0.03696 in generation 98. All four network types consist of 300 reservoir units and all four reported validation scores were averages of 5 networks sampled from the candidate.

of the variable delay model in the four-cluster experiment still show signs of improvement during the final generations. In both experiments we see a larger drop in NRMSE for the variable delay models. Furthermore, in the single cluster experiment we observe that good scores for the best candidates are obtained after few generations, with already adequate best performances at the start of evolution. Obtaining a good average performance takes additional optimization.

To fairly compare the variable delay models with the baselines, we use the best baseline and variable delay parameter sets to randomly generate 40 networks for both conditions. We train these networks using the same NARMA-10 training set as used during evolution, and evaluate them on a test set. We show the results in performance in Table 1. These test scores are in line with our previous observations, as both DDNs perform consistently better than their respective baselines, but also similarly or better compared what is reported in several recent novel ESN implementations [1,5,17].

Note that in our single cluster experiment, sampling neuron location from a GMM with $K = 1$ means that we are in fact sampling from a simple Gaussian distribution, so no extra constraints are added to the connectivity of the weight matrix. As such, we can isolate the use of variable delays as the only alteration compared to conventional ESNs. We can interpret the improvement seen in the $K = 1$ DNNs compared to baseline ESNs as solely the result of variable delays. Although our single-cluster networks achieve the reported scores with random connectivity, in our four-cluster experiment, the within- and between-cluster connectivity is specified and optimized. As such networks are free to evolve into more specific architectures, including deep architectures. Hence, we cannot isolate variable delays as the only cause of improvement. However, we can see that the four-cluster baseline performs better than the single-cluster baseline, but worse than the single cluster DDN. Furthermore, the best four-cluster variable delay candidate achieved the best NRMSE that we report.

Table 1. Average NRMSE on unseen test set of 40 networks sampled from the best candidates from the evolution. Bold font indicates best found approach.

Type	K	NRMSE (test)
Baseline	1	0.1588 ± 0.0124
DDN	1	0.0639 ± 0.0018
Baseline	4	0.0848 ± 0.0056
DDN	**4**	$\mathbf{0.0391 \pm 0.0025}$

As discussed in Sect. 2, inter-neuron delays cause an implicit shift in time between the network activation (which can be seen as features for the linear readout) and the task labels. The exact number of simulation steps this shift amounts to can not be determined beforehand, as it is unclear after how many steps the relevant information is present in the reservoir. Therefore, we introduced the lag parameter. The regression is performed repeatedly, with a lag

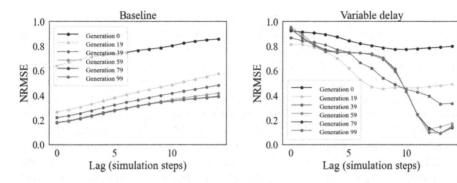

Fig. 2. Best network's average NRMSE on NARMA-10 validation data using different lag parameters, measured for different generations. The left and right plots show performances for baseline candidates and variable delay candidates respectively. The shape of the baseline lag profile remains the same throughout evolution, and only changes in terms of absolute NRMSE. Conversely, the lag profiles of the DDNs do not only change in absolute terms, but the optimal lags are shifted further into the future as evolution progresses.

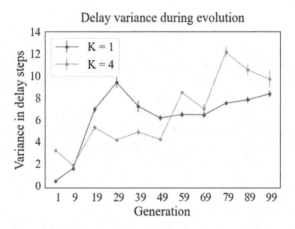

Fig. 3. DDN delay variance in terms of simulation steps throughout evolution. 10 networks are sampled form the best candidate every 10th generation. An average is taken of the variance competed for each of these sample networks.

parameter ranging from 0 to 15 for both the baseline as well as our proposed DDN networks. We can use the performance of DDNs with different lag parameters to gain insight in the effect of the delays. In Fig. 2 we show the effect of the lag parameter with respect to performance on the NARMA-10 task. Shown NRMSE validation scores are averaged across an entire generation as a function of lag for various generations throughout the CMA-ES evolution process. The evolution of this lag profile gives insight in the temporal range where the relevant information is located. Note that Fig. 2 only shows the performance for the conducted single-cluster experiments. Four-cluster lag profiles were qualitatively similar and differed only in terms of absolute NRMSE or absolute lags. In the variable delay case we see that the lag parameter resulting in the lowest NRMSE grows throughout the evolution. This means that processing of task inputs tends to take longer. It is to be expected that baseline networks consistently find an optimum using a lag parameter of 0. Any other value would have suggested that baseline ESNs are superior at predicting future NARMA-10 states without any knowledge of the corresponding input, compared to predicting simply the next value based on known input. On the other hand, Fig. 2 begs the question as to why DDNs always find a non-zero optimal lag, and moreover why the lag parameter grows during evolution. We propose that non-zero delays allow for more variability in delay, and as such we interpret the consistency of non-zero optimal lag throughout evolution as additional support for our hypothesis, namely, that variation in propagation delays can be exploited in echo state networks. To further validate this claim, we plot the variance in delay line length among 10 DDNs generated using the best set of hyperparameters in a single CMA-ES generation in Fig. 3. As the reader can see, the variance of delays indeed grows as the number of generations in the CMA-ES optimisation process increases.

4   Conclusion

We proposed an extension of echo state networks (ESNs), wherein neurons have a stochastically varying physical location. Proximity between neurons guides the variable propagation delay applied to neuron activation, similar to how variations in axon length and propagation speed causes variable delays in biological brains. Similarly, propagation delay is inherent in all physical systems. We hypothesized that this could be exploited to improve the processing of timeseries. We have shown that the resulting distance delay networks (DDNs) proposed in such a way reach much lower validation and test errors on the NARMA-10 task than baseline ESNs. When testing 40 randomly sampled networks based on the best selected parameters on an unseen test set, we observed that the DDNs performed significantly better. As such, we can conclude that the addition of variable delays in simulated rate-based reservoirs improves performance on the NARMA-10 task. Furthermore, inspired by approaches such as DeepESN [7] we have shown that using multiple clusters in a single DDN can be exploited to enhance performance with other means than just the use of delays. However, our single cluster experiments also achieve similar scores with fully random connectivity, hence isolating variable delays as the cause of improvement.

5   Future Work

We see that in our four-cluster experiment, the variable delay models are still improving in performance at the end of evolution, suggesting that there is more performance to be gained, therefore future exploration in this direction seems promising.

Yet, while our findings support our hypothesis, to confirm that this improvement generalises well, our DDNs need to be tested on additional tasks. An especially important question to answer in future work is whether DDNs offer an equally significant improvement in tasks based on natural data (as opposed to artificial tasks such as NARMA-10 task). Additionally, in order to provide better insight into the consistency in results of our methods, results of many evolution runs should be analyzed.

Finally, it should be mentioned that our variable delay implementation runs significantly slower than our baseline. However, as the main objective of this work is to show that variation in propagation speed can be exploited to our benefit, we leave the optimisation of our implementation in order to narrow this performance gap for future work. For newly proposed physical reservoirs, this additional performance cost can usually be avoided since natural substrates implement this varying delay implicitly.

References

1. Akiyama, T., Tanaka, G.: Computational efficiency of multi-step learning echo state networks for nonlinear time series prediction. IEEE Access **10**, 28535–28544 (2022). https://doi.org/10.1109/ACCESS.2022.3158755
2. Atiya, A.F., Parlos, A.G.: New results on recurrent network training: unifying the algorithms and accelerating convergence. IEEE Trans. Neural Netw. **11**(3), 697–709 (2000)
3. Caminiti, R., et al.: Diameter, length, speed, and conduction delay of callosal axons in macaque monkeys and humans: comparing data from histology and magnetic resonance imaging diffusion tractography. J. Neurosci. **33**(36), 14501–14511 (2013). https://doi.org/10.1523/JNEUROSCI.0761-13.2013. https://www.jneurosci.org/content/33/36/14501
4. Dambre, J., Verstraeten, D., Schrauwen, B., Massar, S.: Information processing capacity of dynamical systems. Sci. Rep. **2**(1), 1–7 (2012)
5. Freiberger, M., Bienstman, P., Dambre, J.: A training algorithm for networks of high-variability reservoirs. Sci. Rep. **10**(1), 1–11 (2020)
6. Gallicchio, C., Micheli, A., Pedrelli, L.: Deep echo state networks for diagnosis of Parkinson's disease. arXiv preprint arXiv:1802.06708 (2018)
7. Gallicchio, C., Micheli, A., Pedrelli, L.: Design of deep echo state networks. Neural Netw. **108**, 33–47 (2018). https://doi.org/10.1016/j.neunet.2018.08.002. https://www.sciencedirect.com/science/article/pii/S0893608018302223
8. Hansen, N.: The CMA evolution strategy: a comparing review. In: Lozano, J.A., Larrañaga, P., Inza, I., Bengoetxea, E. (eds.) Towards a New Evolutionary Computation. STUDFUZZ, vol. 192, pp. 75–102. Springer, Heidelberg (2006). https://doi.org/10.1007/3-540-32494-1_4
9. Hansen, N., Akimoto, Y., Baudis, P.: CMA-ES/pycma on Github. Zenodo (2019). https://doi.org/10.5281/zenodo.2559634
10. Hoerl, A.E., Kennard, R.W.: Ridge regression: biased estimation for nonorthogonal problems. Technometrics **12**(1), 55–67 (1970). https://doi.org/10.1080/00401706.1970.10488634. www.tandfonline.com/doi/abs/10.1080/00401706.1970.10488634

11. Holzmann, G., Hauser, H.: Echo state networks with filter neurons and a delay&sum readout. Neural Netw. **23**(2), 244–256 (2010). https://doi.org/10.1016/j.neunet.2009.07.004. www.sciencedirect.com/science/article/pii/S0893608009001580

12. Jaeger, H.: The "echo state" approach to analysing and training recurrent neural networks-with an erratum note. German National Research Center for Information Technology GMD Technical Report, Bonn, Germany, **148**(34), 13 (2001)

13. Jaeger, H.: Tutorial on training recurrent neural networks, covering BPPT, RTRL, EKF and the "echo state network" approach, vol. 5. GMD-Forschungszentrum Informationstechnik, Bonn (2002)

14. Jaeger, H.: Echo state network. Scholarpedia **2**(9), 2330 (2007)

15. Jaurigue, L., Robertson, E., Wolters, J., Lüdge, K.: Reservoir computing with delayed input for fast and easy optimisation. Entropy **23**(12) (2021). https://doi.org/10.3390/e23121560. www.mdpi.com/1099-4300/23/12/1560

16. Jeanson, F., White, A.: Evolving axonal delay neural networks for robot control. In: Proceedings of the 14th Annual Conference on Genetic and Evolutionary Computation, GECCO 2012, pp. 121–128. Association for Computing Machinery, New York (2012). https://doi.org/10.1145/2330163.2330181

17. Li, Z., Tanaka, G.: Deep echo state networks with multi-span features for nonlinear time series prediction. In: 2020 International Joint Conference on Neural Networks (IJCNN), pp. 1–9 (2020). https://doi.org/10.1109/IJCNN48605.2020.9207401

18. Long, J., Zhang, S., Li, C.: Evolving deep echo state networks for intelligent fault diagnosis. IEEE Trans. Ind. Inf. **16**(7), 4928–4937 (2020). https://doi.org/10.1109/TII.2019.2938884

19. Madadi Asl, M., Valizadeh, A., Tass, P.A.: Dendritic and axonal propagation delays determine emergent structures of neuronal networks with plastic synapses. Sci. Rep. **7**(1), 1–12 (2017)

20. Tanaka, G., et al.: Recent advances in physical reservoir computing: a review. Neural Netw. **115**, 100–123 (2019). https://doi.org/10.1016/j.neunet.2019.03.005. www.sciencedirect.com/science/article/pii/S0893608019300784

EduBot: A Proof-of-Concept for a High School Motivational Agent

Hugo Faria[1][iD], Maria Araújo Barbosa[1][iD], Bruno Veloso[1][iD],
Francisco S. Marcondes[1][iD], Celso Lima[2][iD], Dalila Durães[1(✉)][iD],
and Paulo Novais[1][iD]

[1] ALGORITMI Centre, University of Minho, Braga, Portugal
{a81283,pg42844,a78352}@alunos.uminho.pt,
{francisco.marcondes,dalila.duraes}@algoritmi.uminho.pt,
pjon@di.uminho.pt
[2] High School of Caldas das Taipas, Caldas das Taipas, Guimarães, Portugal
diretor@esct.pt

Abstract. Motivation appears to plays a role in student dropout. This paper uses a dataset with high school information (756 entries) for creating a model that relates year failures with the other features on the dataset. The resulting model, based on XGBoost, shows that the top-ranked features are related with motivation issues corroborating with the initial observation. In this sense, a digital assistant embedded with a motivation module may aid on improving motivation and on avoiding dropout. In other words, if the predictor detects year failure possibility it start to act on motivating the student. Considering this paper being direct to underage students, this paper stops on the proof-of-concept level considering the actual dialogues and live tests are expected to performed with the support of educational psychologists.

Keywords: Educational data mining · Motivational assistant · Predict at-risk students · Student's performance

1 Introduction

Educational Data Mining is a research field concerned with find patterns that can be used for improving the education experience. However, due to several difficulties, such a research in compulsory education is yet underdeveloped [1]. This paper is part of a broad research aiming to aid on dropout prevention; it focuses on analysing a high-school dataset, targeting the feature of year failure. In this sense, if there is a risk for a student failing a year, that information can be pushed to the director dashboard that can act for avoiding it.

In addition, it is also suggested that a digital assistant can be used for performing a first support line for the student that is always available and able to provide personalised conversation [14]. This paper also presents a digital assistant proof-of-concept that acts based on the results given by the dataset analysis

© The Author(s), under exclusive license to Springer Nature Switzerland AG 2022
H. Yin et al. (Eds.): IDEAL 2022, LNCS 13756, pp. 223–232, 2022.
https://doi.org/10.1007/978-3-031-21753-1_22

and prediction model. In short, the dataset analysis suggests that motivation plays a major role in year failure, therefore a motivational agent is built for start working if a risk alert is triggered.

As an ethical statement, all information gathered for the dataset have being consented by the student guardian and by the school. Also, the motivational agent development stopped at the proof-of-concept level by considering that further development would require the support of educational psychologists and no experiment was performed upon students. The proof-of-concept is developed willing to open that possibility for the psychologists [10].

The paper is divided into three main sections. First, the analysis of the dataset is presented. Then, the risk predictor is introduced. Lastly, the behaviour of the agent is shown. In the last section, as a final note, all the previous topics are summarised and the main conclusions are drawn, providing some directions for future work.

2   State of the Art

In the study [16], the authors believe that more specialized work is needed for Educational Data Mining to become more mature and suggest some improvements for this: *mining tools that are easier to use by educators or non-experts in data mining, standardization of methods and data*, and then *integration with the e-learning system*. Recent studies seem to support this idea and agree with the need to integrate the extracted data (from educational data mining) with e-learning systems [18] and others go further and suggest to focus on social media as it has already gained a great popularity among students and is suitable to engage students in collaborative learning [17].

Data mining algorithms have been used in educational systems, especially e-learning systems, as these systems are widely used [2–4]. In [20], the authors propose a novel rule-based classification method called S3PSO based on PSO to extract the features that can be used for predicting the students' final score. For the Moodle dataset, this classifier improves the accuracy by 9% compared to others such as SVM, KNN, Naïve Bayes, Neural Network and APSO[5,6]. Another problem that can be addressed in the field of educational data mining, withe good results, is the problem of allocating students to classes [19,21].

3   Dataset Presentation

The ESCT dataset consists of data collected from high school students of one school in Guimarães, Portugal, in April 2022. The dataset is composed by 756 entries and 22 features. The features are described on Table 1:

Before analysing the dataset, some inconsistencies were identified. The values in the columns mother's profession, father's profession and parish had some spelling errors, so researchers manually fixed them. This happened due to those questions being open answer and not multiple choice ones. As for the values on column 13, researchers found that 671 were empty. This occurred because the

Table 1. Feature description of the ESCT dataset

Index	Column	Description
1	Date of registry	Date
2	Student's age	Age
3	Student's Gender	Male; female; other
4	Parish of residence	Parish name
5	County of residence	County name
6	School year	$10°$; $11°$; $12°$
7	Previous school	School name
8	Father's education level	Basic education; secondary education; university education
9	Father's profession	Profession name
10	Mother's education level	Basic education; secondary education; university education
11	Mother's profession	Profession name
12	Did the student fail	Yes; no
13	Why did the student fail a year	Lack of support and accompaniment with the family; lack of interest towards the school; assiduity problems; family problems
14	Student's opinion about the course	The course requires a lot of study hours; the course requires spending many hours in the classroom; most subjects are theorical; most subjects are practical; most subjects allow the developmentof my skills and capacities
15	Who influenced or motivated the course choice	Family; friends; professors; vocational guidance service; example of a success story; personal choice
16	Student's satisfaction with the course	Very satisfied; satisfied; not satisfied
17	Did the course meet student's expectations	Totally; partially; none
18	Is the content in line with the course and professional output	Yes; no
19	Does the course contribute to the student's personal and cultural enrichment	Yes; no
20	Is the hour distribution balanced	Yes; no
21	What the student thinks the conclusion of a training and qualification course will provide	Faster entry into the job market; professional certification; pratical learning; access to higher education for further studies; better future pay; social recognition
22	What the student thinks the conclusion of a scientific-humanistic course will provide	Faster entry into the job market; professional certification; pratical learning; access to higher education for further studies; better future pay; social recognition

questionnaire didn't have an option for those who didn't repeat a year. In this case, the empty values were replaced by the word "None". The columns with index 4, 21 and 22 could have more than one answer in a single entry. In this case, each column was separated in multiple ones, equivalent to the number of options.

4   An Active Motivational Digital Assistant

This module aims to design a prototype of an active motivational virtual assistant that interacts with the student upon receiving alerts from the predictive model. At a high level, the deployment view of the EduBot architecture is described on Fig. 1. In general, the architecture is composed of two main components. The first component is called Education Intelligence (with semantics similar to Business Intelligence), which will contain the predictive model. The second component is the Digital Assistant which, based on educational intelligence information, interacts with the student.

4.1   Education Data Mining

For exploring the ESCT dataset it was proposed three hypothesis for dropout risk factors: 1) the mother's education [13]; 2) the student's place of residence [1]; and 3) previous enrolled school.

Mother's Education as a Risk Factor
    The first hypothesis consisted in checking if a relation between the mother's education level and the type of student existed. Table 2 shows the summation of mother's education level for students that already failed an year and to those that had not. The correlation test using Chi-Squared resulted in $p = 0.0002$, meaning the variables are dependent.

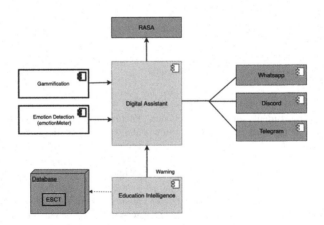

Fig. 1. EduBot high level architecture.

Table 2. Distribution of students mother's education level (%)

Type of students	Basic education	Secondary education	Higher education
Repeating students	72%	26%	2%
Non-repeating students	53%	31%	16%

Place of Residence as a Risk Factor
The second hypothesis consisted in checking if a relation between the parish of residence and the type of student existed. The Chi-Squared resulted in $p = 0.839$, suggesting that they are independent.

Previous School as a Risk Factor
The third hypothesis consisted in checking if a relation between the previous school and the type of student existed. A table with the summation of previous enrolled school for students that already failed an year and to those that had not is presented in Table 3. The Chi-Squared test resulted in $p = 0.089$ suggesting that these features are not dependant.

Table 3. Parish distribution of the students

Type of students	School A	School B	School C	School D
Repeating students	10.0%	17.2%	19.6%	10.3%
Non-repeating students	90.0%	82.8%	80.4%	89.7%

Considering the three hypothesis only the first is corroborated with the data. From a pragmatics perspective that result may suggest that some of the mothers with lower education level may not motive enough their children for education. In this sense a motivational agent would be worthy.

4.2 Education Intelligence Module

Prior to produce the model for triggering the motivational agent behaviour, a pre-processingis need for balancing students that had and had not failed a year (respectively 663 and 93).

The first choice was to use SMOTE-based data augmentation. However it resulted in highly unbalanced features, the skewness test confirmed that some features are not well distributed. Some outliers were then removed for fix the skew. After this, SMOTE was used again but the problem remained, some features became unbalanced after the creation of synthetic entries. As an alternative to SMOTE, The RandomOverSampler was then used, followed by the application of RandomUnderSampler. RandomOverSampler was applied with a sampling_strategy of 0.23. The resulting dataset is then composed by 298 entries, 149 for each type of student.

For creating the model, Support Vector Machine, Random Forest Classifier, and XGBoost were used upon the dataset split with 80% for training and 20% for evaluation. Highlight that XGBoost have being presenting good performance for predicting scholar failure [1]. Grid search cross-validation was used to find optimal parameters for the model targeting accuracy [12]. The results are depicted in Table 4.

Table 4. Results from the algorithms. "A" stands for the students that had never repeat a year (Approved), and "F" stands for the ones that had repeat a year (Failure).

Algorithm	Precision		Recall		F1-score		Accuracy
	A	F	A	F	A	F	
SVC	0.74	0.88	0.90	0.71	0.81	0.79	0.8000
RandomForest	0.80	0.74	0.69	0.84	0.74	0.79	0.7667
XGBoost	0.88	0.78	0.72	0.90	0.79	0.84	0.8167

Whenever is possible it worth to understand the importance of features on the prediction. In this sense, Fig. 2 depicts the importance of each feature for the XGBoost Classifier. The top features are some of the answers a student in a scientific-humanistic course could give when asked about what the conclusion of the course would provide, more specifically students that were looking for professional certification, practical learning and university access, and the mother's education level. Therefore, also motivational issues.

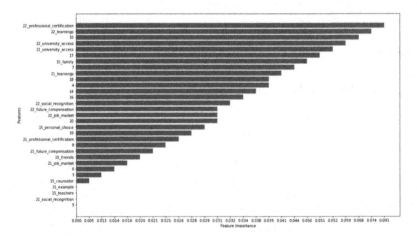

Fig. 2. Feature importance after classification with XGBoost. Each feature number corresponds to the indexes in Table 1

In order to improve the model, XGBoost was used again but with the top eight features resulting in 91.40% of accuracy (considered the best match). In this sense, if the student fill a form with theses eight features, it is possible to the agent to know if is it the case of working or not. Also, considering then that there are both stable and dynamic features, the motivational agent may ask for the dynamic information from time to time in order to re-assess the possible student need for motivation.

4.3   Digital Assistant Motivational Module

A digital assistant is a complex conversational agent that can be composed by several modules working for a good user experience. The focus for this section is the motivation module as an answer to the needs explicated in the previous sections. In short, previous sections had shown that motivation plays an important role in student failure, at least on the school from where that dataset was gathered.

Highlight that gamification is a known approach for motivating students and avoid drop-out [7]. In this sense, it may be important that, in addition to motivational interactions, gamification approaches are put in use [8,11]. Ideally, the agent should also be seen as a friend with personality and propose exciting activities during learning conversation as a break [9], but this is out of the scope of this study.

After defining the basic representation, a proof-of-concept was made to verify some key functional aspects of the intended design. This doesn't present all the pretended functionalities and is just a working prototype. The RASA[1] framework, an open source for building AI assistants and chatbots, was used. The intention was to show two different behaviours for the agent: 1) positive reinforcement in a student showing interest in learning and improving performance; 2) spurring the student to study and reinforcing the idea that they can do better using a gamification approach.

The first part of Rasa is defining a Natural Language Understanding model, which is the transformation of user messages into structured data. The next step is related to the dialogue management system, which means teaching the assistant on how to respond to student messages. After that, it is necessary to create stories that represent the interactions between agent and the student. In this case, the next two stories were created for the two pretended behaviours.

```
- story: positive reinforcement.          - story: challenge Student
  steps:                                    steps:
  - intent: greet                           - intent: greet
  - action: utter_greet                     - action: utter_greet
  - intent: affirm                          - intent: affirm
  - action: utter_ask_homework              - action: utter_ask_study
  - intent: affirm                          - intent: deny
```

[1] https://rasa.com/docs/rasa/.

```
- action: utter_congratulate        - action: utter_challenge
- intent: thanks                     - intent: affirm
- action: utter_goodbye              - action: utter_motivate
                                     - action: utter_goodbye
```

It is important to note that both Intents and Actions are defined in an initial phase before the stories are written. This involves defining a set of examples of possible words and expressions associated for each student's intent and agent actions. After training the model, it was possible to test the agent. For the two stories created the results are showed in Fig. 3. In the Fig. 3a, the agent reinforces positively the student for studying, while in Fig. 3b the agent challenges the student to make him have a reason to study.

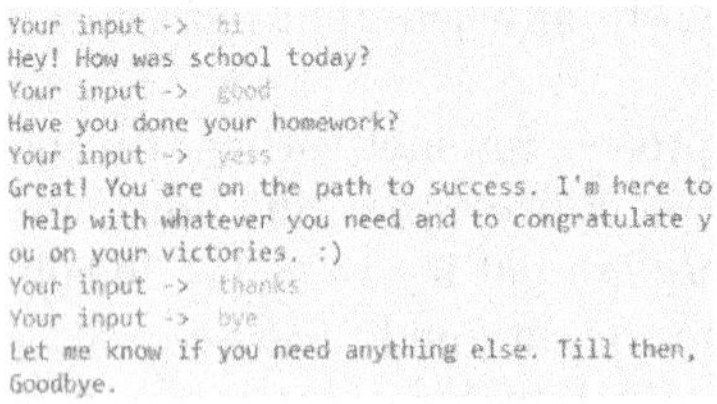

(a) Positive reinforcement action. 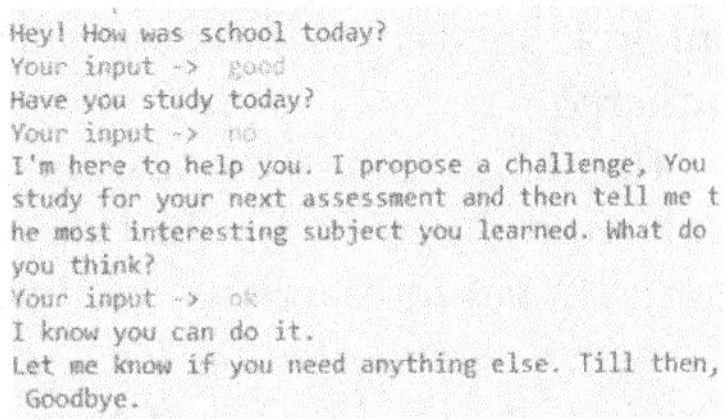   (b) Student challenge for study

Fig. 3. Conversation in the two defined scenarios.

It is worth noticing that after proposing a challenge to the student like in Fig. 3b, the agent should act based on time-spans for asking after the activity suggested. This may be used for the agent to learn which approaches are more compelling for each student. On a long run it will create a personalised toolbox of motivational approaches for each student (an approach already in use, and therefore validated, by the WoeBot [15]).

As already mentioned, this is a simple prototype made for validating key functional aspects. Considering that such agent is intended to be used by under-age people it must follow protocols to be defined by education psychologists in order to reach a proper and comprehensible behaviour. Nevertheless, the proof-of-concept goals are met in the sense that the high-level architecture presented in Fig. 1 is fully validated and is ready to proceed for the next maturity level.

5   Conclusion

The goal of the education system is the student success. For that, it is important to be able to predict the risk of a student's failure at an early stage and create semi-automatic strategies to invert the situation. In this study, it was found that the mother's educational level and the opinion about what concluding a scientific-humanistic course would provide were the most important factors in predicting the student's failure. A prototype digital assistant is also presented to engage and motivate students in the learning process.

Future work includes the development and validation of new predictive models for school performance and the prototype presented for the EduBot. In addition, an investigation in a real school environment with parents permission, due to the students being minors, needs to be conducted to validate the results.

Acknowledgements. This work has been supported by FCT - Fundação para a Ciência e Tecnologia within the R&D Units Project Scope: UIDB/00319/2020.

References

1. Bruno, V., Maria, A.B., Hugo, F., Francisco, S.M., Dalila, D., Paulo, N.: A systematic review on student failure prediction. In: Methodologies and Intelligent Systems for Technology Enhanced Learning, Workshops - 12th International Conference
2. Sokkhey, P., Okazaki, T.: Developing web-based support systems for predicting poor-performing students using educational data mining techniques. Int. J. Adv. Comput. Sci. Appl. **11**, 23–32 (2020). https://doi.org/10.14569/IJACSA.2020.0110704
3. Embarak, O.: Towards an adaptive education through a machine learning recommendation system. In: International Conference on Artificial Intelligence in Information and Communication (ICAIIC) 2021, pp. 187–192 (2021). https://doi.org/10.1109/ICAIIC51459.2021.9415211
4. Siddique, A., Jan, A., Majeed, F., Qahmash, A., Quadri, N.N., Wahab, M.: Predicting academic performance using an efficient model based on fusion of classifiers. Appl. Sci. **11**, 11845 (2021). https://doi.org/10.3390/app112411845
5. Majjate, H., Jeghal, A., Yahyaouy, A.: Predicting factors affecting student's performance in a learning management system. Indian J. Comput. Sci. Eng. **12**, 1771–1779 (2021). https://doi.org/10.21817/indjcse/2021/v12i6/211206015
6. Begum, S., Padmannavar, S.: Genetically optimized ensemble classifiers for multiclass student performance prediction. Int. J. Intell. Eng. Syst. **15**(2), 316–328 (2022). https://doi.org/10.22266/ijies2022.0430.29
7. de la Peña, D., Lizcano, D., Martínez-Álvarez, I.: Learning through play: gamification model in university-level distance learning. Entertain. Comput. **39**, 100430 (2021). https://doi.org/10.1016/j.entcom.2021.100430
8. Morschheuser, B., Werder, K., Hamari, J., Abe, J.: How to gamify? A method for designing gamification. In: Proceedings of the 50th Hawaii International Conference on System Sciences (HICSS 2017), pp. 1298–1307 (2017). www.hdl.handle.net/10125/41308
9. Pérez-Marín, D., Pascual-Nieto, I.: An exploratory study on how children interact with pedagogic conversational agents. Behav. Inf. Technol. **32**(9), 955–964 (2013). https://doi.org/10.1080/0144929X.2012.687774
10. Lee, C.-I., Chen, I.-P., Hsieh, C.-M., Liao, C.-N.: Design aspects of scoring systems in game. Art Des. Rev. **05**(01), 26–43 (2017). https://doi.org/10.4236/adr.2017.51003
11. Kumar, J., Herger, M.: Mechanics. In: Gamification at Work: Designing Engaging Business Software. www.interaction-design.org/literature/book/gamification-at-work-designing-engaging-business-software/chapter-6-58-mechanics. Accessed 05 Sept 2021

12. Krstajic, D., Buturovic, L.J., Leahy, D.E., Thomas, S.: Cross-validation pitfalls when selecting and assessing regression and classification models. J. Cheminformatics **6**(1), 1–15 (2014)
13. Awan, A., Shaheen, N.: The impacts of mother's education on the academic achievements of her child. Glob. J. Manag. Soc. Sci. Human. **6**, 735–756 (2020)
14. Sweeney, C., et al.: Can chatbots help support a person's mental health? Perceptions and views from mental healthcare professionals and experts. ACM Trans. Comput. Healthc. **2**(3), 15 (2021). Article 25. https://doi.org/10.1145/3453175
15. Fitzpatrick, K., Darcy, A., Vierhile, M.: Delivering cognitive behavior therapy to young adults with symptoms of depression and anxiety using a fully automated conversational agent (Woebot): a randomized controlled trial. JMIR Ment. Health **4**(2), e19 (2017). https://doi.org/10.2196/mental.7785. www.mental.jmir.org/2017/2/e19
16. Romero, C., Ventura, S.: Educational data mining: a survey from 1995 to 2005. Expert Syst. Appl. **33**(1), 135–146 (2007)
17. Mohamad, S.K., Tasir, Z.: Educational data mining: a review. Procedia. Soc. Behav. Sci. **97**, 320–324 (2013)
18. Dutt, A., Ismail, M.A., Herawan, T.: A systematic review on educational data mining. IEEE Access **5**, 15991–16005 (2017)
19. Nogareda, A.M., Camacho, D.: Optimizing satisfaction in a multi-courses allocation problem combined with a timetabling problem. Soft. Comput. **21**(17), 4873–4882 (2017). https://doi.org/10.1007/s00500-016-2375-8
20. Hasheminejad, S.M., Sarvmili, M.: S3PSO: students' performance prediction based on particle swarm optimization. J. AI Data Min. **7**(1), 77–96 (2019)
21. Nogareda, A.M., Camacho, D.: A constraint-based approach for classes setting-up problems in secondary schools. Int. J. Simul. Model **16**(2), 253–262 (2017)

A Simulation Model for Predicting the Spread of COVID-19 Virus

Piotr Jastrzębski, Barbara Jagielska, Mateusz Kolasa, Izabela Rejer$^{(\boxtimes)}$ ⓘ,
and Maciej Gabryś

West Pomeranian University of Technology in Szczecin, 70-310 Szczecin, Poland
`izabela.rejer@zut.edu.pl`

Abstract. COVID-19 has shown a high potential of transmission within
the last two years. To interrupt the chain of transmission, it is estimated
that 85% of the population must be immune. Since not all society wants
to take vaccinations, it is very important to predict how the current
precautions will impact the virus development. This paper presents a
simulation model framework that can be used to predict the development of SARS-CoV-2 virus. The model was based on SEIR (Susceptible-
Exposed-Infectious-Removed) model but was significantly extended by
adding a set of additional changeable parameters and a new layer responsible for modelling the virus spread patterns. To test the capability of
the model to predict the virus spread in a hermetic group of people, we
run the 28-days simulation of the spread of the 4-th wave of COVID-
19 in a shopping mall visited by 6500 agents. The simulation results
showed a remarkable relation to the real development of the 4th wave of
COVID-19 in a small hermetic community (the gryfiński district in West
Pomeranian Voivodeship of Poland).

Keywords: COVID-19 · Simulation model · Virus spread

1 Introduction

Coronavirus, COVID-19, SARS-CoV stands for Severe Acute Respiratory Syn-
drome. Posing a serious threat to public health it has originated in the city of
Wuhan, Hubei Province, Central China, and spread quickly all over the world.
As of March 2020, considering the global danger, the World Health Organization
has declared COVID-19 a public health emergency of international concern, with
no vaccines present at the time.

Despite the fact that there are some simulation models for the development of
the next waves of the COVID-19 epidemic (e.g. MOdelling COronavirus Spread
MOCOS [3] or model proposed by Institute for Health Metrics and Evaluation
[2]), they lack some specialized parameters. Therefore the aim of this research

ⓒ The Author(s), under exclusive license to Springer Nature Switzerland AG 2022
H. Yin et al. (Eds.): IDEAL 2022, LNCS 13756, pp. 233–241, 2022.
https://doi.org/10.1007/978-3-031-21753-1_23

was to create a model that would have some of the parameters missing in other simulations. The distinctive elements of the proposed simulation model are hermitized groups of individuals and also the possibility to parametrize the data.

To create an actual model that could correspond to reality, we had to gather information about the pedestrian dynamics and spread of the virus in the previous waves. Furthermore, the research on actual numbers of infected and vaccinated individuals, estimating the percentage of the error that might have been possible within the simulation, was also an important part of the modelling process. Therefore, to simplify the whole process, we focused on the situation in Poland. Since the virus constantly mutates and the healthcare system varies across the country, we decided to base our simulation on the data from one Polish district (gryfiński district). The information used in the model was collected from an official website of The Republic of Poland [4].

We based our simulation model on an SEIR (Susceptible-Exposed-Infectious-Removed) model [9]. It is possible to find many simulation models based on the SEIR model on the Internet. However, they usually do not provide a possibility of changing their parameters. Additionally, they do not allow to focus on hermitized groups. Meanwhile, using the changeable parameters - such as the ability of the virus to spread in the proximity, or parameters of the simulation participants - facilitates a better prediction of a given virus wave for a specific mutation [13].

To provide an environment for a hermetized group of people we focused on groups of six to seven thousand individuals visiting stores in a shopping mall. In the simulation, we used five different health statuses for customers visiting the mall: healthy, infected, transmitting, staying at home or healed. Each customer also conveyed the information about his vaccination status. To simplify the simulation all of the shops used were one-level buildings in theory and 2D maps in practice. As presented in Results section, the predictions provided by the proposed model were very close to numbers reported later (after the end of the simulation) in the official statistical data sources. That proved the algorithm is well suited for both, predicting virus spread and estimating the effects of applied countermeasures.

2   Methods

The backbone of the proposed stimulation model derives from the model described in [8]. The space in the original model is represented by a grid consisting of 40 × 40 cm cells and reflects the typical space occupied by a single pedestrian in the crowd [14].

The basic version of the original model is composed of four layers:

1. The pedestrians position layer - this layer contains pedestrians being part of the simulation. Each cell of this layer can be in one of the two possible states: 'contains a pedestrian', 'is empty'.
2. The floor layer - layer of the 'inanimate' part of the space. The cells of this layer can take one of the three states: 'obstacle' (e.g. wall), 'free space' (floor), and 'exit'.

3. The static layer - each cell of this layer contains value describing distance from the nearest exit.
4. The dynamic layer - each cell of this layer contains value informing how often it is visited by pedestrians. Thanks to the existence of this layer, it is possible to model the patterns of behaviour explaining how the individuals follow the group. This task is accomplished by attracting pedestrians to cells with higher energy. When a pedestrian leaves a cell, he leaves his trace, called boson, on that cell (the value that represents a cell in this layer is a sum of bosons left by the all the clients). At each iteration of the simulation the boson can disappear with a probability α, which is one of the model parameters.

In each iteration pedestrians choose cell to move to, in order to get closer toward the exit, based on the dynamic and static layers. In the case of several pedestrians trying to move into the same cell, only one makes the move, and the rest waits on their cells. To decide which client should move, random numbers, modified by the clients probability of moving to that cell, are assigned to all competing clients.

In order to simulate the spread of the virus, we modified the original model described above introducing the following adjustments. Firstly, the simulated space in our model is a shop or a shopping mall. Accordingly, pedestrians are clients, cashiers or security guards. The exit cells, called now attraction points, are the shop entrances, cash registers, shelves or mall exits. A special type of cell, contained in the floor layer, has been also added to the model - one that acts as wall, preventing pedestrian movement, but at the same time enables virus to spread (it is used for objects like shop counters or benches). Another important amendment of the model structure was the introduction of a new layer - the virus layer. This layer has been added to model the virus spread. Each cell of this layer contains values from 0 to 1, informing about the density of virus particles over this cell. All the layers used in the model and their predestination are shown in Fig. 1.

Secondly, each client entering the mall gets a list of attraction points to visit. After entering, he chooses the first point from the list, approaches it, and then moves to the next point. After visiting all the points from the list, the client moves towards the closest exit, and is removed from the map (but not from the simulation).

Thirdly, in order to reflect realistic behavior of clients in a shop, upon reaching their destination clients stay for a set amount of time in the spot, moving chaotically around that point. To simulate that movement, each point of attraction has its own dynamic layer. Moreover, to simulate an effect of keeping distance between clients 'crowd value' has been added to the model. It expresses how many clients stays within the given radius from the given cell at the given moment.

The proposed construction of the model might lead to the situation when clients block each other. This situation might appear when a group of clients decides to go to the same point of attraction at similar time. In such scenario a client that was first at the attraction point might get stuck because the cells surrounding this point are occupied by other clients trying to move to the same attraction point. In order to avoid such unwelcome model behaviour, an additional constrain, allowing clients to skip the current destination if they have been waiting for too long, was added to the model.

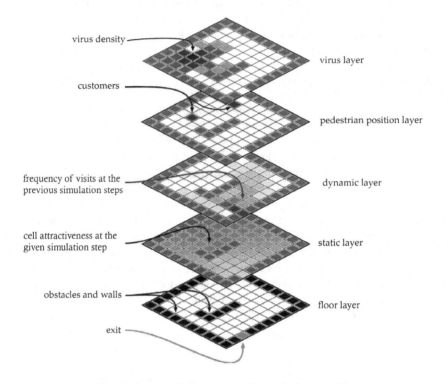

Fig. 1. Layers of the proposed simulation model

As can be seen in Fig. 2, the proposed model is a SEIR based model with a small amendment: the infecting state has been further split into infecting but feeling well, and infecting and feeling ill enough to stay at home. Clients that are staying at home represent people staying in quarantine and hence, they do not appear on the simulation map until recovered but are not removed from the simulation.

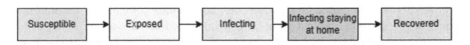

Fig. 2. A set of possible client states employed in the model

The model dynamics can be described as follows. Clients that are infected leave virus in air on the virus layer in the direction they are walking. Virus can only be spread to valid cells, it is to all cells apart from those that are marked as walls. The mechanism of spreading virus particles in the air is represented by 3×3 Gaussian Blur filter (Eq. 1).

$$G(x,y) = \frac{1}{2\pi\rho^2}e^{-\frac{x^2+y^2}{2\rho^2}} \tag{1}$$

where x, y - distances in respective axes, ρ - standard deviation of blur, determining time before virus particles fall on the ground. Argument ρ can be adjusted to match the true virus decay in the air (greater values mean faster decay, ρ must be greater than 0). The track of virus left by a client is different depending on whether the client is wearing a mask or not. While clients wearing mask can spread virus particles only to a single cell just ahead them, clients without masks can spread virus particles also to all cells surrounding this cell.

Each client in the model stores information about virus density in cells he has recently visited. The time, within this information is kept by the client, is one of the model's parameters. At each iteration of the simulation, the density of virus particles assigned to the cell the client stays in at the moment is added to the list, while the virus density from the oldest cell from the list is removed, and the average density of virus particles from all cells in the list is calculated. If this average is greater than a given threshold, the client changes his state from susceptible to exposed. The general threshold for a change in a client's state (from susceptible to exposed) is set at the beginning of the simulation but during the stimulation it is modified for each client depending on the two factors: i) client is/is not wearing a mask, ii) client has been/has not been vaccinated.

In every iteration a client makes decision about his next move. This decision depends on the 3×3 preference matrix, where each element is probability of moving in one direction. The middle element is probability of staying in place. The values of the probability matrix are calculated according to Eq. 2.

$$m_{ij} = N * e^{\beta * J_s * \Delta_s(i,j)} * e^{\beta * J_d * \Delta_d(i,j)} * e^{\beta * J_x * \Delta_x(i,j)} * (1 - n_{ij}) * d_{ij} \tag{2}$$

where N - normalization factor ensuring that $\sum\limits_{i=-1}^{1}\sum\limits_{j=-1}^{1} m_{ij} = 1$, β - parameter determining how strongly cell attractiveness in the given aspect (such as distance from destination, cell popularity etc.) affects its probability, J_s - parameter informing how strongly the distance to the nearest point of attraction affects probability, Δ_s - distance to the nearest point of attraction (based on static layer), J_d - parameter informing how strongly the cell popularity (frequency of visits) affects probability, Δ_d - frequency of visits (based on dynamic layer), J_x - crowd value assigned to the cell, Δ_x - difference in crowd value between neighbouring cells, n_{ij} - cell unavailability (is equal to 1 when a cell is an obstacle or is occupied by other client), d_{ij} - parameter ensuring keeping consistence of client movement, its value is equal to:

- 1 - when, in previous iteration the client did not move or moved in another direction,

– $\exp(\beta * J_0)$ - when, in previous iteration the client moved in the same direction (J_0 is a parameter determining reward for keeping direction),
– $\exp(-\beta * J_d)$ when, in previous iteration the client moved in the opposite direction.

The proposed model was implemented using c++ and qt as two separate applications. The first one, a configuration application, allows for easy adjustment of the model parameters in order to better reflect the current knowledge on COVID-19 virus, or to be used for simulating the spread of other diseases. The second application takes as inputs the parameters, set in the configuration application, and runs the simulation. It provides a 2D map presenting all aspects of the simulation (including the virus spread) step by step (Fig. 3).

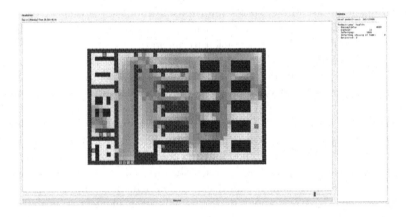

Fig. 3. A visualisation layer of the simulation application

3   Results

The model, described in the previous section, was used to simulate the virus spread in a space resembling a shopping mall, for a population of 6 500 agents. Most of the model parameters, especially those concerning the virus spread and medical aspects of the simulation, were set to the reference levels established on the basis of publications cited in Table 1. The simulation was run on the machine of the following parameters: processor: Intel(R) Core(TM) i5-6500 CPU @ 3.20 GHz 3.2 GHz; memory: 16.0 GB DDR3 2333 Hz; operating system: Windows 10 Enterprise LTSC. The 28-days long simulation took 13 h and 32 min.

Figures 4 and 5 presents the results returned by the proposed stimulation model. Figure 4 shows the distribution of customers by their health status for each day of the simulation. As it can be noticed in the figure, the largest amount of infecting people appeared around the 10th day of the simulation.

Table 1. The list of references used to establish the main simulation parameters

Parameter name	Value	Reference
Initial susceptible	41.5%	[10]
Initial exposed	1.5%	[10]
Initial infectious	2%	[10]
Initial infectious staying at home	0%	[10]
Initial exposed	1.5%	[10]
Recovered	55%	[10]
Vaccinated population	53%	[1]
Vaccine effectiveness	60%	[12]
Days of being infected, feeling good and not infecting	2–3 days	[6]
Days of being infected, feeling good and infecting	2–3 days	[6]
Days of being infected and feeling ill	6–8 days	[5]
ρ determining how long virus stays in the air	0.38	[5]
Percentage of customers wearing masks properly	60%	[11]
Min. amount of virus particles to get infected without mask	90%	[6]
Min. amount of virus particles to get infected wearing mask	97%	[7]

Apart from results specific for SEIR models, the proposed model also returns information about the number of new cases appearing at each simulation day. These results are presented in (Fig. 5). Due to a small scale of our model, sudden changes in count of infected people can be observed in this figure. However, observing the trend of the simulation, it can be noticed that apart from a few first days, when the number of cases grown rapidly, for the rest of the simulation the trend was decreasing.

The simulation results presented in Figs. 4 and 5 have shown a significant similarity to the real development of the 4th wave of COVID-19 in Poland. Figure 6 presents the number of daily cases (and their 3-day average) that appeared in 28 consecutive days (from 20th of December 2021 to 16th January of 2022) in one of the districts in West Pomeranian Voivodeship in Poland - gryfiński district. As can be noticed in Figs. 5 and 6, there are visible similarities between our predictions and gryfiński district. In both figures we can see a high peak on December 23–24 that is followed by a very deep drop on December 26–27. Next, we have one more peak (between December 28 and 31) and then the wave is starting to vanish (very quickly in the case of our simulation, and much slower in the case of gryfiński district). On the contrary to those similarities, one clear difference can be also noticed. In Fig. 6 we can see a high peak at the beginning of the simulation period which is not visible in Fig. 5. This however, is a perfectly normal situation because after introducing virus particles to the model, the simulation needed some time to develop.

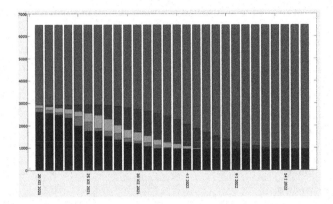

Fig. 4. Distribution of customers by their health status; green - recovered, purple - infecting, staying at home, yellow - infecting, not staying at home, red - exposed, blue - susceptible (Color figure online)

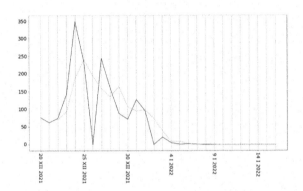

Fig. 5. Simulated infections; blue - new infections, orange - a 3-day average of new infections (Color figure online)

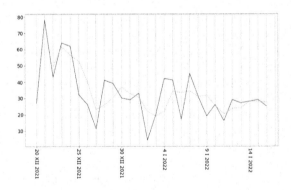

Fig. 6. Reported new cases in gryfiński district from 20th December to 16th January 2022; blue - new infections, orange - 3 day average of new infections (Color figure online)

4   Conclusions

Simulation is a very important tool in fighting epidemic. Simulation models can be used as help to predict the course of virus spread, and to estimate the impact of countermeasures taken to restrict the range of the pandemic, such as keeping social distancing or wearing masks. This paper proposes a new simulation model that provides a unique small scale view on the COVID-19 spread process. As was shown in the paper, the simulation results provided by the proposed model were very close to the number of real new cases reported in the official websites. That shows the practical usability of the model and proves that it can be used for predicting the virus spread. Moreover, since the model is equipped with a number of parameters changing the way the virus is spreading in different situations, it can be also used for evaluating countermeasures taken to fight the pandemic.

References

1. Govermental raport of vaccination status in Poland. https://www.gov.pl/web/szczepimysie/raport-szczepien-przeciwko-covid-19. Accessed 21 Dec 2021
2. Institute for Health Metrics and Evaluation. https://covid19.healthdata.org. Accessed 20 Jan 2022
3. MOdelling Coronavirus Spread. https://mocos.pl/. Accessed 27 Jan 2022
4. The official website of The Republic of Poland Government. www.gov.pl/. Accessed 27 Jan 2022
5. World Health Organisation, Coronavirus disease. www.who.int/news-room/q-a-detail/coronavirus-disease-covid-19. Accessed 10 Jan 2022
6. World Health Organisation, How is it transmitted? www.who.int/emergencies/diseases/novel-coronavirus-2019/question-and-answers-hub/q-a-detail/coronavirus-disease-covid-19-how-is-it-transmitted. Accessed 10 Jan 2022
7. World Health Organisation, Similarities and differences with influenza 2020. www.who.int/news-room/questions-and-answers/item/coronavirus-disease-covid-19-similarities-and-differences-with-influenza. Accessed 10 Feb 2022
8. Burstedde, C., Klauck, K., Schadschneider, A., Zittartz, J.: Simulation of pedestrian dynamics using a two-dimensional cellular automaton. Phys. A **295**(3–4), 507–525 (2001)
9. Carcione, J.M., Santos, J.E., Bagaini, C., Ba, J.: A simulation of a COVID-19 epidemic based on a deterministic SEIR model. Front. Public Health **8**, 230 (2020)
10. Dong, E., Du, H., Gardner, L.: An interactive web-based dashboard to track COVID-19 in real time. Lancet Infect. Dis. **20**(5), 533–534 (2020)
11. Haischer, M.H., et al.: Who is wearing a mask? Gender-, age-, and location-related differences during the COVID-19 pandemic. PLoS ONE **15**(10), e0240785 (2020)
12. Katella, K.: Comparing the COVID-19 vaccines: how are they different? www.yalemedicine.org/news/covid-19-vaccine-comparison. Accessed 10 Jan 2022
13. Liu, H., He, S., Shen, L., Hong, J.: Simulation-based study of COVID-19 outbreak associated with air-conditioning in a restaurant. Phys. Fluids **33**(2), 023301 (2021)
14. Weidmann, U.: Transporttechnik der fussgänger-schriftenreihe ivt-berichte 90. Institut für Verkehrsplanung, Transporttechnik, Strassen-und Eisenbahnbau (1993)

ICU Mortality Prediction Using Long Short-Term Memory Networks

Manel Mili[1,3] ![ORCID], Asma Kerkeni[2,3(✉)] ![ORCID], Asma Ben Abdallah[2,3] ![ORCID],
and Mohamed Hedi Bedoui[3] ![ORCID]

[1] Faculty of Medicine, University of Monastir, Monastir, Tunisia
[2] Higher Institute of Computer Sciences and Mathematics, University of Monastir,
Monastir, Tunisia
{manel.mili,asma.kerkeni,asma.benabdallah}@isimm.u-monastir.tn
[3] Laboratory of Technology and Medical Imaging, Faculty of Medicine,
University of Monastir, Monastir, Tunisia
medhedi.bedoui@fmm.rnu.tn

Abstract. Extensive bedside monitoring in Intensive Care Units (ICUs) has resulted in complex temporal data regarding patient physiology, which presents an upscale context for clinical data analysis. In the other hand, identifying the time-series patterns within these data may provide a high aptitude to predict clinical events. Hence, we investigate, during this work, the implementation of an automatic data-driven system, which analyzes large amounts of multivariate temporal data derived from Electronic Health Records (EHRs), and extracts high-level information so as to predict in-hospital mortality and Length of Stay (LOS) early. Practically, we investigate the applicability of LSTM network by reducing the time-frame to 6-hour so as to enhance clinical tasks. The experimental results highlight the efficiency of LSTM model with rigorous multivariate time-series measurements for building real-world prediction engines.

Keywords: Electronic health record · Multivariate time-series data · MIMIC-III

1 Introduction

An ICU serves patients with severe complications or life-threatening injuries, which involve constant care in order to maintain normal bodily functions. To improve hospital services, it seems important to adequately select patients to be admitted to ICUs early on. In an ICU, the patient is monitored using Electronic Health Record (EHR) systems, entering many medical data a day including physiological measurements. Finding statistic models in these measurements has the potential to provide a high aptitude for more accurate and earlier predictions of future clinical events. This might not only help clinicians make more effective medical decisions but also facilitate an economical allocation of hospital

© The Author(s), under exclusive license to Springer Nature Switzerland AG 2022
H. Yin et al. (Eds.): IDEAL 2022, LNCS 13756, pp. 242–251, 2022.
https://doi.org/10.1007/978-3-031-21753-1_24

resources. Naturally, mortality prediction and Length of Stay (LOS), are mainly performed with an interest in the prediction of possible outcomes, which are the death or survival of the patient, and for how long a patient may remain in the intensive units. Nevertheless, most available mortality and LOS prediction systems [1–4] in the literature were designed for at least 24-hour to provide a real-time or retrospective prediction on patients' mortality. To enhance prediction for early diagnosis, the main objective of this paper is to develop an end-to-end approach based on deep learning models, within a data mining framework, specifically intended for predicting mortality and LOS, based on multivariate time-series physiological measurements from the first few hours of admission, in particular after the first 6 h of a patient's acceptance in the ICU. The rest of the paper is organized as follows: Sect. 2 provides a comprehensive literature review on the state-of-the-art works. Section 3 details the process of dataset collection and preparation. Section 4 discuss the proposed model and presents its configuration and implementation tools. To consider the effectiveness of the proposed method, Sect. 5 deals with experiments. Ultimately, Sect. 6 concludes the paper and highlights the fundamental contributions.

2 Related Works

Over the past few decades, substantial researches are undertaken to affect predicting mortality risk and LOS tasks. A number of the more frequently used mortality prediction models in an ICU setting include SAPS-II [1] and SOFA [2]. SAPS-II was designed to estimate the probability of mortality, while SOFA was wont to describe organ dysfunction. Using the primary 24-hour patient physiological measurements, these scores are only designed to form one prediction. As a result, it's unknown how well each system predicts mortality following the primary day of admission. Moreover, it seems intuitively likely that straightforward clinical judgment also will discriminate more effectively as time passes. Existing tools are therefore slow to succeed in useful discriminatory effectiveness and aren't generally felt by clinicians to be useful to help decision-making once they will discriminate.

Adding to severity scores, several authors have converged on management mortality risk, as an example, Pirracchio et al. [4] aimed to develop a scoring procedure to predict mortality in ICUs supported Super-Learner (SL) model. They have proved that the SL method improved performance. However, the authors evaluated the performance of SL using data recorded within the primary 24-hour. Moreover, Darabi et al. [5] developed a model supported Gradient Boosted Tree (GBT) and Convolutional Neural Network (CNN) to estimate the mortality risk of patients admitted to ICUs. Their results prove usability a smaller number of features which will generate satisfactory outcomes for GBT, unlike, CNN that need a wealthy amount of knowledge for training. However, their model was designed within the period of 30-day after admittance.

In addition to mortality risk prediction, few researchers have converged to estimate LOS. Mentioning Gentimis et al. [6] who explored the utilization of

Neural Network (NN) for predicting the entire LOS of a patient within the hospital. The predictive model outperforms machine learning models. However, the studied scenarios considered time-frames > 5 days, or ≤ 5 days, to validate the potency of the model. Furthermore, Zebin et al. [7] applied an Auto-Encoder (AE) along side a dense neural network technique attempted at identifying short and long stays for patients. The proposed model improved the performance compared to employing a simplistic dense neural network for the classification task. However, their assessment results were validated using recordings observed after 24 h of admission.

To conclude, all the above-mentioned works only focused on predicting the risk of mortality and LOS for patients who required intensive care within a minimum of 24-hour of their ICU admission [3–5,8,9]. The challenge, therefore, lies within the early hours of a patient's admission, for instance, the primary 6 and 12 h. Additionally, not all critically ill patients can enjoy ICU admission. Hence, determining the priority of patients' treatments by the severity of their condition is crucial because the ICU is extremely costly with limited resources. The challenge, therefore, lies in triaging patients consistent with their medical conditions, while estimating their expected time of hospitalization. Adding to the present , most research has centered on the evaluation of the efficiency of their predictive models using univariate time-series data and that they didn't consider the potency of multivariate time-series records for improving the accuracy and therefore the efficiency of time-series modeling [10].

3 Dataset

This effort is conducted over the well-known publicly available, large-scale ICU database, the MIMIC-III [11], which presents a single-center electronic database developed by the MIT Lab for Computational Physiology, comprising health data related to 61.532 ICU admissions of 46.520 distinct de-identified patients admitted between 2012 and 2020.

3.1 Feature Engineering

Every day, different vital signs measurements are computed and analyzed during intensive stays. In this proceeding, we focused primarily, in hidden patterns within ICU time-series data and investigated the hypothesis that there is much useful knowledge in motifs within these data that can aid to improve prediction clinical tasks. This hypothesis is motivated by observations considered within several studies, for example, in [12], we found that in the event of a lack of oxygen transport, measurements in this time-frame of associated variables increase the risk of death. We therefore explored some temporal variables defined in acuity severity scoring systems and added others since they have proven to possess a powerful effect in predicting mortality and hence LOS [13]. These variables include "heart rate", "systolic BP", "diastolic BP", "mean BP", "respiratory rate", "oxygen saturation", "glasgow coma score", "blood urea nitrogen", "temperature", "white blood cells", and last not least "bilirubin".

Some of the foremost pertinent measurements could also be obtained using information available within the earliest phase [3]. So, we've extracted features for the primary 6 hours for every ICU stay. We have also extracted features for the 12 and 24 h so as to verify the effectiveness of the proposed model in maintaining its accuracy for long periods.

3.2 Feature Preprocessing

EHRs contain valuable information for estimating mortality risk and discharge time for ICU patients, but substantial missing and imbalanced data present mutual problems for the development and implementation of a prediction model. Hence, the subsequent two issues were identified and handled accordingly.

Missing Data Imputation. The percent of missing values for certain features is higher than 50%. To manage this problem, data imputation was performed including two strategies: we start by filling them using linear interpolation on each multivariate time-series data. Some observations are still missing after this imputation since there are missing data for certain variables. Hence, we impute missing observations using the Mean as the second strategy.

Imbalanced Data Regulation. The number of patients who passed away inside the intensive department is relatively small in comparison with the number of patients who survived, yielding an imbalanced dataset. To manage this problem, re-sampling methods were adopted since they are less sensitive to outliers than other techniques like Cost-sensitive classifiers [14] and Automatic support vector data description [15]. Two of the most common categories of re-sampling methods are under-sampling and over-sampling strategies. The former remove observations from the training dataset that belong to the dominant class, while the latter duplicate samples that belong to the lesser class, thus increasing its impact within the training process. We have applied the former on the dataset since the latter would make models inflexible in learning during the training process by causing overfitting. As a result, the size of the data was reduced from 33.6 Mo to 7.76 Mo, from 66, 7 Mo to 15 Mo and from 129 Mo to 29 Mo, over the 6-hour, 12-hour and 24-hour time-frames, respectively.

4   Methodology

The idea behind time-series prediction is to predict future events supported past values with reference to historical measurements and associated patterns. Turning to the philosophy of the research methodology, we would like to hold relevant information throughout the processing of medical data sequences, as physiological variables begin to decrease or increase over a period of your time, thus making it possible to predict future outcomes associated with patient conditions in care units. To reach these specific goals, a typical two-stage architecture is presented in Fig. 1.

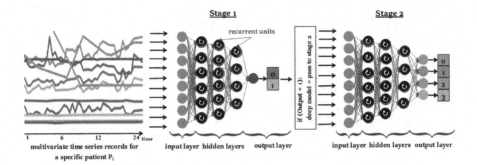

Fig. 1. A summary structure of two-stage architecture: within the first stage, a binary classifier is trained to predict mortality. Then, if the mortality is predicted to be positive, the model would further provide an estimation about LOS.

The philosophy behind the defined architecture is detailed as follows: we start by interpreting multivariate time-series of 11 past clinical records for each patient P_i:

$$P_i : X_{1,t_k}, X_{2,t_k}, ..., X_{11,t_k} \tag{1}$$

with $k = 1, ..., n$ and $n \in \{$6-hour, 12-hour, 24-hour$\}$. In the first stage, a binary classifier is trained to predict the risk of mortality. In a mathematical interpretation we identify:

$$Class = \begin{cases} 0, & \text{survivors group} \\ 1, & \text{non-survivors group.} \end{cases}$$

Therefore, we define a knowledge set of two exclusion criteria: we start by filtering by $16 \leq age \leq 89$ [8]. Then, we exclude ICU stays of but one hour to get rid of obscurity in data due to unusual short stays. After filtering, we observe 49.632 ICU stays of 36.343 patients. While a multi-class classifier is trained at the second stage using a similar vital signs so as to predict LOS for those that are predicted dead in stage 1. Accordingly, we filter the ICU stays with death time ≤ 0. As a result, 5.718 in-hospital mortalities were obtained. We then label each data to at least one of the four classes represented below:

$$Class = \begin{cases} 0, & \text{if } death\_time\_hours < 6, \\ 1, & \text{if } 6 \leq death\_time\_hours < 12, \\ 2, & \text{if } 12 \leq death\_time\_hours < 24, \\ 3, & \text{otherwise.} \end{cases}$$

The proposed model will predict outcomes values by identifying short-term (6 h/12 h) and long-term (24 h) dependencies. For this purpose, we have employed the LSTM architecture [19]. This type of network improves the simple Multilayer Perceptron (MLP) network by including an output that depends on historical learned informations. The LSTM architecture is characterized by hidden units, called memory blocks. These units allow the network to remember information over short/long sequences. Moreover, these gates allow the LSTM model to beat the issues that inhibit the training of other deep models including RNNs and MLPs. This, and therefore the impressive results that may be achieved, are the rationale for its popularity on an outsized sort of problems [16,17].

4.1   Model Configuration

The efficient implementation of deep learning requires the selection and optimization of many hyperparameters, as well as extensive trial and error to find the optimal values. In order to assess the advanced performance, data is divided into training, test and validation sets; The training set is being used to train learning classifier; the validation set is used to fine-tune the parameters and estimate the behavior of the classifier; and the test set is going to be used to determine the efficiency of the classifier. Once data is splitted, we tune models using K-fold cross-validation. In this study, we set K = 3. The implemented LSTM model used Tanh activation function in the hidden layers and Sigmoid activation function in the output layer. Dropout with a rate of 0.2 is used as a regularization technique for weight optimization. In our model, a learning rate of $1e^{-03}$ is used, the number of epochs to train is set to 60 and the batch size is set to 100.

4.2   Model Implementation

In this work, the model was implemented using Keras framework, with TensorFlow backend. The implementation part of the proposed model consists of two stages:

1. Feature Engineering: we chose big data tool like Apache Hive 2.1.0 on Microsoft Azure remote cluster (2 head nodes and 1 worker node, each with 200 GB space, 14 GB RAM, and 4 processors), to perform data preprocessing and feature engineering.
2. Deep Learning using Colaboratory.

We also used Python and several packages for efficient model testing, hyperparameter tuning and model evaluation including: Pandas, NumPy, SciPy, Scikit-learn, Matplotlib, Seaborn.

5   Experimental Results

In this section, we describe the results of our experiments by evaluating the LSTM model against the traditional state of the art acuity scores and machine learning approaches that were used to predict possible future clinical events supported time-series measurements, including SOFA score, SAPS-II score, SL, SVM, LR, NB and CNN. Individual sets of parameters were tuned using 3-fold cross-validation to evaluate the potency of every fixed model. Experiments were conducted under three settings: using temporal physiological measures within 6-hour, 12-hour, and 24-hour time-frames. It's worth noting that SAPS-II and SOFA acuity scores use the primary 24 h of data to evaluate patient severity of illness.

For binary-classifier, we opt for F1-score and MCC metrics to evaluate the effectiveness of the model. In gist, these two metrics were chosen because they provide a more realistic measure of a model's performance, and hence they are robust for binary classification problems [18].

Results outputs of different classifiers are presented in Table 1. In the light of the obtained results, fitting an LSTM model on the multivariate time-series records within a 6-hour time-frame has improved the prediction of early diagnosis of mortality risk for patients who remained in intensive departments. In fact, it is often seen from Table 1 that the LSTM model under the tuned configuration features a higher F1-score and MCC compare to the opposite mortality predictive approaches, which approved that the performance of the LSTM model is more consistent. Although the CNN model has attained a better F1-score and MCC within a 24-hour time-frame, the LSTM model outperformed it within 6-hour and 12-hour time-frames, validating its potency in predicting mortality risk as soon as possible following the admission of patients to the critical units.

Table 1. Mortality prediction performance for binary-classification approaches (The best performing model is highlighted in **bold**).

Classifier	Observation periods					
	6-hour		12-hour		24-hour	
	F1-score	MCC	F1-score	MCC	F1-score	MCC
SAPS-II	–	–	–	–	0.41	0.33
SOFA	–	–	–	–	0.06	0.14
NB	0.44	0.36	0.55	0.49	0.55	0.49
LR	0.09	0.20	0.14	0.25	0.17	0.28
SVM	0.33	0.20	0.30	0.18	0.27	0.17
SL	0.55	0.57	0.65	0.66	0.67	0.68
CNN	0.92	0.91	**0.97**	0.96	**0.99**	**0.99**
LSTM	**0.96**	**0.95**	**0.97**	**0.97**	0.96	0.96

Regarding multi-class classification, the average of the evaluation measures can provides a view on the overall results for the potency of LSTM fitted on the data aggregated over 6-hour within the prediction of LOS compared to those aggregated over 12-hour and 24-hour time-frames. Two major names to refer to averaged results are micro-average and macro-average. In gist, a macro-average will compute the metric independently for every class then take the average, whereas a micro-average will aggregate the contributions of whole classes to compute the average metric. Figure 2 summarizes Micro and Macro-average results for AUROC metrics and confirms that multivariate time-series data aggregated over a 6-hour time-frame offer rigorous multi-classification results compared with 12-hour and 24-hour time-frames that indicate slight improvement results.

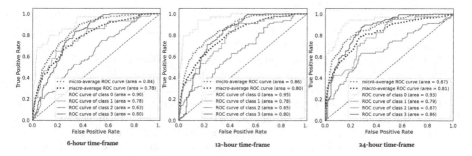

Fig. 2. ROC curves of the LSTM model fitted on data aggregated over 6-hour (in the left), 12-hour (in the middle), and 24-hour time-frames (in the right), applied for the multi-classification problem.

6   Conclusion and Future Works

Enhancing the excellence of care for patients and predicting future outcomes are the foremost important targets in critical care research. In this paper, and by deploying multivariate time-series data obtained from EHR-database MIMIC-III, we reveal that the LSTM model systematically outperforms all opposing predictive models of mortality using physiological measures observed during 6 and 12 h. These positive results recommend that access to the patient's physiological data trajectory as early as possible could enhance the potential in monitoring and predicting possible future events concerning the patient's conditions in ICUs. In future work, we arrange to apply the proposed model in other clinical tasks including early triage and risk assessment, prediction of physiologic decompensation, and identification of high-cost patients.

Acknowledgement. This work was supported by the Ministry of Higher Education and Scientific Research of Tunisia through the PEJC Young Researchers Encouragement Program (Project code 20PEJC 05-16).

References

1. Le Gall, J.R., Lemeshow, S., Saulnier, F.: A new simplified acute physiology score (SAPS II) based on a European/North American multicenter study. JAMA **270**(24), 2957–2963 (1993)
2. Simpson, S.Q.: New sepsis criteria: a change we should not make. Chest **149**(5), 1117–1118 (2016)
3. Awad, A., Bader-El-Den, M., McNicholas, J., Briggs, J., El-Sonbaty, Y.: Predicting hospital mortality for intensive care unit patients: time-series analysis. Health Inform. J. **26**(2), 1043–1059 (2020)
4. Pirracchio, R.: Mortality prediction in the ICU based on MIMIC-II results from the super ICU learner algorithm (SICULA) project. In: Secondary Analysis of Electronic Health Records, pp. 295–313. Springer, Cham (2016). https://doi.org/10.1007/978-3-319-43742-2_20
5. Darabi, H.R., Tsinis, D., Zecchini, K., Whitcomb, W.F., Liss, A.: Forecasting mortality risk for patients admitted to intensive care units using machine learning. Procedia Comput. Sci. **140**, 306–313 (2018)
6. Gentimis, T., Ala'J, A., Durante, A., Cook, K., Steele, R.: Predicting hospital length of stay using neural networks on mimic III data. In: 2017 IEEE 15th Intl Conf on Dependable, Autonomic and Secure Computing, 15th Intl Conf on Pervasive Intelligence and Computing, 3rd Intl Conf on Big Data Intelligence and Computing and Cyber Science and Technology Congress (DASC/PiCom/DataCom/CyberSciTech), pp. 1194–1201. IEEE (2017)
7. Zebin, T., Rezvy, S., Chaussalet, T. J.: A deep learning approach for length of stay prediction in clinical settings from medical records. In: 2019 IEEE Conference on Computational Intelligence in Bioinformatics and Computational Biology (CIBCB), pp. 1–5. IEEE (2019)
8. Purushotham, S., Meng, C., Che, Z., Liu, Y.: Benchmark of deep learning models on large healthcare mimic datasets. arXiv preprint. arXiv:1710.08531 (2017)
9. Johnson, A.E., Dunkley, N., Mayaud, L., Tsanas, A., Kramer, A.A., Clifford, G.D.: Patient specific predictions in the intensive care unit using a Bayesian ensemble. In: 2012 Computing in Cardiology, pp. 249–252. IEEE (2012)
10. Aboagye-Sarfo, P., Mai, Q., Sanfilippo, F.M., Preen, D.B., Stewart, L.M., Fatovich, D.M.: A comparison of multivariate and univariate time series approaches to modelling and forecasting emergency department demand in Western Australia. J. Biomed. Inform. **57**, 62–73 (2015)
11. Johnson, A.E., et al.: MIMIC-III, a freely accessible critical care database. Sci. Data **3**(1), 1–9 (2016)
12. https://physionet.org/content/challenge-2012/1.0.0/
13. Vold, M.L., Aasebø, U., Wilsgaard, T., Melbye, H.: Low oxygen saturation and mortality in an adult cohort: the Tromsø study. BMC Pulm. Med. **15**(1), 9 (2015). https://doi.org/10.1186/s12890-015-0003-5
14. Perry, T., Bader-El-Den, M., Cooper, S.: Imbalanced classification using genetically optimized cost sensitive classifiers. In: 2015 IEEE Congress on Evolutionary Computation (CEC), pp. 680–687. IEEE (2015)
15. Sadeghi, R., Hamidzadeh, J.: Automatic support vector data description. Soft. Comput. **22**(1), 147–158 (2018). https://doi.org/10.1007/s00500-016-2317-5
16. Choi, E., Schuetz, A., Stewart, W.F., Sun, J.: Using recurrent neural network models for early detection of heart failure onset. J. Am. Med. Inform. Assoc. **24**(2), 361–370 (2017)

17. Reimers, N., Gurevych, I.: Optimal hyperparameters for deep lstm-networks for sequence labeling tasks (2017). arXiv preprint. arXiv:1707.06799
18. How to evaluate model performance in Azure Machine Learning Studio. https://docs.microsoft.com/fr-fr/azure/machine-learning/studio/evaluate-model-performance/
19. Lindemann, B., Müller, T., Vietz, H., Jazdi, N., Weyrich, M.: A survey on long short-term memory networks for time series prediction. In: Procedia CIRP, vol. 99, pp. 650–655 (2021)

Intelligent Learning Rate Distribution to Reduce Catastrophic Forgetting in Transformers

Philip Kenneweg(✉) [ID], Alexander Schulz[ID], Sarah Schröder[ID],
and Barbara Hammer[ID]

Bielefeld University, Inspiration 1, 33615 Bielefeld, Germany
pkenneweg@techfak.uni-bielefeld.de

Abstract. Pretraining language models on large text corpora is a common practice in natural language processing. Fine-tuning of these models is then performed to achieve the best results on a variety of tasks. In this paper, we investigate the problem of catastrophic forgetting in transformer neural networks and question the common practice of fine-tuning with a flat learning rate for the entire network in this context. We perform a hyperparameter optimization process to find learning rate distributions that are better than a flat learning rate. We combine the learning rate distributions thus found and show that they generalize to better performance with respect to the problem of catastrophic forgetting. We validate these learning rate distributions with a variety of NLP benchmarks from the GLUE dataset. The source code is open-source and free software, available at https://github.com/TheMody/NAS-CatastrophicForgetting.

Keywords: Natural language processing · BERT · Learning rate · Transformer

1 Introduction

In the real world, data is sequential, tasks may switch unexpectedly and individual tasks may not reoccur for a long time. Thus, the ability to learn tasks in succession is a core component of biological and artificial intelligence [6]. Catastrophic forgetting constitutes one major problem in this context [10]: While learning, there is a high risk of forgetting previously learnt information. There has been a large corpus of work dedicated to dealing with this problem [9]. However the problem of catastrophic forgetting also occurs during fine-tuning in modern transformer models [12], yet it is widely unsolved how to efficiently deal with it, with minimal effort. In this approach we propose a comparatively efficient, yet powerful novel approach how to deal with this issue: Careful and automated tuning of the learning rates of a specific network per layer. It turns out that a highly specific and non-monotonic choice obtained via automated bayesian optimization is the best choice, as we will show for state of the art transformer models.

The transformer architecture pioneered by Vaswani et al. [14] has enabled large pre-trained neural networks to efficiently tackle previously difficult Natural Language Processing (NLP) tasks with relatively few training examples. Most language models are pre-trained on a large corpus of text data (for example Wikipedia, Reddit, etc.) using a variety of different unsupervised pre-training objectives (Masked Language Modeling, Next Sentence Prediction, etc.) [1]. In addition, many common architectures use a

© The Author(s), under exclusive license to Springer Nature Switzerland AG 2022
H. Yin et al. (Eds.): IDEAL 2022, LNCS 13756, pp. 252–261, 2022.
https://doi.org/10.1007/978-3-031-21753-1_25

fine-tuning step for a specific task in parallel with a shallow layer on top of the generated contextualized embeddings to achieve good performance [5].

In this paper we take a closer look at different kinds of data shifts and how they affect the performances of transformer based networks. Typically, neural networks perform poorly when the data they are trained on is input in a sequential fashion. To mitigate this problem, we investigate different learning rate distributions over the layers of the network to decrease the effect of this phenomenon called catastrophic forgetting specifically in transformer based networks.

We hypothesize that different layers of the transformer network represent different abstract concepts and, therefore, should be adapted with different speed when fine-tuning to a new task to reduce catastrophic forgetting. Thereby, we hope to increase the generalization capabilities. We investigate this concept for a few datasets from the NLP domain and compare it to the state of the art from the literature [6].

Based on current best practices, manual effort should be minimized; therefore, we aim for an automation of these optimization steps w.r.t learning rate distribution. We achieve this by three contributions: first, we define a landscape of promising options for learning rate choices. Second, we introduce an automatic hyperparameter optimization process often referred to as a type of AutoML [3], which enables us to automatically search these options to find the best possible learning rate distribution. Third, we find a good way to combine the results of the optimization into a universally applicable solution.

2   Related Work

At present, the most common approach to fine-tuning a language model is to process the outputs of the transformer with a single layer neural network [1, 14] and fine-tune this stack with a flat learning rate. One of the challenges which occur while fine-tuning is the well known problem of catastrophic forgetting, where the network over-fits on new data and/or forgets previously learned information. This is due to its inherent complexity and comparatively small size of the fine-tuning datasets. Some recent papers have attributed large pre-trained networks to be affected less by catastrophic forgetting [12]. However, they are still affected especially when they are smaller or not as extensively pre-trained, as we will exemplary see later.

In the last years, more sophisticated fine-tuning approaches have evolved to improve this baseline approach. One wide spread technique to prevent catastrophic forgetting is elastic weight consolidation (EWC) [6]. Here, an additional loss term is added when fine-tuning the network. This additional loss term prevents the loss of previously learned information by taking the pre-training task into account. During the evaluation section we compare our method to EWC, as it is easy to implement and widely used.

The approach SMART [5] addresses the problem of catastrophic forgetting by introducing a smoothness inducing regularization technique and an optimization method, which prevents aggressive updating of the network weights. While SMART claims to increase performance during fine-tuning with their method, no experiments regarding catastrophic forgetting with consecutive task learning are mentioned in their paper and no code to make evaluations on other settings possible has been published.

Other works have tried to find an optimal learning rate distribution but only for older language models and their respective architectures namely LSTMs. [4]

Before large pre-trained networks like ViT, ResNet50 etc., neural networks were fine-tuned by training the latter layers of the network more as they are expected to be more specialized to the specific task the network was trained on, while the earlier layers are more task agnostic. Following this, the earlier layers are not adapted at all/only adapted slightly during transfer and the latter layers are adapted more freely. For the transformer architecture this practice is not widely adapted as in most papers Transformer based architectures are adapted equally across all layers.

In this paper we particularly introduce hyperparameter optimization technologies in this setting, which enables us to investigate a variety of learning rate distribution choices of the language model, during fine-tuning. To the best of our knowledge, this constitutes one of the first approaches in which the effect of specific learning rate on different parts of the transformer network on the fine-tuning process is extensively investigated.

3   Proposed Approach

The transformer architecture is very powerful and is able to adapt to most classification tasks well, but is still susceptible to the problem of catastrophic forgetting [12].

Here, we propose an approach to automatically select different learning rates for different parts of a transformer model in order to reduce the effect of catastrophic forgetting. Our approach is based on a two-step method: First, we determine such a learning rate distribution for a pair of two datasets with the goal to obtain the best performance for both while training sequentially. Then, we do this for a few such pairs and, in the second step, combine the resulting learning rate distributions such that catastrophic forgetting is also reduced for a novel unseen dataset pair.

For the first step, we consider an original dataset D_o and a shifted one D_s. The indicated model is first fine-tuned on D_o and then fine-tuned on D_s. Subsequently, the performance p_o on D_o is investigated for catastrophic forgetting, while the performance p_s on D_s is also computed to guide the hyperparameter optimization. We combine the performance measures on both datasets to provide the rewards measure used during our hyperparameter optimization $p = p_s + p_o$.

The second step, the combination of different learning rate distributions is detailed in Sect. 3.2 and the search space for the hyperparameter optimization is described in Sect. 3.1.

3.1   Search Space

In this paper we postulate that different learning rates on different parts of the network could provide value by letting these parts adopt at different speeds. The more general parts of the network will be modified less than the parts of the network which are responsible for more task specific computations. With this in mind we determine a space of learning rate distribution choices, which constitute the possible search options. We perform the choice of a different learning rate in the range of $1e-7$ to $1e-3$ for 10 different parts of the network, i.e. $lrs_j = \{lr_{ij}, i = 1, \ldots, 10\}$, $lr_{ij} \in [1e-7, 1e-3]$, for candidates enumerated with j.

This enables the hyperparameter optimization to find a good configuration for every part of the network. We do not expect that 10 vastly different learning rates are needed, but rather choose this number high, as the optimization process has the possibility to converge to similar choices for consecutive parts of the network, thereby reducing the variation in the learning rates chosen.

Since BERT has 12 encoder layers we divide the different layers equally into 8 of these (choice 1 affects layer 1, choice 2 affects layers 2,3, choice 3 affect layer 4, choice 4 affect layers 5,6 ...). Choice 0 is reserved for the embedding weights of the networks. Choice 9 is reserved for the single dense layer which is appended for a specific task.

The optimization strategy used is Bayesian Optimization [13] with Hyperband Scheduling [11], as a particularly promising technology in the domain of hyperparameter optimization. We refer to the output of this procedure, i.e., the best configuration found by this approach in a specific training task, as BERT continual Learner or short $BERTcL$.

3.2 Combining Learning Rate Distributions

During the training on one dataset pair a, the optimization process evaluates different learning rate vectors $lrs_j^a = (lr_{j,1}^a, \ldots, lr_{j,10}^a)$ for layers $1, \ldots, 10$ and experimental runs $j = 1, \ldots$. Each set of lrs_j^a is evaluated and hereby assigned a performance measure p_j^a resulting in a rank r_j^a if sorted according to experimental performance p_j^a for all j. Instead of using just the best performing learning rate set, we utilize a weighted combination of them for the purpose of generalization to new data. They are weighted by their performance ranking and combined with the weighted geometric mean (in correspondence to the exponential search of the learning rates):

$$lr_i^a = \exp\left(\frac{1}{\sum_j b^{-r_j^a}} \sum_j \ln(lr_{ji}^a) * b^{-r_j^a}\right), \text{ for all layers } i \qquad (1)$$

with $b \in [1, \infty]$. We choose $b = 1.8$ (higher values put more emphasis on the best performing samples during averaging).

In practice performing hyperparameter optimization for every dataset pair is infeasible and not desirable. Consequently, we try to find a learning rate distribution which improves the performance for most dataset combinations. To do this, we combine the learning rate distributions lrs^a found for a few dataset pairs a using the geometric mean.

We call the thusly created BERT version $BERTcL\ combined$.

4 Experimental Approach

In this section, we detail our experimental design to investigate the effects of different learning rates for different parts of the BERT network for natural language classification tasks. We utilize AutoGluon [7] and the BertHugginface library [16] for implementation and the pre-trained Bert model ('bert-base-uncased') for all experiments.

We evaluate the two steps of our proposed approach: First we investigate in how far optimizing learning rates for different parts of the model can reduce catastrophic forgetting. For this purpose, we compare the performance drop when fine-tuning standard Bert subsequently to $BERTcL$ and EWC. This comprises a sanity check in how far our proposed modelling is in principle capable of reducing catastrophic forgetting. Then we perform the more interesting evaluation of the second step, whether $BERTcL$ *combined*, i.e. a fixed optimized learning rate distribution for a specific pretrained model, is able to reduce catastrophic forgetting for unseen datasets.

In each step, the training and testing is done on two independent subsets of the dataset. The splits as provided by the data source are used. We follow the EWC [6] paper insofar that we change tasks or data subsets sequentially and evaluate the success of our method by the performance on the combination of performances at the end of training on all datasets. Further details on the data and on hyperparameters are given in the following.

4.1 Datasets

The Glue dataset by Wang et al. [15] is a collection of various popular datasets in NLP, and it is widely used to evaluate common natural language processing capabilites. All datasets used are the version provided by tensorflow-datasets 4.0.1. More specifically, we use the Stanford Sentiment Treebank *SST2*, the Microsoft Research Paraphrase Corpus *MRPC*, the Recognizing Textual Entailment *rte*, the Stanford Question Answering Dataset *QNLI*, the Quora Question Pairs2 Dataset *QQP*, and the Multi-Genre Natural Language Inference Corpus *MNLI*.

4.2 Types of Data Shift

In this paper we look at 2 different kinds of shift to evaluate the problem of catastrophic forgetting in transformer networks.

Dataset shift. With the term dataset shift we refer to a shift where the dataset and thereby task of the network shifts to a different task. By choosing the first and second task carefully we can produce more or less extreme shifts. For a shift with a small impact, we choose the *MRPC* dataset and the *QQP* dataset, here the task of evaluating if two questions are the same shifts to the task of evaluating if two sentences are the same.

For a more substantial shift, we evaluate the *SST2* dataset and the *MRPC* dataset, here the task changes from evaluating sentiment of a single sentence, to predicting entailment of 2 different sentences. This kind of shift shows if the transformer forgets part of the knowledge learned during pre-training and thus is no longer able to generalize to other tasks.

Distribution Shift. Here, we talk about shifts where the task the network is trained upon stays the same, but rather some characteristics of the data distribution change.

Sentence Length Shift. With the term sentence length shift we refer to a shift created by splitting the dataset by sentence length into 2 equally sized smaller datasets. One with sentences smaller than average D_o, the other with sentences larger D_s. Since the Transformer architecture is agnostic towards sequence length, this should have a comparatively small impact for the language model performance, but is still interesting.

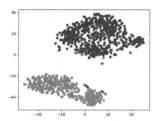

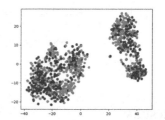

(a) Embeddings of the *SST2* dataset (validation data) colored according to their clusters computed with K-Means (K=2).

(b) Embeddings of the *SST2* dataset (validation data) colored according to their ground truth labels.

Fig. 1. Visualizations of sentences of the *SST2* dataset embedded with BERT and projected to 2-D with TSNE.

Artificial Shift. We call the shift introduced by clustering the embeddings of the dataset generated by the pre-trained BERT into two clusters D_o and D_s with a clustering algorithm (K-means), artificial shift. In the *SST2* dataset the embeddings are naturally split in 2 clusters, see Fig. 1a, 1b.

4.3 Baselines and Implementation Details

As a Baseline comparison we evaluate BERT with a learning rate of 2e-5, henceforth referenced to as $BERTbase$. These values are taken from the original paper [1] and present good values for a variety of classification tasks.

For another comparison we evaluate BERT combined with Elastic weight decay as a state of the art method to counter catastrophic forgetting [6], referenced as $BERT + EWC$. The task used to compute the Elastic weight decay loss is the Masked Language Modeling task used during pre-training of BERT, computed on the Wiki40b [2] dataset. The importance value is set to 675 as indicated in [8].

The metaparameter choice for all experiments are as follows:

- All models are trained for 5 epochs on the smaller (*SST2*, *RTE* and *MRPC*) datasets and for 3 epochs on the larger (*MNLI*, *QNLI* and *QQP*) datasets.
- All models are trained using a cosine decay of their learning rate with warm starting for 10 % of the total training time.
- The pooling operation used in all experiments is [CLS].
- Batch size used for training was 16.
- The Adam optimizer with betas (0.9,0.999) and epsilon 1e-08 was used.
- The maximum sequence length was set to 256 tokens.
- No callbacks for early stopping were used.
- A random seed (999) was used for numpy and python, not pytorch.

Table 1. Classification accuracies, without any shift for later comparison.

Method	SST2	MRPC	MNLI	QNLI	QQP	RTE	Average
$BERTbase$	0.925	0.860	0.829	0.905	0.899	0.700	0.848

5 Results

In Table 1 we display the baseline results on the GLUE tasks without any datashift. Here we use the results obtained by the Huggingface BERT implementation and not the results of the original paper, as the original paper did not provide performances for all GLUE tasks.

5.1 Dataset Shift

In the dataset shift experiments the performance drop off is quite significant depending on the datasets used, as can be seen in Table 2, compared to Table 1. Our method of searching for a good learning rate distribution for every dataset combination called $BERTcL$ can mitigate this performance drop or in some cases, completely negate it. It results in on average 2.4% better performance p_o after training on the second dataset D_s, while having a small performance drop of p_s of 0.6% on the second dataset.

We combine the learning rate distributions found as described in Chap. 3 to create $BERTcL$ *combined*. The resulting learning rate distribution is visualized in Fig. 2. It is on average lower than the standard learning rate of $2e - 5$, but shows a spike for the last (newly initialized dense) layer, as well as a very low learning rate for choice 2 (in the BERT architecture this represents the encoder layers 2 and 3). This suggest that these layers are very general and do not need to be retrained much for a specific task.

For the sake of completeness, we provide $BERTcL$ *combined* scores on the datasets it was trained on in Table 2. The more interesting case of generalization towards unseen data during the hyperparameter optimization process is illustrated in Table 3. In all cases a better performance p_o on the first dataset is achieved compared

Table 2. Classification accuracies, for the dataset shift experiment. The first number denotes the performance p_o on the original dataset after training on the second dataset. The second number denotes the performance p_s on the second dataset after training. Improvements are marked in **bold**.

Method	SST2-MRPC	MRPC-SST2	QQP-MRPC	MRPC-QQP	Average
$BERTbase$	$0.904 - 0.860$	$0.740 - 0.922$	$0.869 - 0.831$	$0.725 - 0.901$	$0.809 - 0.879$
$BERT + EWC$	$0.845 - 0.836$	$0.650 - 0.915$	$0.851 - 0.855$	$0.691 - 0.895$	$0.759 - 0.875$
$BERTcL$	$\mathbf{0.923} - 0.853$	$\mathbf{0.765} - 0.910$	$\mathbf{0.877} - 0.838$	$\mathbf{0.767} - 0.892$	$\mathbf{0.833} - 0.873$
$BERTcL$ combined	$0.908 - 0.836$	$0.755 - 0.916$	$0.877 - 0.823$	$0.738 - 0.891$	$0.820 - 0.867$

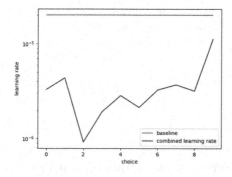

Fig. 2. Combined learning rate as determined by the hyperparameter optimization process over the dataset shift experiments. X-Axis denotes position of learning rate in the transformer architecture as described in Sect. 3.1. Lower numbers indicate earlier layers in the transformer. Y-Axis denotes the learning rate (log scale).

to $BERTbase$ with the combined distributed learning rate. However, a lower performance p_s on the second dataset is also sometimes observed. Overall, on average over the unseen data, $BERTcL$ $combined$ outperforms the flat learning rate, as well as $BERT + EWC$ by about 5%.

In conclusion, we find that $BERTcL$ $combined$ constitutes a robust, flexible, easy to implement and well performing method to mitigate the problem of catastrophic forgetting during dataset shifts.

5.2 Distribution Shift

In this section we depict results of the distribution shift experiments described in 4.2.

Sentence Length Shift. For the sentence length shift experiments we choose the datasets $SST2$ and $QNLI$. The results can be seen in Table 4. They indicate that the original BERT architecture can generalize well from longer to smaller sentences. Our $BERTcL$ is not expected to provide any improvement and we do not perform further evaluation on this kind of data shift.

Table 3. Classification accuracies for the dataset shift experiment on unseen datasets during the hyperparameter optimization process search. The first number denotes the performance on the original dataset after training on the second dataset. The second number denotes the performance on the second dataset after training. Improvements are marked in **bold**.

Method	RTE-MRPC	MRPC-RTE	MNLI-QNLI	QNLI-MNLI	Average
$BERTbase$	$0.523 - 0.836$	$0.745 - 0.632$	$0.693 - 0.896$	$0.517 - 0.833$	$0.620 - 0.799$
$BERT + EWC$	$0.570 - 0.846$	**0.774** $- 0.639$	$0.610 - 0.908$	$0.509 - 0.835$	$0.616 - 0.807$
$BERTcL$ combined	**0.588** $- 0.760$	$0.760 - 0.599$	**0.770** $- 0.893$	**0.595** $- 0.815$	**0.678** $- 0.767$

Table 4. Classification accuracies, for the distribution shift experiment. Two numbers per dataset are given, the first denoting the accuracy on the partial dataset before the data shift and the second denoting the same accuracy after the data shift.

	Sentence length shift		Artifical shift		
Method	SST2	QNLI	SST2	MRPC	MNLI
$BERTbase$	$0.931 - 0.933$	$0.903 - 0.905$	$0.909 - 0.900$	$0.845 - 0.888$	$0.814 - 0.826$

Artifical Shift. The artifical shift experiment was performed as described in Sect. 4.2, on the datasets *SST2*, *MRPC* and *MNLI*. The results of these experiments can be seen in Table 4. The *SST2* dataset has clearly seperable clusters, but the performance only drops slightly when introducing the artifical shift. For the *MRPC* and *MNLI* datasets the performance p_o improves after training on D_s. Following, this we do not expect $BERTcL$ to provide any improvement.

In conclusion, we find that the transformer architecture is robust during distribution shifts, and can in many cases even improve performance, contrary to the catastrophic forgetting paradigm.

6   Conclusion

Transformer networks are surprisingly robust to varying lengths of input sequences during training and testing, as well as to artificially clustered data shifts.

For subsequent learning over consecutive datasets and tasks we present an intelligent learning rate distribution $BERTcL$ *combined* for the BERT sentence embedder which mitigates or in some cases completely solves the problem of catastrophic forgetting. In comparison to other approaches we do not modify the underlying transformer architecture or provide additional regularization, but rather change the learning rate based on the layer. This approach can be applied to many common encoder or decoder models like BERT, RoBERTa, distilBERT or GPT-2.

The exact contribution of the learning rate distribution found in comparison to a flat learning rate is of interest and merits further research. It is difficult to provide meaningful intuition about what changed in a transformer based network during training. While Attention visualizations can show on a case by case basis which words are associated with each other, this can only be done for individual sentences and is not representative of the dataset as a whole. Possible further research also includes other options for the hyperparameter optimization process, like different optimizer choices as well as learning rate schedules, for different parts of the network.

The source code is open-source and free (MIT licensed) software and available at https://github.com/TheMody/NAS-CatastrophicForgetting.

Acknowledgements. We gratefully acknowledge funding by the BMWi (01MK20007E) in the project AI-marketplace.

References

1. Devlin, J., Chang, M., Lee, K., Toutanova, K.: BERT: pre-training of deep bidirectional transformers for language understanding (2018). CoRR abs/1810.0480, http://arxiv.org/abs/1810.04805

2. Guo, M., Dai, Z., Vrandecic, D., Al-Rfou, R.: Wiki-40b: Multilingual language model dataset. In: LREC 2020 (2020). https://www.lrec-conf.org/proceedings/lrec2020/pdf/2020.lrec-1.296.pdf

3. He, X., Zhao, K., Chu, X.: Automl: a survey of the state-of-the-art. Knowl. Based Syst. **212**, 106622 (2021). https://doi.org/10.1016/j.knosys.2020.106622, https://www.sciencedirect.com/science/article/pii/S0950705120307516

4. Howard, J., Ruder, S.: Fine-tuned language models for text classification (2018). CoRR abs/1801.06146, http://arxiv.org/abs/1801.06146

5. Jiang, H., He, P., Chen, W., Liu, X., Gao, J., Zhao, T.: SMART: robust and efficient fine-tuning for pre-trained natural language models through principled regularized optimization (2019). CoRR abs/1911.03437, http://arxiv.org/abs/1911.03437

6. Kirkpatrick, J., et al.: Overcoming catastrophic forgetting in neural networks (2016). CoRR abs/1612.00796, http://arxiv.org/abs/1612.00796

7. Klein, A., Tiao, L., Lienart, T., Archambeau, C., Seeger, M.: Model-based asynchronous hyperparameter and neural architecture search. arXiv preprint. arXiv:2003.10865 (2020)

8. Kutalev, A., Lapina, A.: Stabilizing elastic weight consolidation method in practical ML tasks and using weight importances for neural network pruning (2021). CoRR abs/2109.10021, http://arxiv.org/abs/2109.10021

9. Lange, M.D., et al.: Continual learning: a comparative study on how to defy forgetting in classification tasks (2019). CoRR abs/1909.08383, http://arxiv.org/abs/1909.08383

10. Legg, S., Hutter, M.: Universal intelligence: a definition of machine intelligence (2007). CoRR abs/0712.3329, http://arxiv.org/abs/0712.3329

11. Li, L., Jamieson, K., DeSalvo, G., Rostamizadeh, A., Talwalkar, A.: Hyperband: a novel bandit-based approach to hyperparameter optimization. J. Mach. Learn. Res. **18**(1), 6765–6816 (2017)

12. Ramasesh, V.V., Lewkowycz, A., Dyer, E.: Effect of scale on catastrophic forgetting in neural networks. In: International Conference on Learning Representations (2022). http://openreview.net/forum?id=GhVS8_yPeEa

13. Snoek, J., Larochelle, H., Adams, R.P.: Practical bayesian optimization of machine learning algorithms (2012). 10.48550/ARXIV.1206.2944, http://arxiv.org/abs/1206.2944

14. Vaswani, A., et al.: Attention is all you need (2017). CoRR abs/1706.03762, http://arxiv.org/abs/1706.03762

15. Wang, A., Singh, A., Michael, J., Hill, F., Levy, O., Bowman, S.R.: GLUE: a multi-task benchmark and analysis platform for natural language understanding. In: Proceedings of the ICLR (2019)

16. Wolf, T., et al.: Huggingface's transformers: state-of-the-art natural language processing (2020)

How Image Retrieval and Matching Can Improve Object Localisation on Offshore Platforms

Youcef Djenouri[1(✉)], Jon Hjelmervik[1], Elias Bjorne[2], and Milad Mobarhan[2]

[1] Mathematics and Cybernetics Department, SINTEF Digital, Oslo, Norway
{youcef.djenouri,jon.m.hjelmervik}@sintef.no
[2] Cognite As, Oslo, Norway
{elias.bjorne,milad.mobarhan}@cognite.com

Abstract. Deep learning is gaining popularity in the realm of object localization. Existing deep learning methods have shown good accuracy and inference runtime, but they require a lot of training data. This needs a major investment in resources, especially for offshore industrial sites that lack huge datasets. Furthermore, because the inference set should contain the same types of objects as the training set, deep learning solutions are highly sensitive to object types. To address these two challenging issues, we proposed a novel framework based on image retrieval and matching algorithms. The set of relevant images to the object query is first retrieved using the Bag of Words. Furthermore, we developed two alternative image matching algorithms to localize the object query on the relevant images. The first one is based on generate and test, and the second one is based on geometric verification. Extensive simulation has been carried out to validate the suggest methodology, and the results are highly promising in terms of computing time and accuracy.

Keywords: Image retrieval · Image matching · Object localisation · Industrial platforms

1 Introduction

Computer vision is a multidisciplinary field whose purpose is to extract useful visual knowledge from data acquired by human eyes. The purpose of object localisation is to automatically identify the locations of the objects in the given image and/or scene, which is considered a promising discipline in computer vision [1,2]. Consider a database represented by the set of images, and a given object query which is also represent as an image. Object localisation problem aims to find the location of the object query on images in the database. Deep learning for object localization has had a lot of success. The ground for this is that such methods can use convolution operators to capture spatial features in an image. In this area, two large families of object localisation algorithms have been proposed [3,4]: 1) A region convolution neural network-based solution involves two passes:

© The Author(s), under exclusive license to Springer Nature Switzerland AG 2022
H. Yin et al. (Eds.): IDEAL 2022, LNCS 13756, pp. 262–270, 2022.
https://doi.org/10.1007/978-3-031-21753-1_26

first, the convolution neural network generates bounding box candidates, and then the bounding box refinement is done by solving the regression problem; and 2) you only look once based solutions involve only one pass to localise the objects. The first family of algorithms takes a long time to run because they work in two steps; however, the second family suffers from low accuracy, especially when dealing with complicated data.

This study looks into offshore oil rigs, where previous deep learning approaches based on the two aforementioned families were unable to be applied for two reasons. The first is that deep learning approaches have traditionally required a substantial amount of training data. This necessitates a significant investment in terms of resources, particularly for offshore industrial platforms. Furthermore, they are quite sensitive to the object query, as the types of objects in the training and inference processes should be identical. The current study provides a unique framework referred known as the ARRANGE (imAge RetRieval mAtchiNG objEct), that combines image retrieval and matching to address these two challenges. The image retrieval is first performed to retrieve the similar images to the object query. The image matching is then used to locate the objects on the similar images. The following are the major contributions of this paper:

1. We develop an image retrieval algorithm based on bag of words [5]. The SIFT features are first computed for each image in the database. The image database's bag of words is then produced based on the visual features of each image. Finally, the bag of words is utilized to find images that are similar to the object query.

2. We propose two strategies for bounding box exploration. The first one is based on generate and test where the bounding are first generated from the similar images to the object query and then evaluated based on the matching score. The second strategy is based on geometric verification where a refinement process is performed to exclude outlier features and concentrate on the regions where the object is probably located on.

3. We conduct extensive analysis of the computational time and accuracy using real data of the offshore oil platforms. The results show that the geometric based strategy outperforms the generate and test strategy in both success rate, and computational runtime.

2   Related Work

Garg et al. [6] present an image retrieval technique based on content information of pixel values. The discrete wavelet transformation strategy is used to decompose each image in multi levels, where several feature extractor algorithms such as the scale-invariant feature transform, the speed-up robust feature and the local binary pattern are applied to derive the descriptors of the images in the database and the image query. The Euclidean distance is finally processed to retrieve the relevant images to the image query. Zhang et al. [7] developed a new hash structure to speedup the computation of the image retrieval process. It considered the correlation among the images using semantic reconstruction under the unified Bayesian inference. It also used the sub-space learning to accurately explore the suggested structure for the searching process. Khade et al. [8] proposed an image retrieval for retrieving the similar floor plans to the current floor plan query while exploring the geometric features. It allows to proceed various types of rotation and scale. Jia et al. [9] used the region fully connected layers to determine the features of the images in the database and the image query in indoor environment. The semantic discrimination is also integrated to prune the irrelevant images. Yin et al. [10] attempted in developing a new image retrieval strategy to quickly handle the large databases without losing of the quality of returned images. It integrated both the coarse localization, and the fine localization in the image retrieval process for excluding the non matched images with the image query. Djenouri et al. [11] explored the decomposition process to divide the images into similar clusters, where a new bag of words based structure called vocabulary forest is suggested. During the searching process, only the similar clusters (trees) to the image query are requested. Salazar et al. [12] revealed the benefit of using the convolution neural network in geometric inspection of the offshore platforms from the photogrammetry model, which is featured y a dynamic scenes. Therefore, the use of the YOLOv2 achieved a good performance in terms of detection accuracy of both risers and i-tubes on offshore platforms. Gong et al. [13] collected an archive of activities data from the offshore drilling platform where the extended multi-level convolutional pose machine algorithm is trained to learn and identify the activities of the different workers. In addition, the multi-action region proposal strategy is suggested to allow the separation of the seawater area. From this literature review, solutions to object localisation show great performance in terms of both accuracy when exploring the deep learning. The main issue of these algorithms is that they require large data in the training phase, in addition to that the object labels should be defined a priori. Collecting large number of images of training is costly in time and sometime impossible in Offshore platforms where the access is very limited to specified personal. In addition, learning from fixed-number of labels is not efficient way where different kind of objects may be added in these offshore platforms due to the maintenance and renovation procedures. Motivated by the success of image retrieval and matching algorithms [11,14,15] in solving complex computer vision problems, in the next section, we propose a novel framework for object localisation on offshore platforms.

3   ARRANGE: ImAge RetRieval mAtchiNG ObjEct

3.1   Principle

In this section, we suggested the ARRANGE framework for object localization on offshore industrial platforms. As illustrated in Fig. 1, the process begins by employing the bag of words solution to retrieve the most similar images to the object query. For the localization process, the matching process is then applied to each retrieved image to the object query. The following are the key components of ARRANGE:

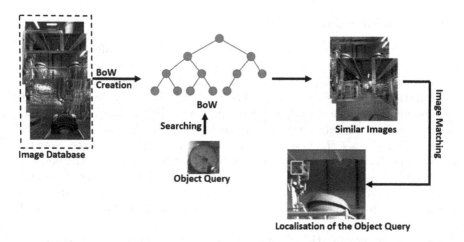

Fig. 1. ARRANGE Framework: The bag of words is first created from the image database, the searching step is then performed to find the similar images to the object query. The image matching is finally executed to localise the object query.

3.2 Image Retrieval

The image retrieval step aims to return the top relevant images according to the object query. This step starts by computing the features of both the images created previously, and the object query. The matching between each image and the object query is determined, where a ranking is performed to extract the top relevant images to the object query. The main stages of this step is given in the following:

1. **Bow Creation**: The feature extraction of the images is performed using the SIFT extractor. The set of keypoint candidates of each image is first detected based on the local minima and the local maxima of each filter of size 3×3. This set is extended using the taylor function. From each keypoint, a filter of size 16×16 is created, where the histogram of oriented gradient is calculated. Based on the histogram of oriented gradient, the descriptor of 128 features of each keypoint is derived. As a result, a pair of (keypoint, descriptor) is generated for every image in the database. Each keypoint is represented by its pixel location in the image, and the descriptor is represented by 128 values. The collection of features extracted from the images is used to compute the visual words. Here, for each image, the visual words from these features are determined using the hierarchical kmeans algorithm, where each center will be considered as one visual word. This results a bag of words that represent the entire images in the database.

2. **Searching Process**: The generated bag of words is explored to find the most similar images to the object query. It is used to compute the visual words from the SIFT features of the object query. The score function based on cosine similarity between the object query and each image in the database is then calculated based on the visual words belonging to both the object query and such an image. The most similar images to the object query are those with highest score values.

3.3   Image Matching

This stage is to locate the correct position of the object query after retrieving its most similar images. As a result, we offer two different image matching strategies. In the following, we will give a full explanation of both strategies:

1. **Generate and test strategy**: The goal of this technique is to consider the bounding box space generated from the most similar images to the object query. The ith bounding box is generated by selecting the values of the pixels $[i \times ws + W_b, i \times ws + H_b]$ in each image from the most similar images. Note that ws, W_b, and H_b are the window size, the bounding box width and the bounding box height respectively. The features of each bounding box is then determined and matched with the features of the object query using the approximate nearest neighbor algorithm. The bounding box with high number of matched with the object query is returned.

2. **Geometric verification based strategy**: By conducting geometric verification between each image in the most similar images and the query object, this technique seeks to reduce the bounding box space. This makes it possible to figure out which parts of each image correlate to which parts of the query object. The RANSAC method [16] is used to keep just the inliers while removing noise and outlier keypoints. The mean of the inlier keypoints is calculated and used to establish the bounding box's center. Based on the bounding box width and bounding box height, the bounding box is constructed from the center.

4   Performance Evaluation

The applicability of ARRANGE on offshore platforms has been validated by extensive numerical simulation. We use real industrial data for offshore platforms provided by our partners *Aker BP*, *Aker Solutions*, *Lundin*, and *Kværner*.

4.1   ARRANGE's Analysis

Figure 2 shows the runtime and accuracy of the matching process with varying matching thresholds and image resolutions. The matching step's runtime was greatly increased from 3 ms to around 30 ms by changing the image resolution from $(32, 32)$ to $(1024, 1024)$. However, when the matching threshold was changed from 0.10 to 0.95, the runtime of the matching step did not change substantially, with less than 5 ms, regardless of the image resolution. These findings are explained by the fact that when an image's resolution is too high, a large number of features are extracted, and the matching process then takes a long time to discover the appropriate images. Nonetheless, the matching process' accuracy is highly dependent on both the image resolution and the matching threshold. The mean average precision is 0.0 and 0.30, respectively, with resolution set to $(32, 32)$ and threshold set to 0.95 and 0.05, and with resolution set to $(1024, 1024)$ and threshold set to 0.05 and 0.95, the mean average precision is 0.67 and 1.00, respectively. These results can be explained by the fact

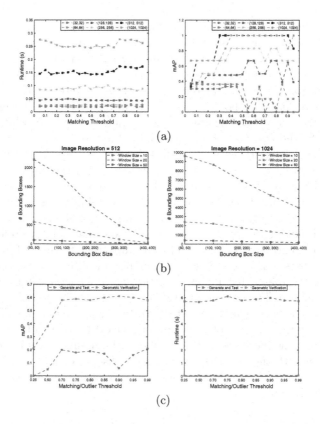

Fig. 2. ARRANGE's Analysis: (a). Runtime and accuracy of the matching step with different matching threshold and image resolution. (b). Number of generated bounding box of the generate and test strategy with different window sizes, bounding box sizes, and image resolution. (c). Accuracy and Runtime of the generate and test strategy, and geometric verification based strategy with different matching and outlier thresholds.

that when the resolution is low, only a small number of features are extracted, lowering the total accuracy of the matching process. In contrast, high resolution allows for the discovery of a large number of features, making matching the object query to the similar images a simple operation. Figure 2 shows how many bounding boxes are created by the generate and test strategy. The number of bounding boxes is greatly increased from 9 to 9604 bounding boxes by raising the image resolution from $(512, 512)$ to $(1024, 1024)$, decreasing the bounding box size from $(400, 400)$ to $(50, 50)$, and decreasing the window size from 50 to 20. The number of bounding boxes generated is 9 when the image resolution is $(512, 512)$, the window size is 50, and the bounding box size is $(400, 400)$. However, with an image resolution of $(1024, 1024)$, a window size of 10, and a bounding box size of $(50, 50)$, the total number of bounding boxes generated is 9604. The reason for these results is that when the image resolution is too high and the window size is too small, the bounding box space becomes too large, resulting in a high number of small bounding boxes . Figure 2 shows the runtime and mAP of the returned object for both the generate and test strategy and the geometric verification strategy, with image resolution set to (1024, 1024) and various matching and outlier thresholds for each strategy. The mAP of the geometric verification based strategy outperforms the generate based strategy when the matching/outlier threshold is changed from 0.25 to 0.99. The mAP of the geometric verification based strategy ranges from 0.20 to 0.60, while the mAP of the generate and test strategy is between 0.0 and 0.20. These results can be explained by the geometric verification-based strategy's capacity to prune outlier features and extract more relevant characteristics from the object query, especially when the outlier threshold value is large. The produce and test strategies, on the other hand, are dependent on how the bounding boxes are created from the similar image to the object query. For example, the object could be situated between two bounding boxes or, in the worst-case scenario, between four bounding boxes. SIFT extractor's ability to locate important features of object query in bounding boxes is harmed as a result of this. In all circumstances, the geometric verification-based technique outperforms the generate and test strategy in terms of runtime. Thus, the geometric verification-based technique is limited to 10 ms, whereas the generate and test strategy is limited to 5 s. The time it takes to extract the features of the bounding boxes has led to these outcomes. The processing time increases as the number of bounding boxes increases. This is not the case with the geometric verification-based technique, which extracts key features by deleting outliers and then uses those features to locate the object query in a similar image.

Table 1. ARRANGE Vs. object localisation solutions.

Measures	ARRANGE	BoW	DCNN-vForest	HDE	SSMV-BoW
Runtime (sec)	0.06	0.12	0.09	0.15	0.19
mAP	0.87	0.76	0.84	0.81	0.82

4.2 ARRANGE Vs. State-of.the-art Object Localisation Algorithms

This section studies the performance of ARRANGE compared to BoW [5], DCNN-vForest [17], HDE [18], and SSMV-BoW [19]. Table 1 presents the average runtime and the average accuracy performance for ARRANGE and baseline solutions for handling 100 user queries. As can be seen, the query time is lower for ARRANGE than the other solutions (DCNN-vForest, BoW, HDE, and SSMV-BoW) implying that being able to disregard huge areas of images saves more time than the overhead cost of evaluating the neural net and selecting the most relevant clusters. Similarly, the results shows that the accuracy is also improved with the ARRANGE algorithm, for all evaluated datasets. These results are explained by the fact that the efficient combination of image search, and bounding box exploration in localize the user queries.

5   Conclusion

This study demonstrated how image retrieval and matching may be used to solve the problem of object localization on offshore industrial platforms. The designed system addresses the challenging issues of the deep learning solutions in acquiring training offshore industrial data as well as localizing different kinds of objects in the inference process. The process begins with the retrieval of the most relevant images to the object query, followed by a match of the retrieved images with the object query to locate it. Furthermore, two different matching strategies based on generate and test and geometric verification have been proposed. Extensive simulation has been performed to validate the proposed methodology, with promising results in terms of computing time and accuracy. As future perspective, we plan to investigate other image retrieval solutions based on vocabulary forest, and pattern mining [11,17,20]. This allows to improve the overall process in terms of runtime as well as accuracy performances.

Acknowledgements. This paper is supported by the Norwegian Research Council funded project Advanced 3D visualization and AR for industrial operations. We would like to thank all project partners, including Aker BP, Lundin, Aker Solutions and Kværner for sharing ideas and data.

References

1. Wang, W., Lai, Q., Fu, H., Shen, J., Ling, H., Yang, R.: Salient object detection in the deep learning era: an in-depth survey. IEEE Trans. Pattern Anal. Mach. Intell. (2021)
2. Aftf, M., Ayachi, R., Said, Y., Pissaloux, E., Atri, M.: Indoor object classification for autonomous navigation assistance based on deep cnn model. In: 2019 IEEE International Symposium on Measurements & Networking (M&N), pp. 1–4. IEEE (2019)
3. Liu, Y., Sun, P., Wergeles, N., Shang, Y.: A survey and performance evaluation of deep learning methods for small object detection. Expert Syst. Appl. **172**, 114602 (2021)
4. Boukerche, A., Hou, Z.: Object detection using deep learning methods in traffic scenarios. ACM Comput. Surv. (CSUR) **54**(2), 1–35 (2021)
5. Kim, J.J.Y., Urschler, M., Riddle, P.J., Wicker, J.: Symbiolcd: Ensemble-based loop closure detection using CNN-extracted objects and visual bag-of-words. In 2021 IEEE/RSJ International Conference on Intelligent Robots and Systems (IROS), pp. 5425–5425. IEEE (2021)
6. Garg, M., Dhiman, G.: A novel content-based image retrieval approach for classification using GLCM features and texture fused LBP variants. Neural Comput. Appl. **33**, 1311–1328 (2021). https://doi.org/10.1007/s00521-020-05017-z
7. Zhang, Z., Zhu, X., Guangming, L., Zhang, Y.: Probability ordinal-preserving semantic hashing for large-scale image retrieval. ACM Trans. Knowl. Discov. Data (TKDD) **15**(3), 1–22 (2021)
8. Khade, R., Jariwala, K., Chattopadhyay, C., Pal, U.: A rotation and scale invariant approach for multi-oriented floor plan image retrieval. Pattern Recogn. Lett. **145**, 1–7 (2021)

9. Jia, S., Ma, L., Yang, S., Qin, D.: Semantic and context based image retrieval method using a single image sensor for visual indoor positioning. IEEE Sens. J. (2021)

10. Yin, X., Ma, L., Tan, X.: A novel image retrieval method for image based localization in large-scale environment. In: 2021 IEEE Wireless Communications and Networking Conference Workshops (WCNCW), pp. 1–5. IEEE (2021)

11. Djenouri, Y., Hjelmervik, J.: Hybrid decomposition convolution neural network and vocabulary forest for image retrieval. In: 2020 25th International Conference on Pattern Recognition (ICPR), pp. 3064–3070. IEEE (2021)

12. Salazar, J.D., et al.: 3d photogrammetric inspection of risers using RPAS and deep learning in oil and gas offshore platforms. Int. Arch. Photogrammetry Remote Sens. Spatial Inf. Sci. **43**, 1265–1272 (2020)

13. Gong, F., Ma, Y., Zheng, P., Song, T.: A deep model method for recognizing activities of workers on offshore drilling platform by multistage convolutional pose machine. J. Loss Prev. Process Ind. **64**, 104043 (2020)

14. Hossein-Nejad, Z., Agahi, H., Mahmoodzadeh, A.: Image matching based on the adaptive redundant keypoint elimination method in the sift algorithm. Pattern Anal. Appl. **24**(2), 669–683 (2021). https://doi.org/10.1007/s10044-020-00938-w

15. Wang, Y., Zhao, R., Liang, L., Zheng, X., Cen, Y., Kan, S.: Block-based image matching for image retrieval. J. Vis. Commun. Image Represent. **74**, 102998 (2021)

16. Wu, J., Zhang, L., Liu, Y., Chen, K.: Real-time vanishing point detector integrating under-parameterized ransac and hough transform. In: Proceedings of the IEEE/CVF International Conference on Computer Vision, pp. 3732–3741 (2021)

17. Djenouri, Y., Hatleskog, J., Hjelmervik, J., Bjorne, E., Utstumo, T., Mobarhan, M.: Deep learning based decomposition for visual navigation in industrial platforms. Appl. Intell. **52**(7), 8101–8117 (2022). https://doi.org/10.1007/s10489-021-02908-z

18. Yang, X., Gao, X., Song, B., Han, B.: Hierarchical deep embedding for aurora image retrieval. IEEE Trans. Cybern. (2020)

19. Giveki, D.: Scale-space multi-view bag of words for scene categorization. Multimedia Tools Appl. **80**(1), 1223–1245 (2021). https://doi.org/10.1007/s11042-020-09759-9

20. Djenouri, Y., Belhadi, A., Fournier-Viger, P., Lin, J.C.W.: Fast and effective cluster-based information retrieval using frequent closed itemsets. Inf. Sci. **453**, 154–167 (2018)

Ethereum Investment Based on LSTM and GRU Forecast

Adrián Viéitez Mariño[1]([⊠]), Matilde Santos Peñas[2] [ID], and Rodrigo Naranjo[1]

[1] Faculty of Informatics, Universidad Complutense de Madrid, 28040 Madrid, Spain
adrian.vmarino@gmail.com

[2] Institute of Knowledge Technology, Universidad Complutense de Madrid, 28040 Madrid, Spain

Abstract. This paper examines how a cryptocurrency price forecast system can be built using supervised learning techniques, not taking into consideration any technical indicators extracted out of the price evolution itself, and solely based on the values of other contextual stock indices, market indicators and online trends. LSTM and GRU models are used and their metrics compared in order to first clarify if any of them offers a better performance. The obtained results show that both are capable of forecasting very closely the actual evolution of the price and not having any of them clearly outperforming the other. Moreover, the viability of its application on the real market is analyzed by creating an investment strategy based on the predictions done by the neural networks. After applying the proposed investment strategy in two different periods of time, it is confirmed that even with a simple approach as the one presented in this work, it is possible to generate an outstanding profit with few operations.

Keywords: Cryptocurrency · Ethereum · Neural networks · LSTM · GRU · Investment

1 Introduction

Thirteen years have passed since the mysterious Satoshi Nakamoto launched his blockchain-based Bitcoin cryptocurrency back in early 2009. Since then, many events have taken place until reaching the current situation where several thousand of different cryptocurrencies absorb a volume of around 2.5 Trillion dollars [1], cryptocurrencies are accepted as payment method by more than 160000 business worldwide [2], and at least 103 countries have been identified, whose governments directed their financial regulatory agencies to develop regulations and priorities for financial institutions regarding cryptocurrencies and their use in AML/CFT, apart from many other countries that allow cryptocurrencies to be used [3].

As the cryptocurrency market has been growing and accumulating value, the study and understanding of its behavior and the variables that might affect it, has gained importance. Along with its growth, an increasingly strong link has been stablished with the traditional stock markets, and this facilitates the use of new forms of prediction that

© The Author(s), under exclusive license to Springer Nature Switzerland AG 2022
H. Yin et al. (Eds.): IDEAL 2022, LNCS 13756, pp. 271–279, 2022.
https://doi.org/10.1007/978-3-031-21753-1_27

can be as accurate, or more, than technical analysis or time series models [4]. It is, thus, of great interest the study of methods that may help us forecast the behaviour of the cryptocurrency market, since they have become another investment field for more and more people and companies.

In this paper, the application of deep learning techniques to forecast the evolution of cryptocurrency prices, in this specific case Ethereum, is presented. Indeed, stock markets trading and investments are areas where artificial intelligence have been proved to be a very useful approach [5, 6]. This work introduces an innovative group of features to feed the proposed neural networks, by first, selecting contextual financial indices and indicators, as well as cryptocurrency-related values, and second, by applying Feature Selection methods over the pre-selected group. Furthermore, the main goal and achievement is to deal with the scarcity of available data, building a system that can accurately predict prices under very different market conditions. In addition, to have a better idea of how reliable the application of the proposed neural networks together with the selected features, to the real market can be, an investment strategy is developed based on the networks predictions and its results are discussed. In summary, after an intense study of the state of art, the work builds all the steps of an investment strategy based on deep learning predictions, using conclusions that have been drawn in different works to understand what the optimal combination of methods is, and supports the conclusions on tangible results extracted from two periods of very different Ethereum price behavior, showing the robustness of the solution.

The structure of the paper is as follows. Section 2 presents the materials used for the study and the processing of the data. Section 3 shows the application of two neural network, LSTM and GRU, to predict the price of the cryptocurrency. Section 4 discusses the results of the investment strategy proposed. The paper ends with conclusions and future works.

2   Materials: Data and Pre-processing

The software used to develop this proposal is Visual Studio 2022 with Python 3.9. The main python libraries used have been: pandas, numpy, urllib, yfinance, pytrends, sklear and tensorflow [7]. The work has been developed using a laptop Medion Erazer X7857 with Intel Core i7–7820 HK, 4 cores with speed 2.90 GHz, 32 Gb RAM and graphic card Nvidia GeForce GTX 1070.

The collected data comes from different sources that might impact the evolution of cryptocurrency prices in general, as the Chinese stock market, American stock market or important financial assets, and the price of Ethereum in this particular case, as the price evolution of the Ethereum Gas, Ethereum transactions volume and indicators that express the interest of the online world on this cryptocurrency as the Wikipedia visits and the Google Trends search volume.

The data collection process has been mostly automated using Python. Therefore, the focus was placed on using services that provide an API to implement requests using this programming language.

For the retrieval of stock market indices as well as cryptocurrencies historical data it has been used the library *yfinance*, which is an open-source tool that offers a threaded and pythonic way to download market data from Yahoo! Finance. This portal, part of the Yahoo! Network, offers news and information of financial nature, online tools for personal finance management and data -also historical- of more than 10000 cryptocurrencies and a myriad of stocks and indices.

In the same way, the goal was to automate the collection of data related to other indicators used as features by the neural networks that will be presented later in this work. Thus:

– To gather data regarding the number of visits to the Wikipedia page of the term "Ethereum", it was used the python library *requests*, which allows us to send HTTP requests in a very easy way directly to the Wikipedia API and get back the information with a convenient JSON format that is easy to parse and save as *.csv in our local disk.
– To be able to fetch historical data for the search volume on Google of the term "Ethereum", the Python library pytrends has been used. This library provides a simple interface to Google backend and enables us to get daily and monthly data that can be used to compose our historical data record.
– EtherScan: Ethereum Gas price and Ethereum transactions volume are variables not available in all platforms. Their historical data has been retrieved manually from the website EtherScan.

Two tasks have been carried out in the pre-processing of the data:

• Fill in missing data: for all market indicators or indices, the data pertaining to weekends and non-working days in general was not available. For those dates, the last known value has been used and the date inserted in our data structure.
• Scale each data source between -1 and 1, using the MinMaxScaler function of the sklearn.preprocessing library.

Due to the daily granularity of all data and the nature of it, no changing drastically from day to day, the elimination of outliers or some filtering have not been performed.

2.1 Feature Selection

As it can be seen, a large number of variables have been collected to be used as inputs of our networks. However, this is not always the most convenient condition to obtain a good performance out of a supervised deep learning algorithm.

Feature selection methods shall be applied in this case, in order to reduce dimensionality of the feature space, gain accuracy in the forecasting results and reduce computational cost. In general, it does not result a simple task to select the best feature selection method, and some works [8, 9] suggest that wrapped methods outperform other techniques for very high dimensional feature spaces. In this case, two methods will be used and compared:

• RFE – Random Forest Regressor
• Pearson's Correlation Coefficient

To implement the feature selection techniques, the Python library Scikit has been used. The results are shown in the Table 1.

Table 1. Selected features

	RFE-RFR	Pearson's
Eth	X	X
Btc	X	X
BtcEth		X
SP500		
EurUsd	X	X
WkViews	X	X
VIX		
GTV		X
DJP		
DJI	X	
IXIC	X	X
MSCI	X	
HSI		X
SSE		X
N100		
CrOil		
Gold		
EthGas	X	
ETV	X	X

Both selections, done for seven days prediction, match in more than half of the features. As it could be expected, some of the most important features, selected by both methods are the price of Ethereum and the price of Bitcoin. At the same level appear to be the Nasdaq Composite (IXIC) and the relation Euro-USD (EurUsd), the visits to the "Ethereum" entry in Wikipedia (WkViews) and the Ethereum transaction volume (ETV).

It is also interesting to observe how one of the differences between methods is the greater importance that Pearson's method gives to Chinese market indices (HIS and SSE) over American indices, while RFE does the opposite (DJI, MSCI). Besides, while for RFE two of the selected features must also be the price of the Gas of Ethereum (EthGas) and the relation of Ethereum price with Bitcoin price (BtcEth), for Pearson's method the selected feature must be the number of searches done of the term "Ethereum" in Google (GTV).

Finally, as wrapped method, the Recursive Feature Elimination technique has been applied, that consists on a backward selection of the predictors.

3   Methods: Neural Networks Application

The selection of the neural network to apply is one of the most complex steps. In this case, based on the literature studied, the use of recurrent neural networks, more specifically the LSTM and GRU models, have been considered the best options. The object of study of this work is focused on supervised learning models, with the goal of forecasting of the cryptocurrency prices as a regression problem with seven days horizon.

3.1   Network Architecture and Parametrization

The selection of the right architecture and the tuning of the parameters is by itself a topic worthy of further study [10]. In this work, some tests have been done in order to find a reasonable solution that balances good performance with small investment in terms of computational time and effort, therefore no hyperparameter optimization has been carried out.

The architecture used in this work consists of one input layer and one hidden layer, both with 128 nodes and Dropout regularization with rate 0.2 that helps to prevent overfitting, plus a Dense layer to reduce the output to only one value. The input and the hidden layers will be LSTM or GRU depending on the implemented model.

The models have been fit using the efficient Adam optimization and the mean squared error loss function, using a batch size of 32 and 60 epocs.

3.2   Metrics

The selected metrics to evaluate the forecast results obtained from the networks are: Mean Squared Error (MSE), Root Mean Squared Error (RMSE) and Mean Absolute Error (MAE).

$$MAE = \frac{1}{N} \sum_{i=1}^{N} |e_i| \tag{1}$$

$$MSE = \frac{1}{N} \sum_{i=1}^{N} (e_i)^2 \tag{2}$$

$$RMSE = \sqrt{\frac{1}{N} \sum_{i=1}^{N} (e_i)^2} \tag{3}$$

3.3   Training and Testing Data

The data collection goes from 09-11-2017 until 13-05-2022. During this period of time, the evolution of the Ethereum price in USD looks as shown in Fig. 1.

Usually, between 70 to 80% of the data should be used for training the network and the remaining 20 to 30% for testing. In this case, since the data are so limited, having the

20% of data for testing would mean to have less than a year time, and using a 30% would only consider the long period of more stability in the price of Ethereum leaving the time where the most significant changes have taken place, out. This has an obvious negative impact on the capability of the network to generalize. Therefore, the goal has been to use a shorter period of time that could give us as much information as possible and represent a good variety of behavior of the Ethereum price evolution. Only 650 samples out of the 1647 have been used for training, what represents a 39% of the total data. The training data is the represented with shadowed background again in Fig. 1.

This selection leaves two very different periods for testing. The first with almost two years of data with mostly small changes and the second with around ten months of big jumps in the price of Ethereum.

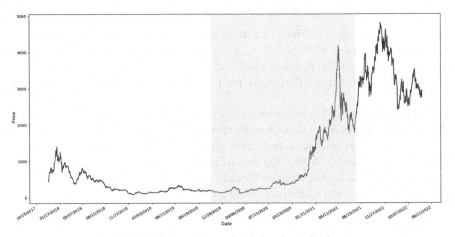

Fig. 1. Ethereum data and training data selection

3.4 Forecasting Results

For the evaluation of the two models, a loop with one hundred iterations has been implemented. Within this loop, the networks have been trained and tested in each iteration and the results collected to calculate the average value of the previously mentioned metrics MSE, RMSE and MAE.

To simplify, the whole set of data is used. The goal here is to obtain an overview of the performance of both networks, in order to see if any of them clearly outperforms the other.

Figure 2 shows the graphical results, giving the chance to compare at a glance the two forecasts with the actual data. Just by looking at the plot, it seems that the GRU model (red line) follows closer the actual evolution of the prices (blue line) in comparison to LSTM forecast (green line).

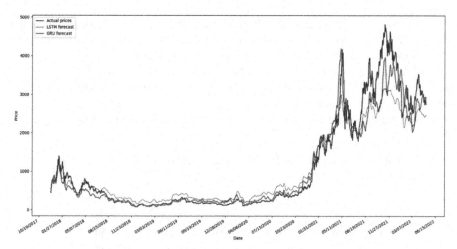

Fig. 2. Graphical comparation between actual data and forecasts

Now, looking at the numbers in Table 2, there is no big difference. Nevertheless, using the features selected by the Pearson's method seem to offer a slightly better performance and therefore, those will be the features used as inputs of our networks to apply the investment strategy. Regarding the model, the difference is so small that is worth to consider both and compare them putting them under test with the same investment strategy.

Table 2. Average values of metrics after one hundred tests

	LSTM			GRU		
	MSE	RMSE	MAE	MSE	RMSE	MAE
RFE	0.73	0.85	0.045	0.69	0.83	0.045
Pearson's	0.51	0.70	0.040	0.56	0.74	0.042

4   Investment Strategy

To have a good investment strategy that suits well the forecast data is a key factor. It could be that even not having good metrics measurements, the investment strategy throws profitable results. As the main purpose of this work is to study if the use of deep learning can somehow help to have a reliable investment system, the focus has been to find a strategy that using the obtained forecast data, would be profitable, keeping it as simple as possible.

The solution implemented detects local maximums or minimums in the forecast data, understanding those peaks as the price that on a specific day is higher or lower than the two previous and the two following days in the case of maximum and minimum, respectively.

When a local minimum is recognized, a purchase order is set for the correspondent date ahead of the mentioned minimum. When a local maximum is recognized, a selling order is set for the correspondent date ahead of the mentioned maximum.

Additionally, a stoploss strategy has been implemented and set to the 7% value of the Ethereum owned at the time. The stoploss is raised every time that the value of the owned Ethereum increases.

In this strategy, there will always be either only USD or only Ethereum.

Table 3. Average metrics after one hundred tests of the investment strategy

	1st period		2nd period	
	LSTM	GRU	LSTM	GRU
Final Capital	22483	24230	24791	24505
Earnings	12483	14230	14791	14505
% Earnings	124.83	142.30	147.91	145.05
Max. Drawdown	9517.58	15595.84	5429.22	4338.51
Profit factor	1.52	1.87	3.64	1.55
Avg. Profit	312.01	360.88	336.67	281.24
Avg. Loss	202.38	192.59	103.35	185.32
Buy Oper.	25.98	37.25	10.04	13.11
Sell Oper.	25.38	36.33	9.75	13.11
Stop-loss	16.42	20.27	7.25	7.79

The results (Table 3) obtained are remarkable when combining this strategy with any of the models. For both periods, before and after the training data set, the outcome is a very high profitability with a low number of operations.

5   Conclusions and Future Works

In this paper, supervised neural networks have been applied to forecast the price evolution of Ethereum. Using an extensive set of inputs that has been refined through Feature Selection methods, it has been shown the importance of choosing the right data to get an accurate prediction. Furthermore, it has been successfully overcome the challenge of obtaining well trained networks with a small amount of data, capable of offering an outstanding performance for all known Ethereum price evolution scenarios known so far. The self-developed investment strategy has confirmed the robustness of the solution and closed a complete sequence of steps followed to cover the full scope of building a reliable investment tool, from the selection of the data until the application of the predictions to the actual day-to-day market evolution.

Indeed, the use of RNN as LSTM and GRU models offers a very good performance and requires low architecture design and hyperparameter tuning effort.

In addition, an implementation of an investment strategy has been developed. Even ruling out complex metrics such as the Maximum Favorable Excursion and Maximum Adverse Excursion or indicators as the Take Profit, the results have shown an extraordinary outcome with earnings that exceed the 120% for all studied cases.

Nevertheless, the investment strategy used in this work has also revealed some weakness in the obtained forecast data, as it can be seen that many times the stoploss has triggered a selling operation, instead of the plan generated after the prediction. This is however understandable taking into consideration that both, the neural networks and the investment strategy, have big room for improvement. It is also expected that as time goes on and more historical data is accumulated, in the future it will be possible to have better trained networks and, together with the growing connection between cryptocurrency markets and traditional stock markets, the cryptocurrency prices should reduce their volatility and therefore become easier to predict with higher level of accuracy.

References

1. Follak, K.P.: Crypto assets: evolution and revolution in international financial markets. Mark. Sci. Technol. J. **1**(1), 1–11 (2022)
2. Holtfort, T., Horsch, A., Schwarz, J.: Economic, technological and social drivers of cryptocurrency market evolution and its managerial impact (2022). http://hdl.handle.net/10419/260544
3. Bajpai, P.: Countries where Bitcoin is legal and illegal, article in Investopedia (2021). www.investopedia.com. Accessed 12 July 2022
4. López, V., Santos, M., Montero, J.: Fuzzy specification in real estate market decision making. Int. J. Comput. Intell. Syst. **3**(1), 8–20 (2010). https://doi.org/10.1080/18756891.2010.9727673
5. Naranjo, R., Arroyo, J., Santos, M.: Fuzzy modeling of stock trading with fuzzy candlesticks. Expert Syst. Appl. **93**, 15–27 (2018). https://doi.org/10.1016/j.eswa.2017.10.002
6. Naranjo, R., Santos, M.: A fuzzy decision system for money investment in stock markets based on fuzzy candlesticks pattern recognition. Expert Syst. Appl. **133**, 34–48 (2019). https://doi.org/10.1016/j.eswa.2019.05.012
7. Python Package Index Projects Homepage. https://pypi.org/. Accessed 12 July 2022
8. Bagherzadeha, F., Mehranib, M.-J., Basirifardc, M., Roostaeid, J.: Comparative study on total nitrogen prediction in wastewater treatment plant and effect of various feature selection methods on machine learning algorithms performance, **41**, 102033 (2021). https://doi.org/10.1016/j.jwpe.2022.103203
9. Pirbazari, A., Chakravorty, A., Chumming, R.: Evaluating feature selection methods for short-term load forecasting (2021). https://doi.org/10.1109/BIGCOMP.2019.8679188
10. Márquez-Vera, M.A., López-Ortega, O., Ramos-Velasco, L.E., Ortega-Mendoza, R.M., Fernández-Neri, B.J., Zúñiga-Peña, N.S.: Diagnóstico de fallas mediante una LSTM y una red elástica. Revista Iberoamericana de Automática e Informática Ind. **18**(2), 160–171 (2021). https://doi.org/10.4995/riai.2020.13611

Generating a European Portuguese BERT Based Model Using Content from Arquivo.pt Archive

Nuno Miquelina(✉) ⓘ, Paulo Quaresma ⓘ, and Vítor Beires Nogueira ⓘ

Universidade de Évora, Évora, Portugal
d37384@alunos.uevora.pt, {pq,vbn}@uevora.pt

Abstract. Building a language model from free available internet information takes several steps and challenges. This new model aims to be a BERT-based language model for European Portuguese, with no specific context. The corpus was built using a web page archive infrastructure provided by Arquivo.pt and restricted to .pt domains. This paper will describe the overall process of building the corpus and training a BERT model.

Keywords: BERT · Vocabulary · Arquivo.pt · Portuguese European

1 Introduction and Motivation

The available text sources on the internet allow the gathering of vast amounts of content for training linguistics models [8]. These massive sources of information still need additional processing techniques like web scraping [5] to guarantee that only text with quality is retrieved, by eliminating pieces of code (HTML, JS, and others) from the actual text. Sentences and words are transformed into vectors (embeddings) and processed in an unsupervised way by the Deep Learning networks, generating language models that can be used for some Natural Language Processing (NLP) tasks like automatic translation among others [16].

Recurrent Neural Networks (RNN) were the main processing method for NLP tasks, but the true nature of this kind of neural networks fail or have less performance in processing long sequences because the first processed tokens get forgotten or lose importance. A novel approach introduced the concept of transformers [17]. This new architecture takes into account the weight of other tokens in the context. Based on this work, investigators presented BERT (Bidirectional Encoder Representations from Transformers) [4].

Recent benchmarks for evaluation of various tasks of natural language understanding (GLUE, MultiNLI, SQuAD v1.1 and SQuAD v2.0 benchmarks) showed that the BERT language representation model improved the state-of-the-art results. This new technique, for creating a language model representation, was designed to have a bidirectional context in all layers. Unlike other models like the OpenAI GPT (Generative Pre-Trained Transformer) that are based only

© The Author(s), under exclusive license to Springer Nature Switzerland AG 2022
H. Yin et al. (Eds.): IDEAL 2022, LNCS 13756, pp. 280–288, 2022.
https://doi.org/10.1007/978-3-031-21753-1_28

in a left-to-right only context. The OpenAI GPT has evolved and originated the new GPT-2 and GPT-3 (bigger training dataset and number of parameters) improving the overall benchmarks [3,14]. Another approach gives attention to the morphology of words, like fastText [2], and allows training models on large corpora and compute word representations for first seen words. ELMo [7] is also an example of a pre-training model with context-sensitive word representations.

BERT good results inspired other investigators to follow their work and to propose new models, like RoBERTa (Robustly Optimized BERT Pretraining Approach) [10], trying to improve the processing robustness changing: training the model longer over more data, removing next sentence prediction objective, training on longer sequences and dynamically changing the masking pattern applied to the training data. BERTimbau [15] and CamemBERT [11] are other projects in Brazilian Portuguese and French respectively, that aim to train mono-lingual models and evaluate their performance in different NLP tasks.

The remainder of this paper is organized as follows: Sect. 2 introduces the archive Arquivo.pt; in the Sect. 3 we provided a description of the process to create the European Portuguese corpus and how to use this corpus to train a BERT language model. Finally, Sect. 4 presents our conclusions together with some pointers for future work.

2   Arquivo.pt

Arquivo.pt[1] is an investigation infrastructure that allows to search and access web pages archived since 1996. Arquivo.pt started in January 2008, but the original idea to create a Portuguese website archive started in 2001 with the scientific project "tumba!" from Faculty of Sciences of the University of Lisbon. This archive makes site content available to researchers, content that could get lost by disappearing from the original sites [6]. The crawling process underneath only considers sites related to Portugal, i.e., sites under the *.pt* domain or embedded on a page hosted in *.pt* or even redirected from a *.pt* domain.

Table 1 describes the volume of data archived and the infrastructure that supports the archive process.

2.1   Arquivo.pt Interfaces

Arquivo.pt provides API interfaces to access the repository, mainly by two ways:

- Text search: passing a query parameter with the target text, the response is a list of files that have that content. This type of query is not our goal, because we don't want to restrict the response to a given text.
- URL search: passing a given URL the server API (Wayback CDX server API) will return a list of captures of that URL. This is the API used to build the set of URLs to retrieve the content. Since there was no temporal filter, the

---

[1] https://arquivo.pt/.

Table 1. Arquivo.pt information (January 2022)

Preserved data volume	Infrastructure
• 13 158 million web files	• 73 servers
• 28 million websites	• 17 TB of RAM memory
• 852 TB compressed content	• 1.816 vCPU
	• 1.186 Hard drives (4,5 PB)

url could have been retrieved at any time. One example of the API calling is the following:

  https://arquivo.pt/wayback/cdx?url=publico.pt

that returns the list of captures for the URL "publico.pt".

3   Content Retrieve Process

The content retrieve process and model train from Arquivo.pt was developed in Python programming language, using several separated applications, one per process action. Fig 1 shows the overall process actions and information saved. Each step of this process is detailed in the following sections.

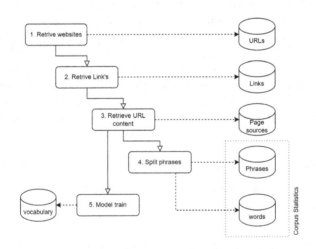

Fig. 1. Content retrieve and model train process

3.1   Step 1: Retrieve Websites

From the millions of available websites in Arquivo.pt we consider a subset of identified Portuguese periodicals. These websites were denoted by dados.gov[2], a Portuguese Government agency. The list has 1.703 websites, of which we consider 1.535 that have correct links for Arquivo.pt archive. The periodicals are Nationwide or Regional, such as "Público" and "Jornal do Fundão".

Our main criteria to select these Portuguese periodical websites is to access well-written and structured Portuguese news/text and with no special or restricted subject. However, we cannot forget that sometimes some of these journals have forums where readers can participate and these forums are more prone to grammatical errors.

3.2   Step 2: Retrieve Links

When we access the Arquivo.pt's archive API services, using the collected websites, we receive a list of links to the content archived over time. Also, here we had to make choices regarding the content retrieved:

- Response Status 200: Each of the returned links has metadata associated that indicates the response status of the retrieved page when it was collected, and we only care about the status 200, that is the HTTP status of a correct accessed page. We discarded pages with status like 301, which is a redirect and assumed that the content is only the necessary information for the browser to follow the new destination.
- Multipurpose Internet Mail Extensions or MIME type: In the HTTP protocol, the response of a request indicates the type of the content returned, and normally a website is composed of several elements (like style sheets, JavaScript, images, . . .) that we didn't want to process. Since Arquivo.pt also has this kind of metadata information, the MIME type, we filtered the following types of content: text/plain, text/html, application/pdf, application/rtf, text/rtf and application/msword.

In Table 2 it is possible to see that the main content type retrieved is the expected one, i.e., text/html. There is also some binary content to be processed (like PDFs). Not all of this content will be saved because the corpus doesn't benefit from repeated text content. It is expected to be repeated because if the site/page is archived for two consecutive days (or near) the content is the same.

The list of links that are extracted through the APIs provided by Arquivo.pt, contains more metadata that allows obtaining a certain page in the various moments in which it was captured. Therefore, we can have links where the address is the same, but the moment they were archived is different. All links are stored in a PostgreSQL database for processing.

---

[2] https://dados.gov.pt/pt/datasets/publicacoes-periodicas-portuguesas-jornais-e-revistas-websites-e-historico-de-versoes-no-arquivo-pt/.

Table 2. Retrieved links statistics

Statistic	Count
Websites processed	1.535
Links	3.487.429
Links Per MIME type	• **application/msword:** 4.880 • **application/pdf:** 275.705 • **application/rtf:** 191 • **text/html:** 3.197.706 • **text/plain:** 8.766 • **text/rtf:** 181

3.3 Step 3: Retrieve URL Content

Due to the fact that were applied filters to receive URLs with text and that succeeded at the time of the crawling process, it is now necessary to obtain the content. Depending on the type of content that was collected (indicated in the link list metadata) different processing is done:

– text/html or text/plain: the text of the request made to the Arquivo.pt repository (Wayback) is extracted, using a Python Trafilatura framework that can interpret the structure of the obtained text (html) and retrieve only the text that is rendered on the page. At this step, the process is also instructed to look only for the text in Portuguese;
– Other MIME types: for these contents, the Apache Tika (Python port) is used. This framework allows extracting text from different formats, including binaries like PDFs.

After extracting the text of the available links, the content is saved in a PostgreSQL database for further processing. Arquivo.pt can retrieve the same page over time (with the same content) but, for our process, there is no advantage in keeping equal content, so a hash of the collected text is generated, which has to be unique in the database, thus avoiding content duplication. A SHA256 hash is calculated and used as a unique constraint in the database. To keep the same encoding for the next steps, is guaranteed to save the extracted content as UTF-8.

The original html is also stored in the database so, if necessary, other frameworks for extracting text from html can be used in the future. Table 3 shows statistics of the content retrieved.

In the following, we described the Python frameworks used for this process:

• **Trafilatura:** [1,9] this library was evaluated as one of the best tools for web scraping, with great performance in retrieving text for web pages. It is used when the URL has a MIME type of text/html or text/plain. This framework has also the possibility to define a target language (that in our case was pt).

Table 3. Text capture statistics

Statistic	Count
Unique content text	428.719
Average text size (bytes)	5.240

- **Apache Tika**[3]: [12,13] proven capacity of extracting text from binary documents, is used to process the content when it is found a MIME type of: application/pdf, application/rtf, text/rtf or application/msword. This will permit the record of the text for the next processing steps. A Python port of the Apache Tika library that makes Tika available using the Tika REST Server.

3.4   Step 4: Split Phrases

For each content collected (text) it is necessary to extract the sentences that make up the text. In the first phase, the blocks are separated (by indicating the line change) and then in each block, the phrases are extracted. For this sentence extraction, is used the Python framework nltk, which allows this separation into sentences. At this stage and for each sentence, it is checked again if it is in Portuguese. The text as a whole may have been classified as being in Portuguese but, we want a second check at the sentence level. Again, when we are going to record the extracted phrases, there is no gain in repeating phrases (regardless of the source text) and we use again the creation of a phrase hash and this hash must be unique in the database. Therefore, in the background, the database insertion can be denied because there is another phrase with the same hash.

Words are also extracted individually from the sentence for:

- Having sentence structure information, indicating the word count, whether classified with stop words. This information per sentence will also allow us to choose sentences for training the model according to its dimension.
- Recording the unique words found.
- Statistics of the words found.

In Table 4 it is possible to see from the collected text, how many unique phrases are retrieved. The phrases also create a unique constraint in the database (hash created on the phrase lower case) to guarantee that there are no replicated phrases.

3.5   Step 5: Model Training

After collecting and building the corpus, it is needed to create a vocabulary that will be used in the training process. The vocabulary defines a set of tokens,

---

[3] https://tika.apache.org/.

Table 4. Phrase capture statistics

Element	Value
Unique phrases	16.198.437
Average phrase size (bytes)	68
Words average	6
Stop words average	4

collected from the corpus or special tokens like "[UNK]", "[PAD]", "[CLS]", "[SEP]" or "[MASK]". BERT can receive two sentences, so at the beginning always receive the "CLS" token and to separate two sentences is used the "SEP" token. "MASK" is a special token used to hide a token and let the algorithm try to find the token that best suits in the context. Since BERT receives a fixed sentence size, the special token "PAD" is used to fill the remain tokens. When a token don't appear in the vocabulary, the special token "UNK" is used to represent it.

- Normalization: used BertNormalizer: this pre-tokenizer splits tokens on spaces, and also on punctuation. Each occurrence of a punctuation character will be treated separately. BertNormalizer used parameters: lowercase = True;
- Pre-Tokenization: used BertPreTokenizer: that takes care of normalizing raw text before giving it to a Bert model. This includes cleaning the text, handling accents, Chinese chars and lowercasing;
- Model: used the WordPiece Tokenization Model, that is the tokenization algorithm Google developed to pretrain BERT. Used the configurations: special tokens=["[UNK]", "[PAD]", "[CLS]", "[SEP]","[MASK]"], min frequency=2
- Post Processor: define the rules for processing inputs with one or two sentences (rules for using "CLS" and "SEP" tokens).

The code was inspired by Hugging Face[4] community. We created vocabularies with 20000, 25000, 30000, 35000, 40000, 45000 and 50000 tokens with the proposal of future evaluations on the results and on the compute performance based on different vocabularies sizes. For the BERT training, we used the vocabulary of size 20000.

For the model training, the corpus was random separated to have a train and a validation corpus (10% of the original corpus). For this first approach, the model was trained for a Masked Language Model - predicting masked tokens in new sentences. The configuration was loaded from a pretrained "bert-base-cased": 12-layer, 768-hidden, 12-heads, 110M parameters. The model run with the following parameters: evaluation strategy = epoch, learning rate= $2e^{-5}$, weight decay=0.01. The training took around 40 h of computation in a high

---

[4] https://huggingface.co/.

performance computing infrastructure. This infrastructure belongs to the University of Évora and is a supercomputer (Vision[5]) made by 2 compute nodes and a management node. Each compute node is an NVIDIA DGX A100 system with the following specifications:

- 8x NVIDIA A100 GPU (40 GB each GPU)
- 2x AMD Rome 7742 (64 cores, 128 threads each CPU)
- 1 TB RAM
- 8x Single-Port Mellanox ConnectX-6 VPI 200Gb/s HDR InfiniBand
- 1x Dual-Port Mellanox ConnectX-6 VPI, 10/25/50/100/200Gb/s Ethernet
- 5 petaFLOPS AI / 10 petaOPS INT 8

4   Conclusions and Future Work

The training procedure was defined and tested to generate a first (that we know) European Portuguese language model based on BERT. The process was run entirely with success, and the future work will be focused on the optimization and evaluation of the model quality and compute performance. The evaluation will be carried throughout NLP tasks, such as NER. The final model and code will be shared with the community for investigation proposes. This work was done using a subset of information retrieved from Arquivo.pt and is expected, when using more content, to have better results in the language model quality.

References

1. Barbaresi, A.: Trafilatura: a web scraping library and command-line tool for text discovery and extraction. In: Proceedings of the Joint Conference of the 59th Annual Meeting of the Association for Computational Linguistics and the 11th International Joint Conference on Natural Language Processing: System Demonstrations, pp. 122–131. Association for Computational Linguistics (2021). https://aclanthology.org/2021.acl-demo.15
2. Bojanowski, P., Grave, E., Joulin, A., Mikolov, T.: Enriching word vectors with subword information. Trans. Assoc. Comput. Linguist. **5**, 135–146 (2017). https://aclanthology.org/Q17-1010
3. Brown, T., et al.: Language models are few-shot learners. In: Larochelle, H., Ranzato, M., Hadsell, R., Balcan, M., Lin, H. (eds.) Advances in Neural Information Processing Systems, vol. 33, pp. 1877–1901. Curran Associates, Inc. (2020). https://proceedings.neurips.cc/paper/2020/file/1457c0d6bfcb4967418bfb8ac142f6 4a-Paper.pdf
4. Devlin, J., Chang, M.W., Lee, K., Toutanova, K.: BERT: pre-training of deep bidirectional transformers for language understanding. In: Proceedings of the 2019 Conference of the North American Chapter of the Association for Computational Linguistics: Human Language Technologies, vol. 1 (Long and Short Papers), pp. 4171–4186. Association for Computational Linguistics, Minneapolis (2019). https://doi.org/10.18653/v1/N19-1423, https://aclanthology.org/N19-1423

---

[5] https://vision.uevora.pt/.

5. Diouf, R., Sarr, E., Sall, O., Birregah, B., Bousso, M., Mbaye, S.: Web scraping: state-of-the-art and areas of application, pp. 6040–6042 (2019). https://doi.org/10.1109/BigData47090.2019.9005594
6. Gomes, D., Nogueira, A., Miranda, J., Costa, M.: Introducing the Portuguese web archive initiative. In: 8th International Web Archiving Workshop. Springer, Heidelberg (2009)
7. Joshi, V., Peters, M., Hopkins, M.: Extending a parser to distant domains using a few dozen partially annotated examples. In: Proceedings of the 56th Annual Meeting of the Association for Computational Linguistics, vol. 1: Long Papers, pp. 1190–1199. Association for Computational Linguistics, Melbourne (2018). https://doi.org/10.18653/v1/P18-1110, https://aclanthology.org/P18-1110
8. Le, H., et al.: Flaubert: unsupervised language model pre-training for French. CoRR abs/1912.05372 (2019). http://arxiv.org/abs/1912.05372
9. Lejeune, G., Barbaresi, A.: Bien choisir son outil d'extraction de contenu à partir du web. In: 6e conférence conjointe Journées d'Études sur la Parole (JEP, 33e édition), Traitement Automatique des Langues Naturelles (TALN, 27e édition), Rencontre des Étudiants Chercheurs en Informatique pour le Traitement Automatique des Langues (RÉCITAL, 22e édition), volume 4: Démonstrations et résumés d'articles internationaux, pp. 46–49. ATALA, AFCP (2020)
10. Liu, Y., et al.: Roberta: a robustly optimized BERT pretraining approach. CoRR abs/1907.11692 (2019). http://arxiv.org/abs/1907.11692
11. Martin, L., et al.: CamemBERT: a tasty French language model. In: Proceedings of the 58th Annual Meeting of the Association for Computational Linguistics, pp. 7203–7219. Association for Computational Linguistics, Online (2020). https://www.aclweb.org/anthology/2020.acl-main.645
12. Mattmann, C.A., Zitting, J.L.: Tika in action (2012)
13. McCandless, M., Hatcher, E., Gospodnetić, O., Gospodnetić, O.: Lucene in Action, vol. 2. Manning Greenwich (2010)
14. Radford, A., Wu, J., Child, R., Luan, D., Amodei, D., Sutskever, I.: Language models are unsupervised multitask learners (2019). https://openai.com/blog/better-language-models/
15. Souza, F., Nogueira, R., Lotufo, R.: BERTimbau: pretrained BERT models for Brazilian Portuguese. In: Cerri, R., Prati, R.C. (eds.) BRACIS 2020. LNCS (LNAI), vol. 12319, pp. 403–417. Springer, Cham (2020). https://doi.org/10.1007/978-3-030-61377-8_28
16. Tripathy, J.K., et al.: Comprehensive analysis of embeddings and pre-training in nlp. Comput. Sci. Rev. **42**(C) (2021). https://doi.org/10.1016/j.cosrev.2021.100433
17. Vaswani, A., et al.: Attention is all you need. In: Guyon, I., Luxburg, U.V., Bengio, S., Wallach, H., Fergus, R., Vishwanathan, S., Garnett, R. (eds.) Advances in Neural Information Processing Systems, vol. 30. Curran Associates, Inc. (2017). https://proceedings.neurips.cc/paper/2017/file/3f5ee243547dee91fbd053c1c4a845aa-Paper.pdf

A Vision Transformer Enhanced with Patch Encoding for Malware Classification

Kyoung-Won Park[1] and Sung-Bae Cho[1,2(✉)]

[1] Department of Artificial Intelligence, Yonsei University, Seoul 03722, Korea
{pkw408,sbcho}@yonsei.ac.kr
[2] Department of Computer Science, Yonsei University, Seoul 03722, Korea

Abstract. With various benefits through software technology development, malicious attacks to steal confidential and company information have constantly been increasing. Recent deep learning models with images converted from malicious code achieve meaningful results, but they have challenges in classifying the same malware family, like Ramnit, Tracur, and Obfuscator. ACY that have similar structures in the image. Instead of observing the overall global features, there is a need for a method of considering the position of local features and learning the relationships between them. In this paper, we propose a vision transformer enhanced with the additional encoding of multiple patches for location information of local features and relationship information between them. For learning considering position information and all relationships between patches, [CLS] tokens that can summarize all information are utilized. 10-fold cross-validation with the Microsoft challenge dataset shows that the proposed model produces better accuracy than comparable studies. The misclassification analysis confirms that the proposed method can detect the same malware family penetrated by the conventional deep learning model. Additional analysis with the activation map emphasizes which structural and sequential features are extracted to detect different codes belonging to the same malware family.

Keywords: Malware detection,·Vision transformer · Location/relation information

1 Introduction

With the recent development of technology, the number of malicious codes attacking vulnerabilities of computers or networks has been increasing exponentially. Various research has been conducted to detect such malicious codes and minimize damage. Recent studies (i.e., static and dynamic analysis) have been steadily conducted. However, code obfuscation and compression technology have been developed, interfering with malicious code analysis. Moreover, there was a problem that the existing static and dynamic analysis could not detect new types of malicious attacks other than previously observed codes. Besides, detection avoidance technology has advanced, neutralizing previous static and dynamic analysis. Consequently, malware images converted from

© The Author(s), under exclusive license to Springer Nature Switzerland AG 2022
H. Yin et al. (Eds.): IDEAL 2022, LNCS 13756, pp. 289–299, 2022.
https://doi.org/10.1007/978-3-031-21753-1_29

assembly code have advanced to discover structural attack patterns concerning each malware type, neutralizing existing avoidance technologies. The image transformed from the byte code contains various information, as shown in Fig. 1. The Portable Executable (PE) assembly code format is used by Windows operating systems for executables, object code, DLLs, and other types of files. According to the PE Format, the information in the PE header corresponds to the header portion of the image and is mainly stored to run the executable files. The part corresponding to the null value corresponds to zero padding on the image. It serves to separate the PE header and the body or to separate sections inside the PE body. The ".text" section is a section containing the program execution code. ".data" of the PE body corresponding to the initialized data in the image has global and static variables initialized into a readable data section. However, the ".rdata" is a read-only data section with constant type and string constants. At the end of the PE file, the ".rsrc" section has information such as icons, cursor, and additional binary. ".reloc" is a section having relocation information and may not exist for each byte code. Since of this, the transformed malware image has these structural features, which can be used in the deep learning approach to classify malware types.

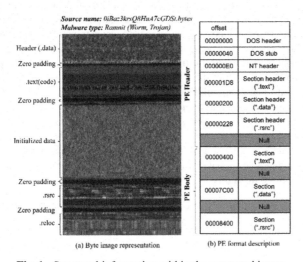

(a) Byte image representation   (b) PE format description

Fig. 1. Structural information within the converted image.

2   Related Works

Various attempts have neutralized the detection avoidance technologies by extracting structural patterns within malware images, starting with malware classification using the Grayscale image developed by Nataraj et al. [2]. The given binary file is read as a vector 8-bit integer to transform binary codes into images. Then each byte is converted into one pixel in the range [0, 255]. In early 2010, before deep learning approaches, various techniques other than decomposition tools such as GIST had extracted features from the images. Thus, the extracted features from Decomposition tools seemed efficient in classifying malicious attacks along with machine learning algorithms, such as KNN and SVM [13, 15, 18]. However, there was a problem that the whole process applied with the additional image decomposition tools in machine learning algorithms was complicated and time-consuming.

Several attempts were made to apply deep learning for easier malicious code detection by simplifying the feature extraction and classification process to solve time-consuming. Spatial features from convolutional neural networks have not only been improved in terms of time but also significantly improved in terms of performance [4, 19, 26]. However, it had shown limitations in achieving improved performance compared to the conventional static and dynamic analysis detection methods Many security experts and researchers have researched image generation methods and models to improve performance by utilizing spatial features from the converted images.

As summarized in Table 1, researchers have expanded the image generation method and considered how to make better images with lots of practical information beyond a grayscale two-dimensional image. After the visualization of the Grayscale image, some works attempted to extend one channel to RGB channels to enrich the amount of information. [4]. However, it was challenging to wish for a significant performance improvement. Besides, 16-bit (2-byte) or 24-bit (3-byte) integer vectors are transformed into images, similar to the process of converting a conventional 8-bit (1-byte) integer vector into one pixel in the malware image. [33]. In addition, a method of converting binary codes into Markov images using bytes transfer probability matrices has also been studied. [35].

In addition, extracting appropriate features from the given images is challenging since the number of actual observed malicious data is insufficient for deep learning. For example, some malware attacks, such as Simda 2015 Microsoft Kaggle competition data, account for less than 5% of the entire dataset. Therefore, extracting valuable features from a small amount of data is nearly impossible. As a second approach to solving the challenge, some works have artificially expanded the dataset's size via data augmentation techniques to extract information from insufficient data. A few attempts to improve performance by oversampling data via the generative adversarial network (GAN) [19, 20, 22, 23, 31, 32, 34]. Besides, other approaches generated data by adding noise-based filters, such as Gaussian and Poisson, to improve performance degradation caused by data imbalance during learning [22, 23, 32, 34].

Table 1. Related works on malware image generation methods.

Category	Reference
8-bit Grayscale image	[1, 2, 3, 5, 6, 7, 9, 10, 11, 12, 13, 18, 19, 20, 23, 25, 26, 27, 28, 30, 31]
8-bit RGB image	[4, 21, 22, 33]
Hash function-based	[14, 15]
Behavior Log Sequence from LSTM	[29]

(*continued*)

Table 1. (*continued*)

Category	Reference
Opcode sequence Matrix	[8]
Spectrogram(signal)	[16]
Hexadecimal Grayscale image	[17]
HTML/JavaScript Grayscale Image	[24]
Markov transfer probability Matrix	[35]
API n-gram to RGB image	[34]

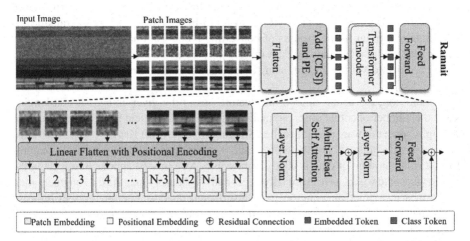

Fig. 2. Vision transformer to learn the positional and sequential attack features within images.

3 Proposed Method

When analyzing malware attacks through images, despite its high performance, the old approaches have some misclassified cases in similar malware-family and cause light degradation in accuracy performance. The location information between the malicious attack patterns and the order of the patterns are essential for precise classification between the same malware types, even though the structural features look similar. When extracting features through a convolution neural network, the network can use only a tiny pixel area in the local receptive field. As a result, information outside the receptive field is not visible and used simultaneously. Thus, the relationship information between different area features within the given data is not used. In addition, the convolution layer randomly changes the location of the filter to extract features, and sequence information between local patterns or location information of extracted features in the image is also not used. Therefore, we improve accuracy performance by making up for the shortcomings of convolution operations through the proposed vision transformer method. Therefore, the proposed model detects the misclassified cases of the CNN-based techniques [36].

3.1 Vision Transformer to Use Location Information of Local Features and Relationship Information

The overall proposed vision transformer structure is depicted in Fig. 2, consisting of four significant steps: (i) The first step is to split the input image into multiple small patches. (ii) The second step is to embed the divided patches through linear computation, generate [CLS] token to summarize all other local information later, and add position information to each linear patch encoding. (iii) The third step summarizes the association between the patches through multi-head attention. (iv) The final step is to classify classes using the [CLS] token of the final output of the transformer encoder.

In detail, we reshape images $x \in R^{C \times H \times W}$ into a sequence of p size of patches $x^p \in R^{N \times (p \times p \times C)}$, where N stands for the total number of patches. Then, 2D patches are flattened to 1D vectors and embedded via linear projection with matrix E:

$$[x_p^1 E; x_p^2 E; \ldots x_p^N E;] \in R^{N \times D} \tag{1}$$

Following the linear projection, we concatenate the flattened embedding vectors and the [CLS] token, which is learnable and will include summarized information from other patches. At the end of the training, the [CLS] token will contain all summarized information from other patches and will use to decide on the final class. Then we add unique position embedding vectors of each patch and create input z_0:

$$[x_{cls}; x_p^1 E; x_p^2 E; \ldots x_p^N E;] \in R^{(N+1) \times D} \tag{2}$$

$$z_0 = \left[x_{cls}; x_p^1 E; x_p^2 E; \ldots x_p^N E; \right] + E_{pos} \in R^{(N+1) \times D} \tag{3}$$

Next, we put z_0 vector into the transformer encoder. The input and output shape of the transformer encoder should be the same to repeat the encoder process L times. The patch embeddings pass through layer normalization (LN) and normalize the variance of the embeddings. Then the vectors pass through multi-head attention (MSA) layer.

$$Z_l = MSA(LN(Z_{l-1})) + Z_{l-1} \tag{4}$$

Before understanding multi-head attention, let's take a look self-attention mechanism first. Self-attention is a method introduced for natural language processing. It solves the long-term dependency problem in the RNN and LSTM in terms of long sequential data in the NLP task. Self-attention receives sequential information as a sentence, uses query(Q), key(K), and value(V) vector for each word in the sentence, and calculates how important the relationship between words in the sentence is. Then it summarizes the information as an attention score:

$$Attention score(Q, K, V) = softmax\left(\frac{QK^T}{\sqrt{d_k}}\right) V \tag{5}$$

According to the perspective of NLP, a patch is the same as a single word, and a sequence of patches is treated as a sentence. Then, the attention score contains relationship information between patches. Multi-head attention consists of several self-attention. It is the same concept that a deep understanding of the given information can be accomplished when viewed from multiple perspectives, not just one. The K number of summary information is returned from numerous self-attentions. Then we sum up those summaries to aggregated information with weight matrix multiplication. Information summarized through multi-head attention helps you not forget previous input information through residual connection.

After passing through the feed-forward network and the same layer normalization, the first transformer encoder completes the learning process. We repeat the encoder process L times to get the final output Z_L after passing through the last transformer encoders.

$$Z_l = MLP(LN(Z'_l)) + Z'_l, l = 1, 2, \ldots L \tag{6}$$

Z_L includes not only class information but also position and order information between patches from vision transformer. At the end of the procedure, only [CLS] token information extracted from Z_L can be utilized for the final classification task.

$$y = \left(LN \left(Z_L^0 \right) \right) \tag{7}$$

4 Experiments

4.1 Data Collection and Implementation Details

In this paper, we use the 2015 Microsoft malware classification challenge dataset to verify the justification of our method [37]. The size of the entire data is 37.89 GB. The total number of samples is 10,898. The distribution by malicious attack consists of Ramnit: 1541, Lollipop: 2478, Kelihos_ver3: 2942, Vundo: 475, Simda: 42, Tracur: 751, Kelihos_ver1: 398, Obfuscator.ACY: 1228, and Gatak: 1013. The assembly code is read as a vector 8-bit integer to transform into an image. Then each 8-bit is converted into one pixel in the range [0, 255] [2].

As seen in Fig. 3, each malware image has its structural characteristics corresponding to the malware class. As look into Simda class, Simda data is relatively less than other classes, about 1/100 up to, and it causes severe data imbalance with performance degradation of training. Besides, the Trojan-type attacks account for six of nine classes and seem to have similar structural characteristics.

We implement a vision transformer as an overall structure to classify using position and sequence information from an image. The entire structure is summarized in Table 2. The number of dimensions used and all details are presented as well. The proposed model receives a two-dimensional image with resolution 64×128 as input, returns a softmax probability for nine classes, and classifies the prediction class as the highest probability.

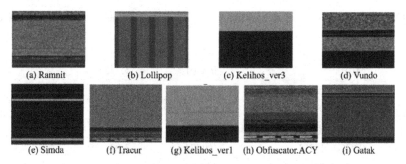

(a) Ramnit   (b) Lollipop   (c) Kelihos_ver3   (d) Vundo

(e) Simda   (f) Tracur   (g) Kelihos_ver1   (h) Obfuscator.ACY   (i) Gatak

Fig. 3. 2015 Microsoft malware classification challenge dataset.

Table 2. Summary of vision transformer structure and training parameters

Operation	Kernel	Size	Dropout	Nonlinearity
Input image: 64 × 128				
Augmented data	N/A	2048	0.0	ReLu
Patch encoding + adding [CLS]	N/A	128	0.0	GeLu
Transformer block (× 8)				
Layer normalization	$\varepsilon = 1e-6$			
Layer multihead attention				
Residual connection				
Layer normalization	$\varepsilon = 1e-6$			
MLP	N/A	256, 128	0.0	GeLu
Residual connection				
Classifier				
MLP with [CLS] token	N/A	2048,1024	0.5	ReLu
Linear (Class)	N/A	10	0.0	Softmax

4.2 Quantitative Experimental Results

As summarized in Table 3, this paper achieves accuracy improvement through a vision transformer compared to the previous image-based method with a resolution of 64 × 128. The old GAN models do not show enhanced performance than the deep neural network due to the resolution problem. In addition, tDCGAN achieves higher performance than CNN due to deep CNN neural networks and weight initialization through AutoEncoder. However, the proposed vision transformer method is better than any other existing deep learning method, and it evaluates the state-of-the-art performance in the image with a resolution of 64 × 128. Besides the evaluation through standard 10-fold cross-validation, we analyze our model through a confusion matrix about the validation dataset with the lowest accuracy among 10-fold cross-validation of the CNN model.

Table 3. 10-fold cross-validation with the recent deep learning methods in terms of images.

Methods	Acc.	Std.
MLP	0.8401	1.54e-03
GAN [19]	0.8781	3.44e-05
tGAN [19]	0.8810	8.05e-05
DCGAN [20]	0.9501	4.60e-05
tDCGAN [20]	0.9574	3.03e-05
CNN: ResNet101v2	0.9493	0.0137
Ours	**0.9683**	**0.0168**

Validation data achieving 0.9851 are used for confusion matrix analysis to support the validity of the proposed method. In the case of CNN (left) in Fig. 4, CNN misclassifies for the Trojan-based Ramnit, Tracur, and Obfuscator.ACY classes. Especially, Obfuscator.ACY, a hidden program, has most characteristics that the Trojan-type attacks should have. Thus, it seems a significant reason for performance degradation for the CNN model. However, as a result of applying the same validation dataset to the proposed method (right), our model shows excellent improvement in Obfuscator.ACY attacks and Ramnit attacks classification.

4.3 Visualization of Activation Area

For a deeper understanding of misclassified cases showing similar structural patterns in the confusion matrix analysis, we visualize the activation area of CNN and the proposed

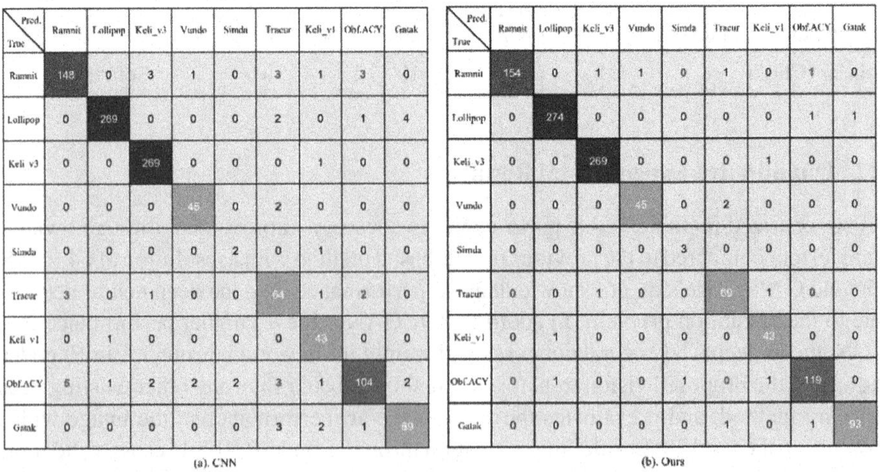

Fig. 4. Confusion Matrix between CNN (left) and the proposed method (right). Red bold represents the number of misclassified cases. Color figure online

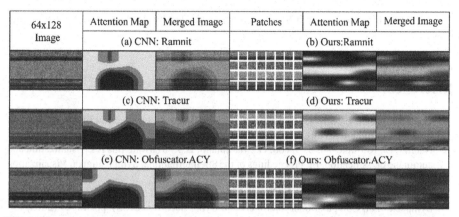

Fig. 5. Comparison of activation maps over the misclassified cases: Ramnit (a and b), Tracur (c and d), and Obfuscator. ACY (e and f). CNN observes an image as a chunk of data, but our model observes the sequence of patch information and put selective attention (red, yellow, and blue) in other patches. Color figure online

method for extracting features, as shown in Fig. 5. For similar-looking Ramnit, Tracur, and Obfuscator. ACY misclassification cases, the CNN extracts feature from mainly the yellow area with the same weight as expressed in a, c, and e. However, as shown in b, d, and f, the proposed method gives different colors for each patch within the image, and the meaning of this color reflects the degree of weights corresponding to the location information. High weights are displayed in red color, medium weights in yellow color, and low weights in blue color. For example, we confirm that the proposed model focuses on the features at the bottom of the Ramnit image and that the Tracur image puts high weight on the information displayed in three red parts. At last, our model focuses on the bottom of the Obfuscator. ACY image, such as Ramnit, extracts features extensively around the base area. As a result, we confirm that the three misclassification cases are accurately predicted in our model.

5   Conclusion and Discussion

The paper proposes a vision transformer-based method to utilize position information of patches and sequential information between the local features of the malware image. In addition, we improve the classification performance through the proposed model. Various experiments evidence that the proposed method resolves the misclassified limit by accurately classifying the Trojan-type malware image. However, preprocessing of resizing images is inevitable because transformer computation exponentially increases the number of parameters for learning high-resolution images. It is likely to occur information loss from the original byte image while resizing. Therefore, as a future study, we will optimize the transformer-based model over high-resolution images, and other research to generate high-resolution images to initialize parameters is still ongoing [38].

Acknowledgement. This work was supported by an IITP grant funded by the Korean government (MSIT) (No. 2020–0-01361, Artificial Intelligence Graduate School Program (Yonsei University)) and Air Force Defense Research Sciences Program funded by Air Force Office of Scientific Research.

References

1. Conti, G., Dean, E., Sinda, M., Sangster, B.: Visual reverse engineering of binary and data files. In: International Workshop on Visualization for Computer Security, pp. 1–17 (2008)
2. Nataraj, L., Karthikeyan, S., Jacob, G., Manjunath, B.S.: Malware images: visualization and automatic classification. In: Proceedings of the 8th International Symposium on Visualization for Cyber Security, pp. 1–7 (2011)
3. Kancherla, K., Mukkamala, S.: Image visualization based malware detection. In: IEEE Symposium on Computational Intelligence in Cyber Security (CICS), pp. 40–44 (2013)
4. Rezende, E., Ruppert, G., Carvalho, T., Ramos, F., De Geus, P.: Malicious software classification using transfer learning of resnet-50 deep neural network. In: 16th IEEE International Conference on Machine Learning and Applications (ICMLA), pp. 1011–1014 (2017)
5. Kalash, M., Rochan, M., Mohammed, N., Bruce, N.D., Wang, Y., Iqbal, F.: Malware classification with deep convolutional neural networks. In: 9th IFIP International Conference on New Technologies, Mobility and Security (NTMS), pp. 1–5 (2018)
6. Rezende, E., Ruppert, G., Carvalho, T., Theophilo, A., Ramos, F., Geus, P.D.: Malicious software classification using VGG16 deep neural network's bottleneck features. In: Information Technology-New Generations, pp. 51–59 (2018)
7. Vasan, D., Alazab, M., Wassan, S., Safaei, B., Zheng, Q.: Image-based malware classification using ensemble of CNN architectures (IMCEC). Comput. Secur. **92**, 101748 (2020)
8. Khan, R.U., Zhang, X., Kumar, R.: Analysis of ResNet and GoogleNet models for malware detection. J. Comput. Virol. Hack. Tech. **15**(1), 29–37 (2019)
9. Cui, Z., Xue, F., Cai, X., Cao, Y., Wang, G.G., Chen, J.: Detection of malicious code variants based on deep learning. IEEE Trans. Ind. Inf. **14**(7), 3187–3196 (2018)
10. Bhodia, N., Prajapati, P., Di Troia, F., Stamp, M.: Transfer learning for image-based malware classification. In: ICISSP (2019)
11. Su, J., Vasconcellos, D.V., Prasad, S., Sgandurra, D., Feng, Y., Sakurai, K.: Lightweight classification of IoT malware based on image recognition. In: IEEE 42Nd Annual Computer Software and Applications Conference (COMPSAC), vol. 2, pp. 664–669 (2018)
12. Yajamanam, S., Selvin, V.R.S., Di Troia, F., Stamp, M.: Deep learning versus gist descriptors for image-based malware classification. In: ICISSP, pp. 553–561 (2018)
13. Cui, Z., Du, L., Wang, P., Cai, X., Zhang, W.: Malicious code detection based on CNNs and multi-objective algorithm. J. Parallel Distrib. Comput. **129**, 50–58 (2019)
14. Jung, B., Kim, T., Im, E.G.: Malware classification using byte sequence information. In: Proceedings of the 2018 Conference on Research in Adaptive and Convergent Systems, pp. 143–148 (2018)
15. Han, K., Lim, J.H., Im, E.G.: Malware analysis method using visualization of binary files. In: Proceedings of the 2013 Research in Adaptive and Convergent Systems, pp. 317–321 (2013)
16. Azab, A., Khasawneh, M.: MSIC: malware spectrogram image classification. IEEE Access **8**, 102007–102021 (2020)
17. Li, L., Ding, Y., Li, B., Qiao, M., Ye, B.: Malware classification based on double byte feature encoding. Alexandria Eng. J. **61**(1), 91–99 (2022)
18. Nataraj, L., Kirat, D., Manjunath, B.S., Vigna, G.: Sarvam: search and retrieval of malware. In: Proceedings of the Annual Computer Security Conference (ACSAC) Worshop on Next Generation Malware Attacks and Defense (NGMAD) (2013)

19. Kim, J.Y., Bu, S.J., Cho, S.B.: Malware detection using deep transferred generative adversarial networks. In: International Conference on Neural Information Proceedings, pp. 556–564 (2017)
20. Kim, J.Y., Bu, S.J., Cho, S.B.: Zero-day malware detection using transferred generative adversarial networks based on deep autoencoders. Inf. Sci. **460**, 83–102 (2018)
21. Catak, F.O., Ahmed, J., Sahinbas, K., Khand, Z.H.: Data augmentation based malware detection using convolutional neural networks. PeerJ Comput. Sci. **7**, e346 (2021)
22. Burks, R., Islam, K.A., Lu, Y., Li, J.: Data augmentation with generative models for improved malware detection: a comparative study. In: IEEE 10th Annual Ubiquitous Computing, Electronics & Mobile Communication Conference (UEMCON), pp. 660–665 (2019)
23. Lu, Y., Li, J.: Generative adversarial network for improving deep learning based malware classification. In: Winter Simulation Conference (WSC), pp. 584–593 (2019)
24. Yoo, S., Kim, S., Kang, B.B.: The image game: exploit kit detection based on recursive convolutional neural networks. IEEE Access **8**, 18808–18821 (2020)
25. Choi, S., Jang, S., Kim, Y., Kim, J.: Malware detection using malware image and deep learning. In: International Conference on Information and Communication Technology Convergence (ICTC), pp. 1193–1195 (2017)
26. Kabanga, E.K., Kim, C.H.: Malware images classification using convolutional neural network. J. Comput. Commun. **6**(1), 153–158 (2017)
27. Hsiao, S.C., Kao, D.Y., Liu, Z.Y., Tso, R.: Malware image classification using one-shot learning with siamese networks. Procedia Comput. Sci. **159**, 1863–1871 (2019)
28. Zhu, J., Jang-Jaccard, J., Watters, P.A.: Multi-loss siamese neural network with batch normalization layer for malware detection. IEEE Access **8**, 171542–171550 (2020)
29. Tobiyama, S., Yamaguchi, Y., Shimada, H., Ikuse, T., Yagi, T.: Malware detection with deep neural network using process behavior. In: IEEE 40th Annual Computer Software and Applications Conf. (COMPSAC), vol. 2, pp. 577–582 (2016)
30. Tran, T.K., Sato, H., Kubo, M.: Image-based unknown malware classification with few-shot learning models. In: International Symposium on Computing and Networking Workshops (CANDARW), pp. 401–407 (2019)
31. Kim, J.Y., Cho, S.B.: Detecting intrusive malware with a hybrid generative deep learning model. In: International Conference on Intelligent Data Engineering and Automated Learning, pp. 499–507 (2018)
32. Moti, Z., et al.: Generative adversarial network to detect unseen internet of things malware. Ad Hoc Netw. **122**, 102591 (2021)
33. Jian, Y., Kuang, H., Ren, C., Ma, Z., Wang, H.: A novel framework for image-based malware detection with a deep neural network. Comput. Secur. **109**, 102400 (2021)
34. Bhaskara, V.S., Bhattacharyya, D.: Emulating malware authors for proactive protection using GANs over a distributed image visualization of dynamic file behavior. arXiv preprint arXiv:1807.07525 (2018)
35. Yuan, B., Wang, J., Liu, D., Guo, W., Wu, P., Bao, X.: Byte-level malware classification based on markov images and deep learning. Comput. Secur. **92**, 101740 (2020)
36. Dosovitskiy, A., et al.: An image is worth 16×16 words: transformers for image recognition at scale. In: ICLR (2021)
37. Ronen, R., Radu, M., Feuerstein, C., Yom-Tov, E., Ahmadi, M.: Microsoft malware classification challenge. arXiv preprint arXiv:1802.10135 (2018)
38. Jiang, Y., Chang, S., Wang, Z.: Transgan: two pure transformers can make one strong gan, and that can scale up. Adv. Neural Inf. Process. Syst. **34**, 14745–14758 (2021)

Association Rules Mining for Reducing Items from Emotion Regulation Questionnaires

Rihab Khadimallah[1,2(✉)], Ilhem Kallel[1,3], and Fadoua Drira[1,4]

[1] REGIM -Lab: Research Groups in Intelligent Machines, Sfax, Tunisia
{rihab_khadimallah,ilhem.kallel}@ieee.org
[2] FSEGS: Faculty of Economics and Management of Sfax, Sfax, Tunisia
[3] ISIMS: Higher Institute of Computer Science and Multimedia of Sfax, Sfax, Tunisia
[4] ENIS: National School of Engineers of Sfax, Sfax, Tunisia
fadoua.drira@enis.tn

Abstract. Regulating learners' negative emotions during a learning session is an important factor in educational settings that aims at enhancing learner's cognitive performance and achievement outcomes. In this context, many researchers in psychology have proposed emotion regulation questionnaires that help to assess the use of emotion regulation strategies by learners in order to regulate their emotions while learning. However, the number of items in each questionnaire is large which may annoy the learner and prevent him/her from completing all of the items; this may lead to inappropriate emotional regulation. Thus, we propose in this paper a machine learning method for mining items in order to reduce the fully associated ones to one item.

First of all, the paper presents a critical overview on statistical methods applied for reducing the large number of items in questionnaires. Then, after the introduction of the selected data set about the emotion regulation questionnaires, we detail the association rules mining method and discuss the obtained results about the significant association rules between the items. This can lead to uphold the items' reduction without loss of reliability.

Keywords: Emotion regulation questionnaire (ERQ) · Reducing items · Association rules mining

1 Introduction

The way we control our emotions has vital impact for our well-being and our social connections. In particular, the field of emotion regulation has become a well known topic over many sub-disciplines [1]. Currently, emotions' handling are considered as the mainspring of any cognitive process including learning activities. In fact, emotions impact the psychological and organic states and they act on learners' attention and on their capacities to get it and memorize knowledge [2]. Hence, many strategies and tools are implemented to allow learners regulate

© The Author(s), under exclusive license to Springer Nature Switzerland AG 2022
H. Yin et al. (Eds.): IDEAL 2022, LNCS 13756, pp. 300–312, 2022.
https://doi.org/10.1007/978-3-031-21753-1_30

their negative emotional states with the aim of enhancing their performance and learning productiveness [3,4].

Human adaptation requires the ability to control emotions. Indeed, neuroimaging research on emotion regulation can explain the relationship between the emotional state, the behavior or the lived experience and the cognitive neural process. Among many tools for emotion regulation, we notice the questionnaire which is a kind of survey composed of many questions/items. However, several issues caused by the use of long surveys have been highlighted. Accordingly, recent studies have revealed that the more there are questions on a survey, the less time a respondent will take to reply it, this means that the respondent spent less time thinking about each survey question, then he spent less time considering the most accurate response. Consequently, it may also indicate that responses are forced and may not reflect a respondent's honest feelings either.

Likewise, a huge number of items in a scale may have a conceivable negative impact on reaction rates and shorter scales minimize reaction predispositions. Accordingly, This paper intend to address the issue of how to come up with an optimal selection of scale items that would produce both focused and dependable scale development responses. Besides, in emotion regulation field, the scarcity of data is the main issue encountered by many researchers in the domain. Hence, this deficiency leads us to propose some questions: Are these available data sets contain well chosen items? Are the items or characteristics in data sets optimal enough for emotion regulation purpose?

2 Review on Some Methods Applied for Reducing Items from Questionnaires

Researchers are interested in using momentary self-report measures (questionnaires) in everyday life for a variety of reasons. In this context, longer questionnaires have been linked to higher perceived burden, reduced participation, and changes in data quality in traditional cross-sectional surveys [5]. Similarly, longer questionnaires have been specified as a common obstacle in many fields in psychology. That's why, in the literature, there exists several methods dealing with reducing the large number of items in questionnaire.

Accordingly, the authors Koczkodaj et al. [6] proposed the "Area Under the Receiver Operator Curve" (AUC ROC) method for reducing the number of rating scale items. This method consists of creating (ROC) curves for each item; where the range of each item under the curve (AUC) was utilized to decide and select the items that best discriminated between participants (the items were reduced from 21 items to 6 items). It is an original and applicable method for reducing the number of items but it is insignificant for a large population. Moreover, the authors in [7] conducted a systematic review about qualitative methods applied to generate questionnaire items in which the items are altered with added or evacuated questions based on subjective information and organized interviews. According to the three forms of self-efficacy test, items with lower discrimination indices were excluded. This method reduces the response bias in the questionnaires but it is used for one sample. In addition, the study of Alvarez et al. [8] used positively and negatively worded items that aims to diminish response predisposition. Item analysis approach is another method applied by [9] which helps reducing scale items without impacting the validity of the questionnaire. This method consists of calculating the mean score of each item by applying a t-test then the items that have the most elevated t-values are chosen. It helps condensing scale items and improving the accuracy of survey results, however; it can not be efficient each and every time used. Besides, the recent study developed by Harel et al. [10] and [11], adopted the Optimal Test Assembly (OTA) as a method for shortening patient-reported outcome measures. The shortened version is composed of only 5 items instead of 16 which has the potential to diminish patient burden and increment information quality, this method consists of sorting items according to their discrimination parameters and selecting them according to the size of their discrimination parameters or by a few blend of their difficulty and discrimination parameters, the scale items were reduced from 16 to 5 items. However, the findings were limited to adults not to the general population. Therefore, we can notice that these methods are not efficient and consistent enough since the questionnaires used have not a huge number of items (i.e., 16 items, 21 items). That's why we propose to contract highly associated items by mining association rules between items as a machine learning technique for reducing the large number of items.

3   Association Rules Mining as an Intelligent Approach for Reducing Items

The rapid growth of data and databases has created an imperative need for new tools and technologies that can quickly and effectively process raw data into usable data information. In the past few years, the development of information technology has promoted at the same time; hence, facilities for storing and managing databases have been added. Since the 1990s, the research field called "data mining" become the central topic of database and Artificial Intelligence (AI).

3.1   ERQ Data Set Description

In this work, and due the scarcity of data sets related to emotion regulation, we consider the SEK-27 data set used in [12] which is composed of emotion regulation questionnaires: ERQ, DERS, NMR, TAS, DASS and BSI.

The SEK-27 data set is provided to evaluate the psychometric properties of each emotion regulation questionnaire. However, in our study we tried to perform the association rules between the items in each emotion regulation questionnaire and then, reducing, a priori, the number of items according to the association rules extracted.

– **ERQ** (Emotion Regulation Questionnaire) is a 10-item self-report measure based on a validated emotion regulation process model involving emotion regulation strategies. It was first introduced by Gross and John [13] and was designed to evaluate particular differences in two emotion regulation strategies: cognitive reappraisal and expressive suppression.

- **DERS** (Difficulties in Emotion Regulation Scale) consists of 36 items and aims at assessing difficulties in emotion regulation (or dysregulation) [14].
- **DASS** (Depression Anxiety Stress Scales) consists of 42 items and aims at evaluating negative affective symptoms on the three named measurements: *Depression, Anxiety* and *Stress* [15].
- **TAS** (Toronto Alexithymia Scale) composed of 20 items and aims to measure alexithymia[1] based on three factors (difficulty identifying feelings, difficulty describing feelings, and externally oriented thinking) [16].
- **NMR** (Negative Mood Regulation) consists of 30 items and aims to measure expectancies for regulation of negative moods [17].
- **BSI** (Brief Symptom Inventory) contains 53 items to assess psychological symptoms in adolescents based on nine symptom dimensions: *Somatization, Obsessive-compulsive symptoms, Interpersonal sensitivity, Depression, Anxiety, Hostility, Phobic anxiety, Paranoid ideation,* and *Psychoticism* [18].

The data set contains 357 instances where the students respond to each item using a specific Likert-Scale especially in what degree the student agree to each questionnaire through a web-based survey.

3.2 Association Rules Mining

Unlike statistic methods which can find a linear correlation between two items, data mining techniques offer significant association rules in order to find any relationship, between the items, hidden in large data sets. Association rules are in the implication form (if/then statements) that can help discover relationships among irrelevant data in the data repository [19].

The basic formula of association rules as follows:

$X \rightarrow Y$, where both X and Y are items or itemsets and $X \cap Y = \emptyset$.

X can be defined as an itemset $\{a_1, a_2, ..., a_n\}$ and Y as an item a_j.

The notation of the association rule $\{a_1, a_2, ..., a_n\} \rightarrow a_j$ means that if X contains all of $\{a_1, a_2, ..., a_n\}$ then it is probable to contain a_j.

X is the antecedent and Y is the consequent.

In order to evaluate the importance and interest of association rules, three basic measures have been proposed:

- *Support of a rule:*
  Is the percentage of transactions containing both X and Y to the total number of items (measure of the frequency of items' occurrence).
  The support formula is :

$$Support(X \rightarrow Y) = \frac{Support(X \cup Y)}{Total\ number\ of\ items} \tag{1}$$

where both X and Y are itemsets and $X \cap Y = \emptyset$.

---

[1] The inability to depict or recognize a person's own emotions.

– *Confidence of a rule:*
Is the percentage of the number of transactions containing both X and Y to the total number of entries that include X (an indication of the number of times the rules are found true). It is expressed by:

$$Confidence(X \rightarrow Y) = \frac{Support(X \cup Y)}{Support\ (X)} \tag{2}$$

where both X and Y are itemsets and $X \cap Y = \emptyset$.
Conventionally, the support and confidence values range between 0 and 1.

– *Lift of a rule:*
Called also interest of a rule, is a measure of the performance of an association rule. It is expressed by:

$$Lift(X \rightarrow Y) = \frac{Support(X \cup Y)}{Support(X) \cdot Support(Y)} \tag{3}$$

where both X and Y are itemsets and $X \cap Y = \emptyset$.

Conventionally, the lift values range between 0 and infinity.

As a first step, we extract the association rules from "WEKA" with minimum support 0.05 and confidence more than 0.9; these values are fixed according to previous research study [20]. In fact, in this study, we are looking for the relationships' strength between the items but not the frequency of rules' occurrence. That's why, we decrease the minimum support value to 0.05 and increase the minimum value of confidence to 0.9. The upper bound for minimum support is set to 1.0 (100%) and the lower bound to 0.05 (5%). The algorithm stops when the lower bound for minimum support is reached. The lift threshold of a rule must be greater than or equal to 1. Then, we extracted the first 10 association rules from each emotion regulation measure.

Association Rules Related to Emotion Regulation Questionnaire
Table 1 reports the best rules extracted from ERQ. According to the extracted rules, the strongest rule is:

R2.{erq1=neutral, erq7=neutral, erq10=neutral} → {erq3 = neutral} which presents the highest lift (4.41), with support value of (0.07) and confidence value of (0.96). When the student responds "neutral" to erq1 and erq7, then; it is highly probable to respond "neutral" to erq3.

Consequently, there is a strong association between the items:

– erq1. *"I control my emotions by changing the way I think about the situation I'm in"*;
– erq7. *"I control my emotions by not expressing them"*;
– erq10. *"When I am feeling positive emotions, I am careful not to express them"*.
– erq3. *"When I want to feel more positive emotion, I change the way I'm thinking about the situation"*;

In this case, the learner controls his emotions by changing the way of thinking about the situation and not expressing them according to the emotion regulation strategies defined by Gross: Reappraisal and Suppression (according to the ERQ items presented in [13]).

One of the goals of this research is reducing a large number of items. Therefore, we extracted a set of useful association rules; after that we can reduce the number of items and develop a new emotion regulation questionnaire containing only the items: erq2, erq3, erq6, erq7 and erq10.

Table 1. Best rules related to Emotion Regulation Questionnaire

Association rule	Sup.	Conf.	Lift
1.{erq1=agree, erq5=agree, erq8=agree} → {erq7=agree}	0.08	0.97	2.61
2.{erq1=neutral, erq7=neutral, erq10=neutral} → {erq3=neutral}	0.07	0.96	4.41
3.{erq1=neutral, erq7=neutral, erq8=neutral} → {erq3=neutral}	0.06	0.96	4.39
4.{erq1=neutral, erq7=neutral, erq8=neutral} → {erq10=neutral}	0.06	0.96	3.68
5.{erq1=agree, erq3=agree, erq5=agree, erq8=agree} → { erq7=agree}	0.06	0.96	2.59
6.{erq1=neutral, erq7=neutral, erq8=neutral, erq10=neutral} → {erq3=neutral}	0.06	0.96	4.38
7.{erq1=neutral, erq3=neutral, erq7=neutral, erq8=neutral} → {erq10=neutral}	0.06	0.96	3.67
8.{erq1=agree, erq5=agree, erq8=agree, erq10=agree} → {erq7=agree}	0.06	0.95	2.58
9.{erq1=agree, erq3=agree, erq5=agree, erq8=agree, erq10=agree} → {erq7=agree}	0.05	0.95	2.57
10.{erq1=neutral, erq8=neutral, erq10=neutral} → {erq3=neutral}	0.08	0.93	4.27

Association Rules Related to Difficulties in Emotion Regulation Scale
In Table 2, the extracted association rules from DERS are presented. The rules (R1 to R7) have the same highest lift (1.56) and same confidence value (0.99). Given the following obtained best rule: R1.{ders27=almost never, ders28=almost never, ders31=almost never } → {ders32 = almostnever}

If the student responds to:

- ders27. *"When I'm upset, I feel like I am weak"*;
- ders28. *"When I'm upset, I feel like I can remain in control of my behaviors"*;
- ders31. *"When I'm upset, I have difficulty controlling my behaviors"*. with "almostnever" then he responds "almost never" to:
- ders32. *"When I'm upset, I believe that there is nothing I can do to make myself feel better"*;

There is a strong association between the factor 1 "Non-acceptance", the factor 3 "Impulse Control Difficulties" and the factor 5 "Strategies" [14]. In this situation, when the learner is upset, he doesn't feel like he is weak or he doesn't remain in control of his behaviors, consequently he almost never believes that there is nothing he can do to make himself feel better. He regulates his emotions by controlling his behaviors according to emotion regulation strategies especially *Suppression*.

After extracting the interesting association rules between the items, we can formulate another questionnaire of Difficulties in Emotion Regulation Scale (DERS) containing only the items number: 1, 2, 5 to 14, 17, 18, 20 to 26, 29, 32 and 34 to 36. So, we can capitalize 26 items instead of 36.

Table 2. Best rules related to Difficulties in Emotion Regulation Scale

Association rule	Sup.	Conf.	Lift
1.{ders27=almost never,ders28=almost never, ders31=almost never} → {ders32=almost never}	0.29	0.99	1.56
2.{ders16=almost never, ders27=almost never, ders31=almost never} → {ders32=almost never}	0.28	0.99	1.56
3.{ders14=almost never, ders27=almost never, ders31=almost never} → {ders32=almost never }	0.27	0.99	1.56
4.{ders14=almost never, ders28=almost never, ders31=almost never} → {ders32=almost never}	0.27	0.99	1.56
5.{ders15=almost never, ders27=almost never, ders31=almost never} → {ders32=almost never}	0.27	0.99	1.56
6.{ders27=almost never, ders30=almost never, ders31=almost never} → { ders32=almost never }	0.25	0.99	1.56
7.{ders16=almost never, ders27=almost never, ders28=almost never, ders31=almost never} → { ders32=almost never }	0.25	0.99	1.56
8.{ders27=almost never, ders31=almost never} → {ders32=almost never}	0.34	0.98	1.55
9.{ders19=almost never, ders27=almost never, ders31=almost never} → {ders32=almost never}	0.27	0.98	1.55
10.{ders27=almost never, ders35=almost never} → {der32=almost never}	0.27	0.98	1.55

Association Rules Related to Depression Anxiety Stress Scale
The best rules found for DASS are depicted in Table 3.

The rules (R4, R5, R6 and R7) have the highest lift (1.15), but they have different support and confidence values.

For the rule R5:

R5.{dass37= "never"} → {dass38= "never"} is the strongest one and has the highest lift (1.15) with support value of (0.76) and confidence value of (0.97). This means that when the student responds "never" to:

- dass37. *"I could see nothing in the future to be hopeful about"*;
Then he responds "never" to:
- dass38. *"I felt that life was meaningless"*.

In the factor of "Depression", there is a strong association between the items: dass37 and dass38 (according to the DASS items presented in [15]). After extracting the association rules between the DASS items, we can a priori, reduce the number of items by excluding the DASS item numbers: 4, 7, 15, 20, 21, 23, 25, 34, 37 and 40.

Table 3. Best rules related to Depression Anxiety Stress Scale

Association rule	Sup.	Conf.	Lift
1.{dass25=never} → {dass23=never}	0.8	0.97	1.1
2.{dass15=never, dass20=never} → {dass23=never}	0.75	0.97	1.09
3.{dass4=never} → {dass23=never}	0.77	0.97	1.09
4.{dass34=never} → {dass38=never}	0.76	0.97	1.15
5.{dass37=never } → { dass38=never}	0.76	0.97	1.15
6.{dass21=never, dass23=never} → {dass38=never}	0.75	0.96	1.15
7.{dass21=never } → { dass38=never}	0.77	0.96	1.15
8.{dass40=never} → {dass23=never}	0.08	0.96	1.08
9.{dass20=never, dass36=never} → {dass23=never}	0.76	0.9	1.08
10.{dass34=never} → {dass23=never}	0.75	0.96	1.08

Association Rules Related to Toronto Alexithymia Scale
Table 4 outlines the 10 association rules related to TAS. The following rules:
R1.{tas3= "strongly disagree", tas6= "strongly disagree", tas14= "strongly disagree"} → { tas7= "strongly disagree"} and R3.{tas3= "strongly disagree", tas6= "strongly disagree", tas13= "strongly disagree" } → { tas7= "strongly disagree"} have the highest confidence (0.99) and lift (1.93), but they have different support values (0.27 and 0.26). When the student responds "strongly disagree" to:

- tas3. *"I have physical sensations that even doctors don't understand"*;
- tas6. *"When I am upset, I don't know if I am sad, frightened or angry"*;
- tas14. *"I often don't know why I am angry"*;

Then he responds "strongly disagree" to:
- tas7. *"I am often puzzled by sensations in my body"*.

In factor 1 "Difficulty identifying feelings" (according to the TAS items presented in [16]), the respondent is "strongly disagree" if he has physical sensations and the doctors don't understand or if he doesn't know whenever he is sad, frightened or angry or if he doesn't know why he is angry. So, he will respond "strongly disagree" about feeling puzzled about the sensations in his body.

For the rule R3, the student responds "strongly disagree" to:

- tas3. *"I have physical sensations that even doctors don't understand"*;
- tas6. *"When I am upset, I don't know if I am sad, frightened or angry"*;
- tas13. *"I don't know what's going on inside me"*;
  Then he responds "strongly disagree" to:
- tas7. *"I am often puzzled by sensations in my body"*.

The participant is "strongly disagree" if he has physical sensations and the doctors don't understand or he doesn't know whenever he is sad, frightened or angry or when he doesn't know what's going on inside him.

In this questionnaire, the item numbers that can be excluded are: 1, 2, 6, 9, 12, 13 and 14 and then, we can formulate a new TAS questionnaire involving the item numbers: 3, 4, 5, 7, 8, 10, 11, 15, 16, 17, 18, 19 and 20.

Table 4. Best rules related to Toronto Alexithymia Scale

Association rule	Sup.	Conf.	Lift
1.{tas3=strongly disagree, tas6=strongly disagree, tas14=strongly disagree} → { tas7=strongly disagree }	0.27	0.99	1.93
2.{tas1=strongly disagree, tas7=strongly disagree} → { tas3=strongly disagree }	0.26	0.99	1.5
3.{tas3=strongly disagree, tas6=strongly disagree, tas13=strongly disagree} → { tas7=strongly disagree }	0.26	0.99	1.93
4.{tas2=strongly disagree} → { tas3=strongly disagree}	0.27	0.98	1.49
5.{tas1=strongly disagree, tas14=strongly disagree, } → {tas3=strongly disagree}	0.25	0.98	1.49
6.{tas3=strongly disagree, tas6=strongly disagree} → { tas7=strongly disagree }	0.3	0.97	1.9
7.{tas7=strongly disagree, tas13=strongly disagree, tas14=strongly disagree} → {tas3=strongly disagree}	0.27	0.97	1.48
8.{tas7=strongly disagree, tas13=strongly disagree} → { tas3=strongly disagree }	0.33	0.97	1.47
9.{tas6=strongly disagree, tas14=strongly disagree} → { tas7=strongly disagree }	0.28	0.96	1.88
10.{tas7=strongly disagree, tas9=strongly disagree} → { tas3=strongly disagree }	0.27	0.96	1.46

Association rules for Negative Mood Regulation

The best rules related to NMR are depicted in Table 5, the rule R1 is the strongest one: {nmr1="agree",nmr6="agree",nmr10="agree", nmr12="agree"} → {nmr2= "agree"}, which has the highest support (0.19), highest confidence (1) and highest lift (1.89). This means that these related answers are totally associated.

If the student answers "agree" to:

- nmr1. *"I can usually find a way to cheer myself up"*;
- nmr6. *"I can feel better by treating myself to something I like"*;
- nmr10. *"It won't be long before I can calm myself down"*;
- nmr12. *"Telling myself it will pass will help me calm down"*;
  Then, he probably responds "agree" to:
- nmr2. *"I can do something to feel better"*.

The items (nmr1, nmr6, nmr10, nmr12 and nmr2) are strongly associated since their lift value > 1.

The learner is "agree" about the way of regulating his negative moods by finding a way to cheer himself up, treating himself to something he likes, calming down himself in short time, consequently he can do something to feel better. In this case, the learner regulates his emotions referring to a particular emotion regulation strategy which is *suppression*.

After extracting the interesting rules related to Negative Mood Regulation (NMR), we can reduce the number of items and help learners save time when responding to the questionnaire. The new questionnaire will be composed of the NMR item numbers: 2, 3, 5, 8, 9, 11, 16, 17, 18, 19, 21, 23, 24, 25, 28 and 20.

Table 5. Best rules related to Negative Mood Regulation

Association rule	Sup.	Conf.	Lift
1.{nmr1=agree, nmr6=agree, nmr12=agree} → { nmr2=agree }	0.19	1	1.89
2.{nmr1=agree, nmr6=agree, nmr10=agree} → { nmr2=agree }	0.18	1	1.89
3.{nmr1=agree, nmr6=agree, nmr15=agree} → { nmr2=agree }	0.18	1	1.89
4.{nmr1=agree, nmr4=agree, nmr10=agree} → { nmr2=agree }	0.17	1	1.89
5.{nmr1=agree, nmr4=agree, nmr12=agree} → { nmr2=agree }	0.17	1	1.89
6.{nmr1=agree, nmr4=agree, nmr15=agree} → { nmr2=agree }	0.17	1	1.89
7.{nmr1=agree, nmr10=agree, nmr15=agree} → { nmr2=agree }	0.16	1	1.89
8.{nmr1=agree, nmr4=agree, nmr6=agree, nmr12=agree} → { nmr2=agree }	0.15	1	1.89
9.{nmr1=agree, nmr4=agree, nmr6=agree, nmr10=agree} → { nmr2=agree }	0.15	1	1.89
10.{nmr1=agree, nmr6=agree, nmr10=agree, nmr12=agree} → { nmr2=agree }	0.15	1	1.89

Association rules related to Brief Symptom Inventory
Table 6 outlines the best rules related to BSI questionnaire.

The rules (R2, R3, R4, R5, R6, R7 and R8) have the highest lift (1.15) and the same confidence (0.97) but different support values.

For the rule:

R6.{bsi12="not at all", bsi28="not at all"} → {bsi8="not at all"}, when the student responds "not at all" to:

- bsi12. *"Suddenly scared for no reason"*;
- bsi28. *"Feeling afraid to travel on buses, subways or trains"*,
  Then he probably responds "not at all" to:
- bsi8. *"Feeling afraid in open spaces"*

In this example, there is a strong association between the symptom dimensions "Phobic anxiety" and "Anxiety" (according to BSI items proposed in [18]). So, if the student is not at all scared for no reason and not at all afraid to travel on buses, subways or trains, then; he is not at all afraid in open spaces.

With the aim of reducing the number of items and creating a new questionnaire by excluding the BSI item numbers: 8, 39, 12, 30, 28, 31, 29, 37, 45, 39, 33, 9, 3, 40 and 13.

Table 6. Best rules related to Brief Symptom Inventory

Association rule	Sup.	Conf.	Lift
1.{bsi8=not at all, bsi39=not at all} → { bsi9=not at all }	0.72	0.98	1.14
2.{bsi12=not at all, bsi30=not at all} → { bsi8=not at all }	0.7	0.97	1.15
3.{bsi28=not at all, bsi31=not at all} → { bsi8=not at all }	0.7	0.97	1.15
4.{bsi8=not at all, bsi31=not at all} → { bsi28=not at all }	0.7	0.97	1.15
5.{bsi12=not at all, bsi29=not at all} → { bsi8=not at all }	0.7	0.97	1.15
6.{bsi12=not at all, bsi28=not at all} → { bsi8=not at all }	0.72	0.97	1.15
7.{bsi8=not at all, bsi37=not at all} → { bsi28=not at all }	0.7	0.97	1.15
8.{bsi8=not at all, bsi45=not at all} → { bsi28=not at all }	0.72	0.97	1.15
9.{bsi28=not at all, bsi39=not at all} → { bsi9=not at all }	0.71	0.97	1.13
10.{bsi28=not at all, bsi33=not at all} → { bsi8=not at all }	0.71	0.97	1.14

4   Conclusion

Emotion regulation measures are very useful in the field of emotion regulation in which the questionnaires are very applied in e-learning environment. In this research exploration, we start from the SEK-27 data set which is composed of emotion regulation measures to find reliable association rules between the questionnaire items. From these rules, we justified the proposed selection of items that can be reduced in order to generate a new questionnaire. This leads to help learners not spending a lot of time when responding to emotion regulation questionnaires.

The obtained results are considered as a training step in the process of reducing items from questionnaires. In one hand, it is necessary to explore the remaining association rules with highest lift and confidence. In another hand, deciding to eliminate items from the questionnaire needs applying test process. In our forthcoming work, we can apply the fuzzy logic to handle the uncertainty of the obtained values of support, confidence and lift. Besides, we can extend the study to construct a dynamic and scalable system that aims to find the right characteristics of the learner while learning with Intelligent Tutoring System (ITS).

References

1. McRae, K., Gross, J.J.: Emotion regulation. Emotion **20**(1), 1 (2020)
2. Oxford, R.L., Bolaños-Sánchez, D.: A tale of two learners: discovering mentoring, motivation, emotions, engagement, and perseverance. In: Gkonou, C., Tatzl, D., Mercer, S. (eds.) New Directions in Language Learning Psychology. SLLT, pp. 113–134. Springer, Cham (2016). https://doi.org/10.1007/978-3-319-23491-5_8
3. Abdelkefi, M., Kallel, I.: Towards a fuzzy multiagent tutoring system for M-learners' emotion regulation. In 2017 International Conference on Information Technology Based Higher Education and Training (ITHET), pp. 1–6. IEEE (2017)
4. Khadimallah, R., Abdelkefi, M., Kallel, I.: Emotion regulation in intelligent tutoring systems: a systematic literature review. In: 2020 IEEE International Conference on Teaching, Assessment, and Learning for Engineering (TALE), pp. 363–370. IEEE (2020)
5. Eisele, G., et al.: The effects of sampling frequency and questionnaire length on perceived burden, compliance, and careless responding in experience sampling data in a student population. Assessment **29**(2), 136–151 (2022)
6. Koczkodaj, W.W., et al.: How to reduce the number of rating scale items without predictability loss? Scientometrics **111**(2), 581–593 (2017)
7. Ricci, L., et al.: Qualitative methods used to generate questionnaire items: a systematic review. Qual. Health Res. **29**(1), 149–156 (2019)
8. Suárez Álvarez, J., et al.: Using reversed items in likert scales: a questionable practice. Psicothema **30** (2018)
9. Pather, S., Uys, C.S.: Using scale reduction techniques for improved quality of survey information. SA J. Inf. Manag. **10**(3) (2008)
10. Harel, D., Baron, M.: Methods for shortening patient-reported outcome measures. Stat. Methods Med. Res. **28**(10–11), 2992–3011 (2019)
11. Harel, D., Mills, et al.: Shortening patient-reported outcome measures through optimal test assembly: application to the social appearance anxiety scale in the scleroderma patient-centered intervention network cohort. BMJ Open **9**(2), e024010 (2019)
12. Grant, M., Salsman, N.L., Berking, M.: The assessment of successful emotion regulation skills use: development and validation of an English version of the Emotion Regulation Skills Questionnaire. PloS One **13**(10), e0205095 (2018)
13. Gross, J.J., John, O.P.: Individual differences in two emotion regulation processes: implications for affect, relationships, and well-being. J. Pers. Social Psychol. **85**(2), 348 (2003)

14. Gratz, K.L., Roemer, L.: Multidimensional assessment of emotion regulation and dysregulation: Development, factor structure, and initial validation of the difficulties in emotion regulation scale. J. Psychopathol. Behav. Assess. **26**(1), 41–54 (2004)
15. Lovibond, P.F., Lovibond, S.H.: The structure of negative emotional states: comparison of the depression anxiety stress scales (DASS) with the beck depression and anxiety inventories. Behav. Research Therapy **33**(3), 335–343 (1995)
16. Parker, J.D., Taylor, G.J., Bagby, R.M.: The 20-item toronto alexithymia scale: III. Reliability and factorial validity in a community population. J. Psychosomatic Res. **55**(3), 269–275 (2003)
17. Catanzaro, S.J., Mearns, J.: Measuring generalized expectancies for negative mood regulation: initial scale development and implications. J. Pers. Assess. **54**(3–4), 546–563 (1990)
18. Derogatis, L.R., Melisaratos, N.: The brief symptom inventory: an introductory report. Psychol. Med. **13**(3), 595–605 (1983)
19. Ji, Y., Liu, S., Zhou, M., Zhao, Z., Guo, X., Qi, L.: A machine learning and genetic algorithm-based method for predicting width deviation of hot-rolled strip in steel production systems. Inf. Sci. **589**, 360–375 (2022)
20. Rekik, R., Kallel, I., Casillas, J., Alimi, A.M.: Assessing web sites quality: a systematic literature review by text and association rules mining. Int. J. Inf. Manag. **38**(1), 201–216 (2018)

Explainable Artificial Intelligence for Improved Modeling of Processes

Riza Velioglu[1,2]($\boxtimes$)[iD], Jan Philip Göpfert[2][iD], André Artelt[1][iD],
and Barbara Hammer[1][iD]

[1] CITEC, Bielefeld University, Bielefeld, Germany
{rvelioglu,aartelt,bhammer}@techfak.de
[2] Recommendy UG, Bielefeld, Germany
j.goepfert@recommendy.dev

Abstract. In modern business processes, the amount of data collected has increased substantially in recent years. Because this data can potentially yield valuable insights, automated knowledge extraction based on process mining has been proposed, among other techniques, to provide users with intuitive access to the information contained therein. At present, the majority of technologies aim to reconstruct explicit business process models. These are directly interpretable but limited concerning the integration of diverse and real-valued information sources. On the other hand, Machine Learning (ML) benefits from the vast amount of data available and can deal with high-dimensional sources, yet it has rarely been applied to being used in processes. In this contribution, we evaluate the capability of modern Transformer architectures as well as more classical ML technologies of modeling process regularities, as can be quantitatively evaluated by their prediction capability. In addition, we demonstrate the capability of attentional properties and feature relevance determination by highlighting features that are crucial to the processes' predictive abilities. We demonstrate the efficacy of our approach using five benchmark datasets and show that the ML models are capable of predicting critical outcomes and that the attention mechanisms or XAI components offer new insights into the underlying processes.

Keywords: Machine learning · Process mining · Transformer · XAI

1 Introduction

Data is collected on anything, at any time, and in any location. A study by IBM [6] found that in 2020, 40 trillion gigabytes (40 zettabytes) were generated. The majority of data in the digital realm, however, is unstructured, making it

This research was supported by the research training group "Dataninja" (Trustworthy AI for Seamless Problem Solving: Next Generation Intelligence Joins Robust Data Analysis) funded by the German federal state of North Rhine-Westphalia, and supported by the European Commission Horizon for *ICU4COVID* project, and the VW-Foundation for the project *IMPACT*.

© The Author(s), under exclusive license to Springer Nature Switzerland AG 2022
H. Yin et al. (Eds.): IDEAL 2022, LNCS 13756, pp. 313–325, 2022.
https://doi.org/10.1007/978-3-031-21753-1_31

impossible for humans to process it in its entirety and leaving businesses struggling to manage such massive amounts of data. As a result, one of today's major issues for businesses is extracting information and value from data contained in information systems.

Process Mining (PM) [2] is a relatively new area of research that lies between process modeling and analysis, machine learning, and data mining. It is an umbrella term for a family of techniques to facilitate the analysis of operational processes by extracting knowledge from event logs in symbolic form, typically in the form of process models. An event log serves as the input to PM algorithms, which is essentially a database of events where each event has (1) a case id: a unique identifier for a particular process instance, (2) an activity: description of the event that is occurring, and (3) a timestamp: timestamp of the activity execution. Resources, expenses, and other event-related attributes may also be integrated by some techniques. PM techniques mostly address three tasks: (1) process discovery: transform the event log into a process model which describes the underlying process generating such data: common techniques include the alpha algorithm, heuristic miner, or inductive miner [1], (2) conformance checking: analyse discrepancies between an event log and an existing process model to detect deviations, for quality checking or risk management [22], (3) process enhancement: improve the existing model's performance in terms of specific process performance metrics [34].

The initial focus of PM was the analysis of *historical* event data to generate a process model. These monitoring systems are reactive in nature, allowing users to detect a violation only after it has occurred. In contrast, Predictive Process Mining (PPM) or Predictive Process Monitoring [18] aims for forward-looking forms of PM with predictive qualities. It is a field that combines machine learning with process mining, aiming for specific tasks such as predicting *next activity*, *activity suffix*, *next timestamp*, *remaining time*, *attributes*, *attribute suffix*, and *outcome* of running cases. Most tasks can be modeled as classification problems, except *next timestamp* and *remaining time* prediction tasks, which are regression problems. Some approaches which have been used in this realm are based on recurrent neural networks (RNNs) or Long-short term memory (LSTMs) [11,23,29]. Alternatives are based on Autoencoders [20] and Convolutional neural networks [24]. More recently, Transformer [32] models have gained a lot attention due to their overwhelming success in computer vision [10] and natural language processing [7]. Inspired by their success, recent approaches proposed a Transformer-based model for process data [5]. A more detailed overview on different models for the aforementioned tasks is covered in the survey [27].

Previous work in PPM faces a few challenges: unclear training/test set splits and pre-processing schemes, which leads to noticeably different findings and challenges w.r.t reproducibility [33]. Since datasets often contain duplicates which are not respected by the evaluation schemes, results are often confounded. In this work, we investigate the capability of important classical ML technologies as well as modern Transformers on a variety of benchmark sets with clear train/test splits, duplicates respected, and different types of representation for classical schemes. More importantly, we do not treat the methods as opaque schemes, but rather rely on feature relevance determination and attention mechanisms to highlight the most important factors for each model. This is in line with previous approaches such as [13], which visualizes the feature relevances of a gated graph neural network or [14] which enhances next activity prediction using LSTMs with counterfactual explanations for wrongly classified samples. Yet both approaches evaluate the behavior on a single and specific dataset only. In our approach, we systematically investigate relevant ML models for different approaches and representations under the umbrella of explainability. In this way, we demonstrate how, in many cases, the presence of specific events can easily fool the system into relying on 'trivial' signals. When these are removed, ML models reveal not only high quality behavior but also intriguing insights into the relevance of more subtle signals or sequences.

Contributions. Our contribution is twofold: a systematic study of the behavior of diverse ML models with train-test split respecting the peculiarities of datasets in PM and appropriate pre-processing, which prevent troubles due to data leakage. Second, we utilize eXplainable Artificial Intelligence (XAI) techniques to compute and visualize the most crucial features. Our code is open source and available at https://github.com/rizavelioglu/ml4prom.

The remainder of this work is structured as follows: In Sect. 2, we present the peculiarities of the datasets and our methodology. In Sect. 3, we present experiments and results. In Sect. 4, we conclude our contribution and discuss potential directions for future work.

2   Methodology

In this section we describe the datasets and how they are pre-processed to be utilized for binary classification, and highlight a few domain-specific peculiarities which have to be taken into account. Then we introduce both classical ML models as well as the state-of-the-art Transformer model used subsequently. Lastly, we present the XAI techniques applied.

2.1   Data

We focus on five widely used datasets which come from a variety of domains including loan applications, road traffic management, and healthcare. These datasets are benchmarks from PM and are also accessible within the well-known ProM tools [31], with the exception of healthcare, which contains personal data. Each dataset has been labeled in order to assess explainability and put them in the ML context. The variety of datasets enables us to check the robustness of our methodology across a wide range of domains. To apply ML classifiers to event data, we first transform the data into a format that resembles a binary classification problem. We present five real-life event logs, each with its own evaluation criterion for a prediction task (referred to as the positive/negative subpart of the logs), and thereafter explain how they are transformed for the task.

BPIC 2017_application+offer [8]: a loan application process of a Dutch financial institute that covers 31 509 cases. Each case represents a loan application and every application has an outcome: positive if the offer is accepted and negative if rejected.

BPIC 2018_application [9]: an agricultural grant-application process of the European Union that contains 43 809 applications over a period of three years together with their duration.

Traffic Fine Management [16]: an event log of an information system managing road traffic fines by the Italian government, with fines being payed or not.

COVID [26]: a dataset of COVID-19 patients hospitalized at an intensive care unit consisting of 216 cases, of which 196 are complete cases (patients have been released either dead or alive) and 20 ongoing cases (partial process traces) that were being treated in the COVID unit at the time data was exported.

Hospital Billing [19]: an event log of the billing of medical services that have been provided by a regional hospital. Each trace in the event log keeps track of the actions taken to bill a group of medical services. A randomly selected sample of 100 000 process instances make up the event log.

2.2 Encoding

Table 1 shows the rules used to label the samples as well as important statistics. One characteristic of typical PM datasets is that they can contain a large number of duplicates, *i.e.* observations of the same process. For example, Hospital dataset contains 41 343 and 58 653 traces in positive and negative classes, respectively. Out of those traces only 306 and 884 are unique. In addition, the top-10 most frequent traces make up the 95.3% and 89.7% of the whole traces. Hence such data are challenging for ML due to a narrow variety. In addition, duplicates need to be accounted for in evaluation to avoid information leakage.

Data consist of sequences of symbolic events of different length. For each of these datasets we compute n-grams for $n \in \{1, 2, 3\}$ to encode traces in vectorial form for classic ML models. Unigrams ($n = 1$) encode occurrence of events, bigrams ($n = 2$) encode two subsequent events and trigrams ($n = 3$) encode three subsequent events. Transformer models can directly deal with sequences. The network learns a vector embedding for each event within a trace. This has the advantage of avoiding problems caused by the high-dimensionality of one-hot encoding method, for example.

2.3 Models

Using these representations, we train a variety of models, including Logistic Regression (LR), Decision Tree (DT), Random Forest (RF), Gradient Boosting (GB), and Transformer models. We only present the LR, DT and Transformer models because the findings do not vary significantly. We selected LR

Table 1. Datasets statistics and the rules used to generate binary classification task (L$^+$/L$^-$ represent positive/negative logs, respectively). The statistics are the number of traces, the number of unique traces, the percentage of unique traces, and lastly the cumulative percentage of the top-10 most frequent traces.

Dataset	Class	Classification criteria	Traces	Uniq.	Uniq./%	Top-10%
BPIC17	L$^+$	Without activity 'A_incomplete'	16506	529	3.2	83.5
	L$^-$	With activity 'A_incomplete'	15003	2101	14.0	51.5
Traffic	L$^+$	End activity 'Payment'	67201	122	0.2	99.1
	L$^-$	End activity not 'Payment'	83169	109	0.1	99.2
COVID	L$^+$	With activity 'Discharge alive'	136	33	24.0	74.3
	L$^-$	With activity 'Discharge dead'	60	20	33.0	83.3
BPIC18	L$^+$	Duration less than 9 months	27966	1081	3.9	57.7
	L$^-$	Duration more than 9 months	15843	477	3.0	82.1
Hospital	L$^+$	Duration less than 3 months	41343	306	0.7	95.3
	L$^-$	Duration more than 3 months	58653	884	1.5	89.7

and DT due to their widespread use and overall high performance for classification problems [12], and we selected Transformer models due to their propensity for learning complex patterns. The type of encoding used in this case limits the amount of information that is accessible because only a portion of the sequential structure is represented. Conversely, transformer models can easily handle sequential data. We use models as proposed in the works [5,32].

2.4 XAI

Since our primary concern is not the classification performance of our trained models, but rather whether ML models can capture underlying regularities, we employ XAI techniques to gain insight into which relationships between events in the data are used by the models [21]. Unlike many explanation approaches such as LIME [28], LRP [3], we are interested in global explanations rather than explanations of single decisions. This is because our goal is not to comprehend the ML model, but rather to gain understanding of the key components of the PM as a whole, which can enable users to enhance processes. As an example, one may anticipate altered process behavior if they come across a specific activity, such as 'Discharge dead' in COVID dataset, which was identified as crucial for the ML model. As a result, we strive for global explanations and, more specifically, we employ a number of well-established feature selection methods for both linear and non-linear settings, including LR with LASSO as a linear model with strong mathematical guarantees [30], DT as a non-linear model with efficient Mean Decrease in Impurity (MDI) and permutation importance [17] that takes into account non-linear relations between features to determine the relevances, and Transformer model equipped with an attention mechanism that highlights complex relationships. Attention mechanisms are technically local explanations.

Table 2. Datasets with features that leak label information.

Dataset	Class	Biased feature
BPIC17	L^-	'A_Incomplete'
Traffic	L^-	'Send for Credit Collection'
	L^+	'Payment'
COVID	L^-	'Discharge dead'
	L^+	'Discharge alive'

We utilize them to see if highly flexible non-linear explanations provide more complex relations than global feature importance measures.

As a first result of the analysis, feature selection methods immediately reveal a potential source of information leakage: Leaving the datasets as they are, the algorithms rely on attributes that directly encode the class label but provide no additional insight into the process. Table 2 shows the detected features that are removed from the logs to avoid such trivial outcomes. All of the listed events exist only in their respective class, e.g. 'A_Incomplete' is present only in L^- but not in L^+, except 'Payment' where the event is present in both classes. Therefore, we removed the event only if it is the last event in a trace from L^+, as that is the source of leakage (see Table 1).

3   Experimental Results

In this section we present the data pre-processing pipeline used to transform all five datasets for training. Then, we present the evaluation metric used as well as the model design and training on each dataset. Finally, we present the scores of different approaches on the binary classification task and the resulting feature relevances. To save space, we limit ourselves to the BPIC17 dataset; results for the other datasets can be found in the GitHub repository.

3.1   Data Split and Pre-processing

For simplicity, we only consider event names to encode traces and no other attributes, e.g. timestamp, or resource of an event. After computing unique traces we randomly sample from them to construct train/test sets with a ratio of 70%/30%, respectively. As the datasets are highly imbalanced, we sample class-wise from data–preserving the proportion of classes both in train and test sets. To account for frequency of traces, we keep the duplicate traces in train set while removing the ones in test set.

We apply minimal pre-processing to data at hand. First, we remove the biased features from traces. Then we add <start> and <end> tokens to the input to explicitly define the beginning and the end of traces. To have a fixed-length input, the traces that are shorter than the longest trace in an event log are padded with the <pad> token. We then build a token dictionary consisting of unique event names in an event log and the aforementioned special tokens. Finally, the tokens in the input are replaced by their unique integer values stored in the dictionary.

3.2 Evaluation Metric

Because the datasets are highly imbalanced, we measure performance using the area under the receiver operating characteristic curve (AUROC) [4]. ROC analysis does not favor models that outperform the majority class while under-performing the minority class, which is desirable when working with imbalanced data [20, p. 27]. The area under the curve of a binary classifier is the probability that a classifier would rank a randomly chosen positive instance higher than a randomly chosen negative one, which is given by the following equation:

$$\text{AUROC} = \int_{x=\infty}^{-\infty} \text{TPR}(T)\text{FPR}'(T)dT \qquad (1)$$

AUROC ranges from 0 to 1, with an uninformative classifier (random classifier) producing a result of 0.5.

3.3 Model Design and Training

For ML models we utilize scikit-learn [25] and initialize the models with default parameters. For logistic regression, we employ regularization by a L1 penalty term(penalty hyper-parameter) with regularization strength (C hyper-parameter). We use repeated stratified k-fold cross-validation (CV) to evaluate and train a model across a number of iterations. The number of repeats is 50, and the number of splits $k = 5$, yielding 250 models being trained. The reported metric is then the average of all scores computed.

For Transformer model, following [5], we chose the embedding dimension as 36, i.e. each trace is represented by a point in 36-dimensional space. Since Transformer disregards the positional information of events in traces, we add positional encoding to token embedding which have the same dimensions. During training, the model learns to pay attention to input embedding as well as positional encoding in an end-to-end fashion. The embedding outputs are then fed to a multi-head attention block with $h = 6$ heads. On the final layer of the attention block, we aggregate features using a global average pooling followed by a dropout at a rate of 0.1. Then, we employ a dense layer with ReLU activation of 64 hidden units and a dropout at a rate of 0.1. Finally, we use a dense layer with sigmoid activation that outputs a value between 0 and 1, which is interpreted as the "probability" of a trace belonging to positive (desirable) class. We train the model for 50 epochs with ADAM optimizer [15], a learning rate of 1e–3, and batch size of 16.

3.4 Results

Predictive Accuracy. We report the experimental results for the binary classification task. Table 3 reports the AUROC scores of models on BPIC17, Traffic, and COVID datasets, where there are some features leaking the label information. Here the perfect score is achieved as the models correctly discover the biased features, as expected. However, the Transformer model and the ML models with 1-gram on Traffic dataset do not achieve the perfect score on the test set. This is due to the fact that the biased feature–'Payment'–appears in both of the classes. Therefore, 1-gram encoding is not capable of leaking the information. The Transformer model also fails to discover the bias. On the other hand, models achieve the expected results with the 2-gram and 3-gram encoding because those encoding types explicitly define the biased feature: as <end> token added to traces, the feature (Payment, <end>) in 2-gram and (Payment, <end>, <PAD>) in 3-gram would leak the label information (positive class if last event is 'Payment', negative class otherwise).

Table 3. AUROC scores on datasets *with* and *without* biased features using different encoding methods and models. The values inside the parenthesis represent the score on the hold-out test set, whereas others represent the score on the training set.

Dataset	Encoding	With biased features			Without biased features		
		LR	DT	Transformer	LR	DT	Transformer
BPIC17	Integer	–	–	100.(100.)	–	–	97.3(97.3)
	1-gram	100.(100.)	100.(100.)	–	89.0(81.3)	89.5(76.0)	–
	2-gram	100.(100.)	100.(99.9)	–	**97.4(97.9)**	97.3(93.5)	–
	3-gram	100.(100.)	99.9(99.8)	–	97.4(97.8)	97.3(91.7)	–
Traffic	Integer	–	–	100.(98.4)	–	–	**63.9(53.5)**
	1-gram	100.(93.3)	100.(92.2)	–	61.5(36.8)	61.5(52.7)	–
	2-gram	100.(100.)	100.(100.)	–	63.8(45.5)	63.9(49.8)	–
	3-gram	100.(100.)	100.(100.)	–	63.8(48.1)	63.9(52.3)	–
COVID	Integer	–	–	90.9(100.)	–	–	**74.8(94.2)**
	1-gram	100.(100.)	100.(100.)	–	67.3(75.0)	69.6(68.3)	–
	2-gram	100.(100.)	99.9(100.)	–	89.3(61.7)	85.4(85.0)	–
	3-gram	99.9(96.7)	98.1(66.7)	–	90.4(48.3)	85.5(76.7)	–
BPIC18	Integer	–	–	–	–	–	98.6(81.2)
	1-gram	–	–	–	97.5(77.9)	97.5(78.5)	–
	2-gram	–	–	–	**98.4(87.3)**	98.2(84.5)	–
	3-gram	–	–	–	**98.4(87.3)**	98.2(81.5)	–
Hospital	Integer	–	–	–	–	–	**92.4(78.7)**
	1-gram	–	–	–	91.8(73.9)	92.3(56.2)	–
	2-gram	–	–	–	92.5(70.5)	92.7(52.5)	–
	3-gram	–	–	–	92.6(66.0)	92.6(48.5)	–

Table 3 also reports the AUROC scores of models on all of the five real-life event logs, where the biased features are removed from BPIC17, Traffic, and COVID datasets. For BPIC17 and COVID datasets the scores worsen but the results are still promising, hinting that the processes, sequences of events, maintain valuable information for the task even though the distinct/biased features are removed. The different amount of information in different encoding is mirrored by the results: 1-gram achieves the worst results compared to 2-gram and 3-gram, as it only incorporates single events, whereas 2-gram and 3-gram incorporates event pairs. None of those methods integrate order information, while Transformer model learns this information during training. We observe that Transformer model successfully captures the relations between events as well as the order information and outperforms other models in Traffic and COVID datasets, whereas in BPIC17 it receives a comparable result. On the other hand, we observe that the scores for BPIC17 and COVID datasets do not fluctuate as much as it does for Traffic dataset when compared to biased scores represented in Table 3. This is due to the fact that after removing the biased feature in Traffic dataset, some traces in both classes become identical.

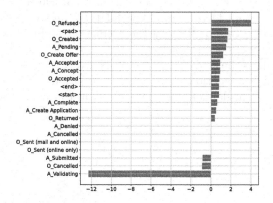

Fig. 1. Relevance of features as considered by the logistic regression model. A high positive/negative value indicates a considerable contribution towards predicting the positive/negative label, *i.e.* desirable/undesirable event trace.

Feature Relevances. We present the relevances of features taken into account by our trained logistic regression model in Fig. 1. We find that 'A_Validating' contributes the most towards predicting an undesirable trace, whereas 'O_Refused' contributes most to predict a desirable trace. Other events have significantly lower impact on the predictions. Interestingly, the special tokens, *i.e.* <pad>, <start>, and <end> affect the predictions when they should not, despite the fact that their influence is negligible. In addition, some features have no effect on the predictions, *e.g.* 'A_Denied', 'A_Cancelled'.

In Fig. 2 we show the feature relevances that our trained decision tree model considers. Based on the MDI value (left figure) the most relevant feature is

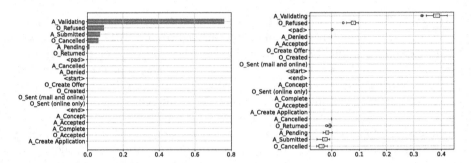

Fig. 2. Relevance of features as considered by decision tree model. **Left**: Relevances according to MDI: A high value indicates a considerable contribution towards deciding between the positive and negative label. **Right**: Relevances according to the permutation importance calculated on the test set, *i.e.* how much shuffling a given feature negatively impacts the performance of the decision tree.

'A_Validating', followed by 'O_Refused', 'A_Submitted', and 'O_Cancelled' with significantly lower impact. Based on the permutation feature importance (right figure) the most relevant features are 'A_Validating', and 'O_Refused', where the rest of the features have no effect on the prediction performance, which aligns well with the relevances of LR model.

Figure 3 visualizes the attention scores of six attention heads for a given trace, as well as normalized attention scores over all attention heads and test samples. We observe that, regardless of where it appears in the trace, the events mostly

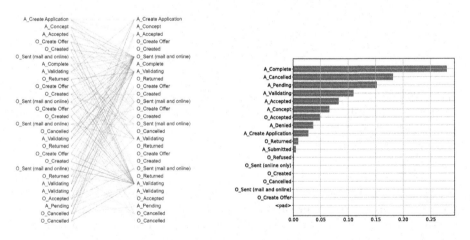

Fig. 3. Left: Relevance of features as considered by transformer model for one randomly selected event trace. The thickness of a line connecting two features indicates the intensity of attention in between, whereas the color represents one of the six different attention heads. **Right**: Normalized attention scores which are averaged among all attention heads and all traces in test set.

attend to 'A_Validating' event for the aforementioned trace. In addition, some events focus on 'O_Sent' in some heads, which differs from the other models. The normalized attention scores, however, demonstrates that not all traces exhibit this behavior. The plot also highlights features whose importances overlap with other models' results. In summary, all models agree on the most relevant features.

4   Conclusion

We have demonstrated how to prepare process data in such a way that it can be used to train classic and modern – state-of-the-art – ML classifiers. All our trained models exhibit high classification performance, *i.e.* they are capable of learning the underlying regularity of the observed processes, whereby Transformers benefit from the fact that the full sequence information is available, unlike *e.g.* 1-gram representations. XAI technologies prevent pitfalls such as information leakage by explicit encoding of the predicted event, and reveal insights into the relevance of events or sequences of events, respectively. These insights enable a further exploration of crucial aspects of the processes, which is useful *e.g.* for the improvement or correction of undesired process outcomes.

Future research could investigate the effects of various model architectures and encoding schemes on the outcomes of feature relevances. Another potential research direction might be to study the impact of adding further event attributes on the learnt representations.

References

1. van der Aalst, W., Weijters, T., Maruster, L.: Workflow mining: discovering process models from event logs. IEEE Trans. Knowl. Data Eng. **16**, 1128–1142 (2004)
2. van der Aalst, W.: Process Mining: Data Science in Action. Springer, Heidelberg (2016). https://doi.org/gjnmht
3. Binder, A., Montavon, G., Bach, S., Müller, K., Samek, W.: Layer-wise relevance propagation for neural networks with local renormalization layers. CoRR abs/1604.00825 (2016)
4. Bradley, A.P.: The use of the area under the roc curve in the evaluation of machine learning algorithms. Pattern Recogn. **30**(7), 1145–1159 (1997)
5. Bukhsh, Z.A., Saeed, A., Dijkman, R.M.: ProcessTransformer: predictive business process monitoring with transformer network. CoRR (2021)
6. Dailey, D.: Netezza and IBM cloud PAK for data: a knockout combo for tough data. https://ibm.co/3xvK4MG. Accessed 17 June 2022
7. Devlin, J., Chang, M.W., Lee, K., Toutanova, K.: Bert: pre-training of deep bidirectional transformers for language understanding. arXiv:1810.04805 (2018)
8. van Dongen, B.: BPI Challenge 2017 (2017). https://doi.org/jcmn
9. van Dongen, B., Borchert, F.: BPI Challenge 2018 (2018). https://doi.org/jcmm
10. Dosovitskiy, A., et al.: An image is worth 16×16 words: transformers for image recognition at scale. arXiv:2010.11929 (2020)
11. Evermann, J., Rehse, J.R., Fettke, P.: Predicting process behaviour using deep learning. Decis. Support Syst. **100**, 129–140 (2017)

12. Fernández, M., Cernadas, E., Barro, S., Amorim, D.: Do we need hundreds of classifiers to solve real world classification problems? JMLR **15**(1), 3133–3181 (2014)
13. Harl, M., Weinzierl, S., Stierle, M., Matzner, M.: Explainable predictive business process monitoring using gated graph neural networks. J. Decis. Syst. **29**(sup1), 312–327 (2020)
14. Hsieh, C., Moreira, C., Ouyang, C.: Dice4el: interpreting process predictions using a milestone-aware counterfactual approach. In: ICPM, pp. 88–95. IEEE (2021)
15. Kingma, D.P., Ba, J.: Adam: a method for stochastic optimization (2014)
16. de Leoni, M.M., Mannhardt, F.: Road traffic fine management process (2015). https://doi.org/jcmk
17. Louppe, G.: Understanding random forests: from theory to practice (2014). https://doi.org/jcms
18. Maggi, F.M., Di Francescomarino, C., Dumas, M., Ghidini, C.: Predictive monitoring of business processes (2013)
19. Mannhardt, F.: Hospital Billing - Event Log (8 2017). https://doi.org/gm85w4
20. Mehdiyev, N., Evermann, J., Fettke, P.: A novel business process prediction model using a deep learning method. Bus. Inf. Syst. Eng. **62**(2), 143–157 (2020). https://doi.org/ggqt7z
21. Molnar, C.: Interpretable Machine Learning, 2 edn(2022). https://www.lulu.com
22. Munoz-Gama, J.: Conformance Checking and Diagnosis in Process Mining, vol. 270, Springer, Heidelberg (2016). https://doi.org/jkmp
23. Nguyen, A., et al.: Time matters: time-aware LSTMs for predictive business process monitoring. In: Leemans, S., Leopold, H. (eds.) ICPM 2020. LNBIP, vol. 406, pp. 112–123. Springer, Cham (2021). https://doi.org/10.1007/978-3-030-72693-5_9
24. Pasquadibisceglie, V., Appice, A., Castellano, G., Malerba, D.: Using convolutional neural networks for predictive process analytics. In: ICPM. IEEE (2019)
25. Pedregosa, F.: Scikit-learn: machine learning in python. JMLR **12**, 2825–2830 (2011)
26. Pegoraro, M., Narayana, M.B.S., Benevento, E., van der Aalst, W.M.P., Martin, L., Marx, G.: Analyzing medical data with process mining: a COVID-19 case study. In: Abramowicz, W., Auer, S., Stróżyna, M. (eds.) BIS 2021. LNBIP, vol. 444, pp. 39–44. Springer, Cham (2022). https://doi.org/10.1007/978-3-031-04216-4_4
27. Rama-Maneiro, E., Vidal, J., Lama, M.: Deep learning for predictive business process monitoring: review and benchmark. IEEE Trans. Serv. Comput. (2021)
28. Ribeiro, M.T., Singh, S., Guestrin, C.: why should I trust you?: explaining the predictions of any classifier. In: ACM SIGKDD, pp. 1135–1144 (2016)
29. Tax, N., Verenich, I., La Rosa, M., Dumas, M.: Predictive business process monitoring with LSTM neural networks. In: Dubois, E., Pohl, K. (eds.) CAiSE 2017. LNCS, vol. 10253, pp. 477–492. Springer, Cham (2017). https://doi.org/10.1007/978-3-319-59536-8_30
30. Tibshirani, R.: Regression shrinkage and selection via the lasso. J. R. Stat. Soc. **58**(1), 267–288 (1996). http://doi.org/gfn45m
31. van Dongen, B.F., de Medeiros, A.K.A., Verbeek, H.M.W., Weijters, A.J.M.M., van der Aalst, W.M.P.: The ProM framework: a new era in process mining tool support. In: Ciardo, G., Darondeau, P. (eds.) ICATPN 2005. LNCS, vol. 3536, pp. 444–454. Springer, Heidelberg (2005). https://doi.org/10.1007/11494744_25

32. Vaswani, A., et al.: Attention is all you need (2017)
33. Weytjens, H., De Weerdt, J.: Creating unbiased public benchmark datasets with data leakage prevention for predictive process monitoring (2021). https://doi.org/jcmp
34. Yasmin, F., Bukhsh, F., De Alencar Silva, P.: Process enhancement in process mining: a literature review. In: CEUR (2018)

Efficient Sensor Selection
for Individualized Prediction Based
on Biosignals

Markus Vieth[1]([✉]) [iD], Nils Grimmelsmann[2] [iD], Axel Schneider[2] [iD],
and Barbara Hammer[1] [iD]

[1] Bielefeld University, Bielefeld, Germany
mvieth@techfak.uni-bielefeld.de
[2] Bielefeld University of Applied Sciences, Bielefeld, Germany

Abstract. Soft sensors combine a hardware component with an intelligent algorithmic processing of the raw sensor signals. While individualization of software components according to a person's specific needs is comparably cheap, individualization of the sensor hardware itself is usually impossible in mass production. At the same time, the number of raw sensors should be minimum to reduce production costs. In this contribution, we propose to model this challenge as a feature selection problem, which optimizes a feature set simultaneously with respect to a family of functions corresponding to individualized post-processing of sensor signals. This concept is integrated into a number of different classical feature selection schemes, and evaluated in the context of the placement of pressure sensors as part of a shoe insole. It turns out that feature selection respecting the class of functions is superior to both placement based on anatomical considerations and classical feature selection methods.

Keywords: Intelligent wearables · Feature selection · Model individualization

1 Introduction

The increasing capability of hardware components and post-processing based on machine learning has led to individualized intelligent wearables supporting, for example, assisted living or demands in healthcare [8]. The term 'soft-sensor' refers to the combination of hardware components and software algorithms, which transform the raw sensor signals into the desired quantity of interest. Mirroring human's individual anatomy and biomechanics, reliable soft sensors often require model individualization to achieve optimum accuracy and robustness for a specific person [6]. While efficient and possibly online or continuous adaptation

The project has been funded by the Ministry of Culture and Science of the Federal State North Rhine-Westphalia in the frame of the project RoSe in the AI-graduate-school https://dataninja.nrw/.

© The Author(s), under exclusive license to Springer Nature Switzerland AG 2022
H. Yin et al. (Eds.): IDEAL 2022, LNCS 13756, pp. 326–337, 2022.
https://doi.org/10.1007/978-3-031-21753-1_32

of the software components can be based on efficient learning schemes [7], the design and placement of the sensors itself is less flexible. Currently, sensors are placed based on anatomical considerations, e.g. in Electrocardiography, or in a grid-like manner, e.g. in Electroencephalography [8]. Optimizing the sensor positions based on data is rare, albeit it has the potential to make the measurements more precise and/or reduce the required number of sensors.

In this contribution, we will deal with an application where sensors can be placed freely on a continuous two-dimensional space, the skin surface.

In many application scenarios, it is not practical to determine an optimal sensor positioning individually for every person. On the other hand, subsequent individualized processing of biosignals to achieve optimum soft sensor measurements is feasible by adapting model parameters for a new person using a calibration procedure or more advanced machine learning processes. In this paper, we address the challenge how to optimize sensor placements such that they are uniformly suited for different individual postprocessing procedures. We phrase the problem as a feature selection problem which takes into account the additional component of soft sensor individualization based on the raw hardware measurements, i.e. a set of subsequent functions, and we compare its per-

Fig. 1. Placement of pressure sensors in an insole (more in Fig. 5)

formance to setups which are based on classical feature selection only.

In the following, we introduce two different modelling approaches:

1. Global optimization-based methods for feature selection (differential evolution and simulated annealing respectively) which are extended to take individualized post-processing into account,
2. approximate and efficient orthogonal matching pursuit, which is extended to individual subsequent weighting schemes.

In addition, we compare to baselines such as suggested by the anatomy, selection using LASSO, or local optimization, as well as the feature selection counterparts without modelling of subsequent individualization of the soft sensor. We explore these methods in a specific application scenario: a shoe insole is to be equipped with few pressure sensors which allow to constantly monitor the total weight put on the foot/leg (see example in Fig. 1). This application is relevant for post-surgery rehabilitation such as femoral neck fracture, where a patient must not overstrain their leg. All methods will be evaluated using realistic data which have been gathered from foot pressure profiles of individuals with full information about all possible sensors.

2   Related Work

The problem of positioning sensors when measuring biosignals has been approached in different ways in the literature:

Regular Grid: In Electroencephalography (EEG) for example, many sensors are placed in a regular grid, typically according to the 10–20 system, the 10–10 system, or systems with even higher resolutions [9]. There may be a basic adaptation to the individual person, in EEG this is realized by taking nasion, inion, and preauricular points as reference, but the positions are usually not optimized regarding the signals to be measured.

Anatomy: In Electrocardiography (ECG), typically 10 electrodes are placed on chest and limbs. They are positioned and connected in a way to measure the electrical activity of the heart in 12 axes, based on anatomical knowledge [14].

Data-Based Optimization: A few approaches base sensor placement on observed data. Ha et al. [4] describe a method to optimize the placement of sensors for Electrocardiography, Photoplethysmography, and body temperature. They determine sensing accuracy and operational robustness at 34 possible sensor locations on the human body. Their goal is to select a single location based on the tested criteria, however. Yoo et al. [17] explore the selection of a subset of 99 pressure sensors in a shoe insole. Choosing n sensors out of the 99 sensors available in the paper gives $\binom{99}{n}$ possibilities (e.g. 3.8 million for 4 sensors and 171 billion for 8 sensors), hence it is impossible to test all. Yoo et al. choose 5 sensor configurations with 4 to 8 sensors based on anatomical considerations, and evaluate them based on the computed center of pressure (COP) during different gait phases. They conclude that only the configurations with 6, 7, and 8 pressure sensors give sufficient accuracy. These approaches deal with the optimization of sensor positions based on data, but the possible positions are pre-selected. In contrast, we propose methods which place the sensors without any constraints and take into account the subsequent individualization of the soft sensor.

3   Methods

Data have the form of a matrix $\mathbf{X}_{s,p}$ containing pressure sensor values of person p at positions s. The objective is to predict the current total weight $\mathbf{y_p}$ as target, which can be explicitly computed if all sensor measures are available. $\mathbf{X_s}$ are the sensor values of *all* persons at sensor positions s and $\mathbf{y}$ are the target values of all persons. To predict the target from a subset of sensor measurements only, an individual parametric function can be used. In this work, linear functions are sufficient to achieve the intended task, parameterized with $\mathbf{w_p}$ for person p, or $\mathbf{w}$ if the soft sensor is not individualized. Extensions to nonlinear functions would be immediate, however.

Formally, we consider the following problem: given a fixed maximum number n of sensor positions, what are the optimum sensor locations, such that individual predictions based on those values yield a minimum overall error? We phrase this problem along several feature selection paradigms [3] in the following:

3.1   Heuristic Placement Based on Foot Anatomy

As a baseline, we consider a sensor placement based on anatomical considerations. We use the configurations suggested in [17]. These sensor positions are not optimized with data. We use the configurations for 4 to 8 sensors as shown in the paper, and derive an additional configuration with 3 sensors by omitting the sensor at the interphalangeal joint from the 4-sensor configuration. The sensor weights $\mathbf{w_p}$ are optimized in the linear least-squares sense.

3.2   Feature Selection as Global Optimization Task

We formulate two different optimization objectives. One objective, which takes into account individualized soft sensors, is given as

$$\mathbf{s}_1^* = \underset{|\mathbf{s}|_1 = n}{\arg\min} \left(\sum_p \underset{\mathbf{w_p}}{\min} \|\mathbf{y_p} - \mathbf{X_{s,p}} \mathbf{w_p}\|_2^2 \right)$$

where $\mathbf{s}$ refers to the selected sensor placement (we enrich it by a fixed implicit bias for convenience), the inner part of the objective constitutes an individual linear least-squares problem mapping sensor values $\mathbf{X_{s,p}}$ of person p at positions s to the individual target, and the outer part minimizes the residuals, summed over all persons, by adapting the sensor positions $\mathbf{s}$.

For comparison, we consider a second objective

$$\mathbf{s}_2^* = \underset{|\mathbf{s}|_1 = n}{\arg\min} \left(\underset{\mathbf{w}}{\min} \|\mathbf{y} - \mathbf{X_s} \mathbf{w}\|_2^2 \right)$$

with uniform weights $\mathbf{w} \in \mathbb{R}^{n+1}$ shared between all persons.

We test two global optimization methods: differential evolution and simulated annealing. We use the implementations from SciPy [15], that are based on [12] and [16]. Note that sensor positions are continuous in these settings.

3.3   Feature Selection as Sparse Approximation

As an alternative, we utilize sparse approximation algorithms for feature selection. Since these rely on a finite set of features, we sample a regular grid of 128×56 sensors from the continuous 2D space (resolution of approx. 2 mm).

OMP: Orthogonal Matching Pursuit [10] is a greedy algorithm for finding the best subset of a set (or dictionary) of features to approximate a target function. In each iteration, the algorithm selects the feature with maximum correlation to the residual (initially to the target function), then updates the model based on the selected feature. These steps are repeated until either a given number of features is selected, or the target function is sufficiently approximated. We use this 'classical' OMP as a baseline. In addition, we propose a modified version, which takes into account subsequent individualization of the soft sensors, as follows: In the classical OMP, each selected feature is weighted with a single weight to approximate the target. Instead, we use an individual weight *per person* which weights the contribution of a selected feature to the personalized prediction model. Residuals are updated according to these individual weights before the next feature is selected. We use and extend the OMP implementation from scikit-learn [11], respectively.

LASSO: The least absolute shrinkage and selection operator (LASSO, [13]) uses the ℓ^1 norm as a regularization on the weights. This leads to a selection of features (for some, the weights are set to 0, like in OMP), but also to smaller weights overall, even among selected features (this is not the case in OMP). We use the implementation from scikit-learn [11].

Adding Local Refinement: Starting with the sensor positions suggested by any of the above mentioned sparse approximation methods, we refine them with a local optimization approach, using the same objective as for the global optimization. This refinement might compensate limitations of the sparse approximation methods, in particular limitations due to a restricted and discrete subset of possible positions.

3.4 Dataset

We record a data set which represents walking/standing of different individuals as measured by a pressure profile, enhanced by the respective total weight. As measuring device, we use the MatSCAN GE sensor plate from Savecomp Megascan. This includes 2288 pressure sensors in a 44 by 52 grid, each sensor with an area of $0.703 \, cm^2$. The sampling rate 50 Hz.

16 test subjects were instructed to walk over the sensor plate, starting a few steps before and ending a few steps behind the plate, to achieve a natural gait. For the left and right foot each, ten stance phases were recorded per person (Fig. 2). Since walking is the most used form of legged locomotion, the resulting data are particularly relevant for practical applications.

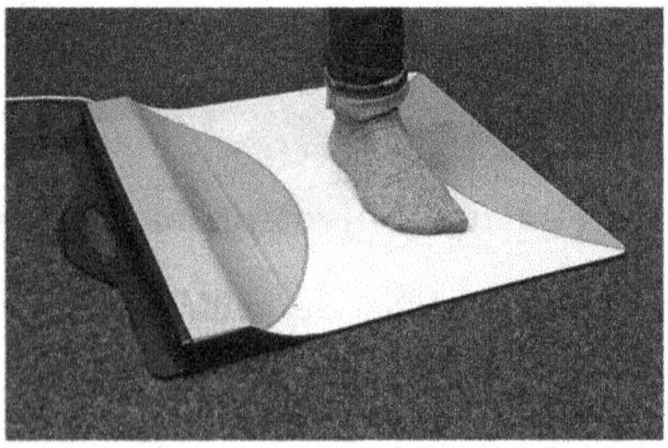

Fig. 2. A person walking over the plate

We remove 3 persons with exceptionally small or large feet (23 cm, 29 cm, and 30 cm). We expect that different sensor placement, which respects an according scaling, would be needed here. In this paper, we focus on people with average foot sizes. The remaining 13 persons have foot sizes between 25.5 cm and 27.5 cm (mean: 26.4 cm), body heights between 171.0 cm and 193.0 cm (mean: 180.4 cm), and body weights between 58.0 kg and 93.8 kg (mean: 70.7 kg). The data amounts to 11393 pressure profile frames.

During preprocessing, pressure profiles are aligned via translation and rotation, by maximising the overlap of sensors with non-zero readings. Left feet were mirrored and then treated the same as right feet. Bicubic interpolation from OpenCV [2] was used to compute continuous sensor measurements from the discrete sensor values.

4 Experiments and Results

Experimental Setup: We evaluate all results based on the R^2 score (i.e. the coefficient of determination [5]) that is equal to 1 for a perfect prediction and smaller for sub-optimal predictions. We evaluate four scenarios with different train and test sets (table 1): **(1)** The first scenario serves as a lower bound as provided by global soft sensors which are not personalized. We use 10 persons for training a global model and 3 persons for testing, in which left and right foot are both in the training set or both in the test set. No individualization of the model to a specific person takes place. This is done to test the generalization of a trained global model to a new person. We run 65 of such splits, so that each person appears in the test set 15 times. The classical OMP and the s_2^* objective are used. **(2)** The second scenario addresses individualized sensor processing and global positioning, in which the latter is not optimized w.r.t. subsequent personalization. We again use 10 persons for training, but only to determine the sensor positions. Then, for each foot of the remaining 3 persons individually, we use 8 of the 10 steps to compute the sensor weights (model coefficients), and the remaining 2 steps for testing (5-fold cross-validation per foot). **(3)** The third scenario is similar to the second one, but takes into account personalization also for sensor positioning. I.e., when determining the sensor positions we now take into consideration that each foot has its own sensor weights and sensor positioning. More concretely, this means that we use the new, modified OMP from Sect. 3.3 and objective s_1^* from Sect. 3.2 (for differential evolution, annealing, and the local refinement of the OMP results). **(4)** In the fourth scenario, everything is individual per person including the sensor positioning: the sensor positions and weights are computed from 8 of the 10 steps per foot, and tested on the remaining 2 steps (5-fold crossvalidation). In this scenario, there is no difference between the classical and the modified OMP, and no difference between objective s_1^* and s_2^*, because the sensor positions are not shared between persons. Scenario (4) serves as an upper bound as provided by completely personalized models. For full results, see Table 2.

Initial Results on Scenario 1: We first look at Fig. 3 to get a general overview of the methods in scenario 1. In general, more sensors result in higher test scores, but the anatomically derived positions are an exception: the 8-sensor configuration has a lower median test score than the configurations with 7 sensors, which can be mostly attributed to one person, who self-reported to have flat feet. For the other persons, the 8th sensor (at the tarsometatarsal joint/the arch of the foot) usually measured only very small values, but for this person with flat feet,

Table 1. Overview over the 4 scenarios

Scenario	Train set		Test set	Methods
	For positions	For sensor weights		
1	10 persons		3 persons	classical OMP, $\mathbf{s}_2^*$ objective
2	10 persons	8 steps	2 steps	
3	10 persons	8 steps	2 steps	Modified OMP, $\mathbf{s}_1^*$ objective
4	8 steps		2 steps	No difference

it had higher readings. When this person was in the test set, the weight for the 8th sensor was set very high based on the persons in the training set, which was too high for the person with flat feet. The OMP gives good results, which are consistently improved with the local refinement. The OMP with refinement gives the best median test score of all tested methods for 3 sensors. The LASSO proves to be an unsuitable method for this task as it only gives acceptable scores with a high number of sensors (at least 7). With refinement, the results for 7 and 8 sensors are quite good, but the percentile ranges remain wide. The global optimization methods give the best results for more than 3 sensors. In all but one case, the annealing method was slightly better than the differential evolution. Due to these initial findings, we will particularly focus on global optimization using differential evolution and partially also other promising variants within other scenarios.

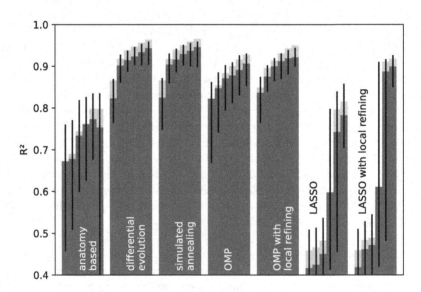

Fig. 3. Scenario 1: Median R^2 scores on test set with whiskers representing the 10 to 90 percentile range (corresponding train scores in pale colour). The scores are grouped by method, and in each group the number of sensors increases from 3 to 8. The y-axis starts at 0.4

Improvement by Model Personalization: Next, we compare the different train/test scenarios in Fig. 4. Comparing scenario 2 with scenario 1 shows a large improvement in the R^2 scores for any number of sensors. This demonstrates the benefit of calibrating the sensor weights for each person. Between scenarios 2 and 3, a (comparably small) difference arises if only few sensors are used. Here, the s_1^* objective gives an advantage. The improvement from scenario 3 to scenario 4 is rather small. Hence a sensor positions which is shared between all persons and customized sensor weights are almost as good as completely personalized sensor positions and weights.

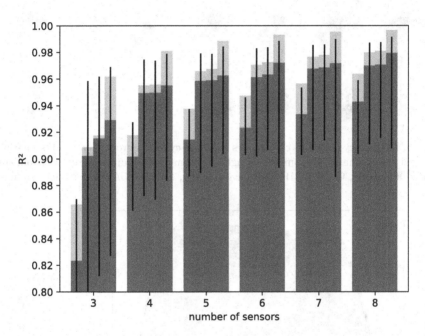

Fig. 4. Differential evolution: Median R^2 scores on the test set (train scores in pale colour). In each group, left to right: scenario 1, 2, 3, and 4. The y-axis starts at 0.8

Exemplary Analysis of Sensor Location: Figure 5 shows exemplary positions of 8 sensors. The OMP gives reasonable results, although often a sensor is placed under the smaller toes or at the edge of the foot, which seems less robust. The images also explain why the LASSO achieves such low R^2 scores: the sensors are often grouped tightly together, which not only wastes sensors but is also impossible to implement in a real insole. Locally refining somewhat alleviates this problem, but the sensors are still concentrated at the ball of the foot. The two global optimization methods almost completely agree with their sensor placement, except for maybe one or two sensors.

Fig. 5. Exemplary positions of 8 sensors shown in cyan. Top row: scenario 1/scenario 2, bottom row: scenario 3. From left to right: anatomy based, differential evolution, simulated annealing, OMP, OMP with local refining, LASSO, LASSO with local refining

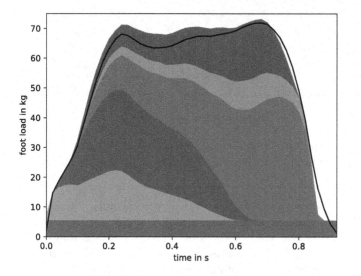

Fig. 6. The contribution of each sensor to the prediction of the foot load during one step (stance phase). The black line is the ground truth leg load. The right side shows the five colour-coded sensor positions.

Table 2. All median R^2 test scores, the highest score per row is marked in bold. Vertically grouped by scenarios 1, 2, 3, and 4. In each scenario, the number of sensors n increases from 3 to 8

n	Anatomy based	Diff. evolution	Simulated annealing	OMP	OMP +refining	LASSO	LASSO +refining
3	0.673	0.823	0.825	0.822	**0.837**	0.416	0.418
4	0.678	0.902	**0.904**	0.847	0.875	0.424	0.461
5	0.734	0.914	**0.916**	0.871	0.900	0.450	0.471
6	0.761	0.923	**0.929**	0.878	0.912	0.598	0.611
7	0.773	0.933	**0.936**	0.891	0.919	0.742	0.887
8	0.753	0.943	**0.946**	0.906	0.921	0.782	0.899
3	0.812	**0.902**	0.897	0.879	0.895	0.501	0.503
4	0.825	0.949	**0.950**	0.901	0.925	0.508	0.528
5	0.859	0.958	**0.960**	0.910	0.937	0.522	0.542
6	0.871	0.961	**0.964**	0.916	0.948	0.638	0.663
7	0.880	**0.968**	**0.968**	0.923	0.953	0.775	0.915
8	0.880	0.970	**0.971**	0.934	0.958	0.847	0.950
3	0.812	0.915	**0.916**	0.879	0.902	0.503	0.503
4	0.825	**0.950**	0.949	0.899	0.929	0.508	0.526
5	0.859	0.959	**0.961**	0.906	0.954	0.520	0.541
6	0.871	0.963	**0.967**	0.918	0.962	0.618	0.635
7	0.880	**0.968**	**0.968**	0.929	0.965	0.791	0.921
8	0.880	0.971	**0.971**	0.938	0.968	0.845	0.950
3	0.812	0.929	**0.931**	0.840	0.888	0.564	0.564
4	0.825	**0.955**	0.952	0.873	0.901	0.584	0.589
5	0.859	0.963	**0.964**	0.888	0.913	0.641	0.626
6	0.871	**0.972**	0.971	0.893	0.929	0.771	0.893
7	0.880	0.972	**0.977**	0.891	0.923	0.861	0.936
8	0.880	**0.979**	0.978	0.882	0.936	0.881	0.930

Figure 6 exemplarily visualizes how the foot load is estimated from 5 sensors. This shows the contribution of each sensor over time. The shifting weight during the movement was considered during the sensor placement. For a model describing the camel-back shape of the curve, see e.g. [1].

5 Conclusion

In this paper, we have formalized the problem of individualized soft-sensors with shared hardware components as a feature selection problem which respects a class of multiple subsequent functional processing, and we have demonstrated its usability for an optimal placement of sensors for measuring biosignals, in the presented case generated from pressure sensors. The experiments on a realistic dataset showed that global optimization methods and Orthogonal Matching Pursuit with local refinement give the best results. It is usually not necessary to compute sensor positions for each person; with only slightly worse performance the positions can be the same for everyone if the sensor weights are individual. In the future, we intend to extend our experiments to non-linear processing models, and to a wider range of persons with different foot sizes or atypical foot anatomies. In the latter, the problem occurs how many global sensor placements are sufficient to serve all different persons based on soft-sensors with individualized post-processing.

References

1. Blickhan, R., Seyfarth, A., Geyer, H., Grimmer, S., Wagner, H., Günther, M.: Intelligence by mechanics. Phil. Trans. R. Soc. Math. Phys. Eng. Sci. **365**(1850), 199–220 (2007). https://doi.org/10.1098/rsta.2006.1911
2. Bradski, G.: The OpenCV library. J. Softw. Tools **25**, 120–123 (2000)
3. Dhal, P., Azad, C.: A comprehensive survey on feature selection in the various fields of machine learning. Appl. Intell. **52**(4), 4543–4581 (2022). https://doi.org/10.1007/s10489-021-02550-9
4. Ha, S., Park, S., Lim, H., Baek, S.H., Kim, D.K., Yoon, S.H.: The placement position optimization of a biosensor array for wearable healthcare systems. J. Mech. Sci. Technol. **33**(7), 3237–3244 (2019). https://doi.org/10.1007/s12206-019-0619-0
5. Hughes, A.J.: Statistics: A Foundation for Analysis. Addison-Wesley Pub. Co., Reading (1971). http://archive.org/details/trent_0116302260611
6. Leite, M., Soares, B., Lopes, V., Santos, S., Silva, M.T.: Design for personalized medicine in orthotics and prosthetics. Procedia CIRP **84**, 457–461 (2019). https://www.sciencedirect.com/science/article/pii/S2212827119309011. https://doi.org/10.1016/j.procir.2019.04.254
7. Losing, V., Hammer, B., Wersing, H.: Incremental on-line learning: a review and comparison of state of the art algorithms. Neurocomputing **275**, 1261–1274 (2018). https://www.sciencedirect.com/science/article/pii/S0925231217315928. https://doi.org/10.1016/j.neucom.2017.06.084
8. Ometov, A., et al.: A survey on wearable technology: history, state-of-the-art and current challenges. Comput. Netw. **193**, 108074 (2021). https://www.sciencedirect.com/science/article/pii/S1389128621001651. https://doi.org/10.1016/j.comnet.2021.108074
9. Oostenveld, R., Praamstra, P.: The five percent electrode system for high-resolution EEG and ERP measurements. Clin. Neurophysiol. **112**(4), 713–719 (2001). https://www.sciencedirect.com/science/article/pii/S1388245700005277. https://doi.org/10.1016/S1388-2457(00)00527-7
10. Pati, Y., Rezaiifar, R., Krishnaprasad, P.: Orthogonal matching pursuit: recursive function approximation with applications to wavelet decomposition. In: Proceedings of 27th Asilomar Conference on Signals, Systems and Computers, vol. 1, pp. 40–44 (1993). https://doi.org/10.1109/ACSSC.1993.342465. ISSN: 1058–6393
11. Pedregosa, F., et al.: SciKit-learn: machine learning in python. J. Mach. Learn. Res. **12**(85), 2825–2830 (2011). http://jmlr.org/papers/v12/pedregosa11a.html

12. Storn, R., Price, K.: Differential evolution – a simple and efficient heuristic for global optimization over continuous spaces. J. Global Optim. **11**(4), 341–359 (1997). https://doi.org/10.1023/A:1008202821328
13. Tibshirani, R.: Regression shrinkage and selection via the lasso. J. R. Stat. Soc. Ser. B (Methodol.) **58**(1), 267–288 (1996). https://onlinelibrary.wiley.com/doi/10.1111/j.2517-6161.1996.tb02080.x.   https://doi.org/10.1111/j.2517-6161.1996.tb02080.x
14. ECG Lead Placement - Normal Function of the Heart - Cardiology Teaching Package - Practice Learning - Division of Nursing - The University of Nottingham. https://www.nottingham.ac.uk/nursing/practice/resources/cardiology/function/placement_of_leads.php
15. Virtanen, P., et al.: SciPy 1.0: fundamental algorithms for scientific computing in python. Nat. Methods **17**(3), 261–272 (2020). https://www.nature.com/articles/s41592-019-0686-2. https://doi.org/10.1038/s41592-019-0686-2
16. Xiang, Y., Gubian, S., Suomela, B., Hoeng, J.: Generalized simulated annealing for global optimization: the GenSA package. R J. **5**(1), 13 (2013). https://journal.r-project.org/archive/2013/RJ-2013-002/index.html. https://doi.org/10.32614/RJ-2013-002
17. Yoo, S., Gil, H., Kim, J., Ryu, J., Yoon, S., Park, S.K.: The optimization of the number and positions of foot pressure sensors to develop smart shoes. J. Ergon. Soc. Korea **36**, 15 (2017). https://doi.org/10.5143/JESK.2017.36.5.395

Understanding the Classes Better with Class-Specific and Rule-Specific Feature Selection, and Redundancy Control in a Fuzzy Rule Based Framework

Suchismita Das$^{(\boxtimes)}$ and Nikhil R. Pal$^{(\boxtimes)}$

Electronics and Communication Sciences Unit, Indian Statistical Institute,
Calcutta 700108, India
suchismitasimply@gmail.com, nrpal59@gmail.com

Abstract. Unlike traditional feature selection methods, the class-specific feature selection methods select an optimal feature subset for each class. Typically class-specific feature selection methods use one-versus-all split of the data set that leads to issues such as class imbalance, decision aggregation, and high computational overhead. We propose a class-specific feature selection method embedded in a fuzzy rule-based classifier, which is free from the drawbacks associated with most existing class-specific methods. Additionally, our method can be adapted to control the level of redundancy in the class-specific feature subsets by adding a suitable regularizer to the learning objective. Our method results in class-specific rules involving class-specific features. We also propose an extension where different rules of a particular class are defined by different feature subsets to model different substructures within the class. The effectiveness of the proposed method is validated through experiments on three synthetic data sets.

Keywords: Class-specific feature selection · Rule-specific feature selection · Redundancy control · Fuzzy rule-based classifiers · Within-class substructures

1 Introduction

Feature selection is an important step for many machine learning tasks. Unlike the traditional feature selection methods that choose a single "optimal" subset of the features for the entire dataset, some studies [6, 8–12, 14, 15] have used class-specific approaches for selecting features, where for each class, a unique subset of the original features is selected. If there are C classes, in the class-specific approach, C subsets are chosen. In a traditional feature selection method, the selected feature subset is chosen based on the global characteristics of the data. It does not take into account any class-specific or local characteristics of the data which may be present. For example, there may exist a group of features

© The Author(s), under exclusive license to Springer Nature Switzerland AG 2022
H. Yin et al. (Eds.): IDEAL 2022, LNCS 13756, pp. 338–347, 2022.
https://doi.org/10.1007/978-3-031-21753-1_33

that follows a distinct distribution for a specific class but varies randomly over the remaining classes. Such a group of features plays a significant role in distinguishing the specific class from other classes but may not be very useful for the C-class problem as a whole. Class-specific characteristics may also exist in the form of class-specific redundancy. Different sets of features could be redundant for different classes. The class-specific feature selection (CSFS) methods in [6,8–12,14,15] have proposed suitable frameworks that exploit class-specific feature subsets to solve classification problems. They have shown that the classifiers built with the subsets chosen by class-specific methods performed better than or comparable to the classifiers built with subsets chosen by global feature selection methods. The CSFS may also enhance the transparency/explainability of the classification process associated with it [6].

The majority of the CSFS methods [6,8,10,11,14,15] follow the one-versus-all (OVA) strategy to decompose a C-class classification problem into C binary classification problems. They choose C class-specific feature subsets optimal for the C binary classification problems. OVA strategy-based CSFS methods have certain drawbacks. Generally, it leads to class imbalance demanding careful design of classifiers. The OVA strategy-based methods are computationally intensive and for testing we need an aggregation mechanism. Even, for the other CSFS methods [9,12] which are not OVA strategy-based, to exploit the obtained class-specific feature subsets in the classification process, C classifiers are employed and require aggregation of C outputs for testing. Moreover, none of these methods consider the presence of - (i) class-specific redundancy, and (ii) within-class substructures in their frameworks.

Here, we propose a CSFS scheme embedded in a fuzzy rule-based classifier (FRBC) that does not use the OVA strategy. Our method selects class-specific feature subsets by learning a single FRBC and hence avoid the issues associated with the CSFS methods involving C classifiers. Moreover, we extend our framework to deal with- (i) CSFS with redundancy control, and (ii) rule-specific feature selection (RSFS) that can exploit the presence of substructures within a class. The rules provided by the FRBC are generally interpretable and more specific. Exploiting the class-specific local features, the proposed FRBC enjoys more transparency and interpretability than a standard classifier exploiting class-specific feature subsets. Our main contributions in this study are summarised as follows.

1. We propose a CSFS method that is not based on the OVA strategy like most of the existing CSFS schemes. Thus, our method is free from the weaknesses of the OVA strategy.
2. We propose a classifier framework that is defined by class-specific fuzzy rules involving class-specific feature subsets.
3. Our method can monitor the level of redundancy in the selected features.
4. We also propose a general version of our class-specific feature selection method that not only chooses different subsets for different classes but also, chooses different subsets for different rules within a class if different substructures are present.

2 Proposed Method

Let the input data be $\mathbf{X} = \{\mathbf{x}^i = (x_1^i, x_2^i, \cdots, x_P^i)^T \in \mathcal{R}^P : i \in \{1, 2, \cdots, n\}\}$. Let us denote the input space by $\mathcal{X} \subseteq \mathcal{R}^P$. The collection of class labels of $\mathbf{X}$ be $\mathbf{y} = \{y^i \in \mathcal{Y} : i \in \{1, 2, \cdots, n\}\}$, where, y^i is the class label corresponding to $\mathbf{x}^i$, and $\mathcal{Y} = \{1, 2, \cdots, C\}$. For our purposes we represent the class label of $\mathbf{x}^i$ as $\mathbf{t}^i \in \{0, 1\}^C$, where $t_k^i = 1$ if $y^i = k$ and $t_k^i = 0$, otherwise. We denote the jth feature by x_j, the class label by y, and the target vector by $\mathbf{t}$. The set of original features is $\mathbf{f} = \{x_1, x_2, \cdots x_P\}$. Let the optimal class-specific subset of features for the kth class be $\mathbf{s}_k$. We need to find out $\mathbf{s}_k$s where, $\mathbf{s}_k \subset \mathbf{f} \; \forall k \in \{1, 2, \cdots, C\}$. Let $\mathcal{X}^{(\mathbf{s}_k)}$ be the projected input space using the features in $\mathbf{s}_k$. The classifier constructed using the class-specific feature subsets $\mathbf{s}_k$s can be defined as $\mathcal{F} : \{\mathcal{X}^{(\mathbf{s}_1)}, \mathcal{X}^{(\mathbf{s}_2)}, \cdots \mathcal{X}^{(\mathbf{s}_C)}\} \mapsto \mathcal{Y}$. We propose a CSFS mechanism embedded in a fuzzy rule-based classifier (FRBC). So next we discuss the FRBC.

2.1 Fuzzy Rule-Based Classifiers

We employ the FRBC framework used in [4,5]. Each class is represented by a set of rules. Let there are N_k rules for the kth class. The lth rule corresponding to the kth class, R_{kl} is given by

$$\mathrm{R}_{kl} : \text{If } x_1 \text{ is } A_{1,kl} \text{ and } x_2 \text{ is } A_{2,kl} \text{ and} \cdots x_P \text{ is } A_{P,kl} \text{then } y \text{ is } k. \quad (1)$$

Here, $k \in \{1, 2, \cdots, C\}$, $l \in \{1, 2, \cdots, N_k\}$, and $A_{j,kl}$ is a linguistic value (fuzzy set) defined on the jth feature for the lth rule of the kth class. Let, α_{kl} be the firing strength of the rule R_{kl}. The rule firing strength is computed using the product T-norm [7] over the fuzzy sets $A_{1,kl}, A_{2,kl}, \cdots, A_{P,kl}$. Let the membership to the fuzzy set $A_{j,kl}$ be $\mu_{j,kl}$. So, α_{kl} is given by, $\alpha_{kl} = \prod_{j=1}^{P} \mu_{j,kl}$. The final output of the FRBC is of the form $\mathbf{o} = (o_1, o_2, \cdots, o_C)$, where, o_k is the support for kth class, computed as $o_k = \max\{\alpha_{k1}, \alpha_{k2}, \cdots, \alpha_{kN_k}\}$. To learn an efficient classifier from the initial fuzzy rule-based system, the parameters defining the fuzzy sets $A_{j,kl}$s can be tuned by minimizing the loss function,

$$E_{cl} = \sum_{i=1}^{n} \sum_{k=1}^{C} (o_k^i - t_k^i)^2. \quad (2)$$

To extract the rules, following [4,5], we cluster the training data of the kth class into N_k clusters. We note here that the kth class may not have n_k clusters in the pattern recognition sense. By clustering we just group the nearby points and then define a rule for each group. Let the centroid of the lth cluster of the kth class be $\mathbf{v}_{kl} = (v_{1,kl}, v_{2,kl}, \cdots, v_{P,kl})$. The cluster centroid $\mathbf{v}_{kl}$ is then translated into P fuzzy sets, $A_{j,kl} = $ "close to" $v_{j,kl} \; \forall j \in \{1, 2, \cdots, P\}$. The fuzzy set '"close to" $v_{j,kl}$' is modeled by a Gaussian membership function with mean $v_{j,kl}$. Although the membership parameters can be tuned to refine the fuzzy rules, in this study, we have not done that. We have used fixed rules defined by the obtained cluster centers and a fixed spread value in the Gaussian membership functions.

2.2   Feature Selection

Following [2–5,13], we use feature modulators which stop the derogatory features and promote useful features to take part in the rules of the FRBC. For each feature, there is an associated modulator of the form $M(\lambda_j) = \exp(-\lambda_j^2)$, where $j \in \{1, 2, \cdots, P\}$ [5]. To select or reject a feature using the modulator function, the membership values associated with the jth feature are modified as

$$\hat{\mu}_{j,kl} = \mu_{j,kl}^{M(\lambda_j)} = \mu_{j,kl}^{\exp(-\lambda_j^2)} \forall k, l \tag{3}$$

Note that, $\lambda_j \approx 0$ makes $\hat{\mu}_{j,kl} \approx \mu_{j,kl}$. Similarly, when λ_j is high (say, $\lambda_j \geq 2$), $\hat{\mu}_{j,kl} \approx 1$. The rule firing strength is now calculated as $\alpha_{kl} = \prod_{j=1}^{P} \hat{\mu}_{j,kl}$. So, when $\hat{\mu}_{j,kl} \approx \mu_{j,kl}$, jth feature influences the rule firing strength α_{kl} and in turn influences the classification process, whereas, if $\hat{\mu}_{j,kl} \approx 1$ then the jth feature has no influence on the firing strength and hence no influence on the predictions by the FRBC. This would be true for any T-Norm as $T(x, 1) = x, x \in [0, 1]$. Thus, for useful features, λ_js should be made close to zero and for derogatory features λ_js should be made high. The desirable values of λ_js are obtained by minimizing E_{cl} defined in (2) with respect to λ_js. The training begins with $\lambda_j = 2+$ Gaussian noise. $M(\lambda_j) \approx 0$ indicates a strong rejection of x_j, while $M(\lambda_j) \approx 1$ suggests a strong acceptance of x_j. However, training may lead λ_js such that $M(\lambda_j)$ takes a value in between 0 and 1. This implies that the corresponding feature influences the classification partially. This is not desirable in our case, as our primary goal is to select or reject features. To facilitate this, we add a regularizer term E_{select} to E_{cl} such that E_{select} adds a penalty if any λ_j allows the corresponding feature partially. In [5], E_{select} is set as follows.

$$E_{select} = (1/P) \sum_{j=1}^{P} \exp(-\lambda_j^2)(1 - \exp(-\lambda_j^2)) \tag{4}$$

So, the overall loss function for learning suitable λ_js becomes

$$E = E_{cl} + c_1 E_{select}. \tag{5}$$

Class-Specific Feature Selection. So far we have not considered selection of class-specific features. In the class-specific scenario, for each class, a different set of P modulators is engaged. So, a total of $C \times P$ feature modulators are employed. Consequently, for each class a different set of features, if appropriate, can be selected. Here, we represent the feature modulator for the jth feature of the kth class as $M(\lambda_{j,k}) = \exp(-\lambda_{j,k}^2)$, where $j \in \{1, 2, \cdots, P\}; k \in \{1, 2, \cdots, C\}$. The modulator value $M(\lambda_{j,k})$ modify the membership values corresponding to the jth feature of the kth class as following:

$$\hat{\mu}_{j,kl} = \mu_{j,kl}^{M(\lambda_{j,k})} = \mu_{j,kl}^{\exp(-\lambda_{j,k}^2)} \forall l \tag{6}$$

For this problem, the E_{select} is changed to

$$E_{select} = (1/(CP)) \sum_{k=1}^{C} \sum_{j=1}^{P} \exp(-\lambda_{j,k}^2)(1 - \exp(-\lambda_{j,k}^2)) \tag{7}$$

We now minimize (5) with respect to $\lambda_{j,k}$s to find the optimal $\lambda_{j,k}$s.

2.3   Monitoring Redundancy

Suppose a data set has three useful features say x_1, x_2, x_3 such that each of x_2 and x_3 is strongly dependent on (say correlated with) x_1 then all the three features carry the same information and only one of them is enough. These three form a redundant set of features. However, if we just use one of them and there is some error in measuring that feature, the system may fail to do the desired job. Therefore, a controlled use of redundant features is desirable. For the global feature selection framework, redundancy control has been realized by adding the regularizer (8) to (5) [3,5,13]:

$$E_r = (1/(P(P-1))) \sum_{j=1}^{P} \sum_{m=1, m \neq j}^{P} \sqrt{\exp\left(-\lambda_j^2\right) \exp\left(-\lambda_m^2\right) \rho^2(x_j, x_m)} \quad (8)$$

Here, $\rho()$ is the Pearson's correlation coefficient (or it could be mutual information also), which is a measure of dependency between two features. When x_j and x_m are highly correlated, $\rho^2(x_j, x_m)$ is close to one (its highest value). In this case, to reduce the penalty E_r, the training process will adapt λ_j and λ_m in such a way that one of $\exp\left(-\lambda_j^2\right)$ and $\exp\left(-\lambda_m^2\right)$ is close to 0 and the other is close to 1. Note the (8) is not suitable for class-specific scenario. Next we change (8) for class-specific redundancy.

Class-Specific Redundancy. For class-specific redundancy, we compute $\rho_k(x_j, x_m)$ between features x_j and x_m considering only instances of the kth class. In the class-specific case, for each class, we have P feature modulators, $M(\lambda_{j,k}) = \exp\left(-\lambda_{j,k}^2\right)$, where $j \in \{1, 2, \cdots, P\}; k \in \{1, 2, \cdots, C\}$. So, (8) is modified as following.

$$E_{r_c} = (1/(CP(P-1))) \sum_{k=1}^{C} \sum_{j=1}^{P} \sum_{m=1, m \neq j}^{P} \sqrt{\exp\left(-\lambda_{j,k}^2\right) \exp\left(-\lambda_{m,k}^2\right) \rho_k^2(x_j, x_m)} \quad (9)$$

Considering the class-specific redundancy, our new loss function for learning the system becomes:

$$E_{tot} = E_{cl} + c_1 E_{select} + c_2 E_{r_c}. \quad (10)$$

2.4   Exploiting Substructures Within a Class

For some real world problems, the data corresponding to a class may have distinct clusters and some of the clusters may lie in different sub-spaces. For example, in a multi-cancer gene expression data set, each cancer may have several sub-types, where each sub-type is characterized by a different set of highly expressed genes/features. This generalizes the concept of class-specific feature selection further. To exploit such local substructures within a class while extracting rules, we need to use rule-specific feature modulators. Each rule of the kth class is assumed to represent a local structure or cluster present in the kth class. So, for the kth class there are $n_k \times P$ feature modulators. For the overall system there are $n_{rule} \times P$ feature modulators where, $n_{rule}(= \sum_{k=1}^{C} n_k)$ is the total number of rules. A modulator function is now represented by $M(\lambda_{j,kl})$ and the corresponding modulated membership is the following.

$$\hat{\mu}_{j,kl} = \mu_{j,kl}^{M(\lambda_{j,kl})} = \mu_{j,kl}^{\exp(-\lambda_{j,kl}^2)} \tag{11}$$

The regularizer, E_{select} is now modified as

$$E_{select} = (1/(CP)) \sum_{k=1}^{C} (1/n_k) \sum_{l=1}^{n_k} \sum_{j=1}^{P} \exp(-\lambda_{j,kl}^2)(1 - \exp(-\lambda_{j,kl}^2)) \tag{12}$$

In this framework, we do not consider redundancy. Using (11) for E_{cl} and (12) for E_{select} we define the loss function $E = E_{cl} + c_1 E_{select}$ for discovering rule-specific feature subset.

3   Experiments and Results

We have done three experiments to validate the three main contributions of our proposed framework. In Experiment 1, we have shown the effectiveness of the proposed CSFS over the usual global feature selection using the proposed FRBC framework. In Experiment 2, we have demonstrated the significance of class-specific redundancy control using our approach. In Experiment 3, a data set having multiple sub-structures in different sub-spaces within a class have been considered to show the utility of our method. We have not tuned the rule base parameters of the FRBC and have tuned only the feature modulators to select/reject features. For clustering, we have used the K-means algorithm. Each fuzzy set is modeled using a Gaussian membership function having two parameters: center and spread. The cluster centers are used as the centers of the Gaussian membership functions and their spreads have been set as 0.2 times the feature-specific range. To minimize the error functions, we use the optimizer, `train.GradientDescentOptimizer` from `TensorFlow` [1]. For all experiments, the learning rate is set to 0.2 and the stopping criterion is 10000 iterations. However, typically the loss function reduces very rapidly at the beginning of the training and it converges within a few hundred iterations. As mentioned in Sec.2 we denote the class-specific feature subset for class 1 as $\mathbf{s}_1$, for class 2 as $\mathbf{s}_2$ and so on.

3.1   Experiment 1

We use a three-class synthetic data set, Synthetic1, with six features having distributions as described in Table 1. Here, $\mathcal{N}(m, s)$ represents a normal distribution with mean, m and standard deviation, s; $\mathcal{U}(a, b)$ represents a uniform distribution over the interval (a, b). Without loss, we have assigned the first 100 points to class 1, next 100 points to class 2, and last 100 points to class 3. From Table 1 we can see that class 2 and class 3 are uniformly distributed over a given interval for features x_1 and x_2. On the other hand, class 1 is clustered around $(0, 0)$ in the feature space formed by x_1 and x_2. Hence the feature space formed by x_1 and x_2, discriminate class 1 from the other two classes. Similarly, (x_3, x_4)

Table 1. Description of the dataset Synthetic1

Instances	Features						Class
	x_1	x_2	x_3	x_4	x_5	x_6	y
$\mathbf{x}^1 \cdots \mathbf{x}^{100}$	$\mathcal{N}(0, 0.5)$	$\mathcal{N}(0, 0.5)$	$\mathcal{U}(-10, 10)$	$\mathcal{U}(-10, 10)$	$\mathcal{U}(-10, 10)$	$\mathcal{U}(-10, 10)$	1
$\mathbf{x}^{101} \cdots \mathbf{x}^{200}$	$\mathcal{U}(-10, 10)$	$\mathcal{U}(-10, 10)$	$\mathcal{N}(0, 0.5)$	$\mathcal{N}(0, 0.5)$	$\mathcal{U}(-10, 10)$	$\mathcal{U}(-10, 10)$	2
$\mathbf{x}^{201} \cdots \mathbf{x}^{300}$	$\mathcal{U}(-10, 10)$	$\mathcal{U}(-10, 10)$	$\mathcal{U}(-10, 10)$	$\mathcal{U}(-10, 10)$	$\mathcal{N}(0, 0.5)$	$\mathcal{N}(0, 0.5)$	3

Table 2. Performance on the dataset, Synthetic1

Run	Features selected		Avg. accuracy of FRBC (%)		
	Class-specific	Global	Class-specific	Global	All features
1–5	$\mathbf{s}_1 : x_1, x_2$; $\mathbf{s}_2 : x_3, x_4$; $\mathbf{s}_3 : x_5, x_6$	x_3, x_6	98.7	34.7	62.08

and (x_5, x_6) discriminate class 2 and class 3 respectively, from the corresponding remaining classes. To understand the importance of CSFS, we perform both global feature selection (GFS) and CSFS, and compare their performances. We have also computed the performance of the FRBC with all features. Number of rules considered per class is one. We have conducted 5 runs for each of the FRBC. We observe from Table 2 that for proposed CSFS, in all five runs, for each class its characteristic features (i.e. x_1, x_2 for class 1 and so on) are selected. The FRBC with the class-specific selected features has achieved an average accuracy of 98.7% - in fact, each run achieved the same accuracy. Whereas, in GFS, the selected subset is x_3, x_6. The FRBC using globally selected feature subset has achieved an accuracy of 34.7% in each of the five runs. One can argue that the class-specific model uses all six features, hence performs better than the global model which uses two features. But, when we learn the FRBC rules using all six features it has achieved an average accuracy of 62.08% over the five runs. Importance of class-specific feature selection is clearly established through this experiment.

3.2   Experiment 2

For Experiment 2, we have considered another synthetic dataset, Synthetic2, which is produced by appending two additional features x_7 and x_8 to Sythetic1 data set. For class 1, x_7 and x_8 are generated as $x_1 + \mathcal{N}(0, 0.1)$ and $x_2 + \mathcal{N}(0, 0.1)$, respectively. For the other two classes, x_7 and x_8 are generated from $\mathcal{U}(-10, 10)$ and $\mathcal{U}(-10, 10)$, respectively. We observe that x_7 is dependent on x_1 and x_8 is dependent on x_2 for class 1 but the remaining two classes are indiscernible among themselves considering features x_7 and x_8. Clearly, features x_7, x_8 are also discriminatory for class 1. However, do x_7, x_8 add any information over x_1, x_2 for class 1? The answer is no, as for class 1, x_7 and x_8 are noisy versions of x_1 and x_2, respectively. This feature-redundancy is specific to class 1. In Table 3 we have described the performances of the FRBCs in the CSFS framework without

Table 3. Class-specific feature selection on Synthetic2 data set

Run	Without redundancy control		With class-specific redundancy control	
	Selected features	Acc.	Selected features	Acc.
1	s_1:x_1, x_2, x_7, x_8; s_2:x_3, x_4; s_3:x_5, x_6	99.3	s_1:x_2, x_7; s_2:x_3, x_4; s_3:x_5, x_6	99.3
2	s_1:x_1, x_2, x_7; s_2:x_3, x_4; s_3:x_5, x_6	99.3	s_1:x_1, x_8; s_2:x_3, x_4; s_3:x_5, x_6	99
3	s_1:x_1, x_2, x_7, x_8; s_2:x_3, x_4; s_3:x_5, x_6	99.3	s_1:x_7, x_8; s_2:x_3, x_4; s_3:x_5, x_6	99.3
4	s_1:x_1, x_2, x_7, x_8; s_2:x_3, x_4; s_3:x_5, x_6	99.3	s_1:x_7, x_8; s_2:x_3, x_4; s_3:x_5, x_6	99.3
5	s_1:x_1, x_2, x_7, x_8; s_2:x_3, x_4; s_3:x_5, x_6	99.3	s_1:x_1, x_7; s_2:x_3, x_4; s_3:x_5, x_6	99

and with class-specific redundancy control. Here also, we have set the number of fuzzy rules per class as one and repeated the experiments five times with each model. The term 'Acc.' in Table 3 refers to accuracy of the FRBC in percentage. For class 1, features x_1 and x_7 are heavily dependent. Hence, to avoid redundancy only one of them should be selected. The same argument is true for features x_2 and x_8. Minimizing (10) which uses (9) to control class-specific redundancy, the FRBC has successfully chosen only one from x_1 and x_7 and one from x_2 and x_8 to include in s_1 in all five runs (Table 3). On the other hand, we observe that without any redundancy control, the class-specific feature selection framework selects all the four discriminatory features to include in s_1 in four runs. The best accuracy achieved by the CSFS framework without redundancy control and that of CSFS with class-specific redundancy control are the same and equal to 99.3% although the later selects only two features. This establishes the benefit of class-specific redundancy control.

3.3 Experiment 3

In experiment 3, we validate our proposed framework for handling the presence of different clusters in different sub-spaces within a class. We have generated Synthetic3, a two class data having four features where each class is composed of two distinct clusters lying in two different sub-spaces. The data set Synthetic3 is described in Table 4. Each class has 200 points. For class 1, instances 1 to 100 create a distinct cluster around $(0, -5)$ in the feature space formed of x_1, x_2 and instances 101 to 200 create a distinct cluster around $(0, 0)$ in the feature space formed of x_3, x_4. Similarly, class 2 also has two distinct clusters in the feature spaces formed of x_2, x_3 and x_1, x_4 respectively. To handle this dataset we employ our proposed rule-specific approach implemented using (11), and (12). The number of rules per class is set to two. As observed from Table 5, the rule-specific feature selection is successful in identifying the two important sub-spaces i.e. x_1, x_2 and x_3, x_4 for class 1 and x_2, x_3 and x_1, x_4 for class 2. We note that for both the classes the selected rule-specific subsets interchange between rules 1 and 2 because the cluster number assignment to different groups of points for a class varies. We also note from Table 5, using the class-specific feature selection method, in different runs, s_1 comprises of x_1 or x_2 and s_2 comprises of x_2 or x_1, x_2. These subsets obviously do not characterize the classes correctly.

Table 4. Description of the dataset Synthetic3

Instances	Features				Class
	x_1	x_2	x_3	x_4	y
$\mathbf{x}^1 \cdots \mathbf{x}^{100}$	$\mathcal{N}(0, 0.5)$	$\mathcal{N}(-5, 0.5)$	$\mathcal{U}(-10, 10)$	$\mathcal{U}(-10, 10)$	1
$\mathbf{x}^{101} \cdots \mathbf{x}^{200}$	$\mathcal{U}(-10, 10)$	$\mathcal{U}(-10, 10)$	$\mathcal{N}(0, 0.5)$	$\mathcal{N}(0, 0.5)$	1
$\mathbf{x}^{201} \cdots \mathbf{x}^{300}$	$\mathcal{U}(-10, 10)$	$\mathcal{N}(0, 0.5)$	$\mathcal{N}(-5, 0.5)$	$\mathcal{U}(-10, 10)$	2
$\mathbf{x}^{301} \cdots \mathbf{x}^{400}$	$\mathcal{N}(5, 0.5)$	$\mathcal{U}(-10, 10)$	$\mathcal{U}(-10, 10)$	$\mathcal{N}(-5, 0.5)$	2

Table 5. Features subsets selected for Synthetic3

Run	Rule-specific		Class-specific	
	Class 1	Class 2	s_1	s_2
1	rule 1:x_1, x_2; rule 2:x_3, x_4	rule 1:x_2, x_3; rule 2:x_1, x_4	x_1	x_1, x_2
2	rule 1:x_1, x_2; rule 2:x_3, x_4	rule 1:x_1, x_4; rule 2:x_2, x_3	x_2	x_1, x_2
3	rule 1:x_3, x_4; rule 2:x_1, x_2	rule 1:x_1, x_4; rule 2:x_2, x_3	x_2	x_2
4	rule 1:x_3, x_4; rule 2:x_1, x_2	rule 1:x_1, x_4; rule 2:x_2, x_3	x_2	x_1, x_2
5	rule 1:x_3, x_4; rule 2:x_1, x_2	rule 1:x_2, x_3; rule 2:x_1, x_4	x_2	x_2

The average accuracy of the FRBC using feature subsets selected by rule-specific, class-specific feature selection and using all features are 100%, 77.4%, and 87.5%, respectively. This demonstrates the usefulness of our proposed RSFS framework.

4   Conclusion

In this work, first, we have proposed a class-specific feature selection scheme using feature modulators embedded in a FRBC. The parameters of the feature modulators are tuned by minimizing a loss function comprising of classification error and a regularizer to make the modulators completely select or reject features. This framework is used in [4,5] for selecting globally useful features. We modified it to make it suitable for CSFS. Our proposed CSFS method does not employ OVA strategy like most of the existing CSFS works and hence is free from the enhanced computational overload and hazards associated with the OVA based methods. We have extended the CSFS scheme so that it can monitor class-specific redundancy by adding a suitable regularizer. Finally, our CSFS framework is generalized to a rule-specific feature selection framework to handle the presence of multiple sub-space-based clusters within a class. The three approaches are validated through three experiments on appropriate synthetic data sets. There are certain limitations of the proposed CSFS. The rule firing strength is computed using product. So, for very high dimensional data it may lead to underflow. However, this issue is true for any FRBC, it is not special to the systems proposed here. Since, we have experimented on synthetic data sets, the parameters like number of rules per class, and the spreads of the Gaussian membership functions are set intuitively. However, for real data sets, these parameters need to be chosen more judiciously with a systematic method.

References

1. Abadi, M., et al.: Tensorflow: a system for large-scale machine learning. In: 12th USENIX Symposium on Operating Systems Design and Implementation (OSDI 2016), pp. 265–283 (2016)
2. Chakraborty, D., Pal, N.R.: A neuro-fuzzy scheme for simultaneous feature selection and fuzzy rule-based classification. IEEE Trans. Neural Networks 15(1), 110–123 (2004)
3. Chakraborty, R., Pal, N.R.: Feature selection using a neural framework with controlled redundancy. IEEE Trans. Neural Netw. Learn. Syst. 26(1), 35–50 (2014)
4. Chen, Y.C., Pal, N.R., Chung, I.F.: An integrated mechanism for feature selection and fuzzy rule extraction for classification. IEEE Trans. Fuzzy Syst. 20(4), 683–698 (2011)
5. Chung, I.F., Chen, Y.C., Pal, N.R.: Feature selection with controlled redundancy in a fuzzy rule based framework. IEEE Trans. Fuzzy Syst. 26(2), 734–748 (2017)
6. Ezenkwu, C.P., Akpan, U.I., Stephen, B.U.A.: A class-specific metaheuristic technique for explainable relevant feature selection. Mach. Learn. Appl. 6, 100142 (2021)
7. Gupta, M.M., Qi, J.: Theory of t-norms and fuzzy inference methods. Fuzzy Sets Syst. 40(3), 431–450 (1991)
8. de Lannoy, G., François, D., Verleysen, M., et al.: Class-specific feature selection for one-against-all multiclass SVMs. In: ESANN. Citeseer (2011)
9. Nardone, D., Ciaramella, A., Staiano, A.: A sparse-modeling based approach for class specific feature selection. PeerJ Comput. Sci. 5, e237 (2019)
10. Panthong, R., Srivihok, A.: Liver cancer classification model using hybrid feature selection based on class-dependent technique for the central region of thailand. Information 10(6), 187 (2019)
11. Pineda-Bautista, B.B., Carrasco-Ochoa, J.A., Martínez-Trinidad, J.F.: General framework for class-specific feature selection. Exp. Syst. Appl. 38(8), 10018–10024 (2011)
12. Qian, Y.: Class-specific guided local feature selection for data classification. In: 2019 IEEE 4th International Conference on Cloud Computing and Big Data Analysis (ICCCBDA), pp. 645–649. IEEE (2019)
13. Wang, J., Zhang, H., Wang, J., Pu, Y., Pal, N.R.: Feature selection using a neural network with group lasso regularization and controlled redundancy. IEEE Trans. Neural Netw. Learn. Syst. 32(3), 1110–1123 (2020)
14. Yuan, L.M., Sun, Y., Huang, G.: Using class-specific feature selection for cancer detection with gene expression profile data of platelets. Sensors 20(5), 1528 (2020)
15. Zhou, W., Dickerson, J.A.: A novel class dependent feature selection method for cancer biomarker discovery. Comput. Biol. Med. 47, 66–75 (2014)

Performance/Resources Comparison of Hardware Implementations on Fully Connected Network Inference

Randy Lozada[1], Jorge Ruiz[1], Manuel L. González[1(✉)], Javier Sedano[1],
José R. Villar[2], Ángel M. García-Vico[3], and E. S. Skibinsky-Gitlin[1]

[1] Instituto Tecnológico de Castilla y León, Burgos, Spain
{randy.lozada,jorge.ruiz,manuel.gonzalez,javier.sedano,
erik.skibinsky}@itcl.es
[2] University of Oviedo, Oviedo, Spain
villarjose@uniovi.es
[3] Andalusian Research Institute in Data Science and Computational Intelligence
(DaSCI), University of Granada, 18071 Granada, Spain
agvico@decsai.ugr.es

Abstract. Fully Connected Network inference is a complex algorithm that can be accelerated using edge devices like Field Programmable Gate Array (FPGA). One commonly known performance improvement for Fully Connected Network inference is quantization. This technique replaces the floating points weights of the network by integers. Frameworks like Open Neural Network Exchange (ONNX) and Tensorflow Lite provide solutions for this procedure. However, these frameworks have different inference algorithms with different operations and data types. In this article inference algorithms of common Fully Connected Networks in ONNX and Tensorflow Lite have been analysed. A performance and resource usage comparison is tested on Xilinx® Zynq UltraScale+™ MPSoC. Results show that to achieve lower latency is better to avoid floating point operations in the inference algorithm. In terms of FPGA resource usage, an increase is observed when the neural network becomes more complex regardless of the algorithm. This growth in resource usage is framework-dependent.

Keywords: FPGA · Fully connected network · Quantization

1 Introduction

Deep neural networks have proven to be a very useful tool for solving a wide variety of real-world problems [1]. However, there are many problems surrounding the development of complex neural network models. Problems such as the large amount of data required for parameter tuning (also known as training), the high computational cost of training or the difficulty of achieving low latency for real-time applications [2].

© The Author(s), under exclusive license to Springer Nature Switzerland AG 2022
H. Yin et al. (Eds.): IDEAL 2022, LNCS 13756, pp. 348–358, 2022.
https://doi.org/10.1007/978-3-031-21753-1_34

A standard solution for accelerating both deep neural networks training and inference is the use of Graphics Processing Units (GPU). However, these are power-hungry, general-purpose devices. In particular, for deep neural network inference, there are other technologies within the edge computing paradigm. These technologies such as FPGA, Tensor Processing Unit (TPU) or Visual Processing Unit (VPU) are characterised by low power consumption while preserving high-performance [3]. On the one hand, TPU and VPU technology is developed by Google and Intel respectively. These technologies are exclusively aimed at deep neural networks inference. On the other hand, FPGA technology is a re-configurable digital circuit for general software-in-hardware implementations.

FPGA technology can achieve lower latency in deep neural network inference than GPU, TPU and VPU [3]. This is due to their intrinsic parallelism and the explicit programming of algorithmic expressions. In addition, FPGA can outperform GPU in terms of energy-efficient and performance per watt [4]. Both low latency and low power consumption make FPGA devices well suited for use in autonomous systems [5]. In these systems, low latency is necessary when making decisions in real time to be able to adapt to the environment. In addition, these systems often use batteries, so an efficient use of energy is necessary.

FPGA are programmed using Hardware Description Languages (HDL). These languages are used to describe digital circuits through the internal components of the device. Examples of HDL are Very High Speed Integrated Circuit HDL (VHDL) or Verilog. However these languages are concurrent in nature and domain-specific, which implies a high cost in implementation time. Moreover, programming with these languages is prone to errors and difficult to debug. In fact, particular debugging methods such as formal verification or universal verification methodology had to be developed, adding up more development time to build solutions with high quality of results. In contrast, High-Level Synthesis (HLS) is a programmer-friendly methodology that allows developers to describe digital hardware with more common programming languages such as C/C++. The code is then transformed into register transfer level constructs which have already been tested by the company responsible for its framework, this makes the description of digital hardware for FPGAs more user-friendly, specially for high performance computing applications.

This work focuses on Fully Connected Network (FCN) inference and their implementation in C++/HLS with algorithmic enhancements such as quantization. The main contribution of this work is the performance/resource comparison between two frameworks, Open Neural Network Exchange (ONNX) and Tensorflow Lite (TFLite). The rest of the article is organized as follows. Different implementations and algorithmic improvements of FCN inference on FPGA devices published in the literature are shown in Sect. 2. In Sect. 3, Floating Point (FP), ONNX and TFLite FCN inference are explained in detail. Thereafter, the hardware implementation of each inference algorithm is explained in Sect. 4. The FCN architectures and the dataset used for experimentation are detailed in Sect. 5. Finally, the article ends with the results, discussion and conclusions.

2   Related Work

FCN inference implementation on FPGA has been studied widely due to its low computational resource usage and low latency. These two goals are typically opposite. The use of more resources allows for greater parallelism, resulting in lower latency. Also, by reusing resources, it is possible to reduce their quantity at the cost of higher latency [6]. To maximize these goals at the same time, different techniques should be applied to FCN. Approaches like quantization, low-bit representation of parameters, and modification of inference algorithms have been proven in the literature. These techniques have a direct impact on the prediction performance metric [7], therefore they must be implemented carefully.

Approaches to the modification of FCN inference algorithms are shown in [6,8]. In [6] the authors test two different FCN and CNN inference designs, N-Fold and Flow. The former refers to instantiating in the FPGA a single layer of each type and updating the weights with the values of each layer during the inference process. The latter refers to instantiating all the layers of the network in the FPGA with their corresponding weights. Results on different FPGA devices show that N-Fold optimizes computational resources and Flow optimizes latency. In [8] FCN with HLS are implemented using Xilinx® Vivado HLS in Xilinx® Zynq ZC702 SoC. The novelty of the implementation is to apply the same operations of the Fully Connected Layer at the same time. Firstly, all multiplications are applied, then all additions and finally activations. With this design, the authors achieve low latency on a regression problem.

Low bit representation and quantization of FCN parameters are explored in [9–12]. In [9] a reusable Single Hidden layer Neural Network architecture compiled in an FPGA overlay is proposed. The input and activation are quantized in power-of-2 values to substitute multipliers of the matrix product operation by bit-shifts. The quantization is optimized using a teacher-student training algorithm and the results show a maximum 0.1% of accuracy loss. Less resource usage is also achieved than standard FCN FPGA implementation. In [10] and [12], a map from FP weights to fixed-point is proposed. These implementations optimize resources and latency. Finally, in [11] a non-linear quantization method on FCN is proposed. This quantization is applied by creating an interval based cumulative distribution of weights. Quantizing to 4-bit fixed-point the authors achieve a 50% memory reduction by worsening the target metric by 2.7%.

These presented works aim to improve the performance or the resources of FCN inference in FPGA. Authors use two main approaches. One is to modify the inference algorithm, and the other is to quantize the model. Combination of these two approaches are implemented in frameworks ONNX and TFLite, which will be detailed in Sect. 3. The novelty of the present work is the comparison between these frameworks in terms of the FCN inference algorithm, the FCN implementation on FPGA and finally, the results of performance and resource usage measured on the FPGA.

3   Inference

A FCN is a concatenation of N Fully Connected layers. Each layer j is composed of I_j inputs and K_j neurons. A Fully Connected layer has four mathematical components: an input tensor, a kernel tensor, a bias tensor and an activation function. The elements of these tensors have real values, represented as FP. The inference of a Fully Connected layer is calculated as follow

$$y = k \cdot i + b \tag{1}$$

$$y' = \mathrm{F}(y) \tag{2}$$

where k stands for the kernel weight's tensor, i for the input tensor, b for the bias vector, y for the intermediate output, F for the activation function and y' for the layer output. Expressions (1) and (2) are known as FP-inference.

3.1   Linear Quantized Inference

Neural Network quantization is a process that transform FP weights to integer ones. Integers can be signed (INT) or unsigned (UINT). This process reduces both memory and computational requirements. In linear quantization, each real tensor is transformed uniformly with the following affine mapping

$$q = \mathrm{clip} \left(\left\lfloor \frac{r}{S} \right\rceil + Z, \alpha_q, \beta_q \right) \tag{3}$$

where q is the quantized value, r is the original real value, S is the scale factor and Z is the zero point offset. There are multiple ways of obtaining S and Z, as stated in [13]. α_q and β_q are the minimum and maximum values that can be represented with a n bit integer.

Quantized Fully Connected Layer. The quantized Fully Connected inference expression (4) can be obtained by applying (3) to (1) and isolating q_y

$$q_y = Z_y + M_1 \left[(q_k - Z_k) \cdot (q_i - Z_i) + M_2(q_b - Z_b) \right] \tag{4}$$

where indexes b, k, i and y stand for the bias, kernel, input, and output tensor. M_1 and M_2 are the scale factors and are defined as a quotient of scales.

Some activation functions (2), like sigmoid ($\sigma(x)$) or hyperbolic tangent (Tanh(x)), can be expensive to compute with quantized layers. These functions require the input q_y to be dequantized into a FP value, calculate its activation function, and quantize again into an integer. To reduce the computational cost of activation function and speed up the inference process, activation results can be pre-calculated and stored in a static activation look-up table (ALUT). ALUTs are a map between the 2^n possible values of y and the 2^n possible values of y'.

Comparison Between FP-Inference and Quantized Inference. Quantization reduces weight sizes. For example, weights from FP32 to INT8 are reduced in size by 75%. Despite this fact, the quantized inference implies a higher number of operations. For instance, a quantized model requires additional input and output layers that quantize and dequantize the input and output data from FP to INT/UINT and vice versa. Furthermore, quantized layers use scales and zero points to avoid integer overflow. Although the inference of a quantized model requires more operations, the use of discrete values allows to replace the activation functions by ALUT, reducing their computational cost.

3.2 UINT8 ONNX Inference

ONNX [14] is an open standard designed to allow interoperability between different machine learning frameworks. The ONNX framework includes a wide variety of libraries, from model converters to ONNX Runtime, a multi-platform hardware-accelerated inference engine. ONNX quantizer is a tool that allows INT and UINT quantization of ONNX models.

ONNX quantizer splits a quantized Fully Connected layer into three different operators: **QLinearMatMul** (5), **QLinearAdd** (6) and **Activation** that is implemented as an ALUT.

$$q_{bias\_in} = Z_{bias\_in} + (M_3(q_k - Z_k) \cdot (q_i - Z_i)) \tag{5}$$

$$q_y = Z_y + [M_4 \cdot (q_{bias\_in} - Z_{bias\_in}) + M_5 \cdot (q_b - Z_b)] \tag{6}$$

By splitting (4) into two separate operators the model becomes less coupled but the overall computational complexity of the layer has increased. An ONNX Fully Connected layer has three FP products M_3, M_4 and M_5, whereas expression (4) only has two FP products, M_1 and M_2.

3.3 INT8 Tensorflow Lite Inference

TFLite is a library that converts Tensorflow and Keras models into a lightweight TFLite model. TFLite includes a post-training INT8 quantization tool designed for the Google Coral Edge TPU hardware accelerator. This hardware-focused design quantization allows the inference engine to infer the model without any FP operations.

In this framework, M_2 in Eq. (4) is set to 1 by forcing $S_b = S_k \cdot S_i$. Setting the S_b as the product of two scales constrains the quantized bias to overflow out of the INT8 quantization range. To overcome this situation, TFLite quantizes the bias in INT32. In order to return to the 8 bit range at the output, M_1 is used to scale down the INT32 values back to the INT8 quantization interval. This operation can be done without any FP by approximating M_1 with an INT32 and bit shift operation by using (7).

$$M \approx \frac{Q}{2^{shift}} \tag{7}$$

The approximation can be accomplished by discretizing the FP interval into an signed integer interval. As M_1 in is always in the interval (0,1), the new INT interval is (0, 2^{shift}). Approximated scale factor Q is an UINT with $shift$ bits of precision.

The main difference between the TFLite and ONNX quantized inference resides in the scaling factors (M_i). Meanwhile ONNX uses three scale factors represented in FP32, TFLite uses a single scale factor represented in INT32 and a bit shift operation, thus removing any FP operation.

4   Hardware Implementation

The above inference algorithms are accelerated using Xilinx® FPGA technology. Vitis HLS [15] is used for the hardware description. This is an HLS tool that allows transforming algorithms described in C/C++ language into synthesizable hardware, called acceleration kernels. In this way, the elements of the C/C++ language are associated with the elements from the programmable logic (PL) of the FPGA: random access memory blocks (BRAM) for mutable memory, like arrays, lookup tables (LUT) for creating custom logic, such as combinatorial elements and ALUT, flip-flops (FF) for the creation of registers and digital signal processing (DSP) blocks for carrying out products and summations [16,17].

4.1   HLS Description

For the implementation of FCN as an acceleration kernel, a design based on a C++ class hierarchy is used. The main objective of this design flow is to obtain an environment that allows different FCN architectures and inference algorithms to be implemented. This standard design allows to compare the results obtained on the FPGA of the inference algorithms described in Sect. 3.

The first class in the design is the **Fully Connected layer**. This class implements the one-layer inference algorithm, which can be FP-inference, ONNX or TFLite. Each instance of this class contains the weights matrix and the bias vector of the layer, as well as the quantization parameters for the quantized variant. These elements are stored inside the FPGA PL using BRAM. This class has two main functionalities: *set_parameters()* and *run()*. The former is used to set the parameters of the layer, copying each element of the weight matrix and the bias vector in the PL while the latter performs the inference operation.

The second class is the **Neural Network**. The main purpose of this class is to create as many Fully Connected layer instances as the number of layers in the FCN. This class is also used to control the parameter setting, as well as the interconnection of instances using the corresponding activation function. In the case of quantized algorithms, the activation function is implemented through an ALUT. Also, for ONNX and TFLite implementation an instance of the initial quantization and the final dequantization functions based on (3) are created.

Finally, the last element is the **Kernel**. This component is a function that describes the acceleration kernel wrapper and creates an instance of the class

Neural Network in the PL that defines the FCN. Through this element, the control of the incoming data flow is handled, controlling when the Neural Network parameterization and inference are performed.

These three elements describe the architecture used for the implementation of all the inference algorithms presented in Sect. 3. Also the HLS optimizations used are equivalent in each variant. This allows a comparison between the different implementations in terms of resource consumption and performance. However, considering the nature of each inference algorithm, there are marked differences between the corresponding hardware implementations.

The main difference is based on the data types used to represent the Fully Connected layer parameters. In the FP-inference this information is represented with FP32, while in the quantized variants INT are used. This implies a performance overhead in FP-inference for each addition and multiplication. For example, a FP addition operation has a maximum latency of 13 clock cycles, while the same operation with INT has 4 clock cycles [15]. Moreover, the size of data types also affect resource consumption since the Fully Connected layers parameters must be stored in the PL. Another important difference is the implementation of the activation functions. The FP-inference uses a combination of FP exponential operations, each with a maximum latency of 30 clock cycles [15]. The quantized variants allow the use of INT8 ALUT, with relatively low resource consumption and a latency of one clock cycle. Among the quantized variants there are several differences. ONNX represents the scale factors using FP32, using fmult [15] type operations. While TFLite replaces this operation by a multiplication between INT32 and a bit-shift. Also, ONNX makes use of UINT8 to represent all weights and TFLite use INT8 for kernel tensor and INT32 for bias. This implies a small increase of memory consumption in PL in TFLite implementations.

5   Experimental Results

5.1   Accelerated Flow

In order to test the implemented algorithms, an accelerated flow application development of the Vitis™ unified software platform [18] is used. The Ultra96v2 development board is used for the hardware implementation, which is an ARM-based Xilinx® Zynq UltraScale+™ MPSoC [17] development board based on the Linaro 96Boards Consumer Edition (CE) specification [19].

The acceleration kernel of FCN inference, regardless of the variant, has the architecture represented in Fig. 1. It is divided into two fundamental elements: Parameter-Settings and Inferences. The former is responsible for the configuration of all the elements that compose the Neural Network while the latter is used to execute the inference. Therefore, the network configuration is performed before the first inference. This implies an initial performance cost.

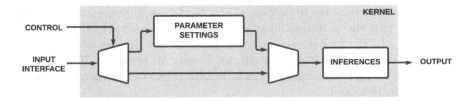

Fig. 1. Acceleration kernel summarized architecture. Parameter-Settings module is responsible for the configuration of all the elements that compose the Neural Network. Inferences module is used to execute the inference using the configuration parameters.

5.2 Model Architecture and Dataset

Two FCN with different topologies are used to test the inference algorithms implementation. The former is composed of 5 layers with 8.6 k-parameters. The latter presents 8 layers with 112.1 k-parameters.

To train and test the different FCN the MNIST dataset [20] is used. It is a small dataset of 28 × 28 handwritten images for pattern recognition. The models were trained with 60000 digits. In ONNX an TFLite the quantization process is done with a small subset of 1000 unlabelled random digits.

5.3 Results and Discussion

The experimental results have three fundamental measures: resource usage, latency and accuracy. The first measure is presented in Table 1 and shows the percentage of each resource used in Ultra96v2. The second and third measures are presented in Table 2 and shows inference time and accuracy lost by the quantization process.

Table 1. Resource consumption of the three FCN algorithms discussed. The table shows the name of the inference algorithms, the number of layers of the FCN and the resources of the FPGA. These resources are: the LUT, LUT used as Memory, BRAM, DSP and Flip-Flops. The best results in the use of FPGA resources have been highlighted.

Algorithm	Layers	LUT (%)	LUT as Mem (%)	BRAM (%)	DSP (%)	Flip-Flops (%)
Non-quantized	5	**27.98**	12.50	12.50	**9.44**	22
	8	**37.61**	13.21	61.11	**6.67**	30.28
ONNX	5	55.85	**7.24**	9.72	35	38.96
	8	86.27	**9.42**	36.57	64.72	61.96
TFLite	5	33.88	14.21	**5.09**	50.83	**19.79**
	8	38.58	10.32	**27.31**	83.06	**29.41**

Analysing the resource utilization (Table 1) of each implementation, it can be observed that the quantized inferences have a higher usage of LUT and DSP. This is because the higher complexity in terms of the number of operations, being the ONNX variant the most affected among all resources. However, the TFLite solution presents a higher increase in DSP than ONNX when the number of layers is incremented.

Table 2. Inference latency and accuracy for 2048 test images of MNIST dataset. The table shows the name of the inference algorithm, the number of layers of the FCN, the accuracy achieved, the initial latency and the average latency. The best results in accuracy and latency have been highlighted.

Algorithm	Layers	Parameters	Accuracy (%)	Initial latency (μs)	Average latency (μs)
Non-quantized	5	8.6k	89.55	1 426.20	616.05
	8	112.1k	95.61	7 475.35	6 630.13
ONNX	5	8.6k	**89.55**	1 697.35	768.29
	8	112.1k	**95.85**	2 749.97	1 772.46
TFLite	5	8.6k	89.50	**994.89**	**316.99**
	8	112.1k	95.70	**2 286.64**	**1 093.45**

To compare the accuracy and performance of the FCN (Table 2), a total of 2048 inferences are computed with a target frequency for the acceleration kernel of 200 MHz. No significant accuracy reduction is observed due to quantization neither in ONNX nor TFLite solutions. The best solution is the one corresponding to TFLite, achieving an average inference per image of 1.1 ms for a 112.1k parameters FCN and an average inference per image of 0.3 ms for an 8.6k parameters FCN. As expected, the FP-inference was the slowest and with the highest latency increase when the number of parameters is augmented. In all solutions, the time of the first inference is much higher than the average. This result is expected due to the initial configuration of the Parameter-Settings process.

6   Conclusions and Future Work

In this article, a comparison between the non-quantized FCN inference and the quantized FCN inference algorithm of TFLite and ONNX is analysed from four main angles or aspects. The first aspect is algorithmic. Quantized FCN inference in ONNX is more decoupled than TFLite, however, it has operations between FP and INT that TFLite manages to eliminate. The second aspect is FPGA resource usage. Non-quantized solution is characterized by high use of BRAM, due to the storage of variables in FP32 as weights, inputs and outputs. Quantized solutions are characterized by high use of LUT, DSP and FF, due to the complex operations of the inference algorithm. The third aspect is the accuracy achieved by each implementation. In this regard, the loss of accuracy due to quantization is negligible in either TFLite or ONNX. Finally, the fourth aspect is latency. TFLite has the lowest latency because its inference algorithm is FP-free. Both quantization frameworks have better latency scalability than the FP-inference when the FCN size is increased. This is because of the use of INT additions, multiplications and ALUT on quantized solutions.

These results show that there is a trade-off between FPGA resource usage and inference latency. This trade-off is due to the increase of the complexity of the quantized inference FCN to be able to operate in INT, thus decreasing the latency while maintaining the accuracy.

The main objective of this work has been to establish a performance/resource comparison between different FCN hardware implementations with the same HLS optimizations. To obtain lower latency results it is necessary to increase the level of parallelism, and even make modifications in the inference algorithms that allow for better exploit these parallel architectures. Future work goes on this way. Also, a reduction of resources will be studied with the implementation of arbitrary precision data instead of common data types like FP32, INT8 or INT32.

Acknowledgments. This work has been founded by the following institutions: the Ministry of Science and Innovation under CERVERA Excellence Network project CER-20211003 (IBERUS), Missions Science and Innovation project MIG-20211008 (INMER-BOT), European Union's Horizon 2020 research and innovation programme (project DIH4CPS) under the Grant Agreement no 872548, CDTI (Centro para el Desarrollo Tecnológico Industrial) under projects CER-20211022, ICE (Junta de Castilla y León) under project CCTT3/20/BU/0002, the Spanish Ministry of Economics and Industry under the grant PID2020-112726RB-I00, the Principado de Asturias under the grant SV-PA-21-AYUD/2021/50994 and the Regional Government of Andalusia, program "Personal Investigador Doctor", reference DOC_00235.

References

1. Abiodun, O.I., et al.: State-of-the-art in artificial neural network applications: a survey. Heliyon **4**(11), e00938 (2018). ISSN: 24058440. https://doi.org/10.1016/j.heliyon.2018.e00938
2. Schwartz, R., et al.: Green AI. Technical report. arXiv:1907.10597, August 2019
3. Gordienko, Y., et al.: "Last mile" optimization of edge computing ecosystem with deep learning models and specialized tensor processing architectures. In: Advances in Computers, vol. 122, pp. 303–341. Elsevier (2021). https://doi.org/10.1016/bs.adcom.2020.10.003
4. Nurvitadhi, E., et al.: Can FPGAs beat GPUs in accelerating next- generation deep neural networks?, pp. 5–14 (2017). https://doi.org/10.1145/3020078.3021740
5. Seng, K.P., Lee, P.J., Ang, L.M.: Embedded intelligence on FPGA: survey, applications and challenges. Electronics **10**(8) (2021). ISSN: 2079-9292. https://doi.org/10.3390/electronics10080895
6. Baptista, D., Sousa, L., Morgado-Dias, F.: Raising the abstraction level of a deep learning design on FPGAs. IEEE Access **8**, 205148–205161 (2020). ISSN: 2169-3536. https://doi.org/10.1109/ACCESS.2020.3036975

7. Nagel, M., et al.: Up or down? Adaptive rounding for post-training quantization. Number. arXiv:2004.10568, June 2020
8. Novickis, R., et al.: An approach of feed-forward neural network throughput-optimized implementation in FPGA. Electronics **9**(12), 2193 (2020). ISSN: 2079–9292. https://doi.org/10.3390/electronics9122193
9. Abdelsalam, A.M., et al.: An efficient FPGA-based overlay inference architecture for fully connected DNNs. In: 2018 International Conference on ReConFigurable Computing and FPGAs (ReConFig). Cancun, Mexico, pp. 1–6. IEEE, December 2018. ISBN: 978-1-72811-968-7. RECONFIG.2018.8641735. https://doi.org/10.1109/RECONFIG.2018.8641735
10. Bjerge, K. Schougaard, J.H., Larsen, D.E.: A scalable and efficient convolutional neural network accelerator using HLS for a system-on-chip design. Microprocess. Microsyst. **87**, 104363 (2021). ISSN: 01419331.104363. https://doi.org/10.1016/j.micpro.2021104363
11. Nicodemo, N., et al.: Memory requirement reduction of deep neural networks for field programmable gate arrays using low-bit quantization of parameters. In: 2020 28th European Signal Processing Conference (EUSIPCO), pp. 466–470. IEEE, Amsterdam, January 2021. ISBN: 978-90-827970-5-3. https://doi.org/10.23919/Eusipco47968.2020.9287739
12. Mukhopadhyay, A.K., Majumder, S., Chakrabarti, I.: 11Systematic realization of a fully connected deep and convolutional neural network architecture on a field programmable gate array. Comput. Electric. Eng. **97**, 107628 (2022). ISSN: 00457906. https://doi.org/10.1016/j.compeleceng.2021.107628
13. Gholami, A., et al.: A survey of quantization methods for efficient neural network inference (2021). https://doi.org/10.48550/ARXIV.2103.13630
14. ONNX Runtime developers. ONNX Runtime. https://onnxruntime.ai/ Version 1.11.0. 2021
15. Xilinx Inc.: Vitis high-level synthesis user guide. Ug1399 2, pp. 1–657 (2020)
16. Xilinx Inc.: UltraScale architecture DSP slice: user guide. Xilinx Tech. Documentation **579**, 1–75 (2018). https://www.xilinx.com/support/documentation/user%7B%5C_%7Dguides/ug579-ultrascale-dsp.pdf
17. Inc, X.: Zynq UltraScale + MPSoC Data Sheet: overview processing system (PS) Arm Cortex-A53 based application dual-core arm Cortex-R5 based on-chip memory. Xilinx Tech. Documentation **891**, 1–42 (2018)
18. Xilinx Inc.: Vitis unified software platform documentation embedded software development. UG1400, p. 667 (2021)
19. Avnet: Ultra96-V2 Board (2022). https://www.avnet.com/wps/portal/us/products/new-product-introductions/npi/aes-ultra96-v2/
20. LeCun, Y., Cortes, C., Burges, C.L MNIST handwritten digit database. http://yann.lecun.com/exdb/mnist/

Gradient Regularization with Multivariate Distribution of Previous Knowledge for Continual Learning

Tae-Heon Kim[1], Hyung-Jun Moon[2], and Sung-Bae Cho[1,2(✉)]

[1] Department of Computer Science, Yonsei University, Seoul 03722, Korea
{thkim0305,sbcho}@yonsei.ac.kr
[2] Department of Artificial Intelligence, Yonsei University, Seoul 03722, Korea
axtabio@yonsei.ac.kr

Abstract. Continual learning is a novel learning setup for an environment where data are introduced sequentially, and a model continually learns new tasks. However, the model forgets the learned knowledge as it learns new classes. There is an approach that keeps a few previous data, but this causes other problems such as overfitting and class imbalance. In this paper, we propose a method that retrains a network with generated representations from an estimated multivariate Gaussian distribution. The representations are the vectors coming from CNN that is trained using a gradient regularization to prevent a distribution shift, allowing the stored means and covariances to create realistic representations. The generated vectors contain every class seen so far, which helps preventing the forgetting. Our 6-fold cross-validation experiment shows that the proposed method outperforms the existing continual learning methods by 1.14%p and 4.60%p in CIFAR10 and CIFAR100, respectively. Moreover, we visualize the generated vectors using t-SNE to confirm the validity of multivariate Gaussian mixture to estimate the distribution of the data representations.

Keywords: Continual learning · Memory replay · Sample generation · Multivariate gaussian distribution · Expectation-maximization

1 Introduction

Continual learning is a learning mechanism that a network learns data sequentially. Sometimes data from all the classes may not be available simultaneously in real-world situations. A new class can be added in the future, or the system does not have enough space to hold all the data. Thus, the research on continual learning is significant. However, as the network learns the tasks, the previous knowledge is lost, and the network is only optimized for the current task, which is the problem called catastrophic forgetting [1]. Recently, many continual learning methods have been proposed to alleviate the catastrophic forgetting. Existing methods can be classified into three categories [2]: regularization [3], parameter isolation [4], and replay approaches [5, 6].

© The Author(s), under exclusive license to Springer Nature Switzerland AG 2022
H. Yin et al. (Eds.): IDEAL 2022, LNCS 13756, pp. 359–368, 2022.
https://doi.org/10.1007/978-3-031-21753-1_35

Replay approach is to replay data from previous tasks. However, this approach causes several problems that disturb its intention to stop forgetting. Overfitting, a problem that happens when the model fits to the data distribution instead of the class distribution, often occurs in the rehearsal methods because the network learns the same data stored in the memory several times [7]. Moreover, since the stored data are fewer than the current task data, the model suffers from the class imbalance problem, where the classes in the current task are more likely to be classified than previously learned classes [6, 8]. Pseudo-rehearsal approach that utilizes the generator is free from the problems mentioned above, but the generator also faces the forgetting problem [9].

In this work, we propose a method that combines gradient regularization and Gaussian mixture estimation to replay representation vectors of all the seen classes generated with the stored means and covariances. Such vectors are extracted by passing an intermediate layer of the CNN, so they contain information about features interpreted by the convolutional layers. It means their distribution is simple enough that multivariate Gaussian can accurately represent the distribution. The distribution of the class representations can change as the model continually learns. So, we utilize a gradient regularization method [10–12] that restricts the learning gradient, which has shown that this method effectively preserves the output vectors of the convolutional layer in continual learning [13]. The proposed method can avoid overfitting and class imbalance by generating the representations and the retraining only with them. Additionally, we implement our method in class incremental learning (CIL) setup, where the network contains a single classifier that can classify all the trained classes. It is more realistic and practical than task incremental learning (TIL) setup, where the network has a classifier for each task and requires a task identifier to select from the multi-head classifiers to perform classification. Our method achieves the best performance on CIFAR10, and CIFAR100 experiments compared to existing continual learning methods.

2   Related Works

This section introduces various related works to this paper. There exist many branches of works in the field of continual learning [2], but two main folds are related to this paper: replay approach and regularization approach. There is also an approach that combines the two categories for better alleviating of the forgetting problem. All the related works are summarized in Table 1.

Replay approach replays the data when learning the next task. iCaRL [5] was the first method in the rehearsal approach that trains the network with current task data and stored data. It uses knowledge distillation loss to prevent forgetting, which is also used in several methods [14, 15], but it was showed that distillation loss interfere the learning [8]. Some methods add a bias-correction layer at the end of the network to solve the class imbalance [6, 8], but they forget more knowledge since they need to use a part of memory for the correction. Several researchers tried to modify data in memory during training, so the network encounters new data every time to avoid overfitting [7, 16–18]. Pseudo-rehearsal approach trains a generator that can create the data in old tasks [9, 19]. It can prevent overfitting and class imbalance, but catastrophic forgetting occurs to the generator, and it is hard to create complicated data. Additionally, there are another branch of methods that uses the stored data to constrain the learning direction of the network by utilizing the gradient of the data [20, 21], but they are only effective in TIL environment.

Table 1. Related works on replay and regularization approach in continual learning.

Category	Sub-Category		Reference
Replay	Rehearsal	Knowledge distillation	[5, 14, 15]
		Bias correction	[6, 8]
		Exemplar selection	[7, 16–18]
	Pseudo-rehearsal		[9, 19]
	Constrained		[20, 21]
Regularization	Parameter regularization		[3, 22]
	Gradient regularization		[10–12]
Replay + Regularization	Sample generation		[13, 23]

Regularization approach restricts the learning direction of the network to prevent knowledge loss. EWC [3] first introduced the regularization method that limits the changes in the important parameters of the model, and similar work was proposed in [22]. Gradient regularization is a branch of this approach that restricts the weight gradient of the model. Gradient regularization approach induces the weight gradient to be orthogonal to input data to prevent changes in the output vectors [10–12]. This was first proposed by OWM [10], which tries to project gradient to the null space of input space.

However, these methods show lower performance in CIL compared to rehearsal methods. Therefore, there was an attempt to combine the two approaches [13, 23]. But their sample generation methods are not precise enough. Also, Shen, et al. uses generated data as knowledge distillation [23], whereas we directly train the network with the generated vectors. Our idea is that by assuming mixture of Gaussian, the distribution of class representations, which ensures to unchanged by gradient regularization, can be accurately represented, and the samples from the estimation can prevent the forgetting.

3 Proposed Method

In this section, we present the working progress of the proposed method. Figure 1 depicts the overall structure of the method. The network is first trained with the input dataset by using gradient regularization. Then, by using expectation-maximization (EM) algorithm, we calculate means and covariances of the representations for each class in the current task and keep those statistics in the memory. Finally, the stored means and covariances are used to generate the representation vectors by assuming multivariate Gaussian, and the generated vectors retrain a classifier part of the network.

3.1 Model Architecture

The network of this method splits into two parts: extractor and classifier. The extractor is a part that learns general characteristics of input data, which is constructed as several convolutional layers. By convolutional layers, irrelevant information in raw input data is removed, and only the relevant features remain in the output vectors that is called class representation. It allows the representations of the same class to cluster together to form a simple distribution. The classifier is a part that classifies the input representations into their classes, which is constructed as several fully-connected layers.

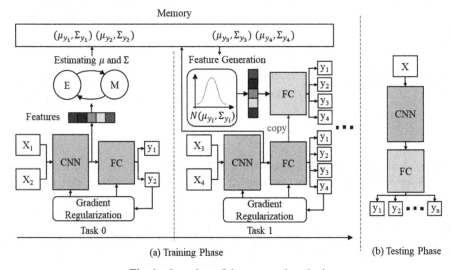

(a) Training Phase   (b) Testing Phase

Fig. 1. Overview of the proposed method.

3.2 Training in Initial Task

In the initial task, the whole network is first trained with input data using gradient regularization method. The idea of gradient regularization is that for previous input x and gradient ∇W computed from a backpropagation of cross entropy loss, if $\nabla W^T x = 0$, then the output will remain the same, which means the learned knowledge is preserved. So, the gradient regularization method computes an orthogonal projector matrix P that projects ∇W to the orthogonal subspace of input space. It is done by a calculation of a matrix that is orthogonal to input space, which is known as least square solution. It is computed by $P = I - A(A^T A - \alpha I)^{-1} A$, where A is the set of all input x, I is Identity Matrix, and α is a hyperparameter. Because the network cannot access every input vector of each layer in continual learning, OWM [10] utilizes recursive least square (RLS) algorithm to approximately compute P by the following equation.

$$P_i = P_{i-1} - k_i x_i^T P_{i-1}$$
$$k_i = \frac{P_{i-1} x_i}{\alpha + x_i^T P_{i-1} x_i} \tag{1}$$

The orthogonal projector P_i of the i-th layer of the network is then multiplied by the weight gradient of the corresponding layer to regularize the learning direction.

We consider two more aspects to help learning continually. As our goal is to approximate the distribution of class representations, so we normalize the output vectors of the extractor to reduce the variance of their distribution to help the estimation. Moreover, we add a self-supervised loss [23] to let the network learn generalized features.

After training the network with OWM, we input the current data to the extractor and collect output vectors. We then estimate the distribution of representations as a mixture of multivariate Gaussian distribution. Unlike univariate Gaussian, which consists of mean and variance, multivariate Gaussian represents data with mean μ and covariance Σ as

$$N(x) = \left(\frac{1}{2\pi}\right)^{\frac{p}{2}} |\Sigma|^{-\frac{1}{2}} exp\left(-\frac{1}{2}\left(x - \mu\right)^T \Sigma^{-1}(x - \mu)\right). \tag{2}$$

For data of multiple random variables, the variance of each variable is not enough to represent the distribution of the data. Covariance shows the correlation between two variables so that multivariate Gaussian can represent the actual distribution accurately. For more precise estimation, we assume the distribution as the mixture of K different multivariate Gaussian. To estimate such distribution, we use EM algorithm to compute means and covariances [24], which is a method to find the maximum log-likelihood of probability function as

$$\mathcal{L}(X;\theta) = \ln p(X|\pi, \mu, \Sigma) = \sum_{n=1}^{N} \ln\left(\sum_{k=1}^{K} \pi_k N(x_n|\mu_k, \Sigma_k)\right) \tag{3}$$

where N is the number of data, K is the number of distributions, $N(\cdot)$ is the Eq. (2) to compute means and covariances, and π is the ratio of data in each distribution. The EM iterates between two steps. The expectation step (E-step) evaluates the responsibility γ, which is the probability of the data point x_n in the k-th distribution, using current means and covariances.

$$\gamma(Z_{nk}) = \frac{\pi_k N(x_n|\mu_k, \Sigma_k)}{\sum_{j=1}^{k} \pi_j N(x_n|\mu_j, \Sigma_j)} \tag{4}$$

γ is then used to compute means and covariances in the maximization step (M-step),

$$\mu_k = \frac{\sum_{n=1}^{N} \gamma(z_{nk})x_n}{\sum_{n=1}^{N} \gamma(z_{nk})}, \Sigma_k = \frac{\sum_{n=1}^{N} \gamma(z_{nk})(x_n - \mu_k)(x_n - \mu_k)^T}{\sum_{n=1}^{N} \gamma(z_{nk})}, \pi_k = \frac{1}{N}\sum_{n=1}^{N} \gamma(z_{nk}) \tag{5}$$

where these equations are derived by partial derivative of Eq. (3) to maximize the likelihood. The computed means and covariances per class are stored in the memory.

3.3 Training from the Next Task

For the next set of data, the network follows the steps of initial stage again. Before moving on to the next stage, the class representation replay is conducted to retrain the classifier. We first generate the representations of all the seen classes from previous and current tasks from the stored means and covariances in the memory. From randomly generated normal vectors S, we generate the representation $\hat{X}$ of class i as

$$\hat{X}_i = (S * \lambda_i)\Gamma_i^T + \mu_i$$
$$\Sigma_i = \Gamma_i\lambda_i\Gamma_i^T \tag{6}$$

where the covariance is decomposed into eigenvalue matrix λ_i and eigenvector matrix Γ_i. The classifier of the network that has learned data of the current task is then learns the classes from the previous tasks $D_{0\sim t-1}$ again. The classifier is trained without the regularization in this step. Because the network only learns part of classes at each task in CIL, the classifier cannot distinguish classes from different tasks that have similar features. By retraining the classifier together with all trained class representations, the network preserves or even improves its ability to classify input data. We generate the same number of representations for each class so that the classifier can learn from the balanced dataset. The network with the retrained classifier is then used to classify input data in testing phase.

4   Experiments

4.1   Datasets and Implementation Details

We conduct CIL experiment on the widely used datasets in related works [5, 23]: CIFAR-10 and CIFAR-100. We split CIFAR-10 dataset into 5 tasks, each contains 2 classes. CIFAR-100 data is split into 2 tasks, 5 tasks, and 10 tasks with 50 classes, 20 classes, and 10 classes per task. We conduct 6-fold cross-validation for each dataset.

The network follows the original model in OWM [10]. Extractor is composed of 3 convolutional layers with 64, 128, 256 output channels and 2x2 filter. Classifier is composed of 3 FC layers with 1000, 1000, n (class number) nodes. For our method, we estimate and store one mean and covariance per class ($K = 1$) and generate 1,000 representations for each class. We compare our method with all the methods from each category in Table 1. For rehearsal methods, we use memory size of 2,000 images for CIFAR10, and 20,000 images for CIFAR100. Note that other branches of continual learning such as parameter isolation methods, are implemented in TIL setup, so they show bad performance in CIL setup. Therefore, we have excluded their results.

4.2   Results

From Table 2, our method shows the highest accuracy compared to other methods except for Joint Learning, where the network is trained with all the data as conventional deep learning. The performance gain in CIFAR10 experiment is 1.14%p, but the performance improves as the number of distributions (K) increases, which is showed in Table 3. Our method achieves 4.60%p improvement on CIFAR100, which is a significant performance gain considering the upper bound, which is 39.97%. Figure 2 shows the task-wise results to visualize the accuracy drop.

We also demonstrate the tSNE result of the generated representations compared to real representations in Fig. 3. It is seen that representations generated multivariate Gaussian distribution can precisely approximate the actual distribution.

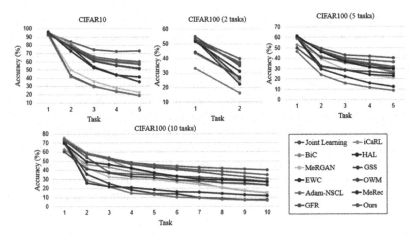

Fig. 2. Task-wise accuracies (%). Joint learning shows the upper bound of the model.

Table 2. 6-fold cross-validation on several continual learning methods. Range represents standard deviation. The bold represents the best accuracy for the task.

Method		CIFAR-10 Accuracy (%)	CIFAR-100 Accuracy (%)		
			2 Tasks	5 Tasks	10 Tasks
Joint learning		73.01 ± 0.74	39.97 ± 0.88		
Replay	iCaRL [5]	58.93 ± 0.19	27.04 ± 0.83	25.43 ± 0.70	23.61 ± 0.51
	BiC [6]	57.08 ± 0.63	36.98 ± 0.13	26.84 ± 0.87	14.95 ± 0.37
	HAL [16]	41.57 ± 0.25	31.10 ± 0.42	29.34 ± 0.49	26.88 ± 0.59
	MeRGAN [19]	22.39 ± 0.83	24.48 ± 0.61	20.52 ± 0.46	15.23 ± 0.62
	GSS [21]	50.95 ± 0.36	27.27 ± 0.30	24.87 ± 0.72	23.63 ± 0.52
Regularization	EWC [3]	18.98 ± 0.10	22.31 ± 0.63	12.53 ± 0.69	7.56 ± 0.25
	Adam-NSCL [11]	18.65 ± 0.85	16.36 ± 0.46	8.69 ± 1.09	6.68 ± 0.88
	OWM [10]	52.11 ± 0.83	34.95 ± 0.66	29.93 ± 0.83	27.40 ± 0.28
Sample Generation	MeRec [13]	35.78 ± 1.36	26.09 ± 0.86	22.75 ± 0.98	11.88 ± 1.02
	GFR [23]	56.88 ± 0.42	35.95 ± 0.38	31.59 ± 0.42	30.20 ± 0.69
Ours		**60.07 ± 0.42**	**37.56 ± 0.34**	**36.11 ± 0.56**	**34.80 ± 0.83**

4.3 Discussions

Our method shows a bigger performance gain in longer tasks, where the performance improves 0.58%p in the 2-task CIFAR100, whereas it improves 4.60%p in the 10-task CIFAR100. This shows the scalability of our method. EWC [3] and Adam-NSCL [11] are originally implemented in TIL, so they show poor performance in this experiment. BiC [6] and HAL [16] try to alleviate the class imbalance and the overfitting, but their result indicates that their methods are not effective. We think that their focus on solving those problems interferes with the learning of the network. In Table 4, we compare the result of our method to the classifier retraining with real representations by storing 1,000 vectors per class in CIFAR10. Our method shows higher accuracy, which implies that (1) generated representations are similar to real representations and that (2) our suggestion improves the performance by avoiding overfitting. Still, considering the gap between our method and joint learning, the method can improve more if we can train the extractor to split the representations of each class to further away.

From Table 3, the accuracy increases as the number of the distribution increases, but a required memory also grows. The memory size needed for storing means and covariances is the same for each class, meaning that the number of classes that our method can learn depends on the system. Thus, we will focus on developing an efficient method that represents the information of different classes into one as a future study.

Table 3. Accuracy on different number of Gaussian distribution number (K) in CIFAR10.

# Distribution (K)	1	2	3	4	5	10
Accuracy (%)	60.07 ± 0.42	62.66 ± 0.24	63.67 ± 0.36	63.55 ± 0.15	64.11 ± 0.08	64.64 ± 0.12

Table 4. Comparison of our method with real representations retraining in CIFAR10.

Method	Accuracy (%)
OWM + real representations	58.95 ± 0.73
Ours	60.07 ± 0.42

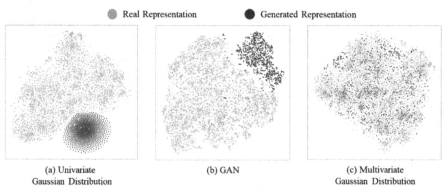

(a) Univariate Gaussian Distribution   (b) GAN   (c) Multivariate Gaussian Distribution

Fig. 3. t-SNE of real representations (light green) and generated representations (red) of class 0 in CIFAR10 by (a) univariate Gaussian, (b) GAN-based generator, and (c) multivariate Gaussian (Color figure online).

5 Concluding Remarks

This paper proposes a method that retrains the classifier with the class representations by assuming the distribution to be a mixture of multivariate Gaussian distribution with the computed means and covariances so that the network can preserve its ability of classification without the catastrophic forgetting for continual learning. It improves the classification performance in CIL setup compared to other existing methods.

Acknowledgment. This work was supported by Institute of Information & Communications Technology Planning & Evaluation (IITP) grant funded by the Korean government (MSIT) (No. 2020-0-01361, Artificial Intelligence Graduate School Program (Yonsei University); No. 2022-0-00113, Developing a Sustainable Collaborative Multi-modal Lifelong Learning Framework).

References

1. McCloskey, M., Cohen, N.J.: Catastrophic interference in connectionist networks: the sequential learning problem. Psychol. Learn. Motiv. **24**, 109–165 (1989)
2. De Lange, M., et al.: A continual learning survey: defying forgetting in classification tasks. In: IEEE Transactions on Pattern Analysis and Machine Intelligence, vol. 44, no. 7, pp. 3366–3385 (2021)
3. Kirkpatrick, J., et al.: Overcoming catastrophic forgetting in neural networks. In: Proceedings of the National Academy of Sciences, vol. 114, no. 13, pp. 3521–3526 (2017)
4. Rusu, A.A., et al.: Progressive neural networks. arXiv preprint. arXiv:1606.04671 (2016)
5. Rebuffi, S.-A., Kolesnikov, A., Sperl, G., Lampert, C.H.: icarl: incremental classifier and representation learning. In: Proceedings of the IEEE Conference on Computer Vision and Pattern Recognition, pp. 2001–2010 (2017)
6. Wu, Y., et al.: Large scale incremental learning. In: Proceedings of the IEEE/CVF Conference on Computer Vision and Pattern Recognition, pp. 374–382 (2019)
7. Buzzega, P., Boschini, M., Porrello, A., Calderara, S.: Rethinking experience replay: a bag of tricks for continual learning. In: 2020 25th International Conference on Pattern Recognition (ICPR), pp. 2180–2187. IEEE (2021)
8. Belouadah, E., Popescu, A.: Il2m: class incremental learning with dual memory. Proceedings of the IEEE/CVF International Conference on Computer Vision, pp. 583–592 (2019)
9. Shin, H., Lee, J.K., Kim, J., Kim, J.: Continual learning with deep generative replay. In: Advances in Neural Information Processing systems, vol. 30, pp. 2994–3003 (2017)
10. Zeng, G., Chen, Y., Cui, B., Yu, S.: Continual learning of context-dependent processing in neural networks. Nature Mach. Intell. **1**(8), 364–372 (2019)
11. Wang, S., Li, X., Sun, J., Xu, Z.: Training networks in null space of feature covariance for continual learning. In: Proceedings. of the IEEE/CVF Conference on Computer Vision and Pattern Recognition, pp. 184–193 (2021)
12. Saha, G., Garg, I., Roy, K.: Gradient projection memory for continual learning. arXiv preprint. arXiv:2103.09762 (2021)
13. Zhang, B., Guo, Y., Li, Y., He, Y., Wang, H., Dai, Q.: Memory recall: a simple neural network training framework against catastrophic forgetting. In: IEEE Transactions on Neural Networks and Learning Systems, vol. 33, no. 5, pp. 2010–2022 (2021)
14. Hou, S., Pan, X., Loy, C.C., Wang, Z., Lin, D.: Learning a unified classifier incrementally via rebalancing. In: Proceedings. of the IEEE/CVF Conference on Computer Vision and Pattern Recognition, pp. 831–839 (2019)

15. Douillard, A., Cord, M., Ollion, C., Robert, T., Valle, E.: Podnet: pooled outputs distillation for small-tasks incremental learning. In: Vedaldi, A., Bischof, H., Brox, T., Frahm, J.-M. (eds.) ECCV 2020. LNCS, vol. 12365, pp. 86–102. Springer, Cham (2020). https://doi.org/10.1007/978-3-030-58565-5_6

16. Chaudhry, A., Gordo, A., Dokania, P., Torr, P., Lopez-Paz, D.: Using hindsight to anchor past knowledge in continual learning. In: Proceedings of the AAAI Conference on Artificial Intelligence, vol. 35, no. 8, pp. 6993–7001 (2021)

17. Shim, D., Mai, Z., Jeong, J., Sanner, S., Kim, H., Jang, J.: Online class-incremental continual learning with adversarial shapley value. In: Proceedings of the AAAI Conference on Artificial Intelligence, vol. 35, no. 11, pp. 9630–9638 (2021)

18. Bang, J., Kim, H., Yoo, Y., Ha, J.W., Choi, J.: Rainbow memory: continual learning with a memory of diverse samples. In Proceedings of the IEEE/CVF Conference on Computer Vision and Pattern Recognition, pp. 8218–8227 (2021)

19. Wu, C., Herranz, L., Liu, X., van de Weijer, J., Raducanu, B.: Memory replay gans: learning to generate new categories without forgetting. In: Advances in Neural Information Processing Systems, vol. 31, pp. 5966–5976 (2018)

20. Lopez-Paz, D., Ranzato, M.A.: Gradient episodic memory for continual learning. In: Advances in Neural Information Processing Systems, vol. 30, pp. 6470–6479 (2017)

21. Aljundi, R., Lin, M., Goujaud, B., Bengio, Y.: Gradient based sample selection for online continual learning. In: Advances in Neural Information Processing Systems, pp. 11816–11825, vol. 32 (2019)

22. Hu, W., Qin, Q., Wang, M., Ma, J., Liu, B.: Continual learning by using information of each class holistically. In Proceedings of the AAAI Conference on Artificial Intelligence, vol. 35, no. 9, pp. 7797–7805 (2021)

23. Shen, G., Zhang, S., Chen, X., Deng, Z.H.: Generative feature replay with orthogonal weight modification for continual learning. In: Proceedings of International Joint Conference on Neural Networks (IJCNN), pp. 1–8. IEEE (2021)

24. Dempster, A.P., Laird, N.M., Rubin, D.B.: Maximum likelihood from incomplete data via the EM algorithm. J. Roy. Stat. Soc.: Ser. B (Methodol.) **39**(1), 1–22 (1977)

Face ReID Method via Deep Learning

Yves Augusto Lima Romero$^{(\boxtimes)}$ ⬤ and Ajalmar Rêgo da Rocha Neto ⬤

Computer Science Department, Federal Institute of Ceará, IFCE, Fortaleza, Brazil
yvesromero1998@gmail.com

Abstract. Recenlty, many face reid strategies have been developped. These strategies create vectorial spaces capable of representing data on reduced dimensions. Such representations are produced by deep learning models that learn to maximize the intra-class similarity and minimize the inter-class similarity. The method described in this article proposes a new face reid strategy based on the Facenet model. It consists of training support vector machines on the euclidean distances calculated between the embeddings of the Facenet model, notably reducing the false positive rate.

Keywords: Face ReID · Machine learning · Deep learning

1 Introduction

In recent years many new technologies focused on biometric systems have been developped. The use of these systems has become quite common, it has become an element of everyday life. One can capture the biometrics of an individual in several ways, for instance, extracting information from face, fingerprints, or voice. Due to the COVID-19 pandemic [16], facial biometrics systems have become more preferred, as the capture of images does not require physical contact with a device, but can be performed at a certain distance. In all types of biometrics, we compare information to check if they were extracted from the same individual. In the case of facial recognition, the information extracted from captures of the face of the same individual will have a high degree of similarity to each other. And their degree of similarity to other individuals' features will be lower. The aim is to find features such that we can minimize inter-class similarity, and maximize intra-class similarity. Depending on the strategy chosen to obtain such information, that is, to extract features from images, this goal will be achieved with greater or lesser success.

In facial recognition, there are basically two main approaches, one versus one, or re-identification, and one versus all, or classification. In the first case, we seek to identify whether two images belong to the same individual, computing their features, and making a comparison through similarity criteria. One versus all is used when we want to search, in a series of enrolled individuals, which one of them most resembles a certain image. One versus all is a classification problem, which consists of pointing out the class whose patterns are closest to the

© The Author(s), under exclusive license to Springer Nature Switzerland AG 2022
H. Yin et al. (Eds.): IDEAL 2022, LNCS 13756, pp. 369–378, 2022.
https://doi.org/10.1007/978-3-031-21753-1_36

received pattern. Generally, when this approach is used, a re-identification step is implemented right after, to validate the classification results. Classification does not guarantee that the suggested individual is the correct one, but only that, among the registered individuals, it is the one with the highest similarity score. Hence the need for a face re-identification stage, which is coupled at the end of the process.

1.1 Related Works

In order to produce techniques capable of extracting relevant information from images, several methods were developed, such as Local Binary Pattern [10], Oriented Gradient Histogram [3], and Eigenfaces, which uses eigenvectors from the data covariance matrix [15]. The information obtained by these techniques can be used to compare human faces [1]. The central objective of feature extractors is to abstract information from the data, generating a vector space of reduced dimensions where the comparison is performed more accurately.

However, with the advent of deep learning algorithms, such as convolutional neural networks [8], researchers became more attentive to this sort of algorithms. Researchers have started to formulate methodologies to solve the problem of face re-identification through neural networks. The great advantage of neural networks, in addition to their great capacity to represent data, is that these structures offer an end-to-end solution, without having to break the problem down into extraction and comparison steps. The neural network already performs these two processes optimally, through automatic parameters update. Convolutional networks play a very important role in extracting features from images. By applying filters along the image pixels, CNN's generate feature maps, which represent patterns in different degrees of abstraction. Using these features, it is possible to train classifiers, of different modalities, in order to categorize the data.

Recently, several efforts have been made in the direction of creating methods for facial recognition, focusing on *re-identification*. The best known studies focus on generating *embeddings* through deep learning algorithms, such as Openface [2], VGG Face [11], the Dlib model [7], and ArcFace [4]. Other methods, such as em [17], do not *embedding* the image characteristics, which makes it difficult to apply them in classification and clustering tasks.

Facenet proposes a strategy that uses image *triplets*, two of which were captured from the same individual(the anchor image and the positive image), the other being a negative pattern, taken from another individual. During training, the neural network learns to produce a vector space that maximizes the Euclidean distance between patterns of different individuals (i.e., minimizes similarity) and minimizes the Euclidean distance between patterns belonging to the same individual (i.e., maximizes the similarity). The cost function that the neural network seeks to optimize is the *triplet loss*, which consists of subtracting the distance calculated from the anchor to the positive image from the distance calculated from the anchor to the negative image. This is shown on Eq. 1.

$$\sum_{i=1}^{N} \|f_i^a - f_i^p)\|_2^2 - \|f_i^a - f_i^n)\|_2^2 + \alpha \qquad (1)$$

On Fig. 1, we want to check whether the anchor image belongs to the same individual portrayed in the positive image, that is, Barack Obama. For this, it is necessary to introduce the images to the input layer of a convolutional network, in order to produce the embeddings of these images, that is, feature vectors, which will be compared with each other.

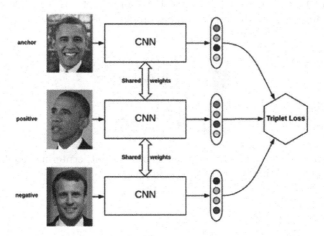

Fig. 1. *Triplet loss*: Anchor images, positive and negative (https://github.com/omoindrot/tensorflow-triplet-loss.).

1.2 Proposal

In this article, we propose a strategy for facial re-identification combining deep learning techniques with minimum margin classifiers. To this end, the neural network trained in the *Facenet* article will be reused to extract *embedding* features and, then, another step, proposed by this work in the methodology section, will be placed, which consists of using vectors (i.e. *embeddings*) extracted with the *Facenet* neural network. Then we calculate Euclidean distances between these vectors and train support vector machines with the obtained distance values. We have created a database to compare this method against another ones.

2 Methodology

2.1 Database

To develop the method proposed in this article, a database of human faces has been created, with a total of 3021 images, belonging to 258 famous individuals, obtained from Google Images through web scraping, facial detection and subsequent data cleaning(to remove repeated and inconsistent images), and face alignment.

Searching the names of celebrities on Google Images using a python code, the images of each individual were stored in separate folders, each with the individual's name. Then, face detection was applied to each of the images, extracting the cutout of the faces, and discarding the other regions of the images. The model used for face detection was a neural network trained on the *Single Shot Detection* strategy [9], using the Caffe framework. Soon after, we applied landmarks detection to obtain the location of parts of the face, such as eyes, nose, eyebrow and mouth, in order to make face alignment. An illustration is shown on Fig. 2. The final database is available on the web[1].

Fig. 2. Human face images with irrelevant parts cut out, centering significant components ((a) Cuba Gooding Jr.; (b) Gerard Butler)

2.2 Architecture

First, we used the Facenet deep neural network, built according to the Inception-ResNet-V1 architecture [13], trained on the LFW database [6]. This architecture aims to combine the advantages of the Inception architectures with the complexity reduction provided by residual connections of the ResNet architecture [5]. The first connections apply convolution operations with unitary filters(1×1) [14]. This is done in order to reduce the number of filters for the following layers, without having to create additional layers to transform data dimensionality. Filters from the previous layer are simply projected onto the dimensions of the next layer, resulting in a layer that has the desired number of filters, without the need to perform computationally expensive operations.

2.3 Feature Extraction

The network used in this article has been saved to h5 format on David Sandberg's github[2], Facenet creator, and its parameters were optimized aiming the triplet loss function, as suggested by the Facenet paper.

Thus, this convolutional network was used to capture embeddings from the images. The embeddings are feature vectors collected in the last layer, in response to an image introduced in the input layer of the network. The embeddings of

---

[1] https://www.kaggle.com/datasets/yveslr/open-famous-people-faces.
[2] https://github.com/davidsandberg/facenet/.

each individual contained in the database were collected. The output layer of the Facenet model has 128 neurons, so is the feature vectors (embeddings) have 128 elements.

2.4   Proposed Strategy

In the strategy proposed by the Facenet article, to find out if a particular image belongs to an individual, the Euclidean, or quadratic, distance between the feature vector of this image and that of an image belonging to that individual is computed; then, the same distance is calculated with respect to the feature vector obtained from an image belonging to another individual. These distances are subtracted and, if the result is less than a certain threshold, the image is considered as belonging to the individual. Otherwise, it is considered not. One of the advantages of this technique is that only three images are needed to make the prediction, the image we want to classify, a positive sample and a negative one, without having to register several images from the same user.

However, it is possible to use more than one image from the same person, in order to improve facial re-identification, reducing both the false negative and false positive rates, as will be shown in the results section.

The method proposed by this article performs the re-identification taking into account the Euclidean distances between the various embeddings belonging to the same individual and the embedding of the image that is to be validated. Thus, for each individual in the database, a support vector machine was trained, totaling 258 models. Before training, the database was divided into four parts, three of which were separated for training these models, and the other part for the testing stage.

During training, the support vector machine is fed with vectors that contain the values of the various euclidean distances, calculated between the embeddings of the training database. Thus, positive samples and negative samples are formed, and the models are trained to issue a binary response, where the value 1 points to the positive case, and 0 to the negative case.

On Fig. 3, there are two schemas of positive sample formation. Each feature vector belonging an individual is compared with feature vectores of other images from the same individual, generating the Euclidean distances circled in the figure. Then these distances are grouped into a vector. All the vectors generated in this step represent the positive samples that will be placed at the input of the support vector machine related to this particular individual. This is made for all individuals.

On Fig. 4, there is a schema of negative sample formation. A face that does not belong to the individual whose model will be trained is taken, one of the positive images is randomly discarded, and the Euclidean distance between the negative image and the remaining positive images is calculated. After that, join these distances in a vector. It is worth noting that, as expected, the negative case distances assume higher values, as the similarity between the data is lower.

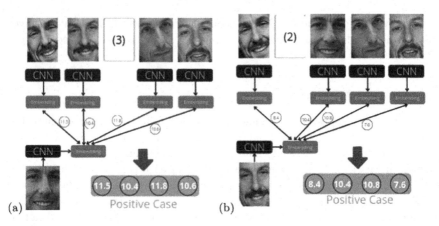

Fig. 3. Sample formation; (a) Formation of a positive sample from the second image. Distance values: 8.4; 10.4; 10.8; 7.6; (b) Formation of a positive sample from the third image. Distance values obtained: 11.5; 10.4; 11.8; 10.6

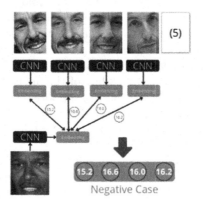

Fig. 4. Formation of a negative sample with random exclusion of one of the five positive images. Distance values: 15.2; 16.6; 16.0; 16.2

Once the samples are generated, the support vector machine related to the individual is trained, using the Grid Search parameter search technique. Each model was individually tested to assess the false negative and false positive rates, using the test database, with a total of 852 images. At the end, we computed the average of false positive and false negative rate through all the models. Each individual model is tested against all the other classes in the test dataset to compute false positive rate, and against the test images belonging to its proper class to compute false negative rate.

Compared Methods. We have tested the proposed strategy(Distance Vectors, on table 1) against four methods: Facenet, Dlib [7], HOG and Embeddings on SVM, which are shown on Table 1. Facenet strategy check if the image belongs to an individual computing the triplet loss cost function between its positive and negative *embeddings*. Dlib strategy goes on the same way, using a different neural network architecture released on 2017 by Dlib team. The Embeddings on SVM strategy consists of training support vector machines using the *embeddings* themselves as samples to model training, instead of using euclidean distances, which is made by our strategy. And HOG strategy uses HOG features extracted from images as samples to the support vector machine model training.

2.5   Results and Discussion

Table 1. Results obtained from each method.

Method	False Positive Rate(%)	False Negative Rate(%)
Distance Vectors	4.93 ± 0.97	0.06 ± 0.004
Embeddings on SVM	4.29 ± 0.92	0.14 ± 0.092
Dlib [7]	5.93 ± 1.40	2.44 ± 0.10
Facenet [12]	8.76 ± 1.31	2.19 ± 0.954
HOG [3]	14.56 ± 0.87	2.13 ± 0.195

Table 1 shows the results of some face reidentification methods on our database. The Distance Vectors method is the strategy that we proposed on this article. The other ones have been explained on the previous sections. Figure 5 shows some face reidentification tasks samples we made using the proposed strategy. This strategy can handle hard negative case such image (a) from Fig. 5, but also leads to false positive when samples are very similar. The false positive rate reduction is due to the fact that we compare the candidate image against all the images of each class, aiming to neutralize the interference of outliers samples(similar samples belonging to different classes, or unsimilar samples belonging to the same class).

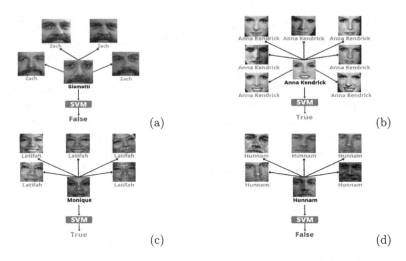

Fig. 5. (a) true negative sample; (b) true positive sample; (c) false positive sample; (d) false negative sample

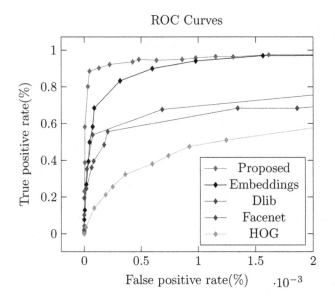

Fig. 6. ROC curves for compared methods

Due to the creation of the distance vector as a sample to the support vector machine model, outlier's influence is desguised and compensated by the presence of other samples distances, which contributes to higher specificity and sensitivity. As we can see on Fig. 3, Adam Sendler used mustache on some images, differing from his other samples. Even if we look to the results table, the comparison between Distance Vectors and Embeddings on SVM methods may be inconclusive, since the first one makes a better performance on false positives but a poorer one on false negatives. But Fig. 6 displays the ROC curves drawn for each tested method, showing that the proposed strategy enbraces a larger area than the other ones.

References

1. Ahonen, T., Hadid, A., Pietikainen, M.: Face description with local binary patterns: application to face recognition. IEEE Trans. Pattern Anal. Mach. Intell. **28**(12), 2037–2041 (2006)
2. Amos, B., Ludwiczuk, B., Satyanarayanan, M.: OpenFace: a general-purpose face recognition library with mobile applications. Tech. rep., CMU-CS-16-118, CMU School of Computer Science (2016)
3. Dalal, N., Triggs, B.: Histograms of oriented gradients for human detection. In: 2005 IEEE Computer Society Conference on Computer Vision and Pattern Recognition (CVPR'05), vol. 1, pp. 886–893. IEEE (2005)
4. Deng, J., Guo, J., Xue, N., Zafeiriou, S.: ArcFace: additive angular margin loss for deep face recognition. In: Proceedings of the IEEE/CVF Conference on Computer Vision and Pattern Recognition, pp. 4690–4699 (2019)
5. He, K., Zhang, X., Ren, S., Sun, J.: Deep residual learning for image recognition. In: Proceedings of the IEEE Conference on Computer Vision and Pattern Recognition, pp. 770–778 (2016)
6. Huang, G.B., Mattar, M., Berg, T., Learned-Miller, E.: Labeled faces in the wild: a database for studying face recognition in unconstrained environments. In: Workshop on faces in'Real-Life'Images: Detection, Alignment, and Recognition (2008)
7. King, D.E.: Dlib-ml: a machine learning toolkit. J. Mach. Learn. Res. **10**, 1755–1758 (2009)
8. Lecun, Y.: Gradient-based learning applied to document recognition. Proc. IEEE **86**, 2278–2324 (1998)
9. Liu, W., et al.: SSD: single shot multibox detector. In: Leibe, B., Matas, J., Sebe, N., Welling, M. (eds.) ECCV 2016. LNCS, vol. 9905, pp. 21–37. Springer, Cham (2016). https://doi.org/10.1007/978-3-319-46448-0_2
10. Ojala, T., Pietikainen, M., Maenpaa, T.: Multiresolution gray-scale and rotation invariant texture classification with local binary patterns. IEEE Trans. Pattern Anal. Mach. Intell. **24**(7), 971–987 (2002)
11. Parkhi, O.M., Vedaldi, A., Zisserman, A.: Deep face recognition (2015)
12. Schroff, F., Kalenichenko, D., Philbin, J.: FaceNet: a unified embedding for face recognition and clustering. In: Proceedings of the IEEE Conference on Computer Vision and Pattern Recognition, pp. 815–823 (2015)
13. Szegedy, C., Ioffe, S., Vanhoucke, V., Alemi, A.A.: Inception-v4, inception-resnet and the impact of residual connections on learning. In: Thirty-first AAAI Conference on Artificial Intelligence (2017)
14. Szegedy, C., et al.: Going deeper with convolutions. In: Proceedings of the IEEE Conference on Computer Vision and Pattern Recognition, pp. 1–9 (2015)
15. Turk, M., Pentland, A.: Eigenfaces for recognition. J. Cogn. Neurosci. **3**(1), 71–86 (1991)

16. V.M. Corman, O. Landt, M.K.E.A.: Detection of 2019 novel coronavirus (2019-nCoV) by real-time RT-PCR. Euro Surveillance (2020)
17. Wu, X.: Learning robust deep face representation. arXiv preprint arXiv:1507.04844 (2015)

Using Design of Experiments to Support the Commissioning of Industrial Assembly Processes

Tim Voigt[1]($\boxtimes$), Marvin Schöne[1], Martin Kohlhase[1], Oliver Nelles[2], and Martin Kuhn[3]

[1] Center for Applied Data Science Gütersloh, Faculty of Engineering and Mathematics, Bielefeld University of Applied Sciences, Bielefeld, Germany
{tim.voigt,marvin.schoene,martin.kohlhase}@fh-bielefeld.de
[2] Automatic Control - Mechatronics, University of Siegen, Siegen, Germany
oliver.nelles@uni-siegen.de
[3] Production Quality, Miele & Cie. KG, Gütersloh, Germany
martin.kuhn@miele.com

Abstract. Ensuring high product quality is an important success factor in modern industry. Data-driven models are increasingly used for this purpose and need to be integrated into industrial processes as early as possible. As these models require high-quality training data, strategies for obtaining such data in the commissioning phase of a process are needed. Design of experiments (DoE) methods can be used for this task, but are not directly applicable for all types of processes. A common class of processes, to which most DoE methods cannot be applied, are assembly processes. The approach described in this paper aims to gather as much information as possible from an assembly process with minimal effort. It makes use of given sets of components from which optimal combinations are created, in order to achieve a space-filling design. Further design improvements are achieved by additionally specifying different mounting position of parts in the design and by using a subset selection procedure to remove less informative design points. The approach was successfully applied to an assembly process of washing machine drums to predict the radial deviation of assembled drums.

Keywords: Design of experiments · Optimal design · Quality prediction · Assembly processes

1 Introduction

Most products manufactured by industrial companies are built from multiple components. The process of assembling these components is important to achieve a desired product quality. In addition to an accurate execution of all assembly steps, the properties of the assembled components play an important role. Modeling the effects of these properties onto quality indicators of the whole assembly makes it possible to enhance product quality and provides additional insights

© The Author(s), under exclusive license to Springer Nature Switzerland AG 2022
H. Yin et al. (Eds.): IDEAL 2022, LNCS 13756, pp. 379–390, 2022.
https://doi.org/10.1007/978-3-031-21753-1_37

into the assembly process. These models and insights are particularly valuable in the commissioning phase of a process, as potential flaws can be detected and optimal component properties can be identified at an early stage in this way.

A common approach for modeling industrial processes is the use of first-principle models. In [17], such an approach with a CAD simulation model is used to virtually commission a small-scale assembly system. However, as real-world industrial processes are typically highly complex, first-principle models are often hard to obtain. Instead, data-based methods based on statistics or *machine learning* (ML) are a common choice for modeling the process behavior. In [18], such an ML-based approach for quality prediction is described using the example of a spot welding process. It helps to avoid quality problems in normal operation, but requires huge data sets for training. During commissioning, the availability of training data is one of the main problems, as these data must be informative and available in sufficient quantity. Methods from the field of *design of experiments* (DoE) can be used to obtain such data. These methods are increasingly used in combination with ML and significantly improve model quality [2]. However, classical DoE methods are often not suitable for experiments with assembled products, as these methods require to realize arbitrary design points, at least with discrete levels for each feature. In experiments with assembled products, each assembly component is described by a fixed set of predefined feature values, leading to constraints concerning the realizable design points.

This paper presents a DoE methodology for assembled products that enables cost-efficient data acquisition for estimating ML models, using given sets of components from which the best possible combinations are assembled. This approach allows for early estimation of quality prediction models that can be used to improve the quality of the assembled products from the beginning of normal process operation. In addition, these models can directly support the commissioning of the assembly process by allowing an evaluation and adjustment of the tolerances of the components. Concerning this methodology, the following research questions are investigated:

- How can informative data from an assembly process be obtained with minimal effort?
- How can multiple mounting positions of components in an assembly be considered in the design?

The remainder of this paper is structured as follows: In Sect. 2, a brief overview of related work in the field of DoE is given. Section 3 gives a formal description of an assembly process in terms of DoE and presents a real-world drum assembly process for washing machines. The proposed DoE methodology is described in Sect. 4 and applied to the drum assembly in Sect. 5. Finally, concluding remarks are given in Sect. 6.

2 State of the Art

DoE methods are widely used in industry to get insights into industrial processes and to model them. Typically, experimental designs like factorial designs or response surface designs are based on geometric patterns [6]. Such designs were successfully used, for example, to optimize a cutting process of titanium [4] or to examine the influence of different process parameters onto a friction stir welding process [13]. However, in order to apply such methods to assembled products, particular components need to be created with combinations of feature values specified by the experimental design. Creating these components might be infeasible or too expensive for practical application. Alternatively, experimental designs can be created from given sets of components. For this purpose, coordinate exchange algorithms that create optimal designs according to different optimization criteria are well suited. Common choices are D-optimal designs for estimating polynomial models [5] or space-filling designs for estimating more complex models like *Gaussian process models* [8]. These algorithms can directly be used for assembled products, if each component is described by exactly one feature. However, allowing only one feature per component is insufficient for most real-world assembly processes.

An algorithm that was specifically developed for experiments with assembled products is presented in [12]. It uses an exchange mechanism of components to create D-optimal designs and incorporates different types of features, some that can be controlled freely and some that are derived from different components. In [1], a genetic algorithm with the same purpose is described. To ensure valid designs, crossover operations like *partially matched crossover* are used. Both algorithms are compared in [11] based on a hydraulic gear pump and an electro-acoustic transducer as real-world industrial assembly processes. For these use cases, the exchange algorithm outperformed the genetic algorithm, which was explained by the strong constraints in this type of DoE problem.

The use case considered in this paper was already described in [16], in which a local search algorithm was used to create a space-filling design by exchanging and rotating the individual components of a washing machine drum. However, a local approach for modeling the functional relationships at certain radial positions of the assembly was chosen ignoring the interactions between these positions.

3 Assembled Products

Experiments on assembled products require special DoE methods to create valid designs. Therefore, a formal description of an assembly task in terms of DoE is given first, followed by the industrial use case considered in this paper.

3.1 Formal Description of the DoE Task

An assembled product consists of multiple components, each of which has properties that influence the resulting assembly. This influence should be investigated in an experiment. Figure 1 gives an overview of the experimental task for an assembly process in terms of DoE. In the experiment, N assemblies are built, each consisting of I different types of components. The individual components $c_i(k)$ with $i = 1, \ldots, I$ and $k = 1, \ldots, N$ are provided as component sets $\mathcal{C}_i = \{c_i(1), c_i(2), \ldots, c_i(N)\}$. An assembly $\mathcal{P}(k) = \{c_1(k), c_2(k), \ldots, c_I(k)\}$ is

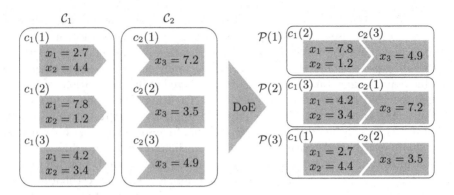

Fig. 1. An exemplary assembly process in terms of DoE for $N = 3$ assemblies $\mathcal{P}(k)$ with $k = 1, 2, 3$, each consisting of $I = 2$ types of components. Components in $\mathcal{C}_1$ are characterized by two features (x_1 and x_2) and components in $\mathcal{C}_2$ by one feature (x_3), leading to assemblies with three features (x_1, x_2 and x_3).

a specific combination of components, drawn from these component sets. Each type of component is characterized by an individual number of m_i features, which together define the $M = \sum_{i=1}^{I} m_i$ features x_j with $j = 1, \ldots, M$ of the assembly. In contrast to classical DoE tasks, the feature values cannot be chosen freely. Instead, each individual component defines one specific combination of values for its m_i features. The aim of the proposed DoE methodology is to systematically match the individual components in order to optimize a design criterion for the overall design that describes all assemblies in an experiment.

3.2 Drum Assembly Use Case

A drum is one of the main parts in a washing machine. It must meet high quality standards in order to withstand mechanical stress during the washing process. It is assembled in a fully automated process and mainly consists of three types of components: cap (c_1), casing (c_2) and base (c_3). These components and the resulting assembly are schematically shown in Fig. 2. The radial deviation of the drum opening, which is influenced by all stacked components, was identified as the most relevant quality indicator that should be predicted. Excessive radial deviation will cause the laundry to be sucked under the drum. The concentricity between the drum opening (marked in blue) and the rotational axis (dashed line) are considered an indicator for radial deviation. This concentricity is calculated from multiple measuring points on the inside of the drum opening, which are determined using a coordinate measuring machine. These measuring points are used to fit a circle by minimizing the mean square distance [15]. The Euclidean distance between the center point of this fitted circle (blue star) and the rotational axis (define by the base) is used as the output value $y = ||[y_x, y_y]^T||$.

A similar procedure with circles fitted to measurement points on the top and bottom is used to characterize each component. The offset between the

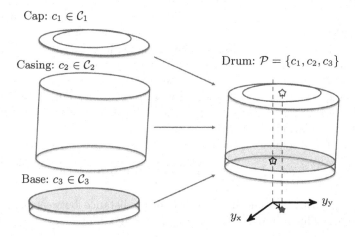

Fig. 2. Main components of a washing machine drum. For better illustration, the deviations of the individual components are exaggerated.

projected center points of the fitted circles is described as a two-dimensional vector $[x_{\mathrm{x},i}\ x_{\mathrm{y},i}]^{\mathrm{T}}$, with elements used as the components features. This leads to $m_1 = m_2 = m_3 = 2$ features per component and makes it possible to take into account different directions of deviations. Since the considered components of the drum are rotationally symmetrical, mounting them in different angular positions is possible. By changing the angular position of a component, the direction of the radial deviation caused by this component is changed too. This mechanism can be used to further improve an experimental design for such types of components.

4 Methodology

The approach described in this paper aims to gather as much information as possible from an assembly process with minimal effort. Therefore, a new DoE procedure for assembled products is described that is able to incorporate different angular positions of components. In addition, a subset selection procedure is proposed to select the best components from a larger set.

4.1 Design Optimization

The DoE problem presented in Fig. 1 is described in a specific way for dealing with assembled products. To enable the usage of standard DoE methods and metrics for evaluating the experimental design, it is transformed into a design matrix $\mathbf{X} \in \mathbb{R}^{N \times M}$. Each column in this matrix contains the feature values x_j with $j = 1, \dots, M$, defined by the available components, whereas each row represents one design point $\mathbf{x}^{\mathrm{T}}(k)$ that is composed of all feature values for one specific assembly. The experimental design $\mathcal{D} = \{\mathbf{x}(1), \mathbf{x}(2), \dots, \mathbf{x}(N)\}$ consists of all these design points. The representation of assemblies as design points enables the application of various optimization criteria. A common choice is the D-optimal design, which minimizes the determinant of its parameter covariance matrix [6]. However, a predefinition of the model structure is required (e.g. polynomial model with a certain degree). A space-filling design aims to cover the input space as uniformly as possible, leading to combinations of components in the assembled products that are as diverse as possible. This type of design is chosen because in the commissioning phase of a process, little knowledge about the functional relationship is given, so that a predefinition of a model structure is not desired. Subsequently, the minimal nearest neighbor distance in a design

$$d_{\min} = \min_{\mathbf{x}(j),\mathbf{x}(n) \,\in\, \mathcal{D},\ j\neq n} \big(d(\mathbf{x}(j),\mathbf{x}(n))\big) \tag{1}$$

is regarded, with $d(\cdot)$ denoting the Euclidean distance between two different design points. This minimal distance should be maximized

$$\max_{\mathcal{D}}(d_{\min}) = \max_{\mathcal{D}} \left(\min_{\mathbf{x}(j),\mathbf{x}(n) \,\in\, \mathcal{D},\ j\neq n} \big(d(\mathbf{x}(j),\mathbf{x}(n))\big) \right), \tag{2}$$

resulting in a type of space-filling design called maximin-design [7].

Several algorithms can be used in order to create optimal designs according to Eq. (2). For example, global optimization methods like the genetic algorithm presented in [1] are suitable. In this paper, an algorithm based on the *Extended Deterministic Local Search* (EDLS) algorithm is described, which makes use of a coordinate exchange mechanism [3]. The greedy local search strategy of EDLS realizes good space-filling properties with a moderate computational demand. In [16], an extension of the EDLS algorithm for experiments on assembled products is presented, referred to below as *local search*. Instead of the coordinate exchange mechanism, a component exchange mechanism was proposed. As a component can be characterized by multiple features, an exchange affects all these features simultaneously, leading to changes in multiple columns of the design matrix $\mathbf{X}$.

4.2   Incorporation of Angular Positions

The exchange of components between two assemblies is an elementary operation in DoE for assembled products. Additional application-specific operations, such as mounting components in different positions, can be added. In this way, the local search algorithm gains flexibility, enabling more uniform designs. In the use case considered here, the concentricity of each component is described by two features. As all assembled components are rotationally symmetrical, a mounting in different angular positions is possible. To incorporate these positions into the feature description, a linear transformation of both features

$$\begin{bmatrix} \tilde{x}_{\mathrm{x},i}(k) \\ \tilde{x}_{\mathrm{y},i}(k) \end{bmatrix} = \begin{bmatrix} \cos(\alpha_i(k)) & -\sin(\alpha_i(k)) \\ \sin(\alpha_i(k)) & \cos(\alpha_i(k)) \end{bmatrix} \cdot \begin{bmatrix} x_{\mathrm{x},i}(k) \\ x_{\mathrm{y},i}(k) \end{bmatrix} \tag{3}$$

that characterize a component $c_i(k)$, with an individual angle of $\alpha_i(k)$, is used. For each possible exchange partner in the local search algorithm, a set of candidate points is created using this transformation. These points, resulting from one

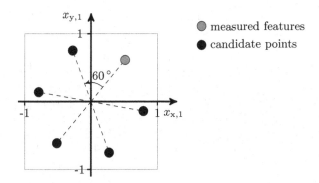

Fig. 3. Feature space of one individual component with additional candidate points resulting from a rotation of the components in steps of $60°$. These candidate points can be used as exchange partners in a component exchange.

exemplary component with $x_{x,1} = 0.51$ and $x_{y,1} = 0.61$ and positions in steps of $60°$, are depicted in Fig. 3. All candidate points are then used as possible exchange partners for a component exchange in the local search algorithm.

Figure 4 demonstrates the effectiveness of the DoE procedure for an artificial assembly process with $I = 2$ types of components. In Fig. 4a, an initial design is shown with design points forming two clusters. All components are initialized with $\alpha_i(k) = 0° \ \forall \, i = 1, \ldots, I$ and $k = 1, \ldots, N$. In this unfavorable design, the concentricity deviation of all components is oriented in the same way, resulting in the worst space-filling properties. An application of the local search algorithm restricted to exchange operations results in the design presented in Fig. 4b, in which combinations of components from both initial clusters are present. The additional consideration of different angular positions, limited for example to steps of $60°$, leads to the design shown in Fig. 4c. In contrast to (a) and (b), large areas of the input space are now covered with design points.

4.3 Subset Selection

In the optimized design shown in Fig. 4c, design points are still present that are close to each other. These points cause measurement without providing much information. Removing these points can improve the design, resulting in both a more uniform design and less measurement effort. For this purpose, a subset selection procedure is proposed. The resulting optimal reduced design with $\tilde{N} = 10$ design points is presented in Fig. 4d. Several strategies for selecting suitable subsets of design points for space-filling designs are presented in [10], including a greedy backward elimination strategy. Such a strategy, in which the worst point in the design is successively removed, is used in this paper. The identification of the worst point is based on minimal Euclidean distances, similar to Eq. 1. For each point in the design $\mathbf{x}(j) \in \mathcal{D}$, the distance to its nearest neighbor

$$d_{\mathrm{NN}}(\mathbf{x}(j)) = \min_{\mathbf{x}(n)\in\mathcal{D}_j} \big(d(\mathbf{x}(j), \mathbf{x}(n))\big) \qquad (4)$$

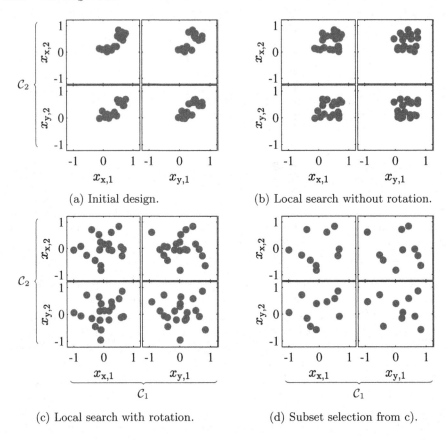

(a) Initial design.

(b) Local search without rotation.

(c) Local search with rotation.

(d) Subset selection from c).

Fig. 4. Comparison of different designs for two types of components $\mathcal{C}_1$ and $\mathcal{C}_2$. $N = 20$ design points are given, that are reduced to $\tilde{N} = 10$ in (d).

is calculated with an initialization $\mathcal{D}_j = \mathcal{D} \setminus \{\mathbf{x}(j)\}$ of the set of potential nearest neighbors. As at least two points are each other's nearest neighbors, an evaluation of this distance is not sufficient to identify a unique worst point. For each point with identical smallest value of d_{NN}, the identified nearest neighbor $\mathbf{x}(n)$ is removed from the candidate set $\mathcal{D}_j = \mathcal{D}_j \setminus \{\mathbf{x}(n)\}$ and Eq. (4) is re-evaluated. In this way, the distances to the next nearest neighbors are hierarchically compared until a unique worst point for elimination is identified. Each elimination clears space for other design points and might allow for further component exchange operations. Therefore, it is advisable to rerun the local search algorithm during subset selection, which can increase the uniformity of the design.

5   Experiments

For the use case presented in Sect. 3.2, an experimental design is created and evaluated. The experimental data are then used to estimate a first process model.

Table 1. Evaluation of different designs for the drum assembly process. For the random designs, mean values over 10 repetitions are given.

No	Combinations	Angular positions	D_{KL}	d_S
1	Random	Same position	10.7	0.82
2	Random		4.1	0.78
3	Local search		3.8	0.79
4	Local search with subset selection		3.1	0.79
5	Rerunning local search during subset selection		2.7	0.74

5.1 Experimental Design

In the experiment performed, $N = 44$ components of each type were regarded. The proposed subset selection mechanism was used to reduce the design $\mathcal{D}$ to $\tilde{N} = 37$ points, leading to increased uniformity. For the remaining $N - \tilde{N} = 7$ components of each type, a second space-filling design was created with the local search algorithm, which was used to gather a first test data set for model evaluation. In order to achieve a reasonable dimensionality of the input space in relation to the given sample size, only the caps and casings were considered in the experimental design ($M = 4$). Since the concentricity of the bases varies only slightly, this type of component was neglected and all bases were mounted in the same angular position. To increase the variance in the concentricity of the caps, half of these components were manually deformed. For practical reasons, the angular positions of the caps were restricted to steps of $60°$.

Evaluation criteria are required in order to compare different designs and demonstrate the proposed DoE methodology's effectiveness. A criterion that describes the uniformity of a design is presented in [16]. It makes use of the *Kullback-Leibler* (KL) divergence D_{KL} to compare the estimated probability density function of a design to the probability density of a uniform distribution. A second evaluation criterion aims to detect the biggest hole in the input space. For this purpose, a space-filling design $\mathcal{S}$ with a huge number of design points $N_S = 2^{16}$ is created using a Sobol set [14]. For each of these Sobol points, the nearest neighbor distance to the design points is identified (Eq. (4)). The used metric is computed by the maximum value of these nearest neighbor distances

$$d_S = \max_{\mathbf{x} \in \mathcal{S}} d_{NN}(\mathbf{x}). \tag{5}$$

Different designs for the drum assembly process are evaluated in Table 1. For both presented criteria, smaller values indicate better space-filling properties of the design. Random combinations of components that are all mounted in the same angular position lead to the least uniform design (No. 1), with the highest D_{KL} value, followed by a design with random combinations and rotations of components (No. 2). The local search algorithm improves this criterion by improving the distance between the design points. Finally, more uniform designs are achieved by the subset selection (Nos. 4–5), which dissolves clusters in the

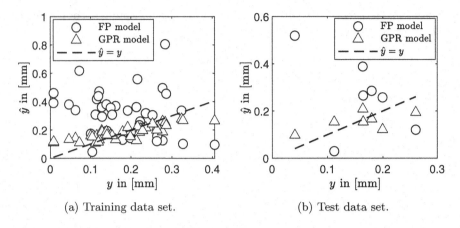

(a) Training data set.   (b) Test data set.

Fig. 5. Predicted radial deviation $\hat{y}$ of the *first-principle* (FP) and *Gaussian process regresssion* (GPR) model versus observed radial deviation y.

input space. Regarding d_S, similar values result from designs Nos. 2- 4. In general, subset selection enlarges holes in the input space, resulting in higher values of d_S. However, even after optimization, the considered data set yields several similar points which can be removed without affecting d_S. A significant improvement of this criterion is achieved by rerunning the local search algorithm during the subset selection process (No. 5), so that beneficial component exchanges are enabled. All in all, design No. 5 provides the best uniformity with the smallest holes in the input space and was therefore used for the real-world experiments.

5.2   Modeling

Based on the experimental data, a process model is estimated. For this purpose, a *Gaussian process regression* (GPR) model is used, which is well suited for such small data sets [9]. The training data consist of the design matrix **X** resulting from design No. 5 and the observed output values $y(k)$ that characterize the radial deviation of each assembled drum. For comparison, a simple first-principle (FP) model $\hat{y}(k) = ||[x_{x,1}(k) + x_{x,2}(k)\ \ x_{y,1}(k) + x_{y,2}(k)]^T||$ is considered, which sums up the radial deviations of the components in an assembly.

The model outputs of both models are compared to the observed radial deviation in Fig. 5, using (a) training and (b) test data. A better fit of the GPR model is visible, which can be explained by the ability of the GPR model to learn influences of the assembly operation from the experimental data. In addition, the *root mean squared error* (RMSE) of the GPR model (0.050 mm) on the test data set is smaller than the standard deviation of y (0.069 mm) and the RMSE of the FP model (0.574 mm). Although this RMSE is quite high compared to the range of output values, this model seems to be suitable for the considered application, since the goal is not to produce drums with predefined values of radial deviations, but drums with radial deviations closer to zero.

To assemble a drum with a specific combination of components, the model can be used directly to determine the best angular position of the cap. A simple strategy is to predict the radial deviations for a discrete number of angular positions and select the one with the minimal value. In this way, the quality of the assemblies can be increased, enabling better adjustments of the assembly process during commissioning and a direct benefit for normal process operation.

6   Conclusions and Future Work

This paper presented a DoE methodology suitable for the commissioning phase of industrial assembly processes. It aims to retrieve as much information as possible from a small number of experiments. In order to achieve low experimental effort, given sets of components are used. From these sets, a space-filling design is created by systematically matching the individual components to each other. For special types of components that are rotationally symmetrical, further improvements of the design are achieved by additionally specifying the angular position of components in an assembly. Furthermore, a subset selection procedure is presented that can be used to reduce the measurement effort and to increase the uniformity of a design by removing the least informative design points.

The proposed methodology was applied to an assembly process of washing machine drums during the commissioning phase, aiming to predict the radial deviation of the assembly, which is one of the most important quality indicators. Based on data from the available components, different designs were created and evaluated. The evaluation revealed the effectiveness of the proposed DoE methodology in increasing the uniformity of the design and reducing holes in the input space. The most suitable design was then used to assemble 37 drums for training and an additional 7 drums for gathering test data. These experimental data sets were used to estimate and validate a first process model. This model can be used to gain process knowledge and support the commissioning of the process. For example, component tolerances can be adjusted based on the model.

Further research could investigate sequential design strategies that make use of multiple small experiments instead of a single huge one. After each step of such a strategy, the need for further experiments can be assessed, avoiding unnecessary experimental runs. In addition, the process model can be improved by incorporating expert knowledge about the process being modeled. To support the incorporation of such expert knowledge, techniques from the field of interpretable ML could be used to provide insights into the model behavior.

Acknowledgements. This work was supported by the EFRE-NRW funding program "Forschungsinfrastrukturen" (grant no. 34.EFRE-0300180). The authors would like to thank Miele & Cie. KG for enabling the experiments.

References

1. Anthony, D.K., Keane, A.J.: Genetic algorithms for design of experiments on assembled products. University of Southampton, Tech. rep. (2004)

2. Arboretti, R., Ceccato, R., Pegoraro, L., Salmaso, L.: Design choice and machine learning model performances. Qual. Reliab. Eng. Int. (2022). https://doi.org/10.1002/qre.3123

3. Ebert, T., Fischer, T., Belz, J., Heinz, T.O., Kampmann, G., Nelles, O.: extended deterministic local search algorithm for maximin latin hypercube designs. In: 2015 IEEE Symposium Series on Computational Intelligence, pp. 375–382. IEEE, Cape Town, South Africa (2015). https://doi.org/10.1109/ssci.2015.63

4. Gariani, S., El-Sayed, M.A., Shyha, I.: Optimisation of cutting fluid concentration and operating parameters based on RSM for turning Ti-6Al-4V. Int. J. Adv. Manuf. Technol. 117(1–2), 539–553 (2021)

5. Mandal, A., Wong, W.K., Yu, Y.: Algorithmic searches for optimal designs. Handbook of design and analysis of experiments, pp. 755–783 (2015)

6. Montgomery, D.: Design and Analysis of Experiments. John Wiley & Sons Inc, Hoboken, NJ (2017)

7. Pronzato, L.: Minimax and maximin space-filling designs: some properties and methods for construction. J. Soc. Française de Statistique 158(1), 7–36 (2017)

8. Pronzato, L., Müller, W.G.: Design of computer experiments: space filling and beyond. Stat. Comput. 22(3), 681–701 (2012)

9. Rasmussen, C.E., Williams, C.K.: Gaussian Processes for Machine Learning. MIT Press, Cambridge (2006)

10. Rennen, G.: Subset selection from large datasets for Kriging modeling. Struct. Multi. Optim. 38(6), 545–569 (2008). https://doi.org/10.1007/s00158-008-0306-8

11. Sexton, C.J., Anthony, D.K., Lewis, S.M., Please, C.P., Keane, A.J.: Design of experiment algorithms for assembled products. J. Qual. Technol. 38(4), 298–308 (2006). https://doi.org/10.1080/00224065.2006.11918619

12. Sexton, C.J., Lewis, S.M., Please, C.P.: Experiments for derived factors with application to hydraulic gear pumps. J. Royal Stat. Soc.: Ser. C (Appl. Stat.) 50(2), 155–170 (2001). https://doi.org/10.1111/1467-9876.00226

13. Shinde, G.V., Arakerimath, R.R.: Multi-response optimization of friction stir welding process of dissimilar AA3003-H12 and C12200–H01 alloys using full factorial method. Proc. Instit. Mech. Eng. Part E: J. Process Mech. Eng. 235(5), 1555–1564 (2021)

14. Sobol, I.M.: On quasi-Monte Carlo integrations. Math. Comput. Simul. 47(2–5), 103–112 (1998)

15. Taubin, G.: Estimation of planar curves, surfaces, and nonplanar space curves defined by implicit equations with applications to edge and range image segmentation. IEEE Trans. Pattern Anal. Mach. Intell. 13(11), 1115–1138 (1991)

16. Voigt, T., Schöne, M., Kohlhase, M., Nelles, O., Kuhn, M.: Space-filling designs for experiments with assembled products. In: 2021 3rd International Conference on Management Science and Industrial Engineering, pp. 192–199. ACM (2021). https://doi.org/10.1145/3460824.3460854

17. Zhang, L., Cai, Z.Q., Ghee, L.J.: Virtual commissioning and machine learning of a reconfigurable assembly system. In: 2020 2nd International Conference on Industrial Artificial Intelligence (IAI), pp. 1–6 (2020)

18. Zhou, B., Pychynski, T., Reischl, M., Kharlamov, E., Mikut, R.: Machine learning with domain knowledge for predictive quality monitoring in resistance spot welding. J. Intell. Manuf. 33(4), 1139–1163 (2022). https://doi.org/10.1007/s10845-021-01892-y

Res-GAN: Residual Generative Adversarial Network for Coronary Artery Segmentation

Rawaa Hamdi[1,2], Asma Kerkeni[1,3(✉)], Mohamed Hedi Bedoui[1],
and Asma Ben Abdallah[1,3]

[1] Laboratory of Technology and Medical Imaging, Faculty of Medicine,
University of Monastir, Monastir, Tunisia
asma.kerkeni@isimm.rnu.tu
[2] Faculty of Sciences of Monastir, University of Monastir, Monastir, Tunisia
[3] Higher Institute of Computer Sciences and Mathematics, University of Monastir,
Monastir, Tunisia

Abstract. Segmentation of coronary arteries in X-ray angiograms is
a crucial step in the assessment of coronary disease. Recently, many
automatic approaches have been proposed to minimize time-consuming
clinicians intervention. However, due to noise and complex vessel structure in this modality, most of those approaches fail to segment thin vessels. In this paper, we introduce a new generative adversarial network
called Res-GAN to obtain accurate vessel segmentation of both thick
and thin vessels. It consists of a Residual-UNet generator following the
encoder-decoder structure; and a Residual CNN discriminator for more
efficient segmentation. Besides, in order to improve the training process,
we adopt a loss function combining both binary cross-entropy and Dice
losses. For the experiment results, we used our private dataset to compare the proposed architecture with others state-of-the-art models. The
results demonstrate that Res-GAN outperforms the others architectures.
It achieves the highest accuracy of 96,55% and Dice metric of 81,18%.

Keywords: Vessel segmentation · Generative networks · X-Ray
Coronary Angiography · U-Net model

1 Introduction

X-Ray Coronary Angiography (XRCA) provides rich information to diagnose
coronary artery diseases [2]. The segmentation of XRCA images is the most
important step to find abnormalities in the vascular structures, which allows
radiologists to detect early signs of these diseases. In order to aid such analysis,
several automatic vessel segmentation methods have been proposed. However,
this task still remains challenging due to the high correspondence between vessels
structure and the background, the complex vessel structure and uneven illumi-
nation. To deal with these issues, many algorithms based on hand-crafted feature

© The Author(s), under exclusive license to Springer Nature Switzerland AG 2022
H. Yin et al. (Eds.): IDEAL 2022, LNCS 13756, pp. 391–398, 2022.
https://doi.org/10.1007/978-3-031-21753-1_38

extraction have been proposed such as filtering based methods, rules based methods and machine learning based methods [6]. However, they are time-consuming and heavily rely on empirical parameter adjustment.

In recent years, with the advance of deep learning methods, more improved results were achieved based on an automatic feature extraction. For instance, the U-Net architecture [9] has shown outstanding performance for medical image segmentation. It has an U-shape encoder-decoder topology. The encoder path is a sequence of convolutions and maxpooling operations. In order to capture contextual information, this path increases the number of features maps and reduces the spatial size of the input image. However, the decoder path is a sequence of up-convolutions and concatenations operations with features maps from the encoder to obtain the original image size and precise the location.

So far, several works based on U-Net and its variants have been proposed such as R2UNet [1], T-Net [3], CSNet [7], etc. Although these models achieve improved performance in segmenting main and large vessels, they are less efficient for segmenting small vessels [4]. This is mainly because the successive downsampling in the encoder path that leads to thin features loss.

Recently, GAN-based architectures have been used in image segmentation. GAN is the abbreviation of Generative Adversarial Networks, which is originally designed to generate synthetic images. It consists of two models, named generator and discriminator, which are trained at the same time. With such networks, the image segmentation is considered as an image translation task.

To have better performance, many researchers have included U-Net as generator type in GAN architecture. Popescuet al. [8] introduced pix2pix GAN architecture for retinal blood segmentation, where the generator is the U-Net model and the discriminator is a patchGAN, which classifies patches instead of the total image. In [11], the authors proposed V-GAN architecture for Retinal Vessel Segmentation in Fundoscopic. It is based on U-Net model as a generator and several discriminative which differ in output size. Moreover, Son et al. [12] proposed RetinaGAN for blood vessels and optic disc segmentation in retinal images, with a similar generator model as pix2pix GAN and several discriminators are explored. In another work, Wu et al. [13] introduced U-Net with attention mechanism as a generator and the discriminator is a convolution neural network(CNN) with dense block layers.

To the best of our knowledge, the only work based on GAN networks for XRCA image segmentation was proposed by Shi et al. [10]. In this work, the authors developed a GAN-based architecture named UENet. As a generator, a simple U-Net model have been used, and for the discriminator, it is based on the average result of three layers which have similar convolutional network structure but with several input image scales. The first layer input is the original image, the next layers inputs are obtained by the down sampling operation of the original image by a factor of 2 and 4 respectively. However this method was proposed to only segment the major arteries.

To overcome the aforementioned limits, we propose in this paper a new GAN-based architecture for coronary artery segmentation, consisting of Residual U-

Net generator and Residual CNN discriminator, thus we call it Res-GAN. This model requires fewer images for the training process and produces more accurate segmentation maps. Additionally, we adopt a loss function combining both binary cross-entropy loss (L_{BCE}) and Dice coefficient loss (L_{Dice}) to ensure better segmentation of small vessels.

2 Proposed Method

In the following subsections, we will detail the proposed architecture and the objective loss function used in this work.

2.1 Network Architecture

Res-GAN architecture includes two main parts: generator model and discriminator model. An overview of the proposed architecture is depicted in Fig. 1. The generator takes as input original X-ray coronary angiography then produces a probability values ranging from 0 to 1 indicating the probability of each pixel belonging to the vessel class. Besides, the discriminator takes the generated image and the ground truth image as input and distinguishes between real image and fake image.

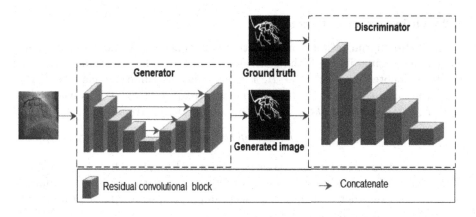

Fig. 1. Overview of the Res-GAN architecture.

The generator is based on Residual U-Net where it takes advantage of residual units as basic blocks and the shape of U-Net with skip connection. The main benefits of this combination: First, the residual unit will facilitate the training process of the network. Second, the skip connection operation between low levels and high levels features will achieve better performance on semantic segmentation. Similarly, in the discriminator model, we exploit the advantage of using residual units with convolution neural networks for the classification task.

The architecture of a residual unit is detailed in Fig. 2 where each unit includes two sequences of convolution layer, batch normalization layer and ReLU activation function. An additional 1 × 1 convolution layer is used to transform the input feature maps into the desired shape before it is entered for the additional operation. Then, it is followed by a maxpooling operation to decrease the spatial dimensions.

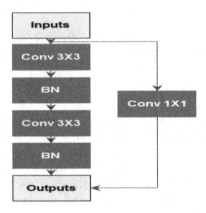

Fig. 2. Residual convolutional block.

2.2   Objective Loss Function

Loss function optimization is an important part of network training performance. Given the original image I, the generator G tries to generate segmented image I', it is denoted by $G : I \rightarrow I'$. However the discriminator classify I' and the ground truth image I'' into two classes (real or fake); it is considered as $D : \{I', I''\} \rightarrow \{Real, Fake\}$. Thus, the loss function is given by:

$$\min_G \max_D L_{GAN}(G, D) = E_{I', I''}[log(D(I', I''))] + E_I[log(1 - D(G(I)))]$$

Note that the discriminator tries to maximize $L_{GAN}(G, D)$ to better categorize the pair of real and fake images. In contrast, the generator tries to minimize $L_{GAN}(G, D)$ In order to generate a close samples as possible to the target data in order to confuse the discriminator.

In fact, for segmentation tasks, we can incorporate ground truth images at the loss function level such as in [11], where authors introduced BCE loss. This loss function is generally used in binary segmentation task; it computes the dissimilarity between the predicted sample and its corresponding ground truth. It is defined as:

$$L_{BCE}(G) = E_{I', I''} - I'' \cdot log(G(I)) - (1 - I'') \cdot log(1 - G(I))$$

However, binary cross-entropy treats each pixel independently of its neighbors; due to imbalanced classes, loss value can be dominated by the most frequent class. To deal with this, we added the Dice loss function which penalizes the overlap mismatch between the predicted image and its corresponding ground truth. It is based on the following equation:

$$L_{Dice}(G) = 1 - \frac{2 \cdot G(I) \cdot I'' + \gamma}{G(I)^2 + I''^2 + \gamma}$$

where γ is the Laplace smoothing parameter which helps to accelerate the training process. It is set to 1 in this work.

Combining the above losses (L_{GAN}, L_{BCE} and L_{Dice}), the full objective function can be formulated as follows:

$$L = \min_G \max_D L_{GAN}(G, D) + \beta L_{BCE}(G) + (1 - \beta)L_{Dice}(G)$$

where β is the weight balance of the objective functions.

3   Experiments and Results

3.1   Materials and Experimental Setup

For the training and testing process, we created our own private database by extending the dataset described in our previous work [5]. It contains 150 X-Ray Coronary Angiography images of size 256×256 pixels and 256 gray levels with a resolution of 0.3×0.3 mm. Each image has a ground truth image verified by clinicians.

To overcome the limited amount of data and improve the performance, we used data augmentation techniques in the training phase, which includes random flipping and random rotation from $-40°$ to $40°$.

The training is based on Adam optimizer with a mini batch size of 8 and a learning rate of 10^{-4}. In addition we adopted the 5-folds cross validation method to split the training and validation sets. For each fold, 80% were used for the training process and 20% for the validation process. Besides, to have an optimal number of epochs and avoid overfitting, we adopted early stopping method.

3.2   Evaluation Metrics

In order to evaluate our approach, we estimated the segmentation quality based on five often used metrics in the literature, namely: Accurary, Sensitivity, Specificity, Precision and Dice coefficient.

The Accurary (Acc) represents the ratio of positive pixels prediction to all pixels' predictions. The Sensitivity (Sen) is the proportion of positive predicted vessel pixels. The Specificity (Sp) is the proportion of positive predicted non-vessel pixels. The Precision (Pre) can be seen as a measure of exactness or fidelity. Finally the Dice coefficient is a measure of region overlap between the ground truth and the segmented image. The equations of these measures are shown in Table 1. All the metrics are in the range $[0, 1]$ with higher values indicating better performances.

Table 1. Performance metrics equations with: **TP**: True Positive, the number of pixels correctly labeled as foreground (value = 1). **FP**: False Positive, the number of pixels wrongly labeled as foreground. **TN**: True Negative, the number of pixels correctly labeled as background (value = 0). **FN**: False Negative, the number of pixels wrongly labeled as background.

Performance Metric	Equation
Accuracy	$Acc = \dfrac{TP + TN}{TP + TN + FP + FN}$
Sensitivity	$Sen = \dfrac{TP}{TP + FN}$
Specificity	$SP = \dfrac{TN}{FP + TN}$
Precision	$Pre = \dfrac{TP}{TP + FP}$
Dice	$Dice = \dfrac{2TP}{2TP + FP + FN}$

3.3 Results

To demonstrate the efficiency of Res-GAN, we compare it with three architecture: U-Net [9], VGAN [11] and Residual U-Net (Res-UNet), which is Res-GAN without discriminator model. The performance of different models are shown in Table 2. VGAN achieves better specificity and precision. However, the specificity value reflects the true negative term; that corresponds to the background pixels segmentation which is less important compared to the vessels pixels segmentation. We want a compromise between sensitivity and precision which is given by Dice metric. Thus, our architecture shows better performance in terms of accuracy, sensitivity and Dice metric. Note that higher Dice metric means better sensitivity and precision. Compared to Res-UNet, Res-GAN provides the higher Dice metric (0.8066 vs 0.8118), this indicates the importance of incorporating the discriminator part.

For the qualitative results, we showed four examples images from the test set in Fig. 3, where several columns present respectively from the left to the right: original images, ground truth images, V-GAN segmentation results, U-Net segmentation results, Res-Unet segmentation results, Res-GAN segmentation results. We found that our proposed achieved better results with less noise-sensitivity. Moreover, Res-GAN architecture improved the connectivity of vessels.

Table 2. Experimental results of the Res-GAN architecture and comparison against U-Net, VGAN and Residual U-Net.

Methods	Accuracy	Sensitivity	Specificity	Precision	Dice
V-GAN	0.9636	0.7494	**0.9861**	**0.8489**	0.7960
U-Net	0.9640	0.7658	0.9847	0.8400	0.8012
Res-UNet	0.9645	0.7812	0.9837	0.8337	0.8066
Res-GAN	**0.9655**	**0.8109**	0.9811	0.8126	**0.8118**

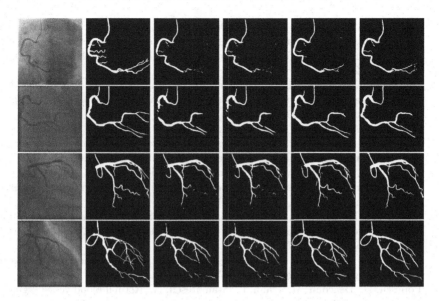

Fig. 3. Qualitative assessment of different methods performance. From left to right: original images, ground truth, V-GAN, U-Net, Res-UNet and Res-GAN outputs.

4 Conclusion

In this paper, we built a new GAN-based architecture called Res-GAN for accurate coronary artery vessels segmentation. Res-GAN consists of a residual-Unet generator using deep residual units and a residual-CNN discriminator. The generator consists of U-Net model with residual units in each block. Furthermore, we adopt a loss function combining both binary cross-entropy loss and Dice coefficient loss to improve training process and to obtain robust segmentation.

To validate the robustness of our architecture, we compared it with three different state-of-the-art models using our private dataset. For quantitative analyse, we measured accuracy, sensitivity, specificity, precision and Dice metric. Referring to the comparison analyses, Res-GAN architecture demonstrated best performance of most measurements compared to the other architectures. In the future work, we will expand the proposed Res-GAN and apply it to various curvilinear structures segmentation from different imaging modalities.

Acknowledgement. This work was supported by the Ministry of Higher Education and Scientific Research of Tunisia through the PEJC Young Researchers Encouragement Program (Project code 20PEJC 05-16).

References

1. Alom, M.Z., Hasan, M., Yakopcic, C., Taha, T.M., Asari, V.K.: Recurrent residual convolutional neural network based on u-net (r2u-net) for medical image segmentation. arXiv preprint arXiv:1802.06955 (2018)

2. Banerjee, A., Galassi, F., Zacur, E., De Maria, G.L., Choudhury, R.P., Grau, V.: Point-cloud method for automated 3d coronary tree reconstruction from multiple non-simultaneous angiographic projections. IEEE Trans. Med. Imaging **39**(4), 1278–1290 (2019)

3. Jun, T.J., Kweon, J., Kim, Y.H., Kim, D.: T-net: encoder-decoder in encoder-decoder architecture for the main vessel segmentation in coronary angiography. arXiv preprint arXiv:1905.04197 (2019)

4. Kamran, S.A., Hossain, K.F., Tavakkoli, A., Zuckerbrod, S.L., Sanders, K.M., Baker, S.A.: RV-GAN: retinal vessel segmentation from fundus images using multiscale generative adversarial networks. arXiv preprint arXiv:2101.00535 (2021)

5. Kerkeni, A., Benabdallah, A., Manzanera, A., Bedoui, M.H.: A coronary artery segmentation method based on multiscale analysis and region growing. Comput. Med. Imaging Graph. **48**, 49–61 (2016)

6. Moccia, S., De Momi, E., El Hadji, S., Mattos, L.S.: Blood vessel segmentation algorithms-review of methods, datasets and evaluation metrics. Comput. Methods Programs Biomed. **158**, 71–91 (2018)

7. Mou, L., et al.: Cs2-net: deep learning segmentation of curvilinear structures in medical imaging. Med. Image Anal. **67**, 101874 (2021)

8. Popescu, D., Deaconu, M., Ichim, L., Stamatescu, G.: Retinal blood vessel segmentation using pix2pix gan

9. Ronneberger, O., Fischer, P., Brox, T.: U-Net: convolutional networks for biomedical image segmentation. In: Navab, N., Hornegger, J., Wells, W.M., Frangi, A.F. (eds.) MICCAI 2015. LNCS, vol. 9351, pp. 234–241. Springer, Cham (2015). https://doi.org/10.1007/978-3-319-24574-4_28

10. Shi, X., Du, T., Chen, S., Zhang, H., Guan, C., Xu, B.: Uenet: a novel generative adversarial network for angiography image segmentation. In: 2020 42nd Annual International Conference of the IEEE Engineering in Medicine & Biology Society (EMBC), pp. 1612–1615. IEEE (2020)

11. Son, J., Park, S.J., Jung, K.H.: Retinal vessel segmentation in fundoscopic images with generative adversarial networks. arXiv preprint arXiv:1706.09318 (2017)

12. Son, J., Park, S.J., Jung, K.H.: Towards accurate segmentation of retinal vessels and the optic disc in fundoscopic images with generative adversarial networks. J. Digit. Imaging **32**(3), 499–512 (2019)

13. Wu, C., Zou, Y., Yang, Z.: U-GAN: generative adversarial networks with u-net for retinal vessel segmentation. In: 2019 14th International Conference on Computer Science & Education (ICCSE), pp. 642–646. IEEE (2019)

Using GANs to Improve the Accuracy of Machine Learning Models for Malware Detection

Ciprian-Alin Simion[1,2]([✉]), Gheorghe Balan[1,2], and Dragoş Teodor Gavriluţ[1,2]

[1] Faculty of Computer Science, "Al.I. Cuza" University, Iaşi, Romania
{asimion,gbalan,dgavrilut}@bitdefender.com
[2] Bitdefender Laboratory, Iaşi, Romania

Abstract. The increase of cyber-attacks and new malware in the last decade led to the usage of various machine learning techniques in security products. While these techniques are designed to improve accuracy, some practical constraints (such as lowering the false positive rate) often influence the selected model.

This paper focuses on how various generative adversarial networks can be used to improve the average detection rate and reduce the false positives for a given neural network, by altering the training set. The result of this paper is a technique that can be used to reduce the number of false positives while preserving or in some cases increasing the detection rate.

Keywords: Malware · Threats · Adversarial machine learning · False positives · Detection · GAN

1 Introduction

During the last decade, the number of known malware files has increased constantly from year to year. With this, so did the countermeasures the security products took to improve detection, the most common one being the usage of machine learning techniques. According to AV-Test[1] the number of know malware files in 2022 is around 1.3 billion. In contrast, 9 years ago (in 2013) there were only 180 million malicious files known. These values emphasize that malware detection is a problem that requires a statistical approach rather than a simple signature-based method. Machine learning was the obvious approach, and nowadays every security product makes use of these techniques to detect malware.

Even though machine learning methods used for malware detection have evolved significantly during the last years, there are a couple of practical constraints (improve the detection rate and reduce false positive rate) that every such approach has to take into consideration.

The development of generative adversarial networks marked a new step in the constant cat-and-mouse game between the security vendors and malware writers.

---

[1] https://www.av-test.org/en/statistics/malware/.

© The Author(s), under exclusive license to Springer Nature Switzerland AG 2022
H. Yin et al. (Eds.): IDEAL 2022, LNCS 13756, pp. 399–410, 2022.
https://doi.org/10.1007/978-3-031-21753-1_39

And, while malware writers are using GANs[2] to develop new techniques that will allow them to avoid detection, security vendors can use the same techniques to evaluate and improve the resilience and accuracy of detection models.

Historically, GANs have been mostly used with images. However, a file can be easily converted into an image representation if the features extracted from that file are numerical values (and in particular 32 bits values that are simply converted into a RGBA[3] format. As a result, various types of GANs can be used to validate the resilience of a machine learning model. However, in this research, we focus on how databases can be augmented with samples generated using GANs. The result is a new database that is more representative from a statistical point of view and has a better chance of obtaining a better model.

The paper is organized as follows: Sect. 2 presents similar research related to this topic, Sect. 3 explains the security vendor/malware writer ecosystem and its challenges, Sect. 4 focuses on the databases used for this study, Sect. 5 shows the experiments conducted and the final chapters reflects on the results and several conclusions that can be drawn from our research.

2   Related Work

Introduced by Szegedy and others in [18], the adversarial machine learning techniques showed for the first time that almost any machine learning model may be deceived by simply adding non-random perturbation to a sample to lead to a wrong classification. These non-random perturbations are called "noise" and different methods to create them have been intensively studied in the last years.

Over the years, different Generative Adversarial Nets architectures were proposed. Each of them came tried to improve the behavior of such a network by maximizing the number of misclassified crafted samples while keeping a similar distribution to a genuine one. Important to note are Conditional Adversarial Networks - CGAN [14], Wasserstein Generative Adversarial Networks - WGAN [3], Maximum-Likelihood Augmented Discrete Generative Adversarial Networks - MaliGAN [4], Coupled Generative Adversarial Networks - CoGAN [11], Adversarial Autoencoders [12], Boundary Seeking GAN - BGAN [8] and Least Squares GAN - LSGAN [13]. Each of these architectures is scoring good results on image processing (like MNIST, TFD[4], CIFAR) and Chinese poem generation problems.

Starting with 2017, these concepts were naturally applied in Cyber-Security. A method to produce functional windows PE Files based on Deep Reinforcement Learning Method called ACER [19] was proposed by Anderson and others in [2]. Their target was to make a malicious file undetected by choosing a finite number of actions to perform when modifying the sample. Some of these methods aimed to create new sections, pack or unpack the file, append bytes to the end of a section, add unused functions to the import address table, alter section names or create a fake entry point that will point to the original one. However, not

---

[2] Generative adversarial networks.

[3] https://en.wikipedia.org/wiki/RGBA_color_model.

[4] Toronto Face Dataset.

surprisingly, the most efficient actions were UPX packing, section renaming, and imports appending.

The robustness of machine learning algorithms that blindly analyze PE file bytes was challenged by Kolosnjaji et al. in [9]. They made use of *MalConv* [17] and achieved a maximum evasion rate of 60% by using a gradient-based strategy when adding 10,000 bytes to the end of the original file. Even if the result is astonishing, we find their solution not suitable for a real-world attack where the majority of state-of-the-art machine learning algorithms are built on more complex and time-efficient feature extraction methods (such as the one mentioned in the previously cited paper [2]). However, we consider this research as a proof of concept for the inefficiency of convolution neural network detection systems which are applied to raw binary files. This idea is also supported by Kreuk and others in [10]. They used the well-known FGSM method to generate adversarial examples for a convolution neural network malware detection. They pose a similarity with [9] in terms of raw binary files fed to the malware detector, but they come with a slightly improved approach which consists of two strategies (mid-file injection and end-of-file injection) that led to a maximum of 99.21% evasion rate.

Maybe the most sophisticated framework to evade black-box detectors developed in the last years is MalFox [21]. By feeding a binary feature vector (based on PE files information and API Calls) to a Generator and applying Obfusmal, Stealmal, or Hollowmal on the resulted perturbation vector (before sending it to the Discriminator) they achieved a very good average evasion rate (56.2%).

Moreover, GANs may be used to balance a given dataset. In this way, Yilmaz et al. showed in [20] that by using a GAN augmented dataset, the value of Accuracy, Recall, Precision, and F1 Score is genuinely improved while at the same time canceling the overfitting effect.

Another interesting usage of the Fast Gradient Sign Method took birth in 2018 with the introduction of SLEIPNIR [1]. This framework is capable of generating functional adversarial examples which may be used both in the training and validation phase to generate robust malware detection models. From all of their research, the most interesting method is $rFGSM^{k5}$ which yields the best results.

One of the most known state-of-the-art frameworks in the malware defense industry is PEberus [16]. It won first place in the *Microsoft Evasion Competition*[6] in the defender challenge. This complex framework consists of different subsystems such as Semantic gap detector rules (Slack space, Overlay, Duplicates), Classification systems (Signature-based models, Skipgram models, Ember-based GBDT models), and Stateful defense. They ranked the best in the competition (only 23% of the attacking samples evaded their model), the only attacks that were not caught by their system being made with obfuscated encrypted samples.

---

[5] Randomized rounding Fast Gradient Sign Method.

[6] https://msrc-blog.microsoft.com/2020/06/01/machine-learning-security-evasion-competition-2020-invites-researchers-to-defend-and-attack/.

3   Problem Description and Proposed Approach

In practice, using a machine learning model for malware detection is not an easy problem. Let's consider the following scenario: an ML model, validated with a cross-validation technique and a test data-set has an accuracy of 99% and a false positive rate of 1%. Strictly from an academic point of view, these metrics reflect a good model. However, a regular Windows 11 machine has more than 100.000 binaries on its main partition (usually the C folder) out of which more than 35.000 are system binaries (located in the C:/Windows folder). This means that 1% false positives are around 1000 binaries (and somewhere close to 350 system binaries). Taking into consideration that most security products have drivers that allow them to delete malicious files, this amount of false positives might have serious consequences on an operating system.

As such, one practical problem security vendors have when dealing with was to divide ways to reduce the number of false positives, such as:

– use a large benign data-set (as a cloud service) and whenever an ML model produces a detection, interrogate that service and see if the detected file is not part of that data-set.
– use telemetry and various anomaly detection to identify when such a false positive appears. This is more of a reactive solution as some customers will still be impacted. It also implied that the false positive is widely spread among a security vendor's customers; a false positive on a not-so-popular software might not be identified.
– modify the algorithms and training methodology to reduce the number of false positives. In [7] authors divide a method that significantly reduces the false positive rate, with the cost of detection rate.

This paper focuses on a similar target for the false positive problem - reduce the false positive rate via a modified training procedure. The approach also follows some practical limitations such as:

– it is virtually impossible to have access to all existing malware files. As such, it is likely that any model will eventually be limited by the sample set used for training.
– for benign files, having a large collection is possible but it implies a large data-set and complex procedure that collect these files during a large period (years). It is also important to understand that some benign files might have geographic characteristics (e.g. a software written for a specific language might be available to download on sites developed for a specific country - making them less likely to be part of a benign data-set).

This research attempts to increase the size of a data set by generating additional samples via various GANs methods.

4   Databases

4.1   Sample Collection

The process of creating a database suited for an industry-ready malware detecting solution is comprised of a series of well-defined steps.

One of the most complex tasks is the collection of malicious samples as they are hard to find and are not publicly available (for understandable reasons). One who wants to collect this type of samples has to made us of different techniques such as *Sandboxes, HoneyPots, Sample pulling, User submission* and tracing of the *Malicious URLs*. Each of these processes comes with its unique advantages and limitations as a malicious sample usually performs an environment fast check before running. On the other hand, the collection of benign files is not as difficult. There are plenty of trusted, verified applications and operating system files which are easy to harvest. However, some filters have to be in place to ensure the purity of the dataset and the correctness of the assigned labels (benign or malicious). Therefore, we chose to remove *Potentially Unwanted Applications, Unethical Software, Adware, Interpreted languages, Installers* and *Damaged files*, as they may induce a bias in the dataset (for example, many of the websites known to serve clean files may also be unknowingly hosting *gray*[7] samples).

Following all the above requirements we collected 500.000 benign files and 500.000 malicious ones in a period of three months (March to May 2019).

4.2   Feature Mining

Each of the 1.000.000 files was analyzed and a total of 25.193 features were extracted. These features are described by different values types (boolean, numeric, and strings) and refer to:

- API[8] calls
- file structure (e.g. section attributes, file entropy)
- information obtained from static analysis (e.g. compiler type, obfuscation methods)
- content-related data (e.g. digital signature, icon)
- executable capabilities (mined from sequences of API libraries, e.g. the capability of a sample to download a file)

4.3   Feature Processing and Selection

All of the extracted features were converted to a numerical value. Strings were either generalized (for example, " http://google.com" may be transformed into a boolean feature that asserts the presence of an URL in the given sample) or transformed into a numerical value (by computing a CRC32 over their content). Next, all boolean features were converted to numerical ones (1 stands for

---

[7] A unclear behaviour which is not recognized neither clean nor malware.

[8] Application program interface.

true, 0 for false). Finally, we obtained the final features by using the formula
$Resulted_{value} = \tanh(\dfrac{Feat_{value} - Feat_{mean}}{Feat_{stddev}})^9$.

The final step of our feature selection algorithm was to apply an F-Score ranking algorithm [5]. Sorting all features and plotting the obtained values we decided to use the first 1536 features as we consider this value the plateau starting point.

5   Experimental Setup

Our experiment made use of the entire database (described in the above chapter). To measure detection rate and FPR[10] we used a Multi-Layer Perceptron further referred to as *Oracle*. Its architecture is comprised of an input layer of size 1536, hidden layers of Batch Normalization, a Leaky ReLu activation function, and a Dropout Layer with 0.3 rate. We choose for the output layer the Sigmoid function as its output would yield the probabilities of the two classes. In terms of scaling to large problem instances, the Adam optimizer is known to be both accurate including FP rates, and time-wise efficient [6]. This strongly supports the usage of *Binary Cross-Entropy* as a loss function.

We start by feeding the Oracle with pairs of malicious and clean samples in order to obtain a baseline Accuracy and FPR. Next, we use GANs **to augment the malicious train dataset**. Hence, we implemented a suite of generative methods: CGAN [14], WGAN [3], CoGAN [11], Adversarial Autoencoders (AAE) [12], BGAN [8] and LSGAN [13]. To achieve our goal of lowering the FPR while keeping a good accuracy, we propose a method to generate a dataset that respects a ratio of 1:10 malicious to benign samples. This dataset will be fed to the Oracle, obtaining new values for Accuracy and FPR.

In order to better explain our algorithms, we introduce the following notations:

- D the entire data set
- Alg - a generative method, one of the following: WGAN, CGAN, CoGAN, AAE, BGAN, LSGAN
- db - dataset used to train the generative method
- k - the number of generated samples ($k = |GD|$).
- $Train$ - is a training function that takes an algorithm, a dataset and a number of iterations
- G - the trained Generator Model of that algorithm
- x - a feature vector
- $part$ - represents which part of the feature vector to start from
- $count$ - represents the number of samples to include in the returned part

---

[9] $Feat_{value}$ is the numerical value of a feature, $Feat_{mean}$ represents the average value of that feature across all samples from the database, $Feat_{stddev}$ is the standard deviation (for the values of that feature).

[10] False positive rate.

- tp - True Positive, tn - True Negative
- fp - False Positive, fn - False Negative
- Acc - accuracy, FPR - false positive rate
- New_D - the newly created dataset for each split
- $Split(d, p)$ - a function that splits the dataset d into p equal parts
- $Next(d, i, c)$ - a function that yields the next c samples from d starting from index $c \times i$

Next, we define an algorithm-agnostic generation method (Algorithm 1).

Algorithm 1. Algorithm-Agnostic generation method

1: **function** $Generate(Alg, db, k, numIter)$
2:     $GD \leftarrow \emptyset$
3:     $G = Train(Alg, db, numIter)$
4:     $GD \leftarrow G(k)$
5:     **return** GD
6: **end function**

For each generative algorithm (WGAN, CGAN, CoGAN, AAE, BGAN, LSGAN) we performed the following steps:

1. the real malware sample set was split into 10 50k parts
2. for each split part:
   (a) we generate 50k malicious samples using the Algorithm 1 (Each model was trained for 5000 iterations), hence a total of 100.000 malicious files (See Lines 3 to 5 in Algorithm 2)
   (b) we double the clean dataset to ensure a ratio of 1 to 10 to better reflect a real user environment, hence 1.000.000 clean files (See Line 6 in Algorithm 2)
   (c) we train the *Oracle* on the resulted dataset (a total of 1.100.000 files) (See Lines 8 to 11 in Algorithm 2)
   (d) for each fold we compute True Negatives, True Positives, False Negatives and False Positive values using *sklearn.metrics.confusion_matrix* [15] (See Lines 12 to 14 in Algorithm 2)
   (e) we compute Accuracy and False Positive Rate using the following formulas: $FPR = \frac{TP}{TN+FP}$, $Acc = \frac{TP+TN}{TN+TP+FN+FP}$ (See Lines 15 to 21 in Algorithm 2)

Algorithm 2. Calculating FPR and Accuracy for Oracle trained with 1-10 augmented dataset - Single

---

1: **function** CALCFPRANDACCFORGAN(Alg)
2:     **for** $i = 1 \rightarrow 10$ **do**
3:         $x_{mal_{real_i}} = Next(x_{mal_{real}}, i, 50000)$
4:         $x_{mal_{gen}} = Generate(Alg, x_{mal_{real_i}}, 50000, 5000)$
5:         $x_{mal} = x_{mal_{real_i}} \cup x_{mal_{gen}}$
6:         $x_{bgn} = x_{bgn_{real}} \cup x_{bgn_{real}}$
7:         $New_D = x_{mal} \cup x_{bgn}$
8:         $New_{D_1}, New_{D_2}, New_{D_3} = Split(New_D, 3)$
9:         $Model_1 = Train(New_{D_1} \cup New_{D_2})$
10:       $Model_2 = Train(New_{D_1} \cup New_{D_3})$
11:       $Model_3 = Train(New_{D_2} \cup New_{D_3})$
12:       $tn_1, tp_1, fp_1, fn_1 = Test(Model_1, New_{D_3})$
13:       $tn_2, tp_2, fp_2, fn_2 = Test(Model_2, New_{D_2})$
14:       $tn_3, tp_3, fp_3, fn_3 = Test(Model_3, New_{D_1})$
15:       $Acc_1, FPR_1 = FPRandAcc(tn_1, tp_1, fp_1, fn_1)$
16:       $Acc_2, FPR_2 = FPRandAcc(tn_2, tp_2, fp_2, fn_2)$
17:       $Acc_3, FPR_3 = FPRandAcc(tn_3, tp_3, fp_3, fn_3)$
18:       $ACC = \frac{Acc_1 + Acc_2 + Acc_3}{3}$
19:       $FPR = \frac{FPR_1 + FPR_2 + FPR_3}{3}$
20:       $ACCs = ACCs \cup ACC$
21:       $FPRs = FPRs \cup FPR$
22:     **end for**
23:     **return** $\frac{ACC}{10}, \frac{FPR}{10}$
24: **end function**

In our last test, we used all generative methods in the same test to augment the malware set. Similar to the previous method, we iterated over the malware set and picked each 50K part of it. We used it to generate another 50K with all the generative methods. Putting them all together, the malware set was comprised of 350K samples (50K real, 300K generated). At the same time, for the clean set, we appended a copy of it to itself for every generated sample set, totaling 3.500.000 clean samples. For each 50K malware subset picked we increased the dataset in the aforementioned manner and proceeded to perform a three-fold cross-validation.

6   Results

As we mentioned in the previous section, we will use an unbalanced dataset scenario with a ratio of 1:10 (1 malicious, 10 benign). We first computed the confusion matrix values for the *Oracle* by splitting the malicious samples into 10 parts of 50k samples with all 500k clean samples. With these datasets, we performed a three-fold cross-validation. The values were averaged (Table 1).

Next, we use the Algorithm 2 with all GANs. For each, we compute True Negatives, False Negatives, False Positives, False Positives Rate, and Accuracy.

Table 1. Baseline FPR and Accuracy for Oracle

Malset Part	TN	TP	FN	FP	FPR	ACC
1–50K	166384,33	12986,33	3680,33	282,33	0,169%	97,839%
50K–100K	166413,33	13878,67	2788,00	253,33	0,152%	98,341%
100K–150K	166592,00	15985,00	681,67	74,67	0,045%	99,587%
150K–200K	166645,67	15860,67	806,00	21,00	0,013%	99,549%
200K–250K	166657,33	16065,00	601,67	9,33	0,006%	99,667%
250K–300K	166606,33	16018,67	648,00	60,33	0,036%	99,614%
300K–350K	166488,33	14352,00	2314,67	178,33	0,107%	98,640%
350K–400K	166077,67	14685,33	1981,33	589,00	0,353%	98,598%
400K–450K	166482,00	15218,00	1448,67	184,67	0,111%	99,109%
450K–500K	166453,33	15759,67	907,00	213,33	0,128%	99,389%
Average	**166480,03**	**15080,93**	**1585,73**	**186,63**	**0,112%**	**99,033%**

In a similar way to the baseline procedure, we selected all 50k for the malware set, generated an additional 50k and we add them together to form the malware set. We also doubled the clean samples set by appending a copy of it to itself to maintain the 1 to 10 ratio. For each of the datasets, we performed a three-fold cross-validation using the *Oracle*. The obtained values were averaged (Table 2).

Table 2. FPR and Accuracy for Oracle trained on GAN augmented malware set

Method	TN	TP	FN	FP	FPR	ACC
Baseline	166480,03	15080,93	1585,73	186,63	0,112%	99,033%
BGAN	333096,80	31662,73	1670,60	236,53	0,071%	99,480%
CGAN	333160,00	31520,37	1812,97	**173,33**	0,052%	99,458%
AAE	333153,63	31650,77	1682,57	**179,70**	0,054%	99,492%
LSGAN	333084,60	31757,97	1575,37	248,73	0,075%	**99,503%**
WGAN	333163,33	31033,70	2299,63	**170,00**	0,051%	99,326%
CoGAN	333045,57	31752,20	1581,13	287,77	0,086%	99,490%

Lastly, we used all the generative methods to augment the malware set for a single test while at the same time increasing the clean set by appending a copy of it to itself. The rest of the test was conducted in a similar manner to the previous one. The results for this test are shown in Table 3.

Table 3. FPR and Accuracy for Oracle trained on 1 to 10 unbalanced dataset augmented with all generative methods

Mal Part	TN	TP	FN	FP	FPR	ACC
Baseline	166480,03	15080,93	1585,73	186,63	0,112%	99,033%
1-50K	1166564,00	112246,00	4420,67	102,67	0,009%	99,648%
100K-150K	1166657,67	113283,67	3383,00	9,00	0,001%	99,736%
150K-200K	1166637,33	114291,00	2375,67	29,33	0,003%	99,813%
200K-250K	1166621,00	114303,67	2363,00	45,67	0,004%	99,812%
250K-300K	1165352,00	114654,00	2012,67	1314,67	0,113%	99,741%
300K-350K	1166478,00	113902,67	2764,00	188,67	0,016%	99,770%
350K-400K	1166567,67	113576,67	3090,00	99,00	0,008%	99,752%
400K-450K	1166546,33	114263,00	2403,67	120,33	0,010%	99,803%
450K-500K	1166617,00	115273,67	1393,00	49,67	0,004%	99,888%
50K-100K	1166403,33	113203,67	3463,00	263,33	0,023%	99,710%
Average	**1166444,43**	**113899,80**	**2766,87**	**222,23**	**0,019%**	**99,767%**

Each of the previous experiments was performed in a cloud-based infrastructure, with 32 CPU cores and 128 GB of RAM. The time format displayed is "hours:minutes:seconds" (Table 4).

Table 4. Training times for each method of augmenting the malware training set for the 1 to 10 ratio between malware and clean

Method	1 fold	1 50K part	Total
Baseline	00:14:13	00:42:40	07:06:39
BGAN	00:28:02	01:24:06	14:01:01
CGAN	00:29:12	01:27:35	14:35:47
AAE	00:27:40	01:22:59	13:49:46
LSGAN	00:31:31	01:34:33	15:45:32
WGAN	00:29:48	01:29:23	14:53:47
CoGAN	00:31:00	01:33:01	15:30:07
GANsemble	01:58:56	05:56:47	59:27:47
All methods			155:10:26

7   Conclusion and Future Work

The results of this study can be summarized as follows:

1. augmenting a data-set with samples build via GANs method improves both the detection rate and the false positive rate. The baseline obtained with

a three-fold cross-validation led to an accuracy of 99.003% and an FPR of 0.112%. WGAN, CGAN, and AAE obtained better results in terms of FPR and Accuracy. It is also worth mentioning that in all of these cases, even though the number of samples increases, the actual number of false positives was smaller than the one obtained with the Oracle (e.g. from 186 false positives obtained to 170 obtained via WGAN methods).

2. best accuracy was obtained by AAE (from a baseline of 99.003% to 99.492%)
3. best FPR was obtained by WGAN (from a baseline of 0.112% to 0.051%)
4. from the practical point of view, the risk appetite should be a factor when choosing a GAN method from the above presented one. If a security vendor is more concerned about false positives, then WGAN is a better choice, while if its main target is a better detection rate, AAE seems to be better. In both cases, it is recommended that this method will be augmented by other techniques (like cloud white-listing).
5. we should also mention that even if the experiment took an important amount of time (as it is shown in Table 4), in practice the overhead should be considered for the GAN training (around 5 min) and sample generation (around 5 min), the rest of the time representing the model training.

In the future, we plan to evaluate the overfitting that these techniques might obtain. At the same time, we plan to evaluate the time-resilience (the drop of detection rate and possible increase of false positive rate over time) of these models over a long period of time and compare it with how a regular model (one that was trained without any sample data-set augmentation) behaves.

References

1. Al-Dujaili, A., Huang, A., Hemberg, E., O'Reilly, U.M.: Adversarial deep learning for robust detection of binary encoded malware (2018). https://doi.org/10.48550/ARXIV.1801.02950, https://arxiv.org/abs/1801.02950
2. Anderson, H.S., Kharkar, A., Filar, B., Evans, D., Roth, P.: Learning to evade static pe machine learning malware models via reinforcement learning (2018). 10.48550/ARXIV.1801.08917, https://arxiv.org/abs/1801.08917
3. Arjovsky, M., Chintala, S., Bottou, L.: Wasserstein GAN (2017). https://doi.org/10.48550/ARXIV.1701.07875, https://arxiv.org/abs/1701.07875
4. Che, T., et al. Maximum-likelihood augmented discrete generative adversarial networks (2017). https://doi.org/10.48550/ARXIV.1702.07983, https://arxiv.org/abs/1702.07983
5. Chen, Y.W., Lin, C.J.: Combining SVMs with various feature selection strategies. In: Guyon, I., Nikravesh, M., Gunn, S., Zadeh, L.A. (eds.) Feature Extraction. Studies in Fuzziness and Soft Computing, vol. 207, pp 315–324. Springer, Heidelberg (2006). https://doi.org/10.1007/978-3-540-35488-8_13
6. Kingma, D., Ba, J.: Adam: a method for stochastic optimization. In: International Conference on Learning Representations, December 2014
7. Gavrilut, D., Benchea, R., Vatamanu, C.: Optimized zero false positives perceptron training for malware detection. In: 2012 14th International Symposium on Symbolic and Numeric Algorithms for Scientific Computing, pp. 247–253 (2012). https://doi.org/10.1109/SYNASC.2012.34

8. Hjelm, R.D., Jacob, A.P., Che, T., Trischler, A., Cho, K., Bengio, Y.: Boundary-seeking generative adversarial networks (2017). https://doi.org/10.48550/ARXIV.1702.08431, https://arxiv.org/abs/1702.08431

9. Kolosnjaji, B., et al.: Adversarial malware binaries: evading deep learning for malware detection in executables (2018). https://doi.org/10.48550/ARXIV.1803.04173, https://arxiv.org/abs/1803.04173

10. Kreuk, F., Barak, A., Aviv-Reuven, S., Baruch, M., Pinkas, B., Keshet, J.: Deceiving end-to-end deep learning malware detectors using adversarial examples (2018). https://doi.org/10.48550/ARXIV.1802.04528, https://arxiv.org/abs/1802.04528

11. Liu, M.Y., Tuzel, O.: Coupled generative adversarial networks (2016). https://doi.org/10.48550/ARXIV.1606.07536, https://arxiv.org/abs/1606.07536

12. Makhzani, A., Shlens, J., Jaitly, N., Goodfellow, I., Frey, B.: Adversarial autoencoders (2015). https://doi.org/10.48550/ARXIV.1511.05644,https://arxiv.org/abs/1511.05644

13. Mao, X., Li, Q., Xie, H., Lau, R.Y.K., Wang, Z., Smolley, S.P.: Least squares generative adversarial networks (2016). https://doi.org/10.48550/ARXIV.1611.04076, https://arxiv.org/abs/1611.04076

14. Mirza, M., Osindero, S.: Conditional generative adversarial nets (2014). https://doi.org/10.48550/ARXIV.1411.1784, https://arxiv.org/abs/1411.1784

15. Pedregosa, F., et al.: Scikit-learn: machine learning in Python. J. Mach. Learn. Res. **12**, 2825–2830 (2011)

16. Quiring, E., Pirch, L., Reimsbach, M., Arp, D., Rieck, K.: Against all odds: winning the defense challenge in an evasion competition with diversification (2020). https://doi.org/10.48550/ARXIV.2010.09569, https://arxiv.org/abs/2010.09569

17. Raff, E., Barker, J., Sylvester, J., Brandon, R., Catanzaro, B., Nicholas, C.: Malware detection by eating a whole exe (2017). https://doi.org/10.48550/ARXIV.1710.09435, https://arxiv.org/abs/1710.09435

18. Szegedy, C., et al.: Intriguing properties of neural networks (2013). https://doi.org/10.48550/ARXIV.1312.6199, https://arxiv.org/abs/1312.6199

19. Wang, Z., et al.: Sample efficient actor-critic with experience replay (2016). https://doi.org/10.48550/ARXIV.1611.01224, https://arxiv.org/abs/1611.01224

20. Yilmaz, I., Masum, R.: Expansion of cyber attack data from unbalanced datasets using generative techniques (2019). https://doi.org/10.48550/ARXIV.1912.04549, https://arxiv.org/abs/1912.04549

21. Zhong, F., Cheng, X., Yu, D., Gong, B., Song, S., Yu, J.: Malfox: camouflaged adversarial malware example generation based on conv-GANs against black-box detectors (2020). https://doi.org/10.48550/ARXIV.2011.01509,https://arxiv.org/abs/2011.01509

Randomized K-FACs: Speeding Up K-FAC with Randomized Numerical Linear Algebra

Constantin Octavian Puiu$^{(\boxtimes)}$

Mathematical Institute, University of Oxford, Oxford, UK
constantin.puiu@maths.ox.ac.uk

Abstract. K-FAC is a successful tractable implementation of Natural Gradient for Deep Learning, which nevertheless suffers from the requirement to compute the inverse of the Kronecker factors (through an eigendecomposition). This can be very time-consuming (or even prohibitive) when these factors are large. In this paper, we theoretically show that, owing to the exponential-average construction paradigm of the Kronecker factors that is typically used, their eigen-spectrum must decay. We show numerically that in practice this decay is very rapid, leading to the idea that we could save substantial computation by only focusing on the first few eigen-modes when inverting the Kronecker-factors. Randomized Numerical Linear Algebra provides us with the necessary tools to do so. Numerical results show we obtain $\approx 2.5\times$ reduction in per-epoch time and $\approx 3.3\times$ reduction in time to target accuracy. We compare our proposed K-FAC sped-up versions with a more computationally efficient NG implementation, SENG, and observe we perform on par with it.

Keywords: Practical natural gradient · K-FAC · Randomized NLA · Deep nets

1 Introduction

Research in optimization for DL has lately focused on Natural Gradient (NG), owing to its desirable properties when compared to standard gradient [1,2]. K-FAC ([3]) is a *tractable* implementation which nevertheless suffers from the drawback of requiring the actual inverses of the Kronecker Factors (as opposed to just a linear solve with them). When these K-Factors are large (eg. for very wide fully-connected layers), K-FAC becomes very slow. A fundamentally different practical implementation of NG which does not have this problem has been proposed: SENG [4] (uses matrix sketching [5] and empirical NG [2]). SENG substantially outperforms K-FAC when the latter suffers from its outlined problems.

In this paper, we provide a way to alleviate K-FAC's issue and make it competitive with SENG. We begin by theoretically noting that the eigenspectrum of the K-Factors must decay rapidly, owing to the exponential-average (EA) construction paradigm of the K-Factors. Numerical results of practically obtained eigenspectrums show that in practice, the decay is much faster than the one implied by

© The Author(s), under exclusive license to Springer Nature Switzerland AG 2022
H. Yin et al. (Eds.): IDEAL 2022, LNCS 13756, pp. 411–422, 2022.
https://doi.org/10.1007/978-3-031-21753-1_40

our worst-case scenario theoretical analysis. Using these observations, we design highly time-efficient approximation routes for K-Factors inversion, with minimal accuracy reduction - by employing randomized Numerical Linear Algebra (rNLA) [6]. Numerically, our proposed methods speed up K-FAC by 2.5× and 3.3× in terms of *time per epoch* and *time to target accuracy* respectively. Our algorithms outperform SENG [4] (in terms of wall time) for moderate and high target test accuracy, but slightly underperform for very high test accuracy.

Related Work. The work of Tang et. al. (2021, [7]) is most related. However, their main approach is to construct a more efficient inversion of the regularized low-rank K-factors, without any rNLA. To make their approach feasible, they have to perform an EA over $A_k^{(l)}$ and $G_k^{(l)}$ rather than over $\bar{\mathcal{A}}_k^{(l)}$ and $\bar{\Gamma}_k^{(l)}$, as is standard (see *Sect.* 2.1). Our approach avoids this issue. Osawa et. al. (2020, [8]) presents some ideas to speed-up K-FAC, but they are orthogonal to ours.

2   Preliminaries

Neural Networks (NNs) are assumed knowledge. We only briefly define related quantities for future reference. Our learning problem is

$$\min_{\theta} f(\theta) := \frac{1}{|\mathcal{D}|} \sum_{(x_i, y_i) \in \mathcal{D}} \left(-\log p(y_i | h_\theta(x_i)) \right), \tag{1}$$

where $\mathcal{D}$ is the dataset containing input-target pairs $\{x_i, y_i\}$, θ are the aggregated network parameters, $h_\theta(\cdot)$ is the neural network function (with n_L layers), and $p(y|h_\theta(x_i))$ is the predictive distribution of the network (over labels - e.g. over classes), which is parameterized by $h_\theta(x_i)$. We let $p_\theta(y|x) := p(y|h_\theta(x))$, $g_k := \nabla_\theta f(\theta_k)$, and note that we can express $g_k = [g_k^{(1)}, ..., g_k^{(n_L)}]$, where $g_k^{(l)}$ is the gradient of parameters in layer l. We will always use a superscript to refer to the *layer* index and a subscript to refer to the *optimization iteration* index.

2.1   Fisher Information, Natural Gradient and K-FAC

The Fisher information is defined as

$$F_k := F(\theta_k) := \mathbb{E}_{\substack{x \sim \mathcal{D} \\ y \sim p_\theta(y|x)}} \left[\nabla_\theta \log p_\theta(y|x) \nabla_\theta \log p_\theta(y|x)^T \right]. \tag{2}$$

A NG descent (NGD) algorithm with stepsize α_k takes steps of the form $s_k^{(\mathrm{NGD})} = -\alpha_k \nabla_{\mathrm{NG}} f(\theta_k)$, where $\nabla_{\mathrm{NG}} f(\theta_k)$ is the natural gradient (NG), defined as [1]

$$\nabla_{NG} f(\theta_k) := F_k^{-1} g_k. \tag{3}$$

In DL, the dimension of F_k is very large, and F_k can neither be stored nor used to complete a linear-solve. K-FAC ([3]) is a practical implementation of the NGD algorithm which bypasses this problem by approximating F_k as

$$F_k^{(\mathrm{KFAC})} := \mathrm{blockdiag}(\{\mathcal{A}_k^{(l)} \otimes \Gamma_k^{(l)}\}_{l=1,...,n_L}), \tag{4}$$

where $\mathcal{A}_k^{(l)} := A_k^{(l)}[A_k^{(l)}]^T$ and $\Gamma_k^{(l)} := G_k^{(l)}[G_k^{(l)}]^T$ are the *forward K-factor* and *backward K-factor* respectively (of layer l at iteration k) [3]. Each block corresponds to a layer and $\otimes$ denotes the Kronecker product. The exact K-Factors definition depends on the layer type (see [3] for FC layers, [9] for Conv layers). For our purpose, it is sufficient to state that $A_k^{(l)} \in \mathbb{R}^{d_{\mathcal{A}}^{(l)} \times n_{\mathcal{A}}^{(l)}}$ and $G_k^{(l)} \in \mathbb{R}^{d_{\Gamma}^{(l)} \times n_{\Gamma}^{(l)}}$, with $n_{\mathcal{A}}^{(l)}, n_{\Gamma}^{(l)} \propto n_{\mathrm{BS}}$, where n_{BS} is the batch size (further size details in [3,9]).

Computing $(F_k^{(\mathrm{KFAC})})^{-1} g_k$ can be done relatively efficiently in a block-wise fashion, since we have $(\mathcal{A}_k^{(l)} \otimes \Gamma_k^{(l)})^{-1} g_k^{(l)} = \mathrm{vec}([\Gamma_k^{(l)}]^{-1} \mathrm{Mat}(g_k^{(l)})[\mathcal{A}_k^{(l)}]^{-1})$, where $\mathrm{vec}(\cdot)$ is the matrix vectorization operation and $\mathrm{Mat}(\cdot)$ is its inverse. Note that since $\mathrm{Mat}(g_k^{(l)})$ is a matrix, we need to *compute* the inverses of $\bar{\mathcal{A}}_k^{(l)}$ and $\bar{\Gamma}_k^{(l)}$ (eg. through an eigen-decomposition - and not just linear-solve with them). This is point is essential.

K-FAC pseudo-code is shown in *Algorithm* 1. Note that in practice, instead of assembling $F_k^{(\mathrm{KFAC})}$ as in Eq. (4), with the K-factors local to θ_k ($\mathcal{A}_k^{(l)}$ and $\Gamma_k^{(l)}$), we use an exponential average (EA) ($\bar{\mathcal{A}}_k^{(l)}$ and $\bar{\Gamma}_k^{(l)}$; see *lines 4* and *8* in *Algorithm* 1). This aspect is important for our discussion in *Sect.* 3. In *Algorithm* 1 we initialize $\bar{\mathcal{A}}_{-1}^{(l)} := I$ and $\bar{\Gamma}_{-1}^{(l)} := I$. θ_0 is initialized as typical [10].

Algorithm 1: K-FAC [3]

---

1   for $k = 0, 1, 2, ...$, *with sampled batch* $\mathcal{B}_k \subset \mathcal{D}$ do
2   for $l = 0, 1, ..., N_L$ do // `Perform forward pass`
3   Get $a_k^{(l)}$ and $A_k^{(l)}$
4       $\bar{\mathcal{A}}_k^{(l)} \leftarrow \rho \bar{\mathcal{A}}_{k-1}^{(l)} + (1 - \rho) A_k^{(l)}[A_k^{(l)}]^T$ // `Update fwd. EA K-factors`
5   Get $\tilde{f}(\theta_k)$;   // `The batch-estimate of` $f(\theta_k)$, `from` $a_k^{(l)}$
6   for $l = N_L, N_{L-1}, ..., 1$ do // `Perform backward pass`
7   Get $g_k^{(l)}$ and $G_k^{(l)}$
8       $\bar{\Gamma}_k^{(l)} \leftarrow \rho \bar{\Gamma}_{k-1}^{(l)} + (1 - \rho) G_k^{(l)}[G_k^{(l)}]^T$ // `Update bwd. EA K-factors`
9   Get gradient $g_k = [(g_k^{(1)})^T, ...(g_k^{(N_L)})^T]^T$
10   for $l = 0, 1, ..., N_L$ do // `Compute K-FAC step:`
11   // `Get Eig of` $\bar{\mathcal{A}}_k^{(l)}$ `and` $\bar{\Gamma}_k^{(l)}$ `for inverse application`
12      $U_{A,k}^{(l)} D_{A,k}^{(l)} (U_{A,k}^{(l)})^T = \mathrm{eig}(\bar{\mathcal{A}}_k^{(l)})$; $U_{\Gamma,k}^{(l)} D_{\Gamma,k}^{(l)} (U_{\Gamma,k}^{(l)})^T = \mathrm{eig}(\bar{\Gamma}_k^{(l)})$
13   // `Use Eigs to apply K-FAC EA matrices inverses to` $g_k^{(l)}$
14      $M_k^{(l)} = \mathrm{Mat}(g_k^{(l)}) U_{A,k}^{(l)} (D_{A,k}^{(l)} + \lambda I)^{-1} (U_{A,k}^{(l)})^T$
15      $S_k^{(l)} = U_{\Gamma,k}^{(l)} (D_{\Gamma,k}^{(l)} + \lambda I)^{-1} (U_{\Gamma,k}^{(l)})^T M_k^{(l)}$; $s_k^{(l)} = \mathrm{vec}(S_k^{(l)})$
16    $\theta_{k+1} = \theta_k - \alpha_k [(s_k^{(1)})^T, ..., (s_k^{(N_L)})^T]^T$ // `Take K-FAC step`

Key Notes on Practical Considerations. In practice, we update the Kronecker-factors and recompute their eigendecompositions ("inverses") only every few tens/hundreds of steps (update period $T_{K,U}$, inverse computation period $T_{K,I}$) [3]. Typically, we have $T_{K,I} > T_{K,U}$. As we began in *Algorithm* 1, we formulate our discussion for the case when $T_{K,I} = T_{K,U} = 1$. We do this purely for simplicity of exposition[1]. Extending our simpler discussion to the case when these operations happen at a smaller frequency is *trivial*, and does not modify our conclusions. Our practical implementations use the standard practical procedures.

2.2   Randomized SVD (RSVD)

Before we begin diving into rNLA, we note that whenever we say RSVD, or QR, we always refer to the thin versions unless otherwise specified. Let us focus on the arbitrary matrix $X \in \mathbb{R}^{m \times n}$. For convenience, assume for this section that $m > n$ (else we can transpose X). Consider the SVD of X

$$X \stackrel{\text{SVD}}{=} U_X \Sigma_X V_X^T, \tag{5}$$

and assume Σ_X is sorted decreasingly. It is a well-known fact that the best[2] rank-r approximation of X is given by $U_X[:,: r]\Sigma_X[: r,: r]V_X[:,: r]^T$ [6]. The idea behind randomized SVD is to obtain these first r singular modes without computing the entire (thin) SVD of X, which is $\mathcal{O}(m^2 n^2)$ time complexity. *Algorithm 2* shows the RSVD algorithm alongside with associated time complexities. We omit the derivation and error analysis for brevity (see [6] for details).

Algorithm 2: Randomized SVD (RSVD) [6]

---

1 **Input:** $X \in \mathbb{R}^{m \times n}$, target rank $r < \min(m, n)$, oversampling param. $r_l \leq n - r$
2 **Output:** Approximation of the first r singular modes of X
3 Sample Gaussian Matrix $\Omega \in \mathbb{R}^{n \times (r+l)}$ // $\mathcal{O}(n(r + l))$ `flops`
4 Compute $X\Omega$ // $\mathcal{O}(mn(r + l))$ `flops`
5 $QR = \text{QR\_decomp}(X\Omega)$ // $\mathcal{O}(m(r + l)^2)$ `flops`
6 $B := Q^T X \in \mathbb{R}^{(r+l) \times n}$ // $\mathcal{O}(nm(r + l))$ `flops`
7 Compute Full SVD (i.e. not the thin one) of B^T, and transpose it to recover
      $B \stackrel{\text{FULL-SVD}}{=} U_B \Sigma_B V_B^T$ // $\mathcal{O}(n^2(r + l)^2)$ `flops`
8 $\tilde{U}_X = QU_B$; $\tilde{\Sigma}_X = \Sigma_B[: r,: r]$; $\tilde{V}_X = V_B[:,: r]$ // $\mathcal{O}(m(r + l)^2)$ `flops`
9 **Return** $\tilde{U}_X \in \mathbb{R}^{m \times r}$, $\tilde{\Sigma}_X \in \mathbb{R}^{r \times r}$, $\tilde{V}_X \in \mathbb{R}^{n \times r}$

The returned quantities, $\tilde{U}_X \in \mathbb{R}^{m \times r}$, $\tilde{\Sigma}_X \in \mathbb{R}^{r \times r}$, $\tilde{V}_X \in \mathbb{R}^{n \times r}$ are approximations for $U_X[:,: r]$, $\Sigma_X[: r,: r]$ and $V_X[:,: r]$ respectively - which is what we were after. These approximations are relatively good with high probability, particularly when the singular values spectrum is rapidly decaying [6]. The total complexity of RSVD is $\mathcal{O}(n^2(r + r_l)^2 + mn(r + r_l))$ - significantly better than the complexity of SVD $\mathcal{O}(m^2 n^2)$ when $r + r_l \ll \min(m, n)$. We will see how we can use this to speed up K-FAC in *Sect.* 4.1. Note the presence of the over-sampling parameter r_l, which helps with accuracy at minimal cost. We will see this r_l appear in many places. Finally, we note that Q is meant to be a skinny-tall orthonormal matrix s.t. $\left\| X - QQ^T X \right\|_F$ is "small". There are many ways to obtain Q, but in *lines 3–4* of *Algorithm 2* we presented the simplest one for

---

[1] To avoid *if* statements in the presented algorithm.
[2] As defined by closeness in the "(p, k)-norm" (see for example [11]).

brevity (see [6] for details). In practice we perform the power iteration in *line 4* $n_{\text{pwr-it}}$ times (possibly more than once).

RSVD Error Components. Note that there are two error components when using the returned quantities of an RSVD to approximate a matrix. The first component is the *truncation error* - which is the error we would have if we computed the SVD and then truncated. The second error is what we will call *projection error*, which is the error between the rank-r SVD-truncated X and the RSVD reconstruction of X (which appears due to the random Gaussian matrix).

RSVD for Square Symmetric PSD Matrices. When our matrix X is square-symmetric PSD (the case we will fall into) we have $m = n$, $U_X = V_X$, and the SVD (5) is also the eigen-value decomposition. As RSVD brings in significant errors[3], *Algorithm* 2 will return $\tilde{U}_X \neq \tilde{V}_X$ even in this case. Thus, we have to choose between using $\tilde{V}_X$ and $\tilde{U}_X$ (or any combination of these). A key point to note is that $\tilde{V}_X$ approximates $V_X[:, : r]$ better than $\tilde{U}_X$ approximates $U_X[:, : r]$ [12]. Thus using $\tilde{V}_X \tilde{\Sigma}_X \tilde{V}_X^T$ as the rank-r approximation to X is preferable. This is what we do in practice, and it gives us virtually zero *projection error*.

2.3 Symmetric Randomized EVD (SREVD)

When X is square-symmetric PSD we have $m = n$, $U_X = V_X$, and the SVD (5) is also the eigen-value decomposition (EVD). In that case, we can exploit the symmetry to reduce the computation cost of obtaining the first r modes. SREVD is shown in *Algorithm* 3. The returned quantities, $\tilde{U}_X \in \mathbb{R}^{m \times r}$ and $\tilde{\Sigma}_X \in \mathbb{R}^{r \times r}$ are approximations for $U_X[:, : r]$ and $\Sigma_X[: r, : r]$ respectively - which is what we were after. The same observations about Q that we made in *Sect.* 2.2 also apply here.

Algorithm 3: Symmetric Randomized EVD (SREVD) [6]

---

1 **Input:** Square, Symmetric PSD matrix $X \in \mathbb{R}^{n \times n}$
2 **Output:** Approximation of the first r eigen-modes of X
3 Sample Gaussian Matrix $\Omega \in \mathbb{R}^{n \times (r+l)}$ // $\mathcal{O}(n(r+l))$ `flops`
4 Compute $X\Omega$ // $\mathcal{O}(n(r+l))$ `flops`
5 $QR = \text{QR\_decomp}(X\Omega)$ // $\mathcal{O}(n(r+l)^2)$ `flops`
6 Compute $C = Q^T X Q$ // $\mathcal{O}(n^2(r+l))$ `flops`
7 $P_C D_C P_C^T = \text{Eigen\_decomp}(C)$ // $\mathcal{O}((r+l)^4)$ `flops`
8 $\tilde{U}_X = QP_C$; $\tilde{\Sigma}_X \in \mathbb{R}^{r \times r}$ // $\mathcal{O}(n(r+l)^2)$ `flops`
9 **Return** $\tilde{U}_X \in \mathbb{R}^{n \times r}$, $\tilde{\Sigma}_X \in \mathbb{R}^{r \times r}$

The complexity is now reduced to[4] $\mathcal{O}(n^2(r+l))$ from the $\mathcal{O}(n^2(r+l)^2)$ of RSVD. However, note that by projecting both the columnspace and the rowspace of X onto Q, we are losing accuracy because we are essentially not able to obtain the more accurate $\tilde{V}_X$ as we did with RSVD. That is because we have $P_C = Q^T U_X$, and thus we can only obtain $\tilde{U}_X = QQ^T U_X$ but *not* $\tilde{V}_X$. Consequently, our *projection error* is larger when using SREVD than when using RSVD, even though the truncation error is the same.

---

[3] Relatively small, but higher than machine precision - as SVD would have.
[4] Dropping $+(r+l)^4$ as $r + l \ll n$ will hold for us.

3   The Decaying Eigen-Spectrum of K-Factors

Theoretical Investigation. Let λ_M be the max. eigenvalue of the arbitrary EA Kronecker-factor

$$\bar{\mathcal{M}}_k = (1-\rho) \sum_{i=-\infty}^{k} \rho^{k-i} M_i M_i^T, \tag{6}$$

with $M_i \in \mathbb{R}^{d_M \times n_M}$, $n_M \propto n_{BS}$. We now look at an upperbound on the number of eigenvalues that satisfy $\lambda_i \geq \epsilon \lambda_M$ (for some assumed $\alpha \in (0,1)$, chosen $\epsilon \in (0,1)$, and "*sufficiently large*" given d_M). Proposition 3.1 gives the result.

Proposition 3.1: Bounds describing eigenvalue decay of $\bar{\mathcal{M}}_k$. *Consider the $\bar{\mathcal{M}}_k$ in (6), let λ_M be its maximum eigenvalue, and let us choose some $\epsilon \in (0,1)$. Assume the maximum singular value of M_i is $\leq \sigma_M$ $\forall k$, and that we have $\lambda_M \geq \alpha \sigma_M^2$ for some fixed $\alpha \in (0,1)$. Assume that $d_M \geq \lceil \log(\alpha \epsilon)/\log(\rho) \rceil$. Then, we have that at most $r_\epsilon n_M$ eigenvalues of $\bar{\mathcal{M}}_k$ are above $\epsilon \lambda_M$, with*

$$r_\epsilon = \lceil \log(\alpha\epsilon)/\log(\rho) \rceil. \tag{7}$$

Proof. We have $\bar{\mathcal{M}}_k = \bar{\mathcal{M}}_{\text{old}} + \bar{\mathcal{M}}_{\text{new}}$ with $\bar{\mathcal{M}}_{\text{old}} := (1-\rho)\sum_{i=-\infty}^{k-r} \rho^{k-i} M_i M_i^T$, and $\bar{\mathcal{M}}_{\text{new}} := (1-\rho)\sum_{i=k-r+1}^{k} \rho^{k-i} M_i M_i^T$.

First, let us find r s.t. the following desired upper-bound holds:

$$\lambda_{\text{Max}}(\bar{\mathcal{M}}_{\text{old}}) \leq \alpha\epsilon\sigma_M^2. \tag{8}$$

Let $\rho_C := (1-\rho)$. we have

$$\lambda_{\text{Max}}(\bar{\mathcal{M}}_{\text{old}}) \leq \rho_C \sum_{i=-\infty}^{k-r} \rho^{k-i} \lambda_{\text{Max}}(M_i M_i^T) \leq \rho_C \sigma_M^2 \rho^r \sum_{i=0}^{\infty} \rho^i = \sigma_M^2 \rho^r. \tag{9}$$

Thus, in order to get (8) to hold, we can set $\sigma_M^2 \rho^r \leq \epsilon\alpha\sigma_M^2$ from (9). That is, we must have $r \geq \log(\alpha\epsilon)/\log(\rho)$. Thus, choosing

$$r := r_\epsilon := \lceil \log(\alpha\epsilon)/\log(\rho) \rceil \tag{10}$$

ensures (8) holds. Now, clearly, $\text{rank}(\bar{\mathcal{M}}_{\text{new}}) \leq n_M r$, so $\bar{\mathcal{M}}_{\text{new}}$ has at most $n_M r$ non-zero eigenvalues. Using $\bar{\mathcal{M}}_k = \bar{\mathcal{M}}_{\text{old}} + \bar{\mathcal{M}}_{\text{new}}$ and the upperbound (8) (which holds for our choice of $r = r_\epsilon$) gives that $\bar{\mathcal{M}}_k$ has at most $n_M r_\epsilon$ eigenvalues above $\alpha\epsilon\sigma_M^2$. But by assumption the biggest eigenvalue of $\bar{\mathcal{M}}_k$ satisfies $\lambda_M \geq \alpha\sigma_M^2$. Thus, at most $n_M r_\epsilon$ of $\bar{\mathcal{M}}_k$ satisfy $\lambda_i \geq \epsilon\lambda_M$. This completes the proof. $\square$

The assumption about λ_M may seem artificial, but holds well in practice. A more in depth analysis may avoid it. Note that the assumption about d_M is for simplicity - if r_ϵ turns out to be larger than d_M in practice, we merely set $r_\epsilon \leftarrow d_M$. Plugging realistic values of $\epsilon = 0.03$, $\alpha = 0.1$ and $\rho = 0.95$, $n_M = n_{BS} = 256$ (holds for FC layers) in *Proposition 3.1* tells us we have to retain at least $n_M r_\epsilon = 29184$ eigenmodes to ensure we only ignore eigenvalues satisfying $\lambda_i \leq 10^{-1.5}\lambda_M$. Clearly, 29184 is very large, and *Proposition 3.1* is not directly useful in practice. However, it does ensure us that the eigenspectrum of the EA K-Factors must have some form of decay. We now show numerically that this decay is much more rapid than inferred by our worst-case analysis here.

Numerical Investigation of K-Factors Eigen-Spectrum. We ran K-FAC for 70 epochs, with the specifications outlined in *Sect.* 5 (but with $T_{K,U} = T_{K,I} = 30$). We saved the eigen-spectrum every 30 steps if $k < 300$, and every 300 steps otherwise. Only results for layers 7 and 11 are shown for the sake of brevity, but they were virtually identical for all other layers. We see that for low k, all eigenvalues are close to unity, which is due to $\bar{\mathcal{A}}$ and $\bar{\Gamma}$ being initialized to the identity. However, the spectrum rapidly develops a strong decay (where more than 1.5 orders of magnitude are decayed within the first 200 eigenvalues). It takes $\bar{\mathcal{A}}$ about 500 steps (that is about 2.5 epochs) and $\bar{\Gamma}$ about 5100 steps (26 epochs) to develop this strong spectrum decay. We consider 1.5 orders of magnitude a strong decay because the K-Factors regularization that we found to work best is around $\lambda_{\max}/10$ (for which any eigenvalue below $\lambda_{\max}/33$ can be considered zero without much accuracy loss). Thus, truncating our K-Factors to an $r \approx 220$ worked well in practice (Fig.1).

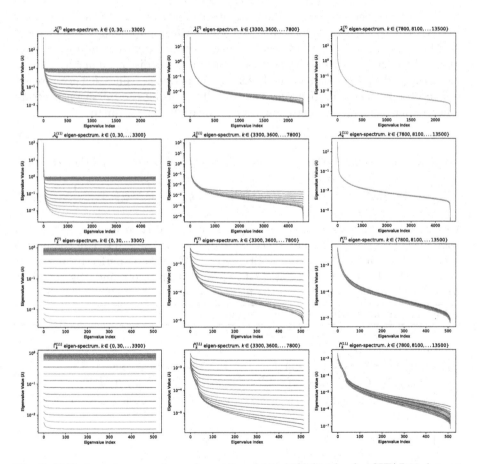

Fig. 1. K-Factors eigen-spectrum: layers 7 and 11 of VGG16_bn for CIFAR10 dataset. Each curve represents the spectrum for a specific step k.

4   Speeding Up EA K-Factors Inversion

We now present two approaches for speeding up K-FAC, which avoid the typically used EVD of the K-factors through obtaining approximations to the low-rank truncations of these EVDs. The ideas are similar in spirit and presented in the order of increasing computational saving (and reducing accuracy).

4.1   Proposed Optimizer: RSVD K-FAC (RS-KFAC)

Instead of computing the eigen-decompositions of the EA-matrices (K-Factors) $\bar{A}$ and $\bar{\Gamma}$ (in *line 12* of *Algorthm 1*; of time complexity $\mathcal{O}(d_{\mathcal{A}}^4)$ and $\mathcal{O}(d_{\Gamma}^4)$), we could settle for using a rank r RSVD approximation:

$$\bar{A} \overset{\text{RSVD}}{\approx} \tilde{U}_A \tilde{D}_A \tilde{U}_A^T, \text{ and } \bar{\Gamma} \overset{\text{RSVD}}{\approx} \tilde{U}_\Gamma \tilde{D}_\Gamma \tilde{U}_\Gamma^T, \qquad (11)$$

where $\tilde{U}_A \in \mathbb{R}^{d_A \times r}$, $\tilde{U}_\Gamma \in \mathbb{R}^{d_\Gamma \times r}$, and $\tilde{D}_A, \tilde{D}_\Gamma \in \mathbb{R}^{r \times r}$.

Using this trick, we reduce the computation cost of *line 12* in *Algorithm* 1 from $\mathcal{O}(d_{\mathcal{A}}^4 + d_\Gamma^4)$ to $\mathcal{O}((d_{\mathcal{A}}^2 + d_\Gamma^2)(r + r_l)^2)$ when using an oversampling parameter r_l. This is a dramatic reduction since we can choose $(r + r_l) \ll \min(d_{\mathcal{A}}, d_\Gamma)$ with minimal truncation error, as we have seen in *Sect. 3*. As discussed in *Sect. 2.3*, for RSVD the projection error is virtually zero, and thus small truncation error means our RSVD approach will give very close results to using the full eigenspectrum. Once we have the approximate low-rank truncations, we estimate

$$(\bar{\Gamma} + \lambda I)^{-1} V \approx (\tilde{U}_{\Gamma,r} \tilde{D}_{\Gamma,r} \tilde{U}_{\Gamma,r}^T + \lambda I)^{-1} V, \qquad (12)$$

where λ is the regularization parameter (applied to K-factors), and then compute

$$(\tilde{U}_{\Gamma,r} \tilde{D}_{\Gamma,r} \tilde{U}_{\Gamma,r}^T + \lambda I)^{-1} V = \tilde{U}_{\Gamma,r} \left[(\tilde{D}_{\Gamma,r} + \lambda I)^{-1} - \frac{1}{\lambda} I \right] \tilde{U}_{\Gamma,r}^T V + \frac{1}{\lambda} V. \qquad (13)$$

We use (13) because its r.h.s. is cheaper to compute than its l.h.s. Note that computing (13) has complexity $\mathcal{O}(rd_\Gamma + 2rd_\Gamma^2)$, which is better than computing *line 15* of *Algorithm 1* of complexity $\mathcal{O}(d_\Gamma^3)$. We take a perfectly analogous approach for $V(\bar{A} + \lambda I)^{-1}$. The RS-KFAC algorithm is obtained by replacing *lines 10–15* in *Algorithm 1* with the for loop shown in *Algorithm 4*.

Algorithm 4: RS-KFAC (our first proposed algorithm)

---

1   Replace *lines 10 - 15* in *Algorithm 1* with:
2   **for** $l = 0, 1, ..., N_L$ **do**
3   $\quad$ // Get RSVD of $\bar{A}_k^{(l)}$ and $\bar{\Gamma}_k^{(l)}$ for inverse application
4   $\quad$ $\tilde{U}_{A,k}^{(l)} \tilde{D}_{A,k}^{(l)} (\tilde{V}_{A,k}^{(l)})^T = \text{RSVD}(\bar{A}_k^{(l)}); \ \tilde{U}_{\Gamma,k}^{(l)} \tilde{D}_{\Gamma,k}^{(l)} (\tilde{V}_{\Gamma,k}^{(l)})^T = \text{RSVD}(\bar{\Gamma}_k^{(l)})$
5   $\quad$ // Use RSVD factors to approx. apply inverse of K-FAC matrices
6   $\quad$ $J_k^{(l)} = \text{Mat}(g_k^{(l)})$
7   $\quad$ $M_k^{(l)} = J_k^{(l)} \tilde{V}_{A,k}^{(l)} \left[(\tilde{D}_{A,k}^{(l)} + \lambda I)^{-1} - \frac{1}{\lambda} I \right] (\tilde{V}_{A,k}^{(l)})^T + \frac{1}{\lambda} J_k^{(l)}$
8   $\quad$ $S_k^{(l)} = \tilde{V}_{\Gamma,k}^{(l)} \left[(\tilde{D}_{\Gamma,k}^{(l)} + \lambda I)^{-1} - \frac{1}{\lambda} I \right] (\tilde{V}_{\Gamma,k}^{(l)})^T M_k^{(l)} + \frac{1}{\lambda} M_k^{(l)}$
9   $\quad$ $s_k^{(l)} = \text{vec}(S_k^{(l)})$

Note that the RSVD subroutine in *line 4* of *Algorithm* 4 may be executed using the RSVD in *Algorithm* 2, but using different RSVD implementations would not significantly change our discussion. As we have discussed in *Section 2.3.1*, even though $\tilde{U}_{A,k}^{(l)}$ should equal $\tilde{V}_{A,k}^{(l)}$ since $\bar{\mathcal{A}}_k^{(l)}$ is square s.p.s.d., the RSVD algorithm returns two (somewhat) different matrices, of which the more accurate one is the "V-matrix". The same observation also applies to Γ-related quantities.

4.2   Proposed Optimizer: SREVD K-FAC (SRE-KFAC)

Instead of using RSVD in *line 4* of *Algorithm* 4, we can exploit the symmetry and use SREVD (e.g. with *Algorithm* 3). This would reduce the computation cost of that line from $\mathcal{O}\big((d_{\mathcal{A}}^2+d_{\Gamma}^2)(r+r_l)^2\big)$ to $\mathcal{O}\big((d_{\mathcal{A}}^2+d_{\Gamma}^2)(r+r_l)\big)$, but at the expense of reduced accuracy, because SREVD has significant *projection error* (unlike RSVD; recall *Sect.* 2.3). We refer to this algorithm as SRE-KFAC and briefly present it in *Algorithm* 5. Note that in *line 4* of *Algorithm* 5 we are assigning $\tilde{V} \leftarrow \tilde{U}$ to avoid rewriting *lines 7–8* of *Algorithm* 4 with $\tilde{V}$'s replaced by $\tilde{U}'s$.

Algorithm 5: SRE-KFAC (our second proposed algorithm)

---

1 Replace lines *lines 3 - 4 in Algorithm 4* with:

2 // Get SREVD of $\bar{\mathcal{A}}_k^{(l)}$ and $\bar{\Gamma}_k^{(l)}$ for inverse application

3 $\tilde{U}_{A,k}^{(l)} \tilde{D}_{A,k}^{(l)} (\tilde{U}_{A,k}^{(l)})^T = \text{SREVD}(\bar{\mathcal{A}}_k^{(l)}); \ \tilde{U}_{\Gamma,k}^{(l)} \tilde{D}_{\Gamma,k}^{(l)} (\tilde{U}_{\Gamma,k}^{(l)})^T = \text{SREVD}(\bar{\Gamma}_k^{(l)})$

4 $\tilde{V}_{A,k}^{(l)} = \tilde{U}_{A,k}^{(l)}; \ \tilde{V}_{\Gamma,k}^{(l)} = \tilde{U}_{\Gamma,k}^{(l)}$

4.3   Direct Idea Transfer to Other Applications

Application to EK-FAC: We can apply the method directly to EK-FAC (a K-FAC improvement; [13]) as well.

Application to KLD-WRM Algorithms: Our idea can be directly applied to the KLD-WRM family (see [14]) when K-FAC is used as an implementation *"platform"*. Having a smaller optimal ρ (0.5 as opposed to 0.95), KLD-WRM instantiations may benefit more from our porposed ideas, as they are able to use even lower target-ranks in the RSVD (or SREVD) for the same desired accuracy. To see this, consider setting $\rho := 0.5$ (instead of $\rho = 0.95$) in the practical calculation underneath *Proposition 3.1*. Doing so reduces the required number of retained eigenvalues down to 2304 from 29184.

5   Numerical Results: Proposed Algorithms Performance

We now numerically compare RS-KFAC and SRE-KFAC with K-FAC (the baseline we improve upon) and SENG (another NG implementation which typically outperforms K-FAC; see [4]). We did not test SGD, as this underpeforms SENG (see *Table 4* in [4]). We consider the CIFAR10 dataset with a modified[5] version of batch-normalized VGG16 (VGG16_bn). All experiments ran on a single *NVIDIA Tesla V100-SXM2-16GB* GPU. The accuracy we refer to is always *test accuracy*.

---

[5] We add a 512-in 512-out FC layer with dropout ($p = 0.5$) before the final FC layer.

Table 1. CIFAR10 VGG16_bn results summary. All solvers reached 91.5% accuracy within the allocated 50 epochs for 10 out of 10 runs. Some solvers did not reach 92% accuracy on all of their 10 runs, and this is shown in the sixth column of the table. Columns 2–4 show the time to get to a specific test accuracy. The fifth column shows time per epoch. All times are in seconds and presented in the form: mean ± standard deviation. For time per epoch, statistics are obtained across 500 samples (50 epochs × 10 runs). For times to a specific accuracy, statistics are obtained based only on the runs where the solver indeed reached the target accuracy (eg. for 92%, the 5 successful runs are used for K-FAC). The last column of the table shows number of epochs to get to 92% accuracy (results format is analogous to the ones of column 4).

	$t_{acc \geq 90\%}$	$t_{acc \geq 91.5\%}$	$t_{acc \geq 92\%}$	t_{epoch}	Runs hit 92%	$\mathcal{N}_{acc \geq 92\%}$
SENG	673.6 ± 34.4	693.2 ± 28.2	718.1 ± 26.0	16.6 ± 0.4	10 out of 10	43.3 ± 0.9
K-FAC	1449 ± 8.7	1971 ± 225	2680 ± 636	75.5 ± 3.4	5 out of 10	35.4 ± 8.3
RS-KFAC	445.8 ± 10.9	600.7 ± 4.9	732.6 ± 153.1	32.6 ± 0.9	10 out of 10	23.0 ± 4.7
SRE-KFAC	439.4 ± 28.5	582.2 ± 24.1	785.3 ± 155.6	30.0 ± 0.4	7 out of 10	26.3 ± 5.1

Implementation Details. For SENG, we used the implementation from the *official github repo* with the hyperparameters[6] directly recommended by the authors for the problem at hand (via email). K-FAC was slightly adapted from *alecwangcq's github*[7]. Our proposed solvers were built on that code. For K-FAC, RS-KFAC and SRE-KFAC we performed manual tuning. We found that no momentum, weight_decay = 7e-04, $T_{K,U} = 10$, and $\rho = 0.95$, alongside with the schedules $T_{K,I}(n_{ce}) = 50 - 20\mathbb{I}_{n_{ce} \geq 20}$, $\lambda_K(n_{ce}) = 0.1 - 0.05\mathbb{I}_{n_{ce} \geq 25} - 0.04\mathbb{I}_{n_{ce} \geq 35}$, $\alpha_k(n_{ce}) = 0.3 - 0.1\mathbb{I}_{n_{ce} \geq 2} - 0.1\mathbb{I}_{n_{ce} \geq 3} - 0.07\mathbb{I}_{n_{ce} \geq 13} - 0.02\mathbb{I}_{n_{ce} \geq 18} - 0.007\mathbb{I}_{n_{ce} \geq 27} - 0.002\mathbb{I}_{n_{ce} \geq 40}$ (where n_{ce} is the number of the current epoch) worked best for all three K-FAC based solvers. The hyperparameters specific to RS-KFAC and SRE-KFAC were set to $n_{pwr-it} = 4$, $r(n_{ce}) = 220 + 10\mathbb{I}_{n_{ce} \geq 15}$, $r_l(n_{ce}) = 10 + \mathbb{I}_{n_{ce} \geq 22} + \mathbb{I}_{n_{ce} \geq 30}$. We set $n_{BS} = 256$ throughout. We implemented all our K-FAC-based algorithms in the empirical NG spirit (using y from the given labels when computing the bwd. K-factors rather than drawing $y \sim p(y|h_\theta(x))$; see [2] for details). We performed 10 runs of 50 epochs for each {solver, b-size} pair[8].

Results Discussion. *Table* 1 shows important summary statistics. We see that the time per epoch is $\approx 2.4\times$ lower for our solvers than for K-FAC. In accordance with our discussion in *Sect.* 4.2, we see that SRE-KFAC is slightly faster per epoch than RS-KFAC. Surprisingly, we see that the number of epochs to a target accuracy (at least for 92%) is also smaller for RS-KFAC and SRE-KFAC than for K-FAC. This indicates that dropping the low-eigenvalue modes does not seem to hinder optimization progress, but provide a further benefit instead. As a result, the time to a specific target accuracy is improved by a factor of 3 - 4× when

---

[6] **Repo:**   https://github.com/yangorwell/SENG.   **Hyper-parameters:**   *label_smoothing = 0, fim_col_sample_size = 128, lr_scheme = 'exp', lr = 0.05, lr_decay_rate = 6, lr_decay_epoch = 75, damping = 2, weight_decay = 1e-2, momentum = 0.9, curvature_update_freq = 200. Omitted params. are default.*.

[7] **Repo:**https://github.com/alecwangcq/KFAC-Pytorch.

[8] **Our codes repo:** https://github.com/ConstantinPuiu/Randomized-KFACs.

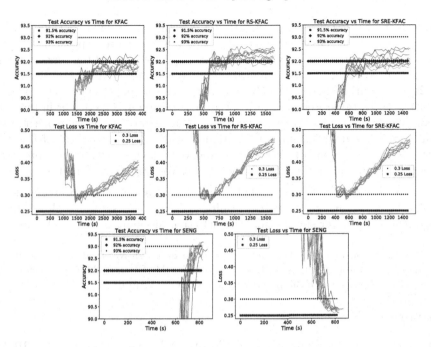

Fig. 2. CIFAR10 with VGG16_bn test loss and test accuracy results.

using RS-KFAC or SRE-FAC as opposed to K-FAC. Note that SRE-KFAC takes more epochs to reach a target accuracy than RS-KFAC. This is due SRE-KFAC further introducing a *projection error* compared to RS-KFAC (see *Sect.* 4.2). For the same reason, RS-KFAC always achieves 92% test accuracy while SRE-KFAC only does so 7 out of 10 times. Surprisingly, K-FAC reached 92% even fewer times. We believe this problem appeared in K-FAC based solvers due to a tendency to overfit, as can be seen in *Fig.* 2.

When comparing to SENG, we see that our proposed K-FAC improvements perform slightly better for 91% and 91.5% target test accuracy, but slightly worse for 92%. We believe this problem will vanish if we can fix the over-fit of our K-FAC based solvers. Overall, the numerical results show that our proposed speedups give substantially better implementations of K-FAC, with time-to-accuracy speed-up factors of $\approx 3.3\times$. *Figure* 2 shows an in-depth view of our results.

6   Conclusion

We theoretically observed that the eigen-spectrum of the K-Factors must decay, owing to the associated EA construction paradigm. We then looked at numerical results on CIFAR10 and saw that the decay was much more rapid than predicted by our theoretical worst-case analysis. We then noted that the small eigenvalues are "washed away" by the standard K-Factor regularization. This led to the idea that, with minimal accuracy loss, we may replace the full eigendecomposition performed by K-FAC with rNLA algorithms which only approximate the strongest few modes. We discussed theoretically that RSVD is more expensive but also more accurate than SREVD, and the numerical performance of the corresponding optimizers confirmed this. Numerical results show we speed up K-FAC by a factor of 2.3× in terms of time per epoch, and even had a gain in per-epoch performance. Consequently, target test accuracies were reached about 3.3× faster in terms of wall time. Our proposed K-FAC speedups also outperformed the state of art SENG (on a problem where it is much faster than K-FAC; [4]) for 91% and 91.5% target test accuracy in terms of both epochs and wall time. For 92.0% our proposed algorithms only mildly underperformed SENG. We argued this could be resolved.

Future work: Developing probabilistic theory about eigenspectrum decay which better reconciles numerical results, refining the RS-KFAC and SRE-KFAC algorithms, and layer-specific adaptive selection mechanism for target rank.

Acknowledgments. Thanks to *Jaroslav Fowkes* and *Yuji Nakatsukasa* for useful discussions. I am funded by the EPSRC CDT in InFoMM (EP/L015803/1) together with Numerical Algorithms Group and St. Anne's College (Oxford).

References

1. Amari, S.I.: Natural gradient works efficiently in learning. Neural Comput. **10**(20), 251–276 (1998)
2. Martens, J. New insights and perspectives on the natural gradient method. arXiv:1412.1193 (2020)
3. Martens, J., Grosse, R.: Optimizing neural networks with Kronecker-factored approximate curvature. arXiv:1503.05671 (2015)
4. Yang, M., Xu, D., Wen, Z., Chen, M., Xu, P.: Sketchy empirical natural gradient methods for deep learning. arXiv:2006.05924 (2021)
5. Tropp, J.A., Yurtsever, A., Udell, M., Cevher, V.: Practical Sketching Algorithms for Low-Rank Matrix approximation. arXiv:1609.00048 (2017)
6. Halko, N., Martinsson P.G., Tropp J.A.: Finding structure with randomness: probabilistic algorithms for constructing approximate matrix decompositions (2011)
7. Tang, Z., et al.: SKFAC: Training Neural Networks with Faster Kronecker-Factored Approximate Curvature. In: IEEE/CVF Conference on Computer Vision and Pattern Recognition (2021)
8. Osawa, K., Yuichiro Ueno, T., Naruse, A., Foo, C.-S., Yokota, R.: Scalable and practical natural gradient for large-scale deep learning. arXiv:2002.06015 (2020)
9. Grosse, R., Martens, J.: A Kronecker-factored approximate Fisher matrix for convolution layers. arXiv:1602.01407 (2016)
10. Murray, M., Abrol, V., Tanner, J.: Activation function design for deep networks: linearity and effective initialisation. arXiv:2105.07741 (2021)
11. Mazeika M.: The Singular Value Decomposition and Low Rank Approximation (2016)
12. Saibaba, A.K.: Randomized subspace iteration: Analysis of canonical angles and unitarily invariant norms. arXiv:1804.02614 (2018)
13. Gao, K.-X., et al.: Eigenvalue-corrected NG Based on a New Approximation. arXiv:2011.13609 (2020)
14. Puiu, C.O.: Rethinking Exponential Averaging of the Fisher. arXiv:2204.04718 (2022)

Guide-Guard: Off-Target Predicting in CRISPR Applications

Joseph Bingham$^{(\boxtimes)}$, Netanel Arussy, and Saman Zonouz

Rutgers University, New Bruswick, NJ 08901, USA
{joseph.bingham,netanel.arussy,saman.zonouz}@rutgers.edu

Abstract. With the introduction of cyber-physical genome sequencing and editing technologies, such as CRISPR, researchers can more easily access tools to investigate and create remedies for a variety of topics in genetics and health science (e.g. agriculture and medicine). As the field advances and grows, new concerns present themselves in the ability to predict the off-target behavior. In this work, we explore the underlying biological and chemical model from a data driven perspective. Additionally, we present a machine learning based solution named *Guide-Guard* to predict the behavior of the system given a gRNA in the CRISPR gene-editing process with 84% accuracy. This solution is able to be trained on multiple different genes at the same time while retaining accuracy.

1 Introduction

The newly found simplicity and versatility of gene editing systems such as the Clustered Regularly Interspaced Short Palindromic Repeats (CRISPR) [2] have spawned interdisciplinary research that explores the power and applicability of sequencing, modifying and editing genomes of model organisms across different domains of life. These promising results have not only prompted the discussion of clinical trials in humans, but have already been tested on humans. As the number of viable genes therapies increases, so too does the need to be a way for practitioners to be able to understand the underlying biological and chemical models that determine the safety of the treatment.

CRISPR is the defense mechanisms of prokayotic organisms, such as some bacteria, against infecting viruses. It does this by capturing a sub-sequence of the infecting viruses' DNA and insert it into the cell's DNA in a particular pattern to create segments. These patterns are then used as guides, for future CRISPR to seek the sequence and cut sequences that contain them. In this capacity, it is the organisms highly adaptive immune system [1].

CRISPR requires a guide RNA (gRNA) [6] sequence, which determines where it will cut the given transcriptome [5]. In the prokayotic organism, this cut in the genome would kill the would be infectious virus. In modern gene therapies, the guide sequence would be selected to be a part of the gene that is problematic. This would either kill the sequence, or induce a closure of the shortened gene sequence or to include a new sequence.

© The Author(s), under exclusive license to Springer Nature Switzerland AG 2022
H. Yin et al. (Eds.): IDEAL 2022, LNCS 13756, pp. 423–431, 2022.
https://doi.org/10.1007/978-3-031-21753-1_41

As these technologies become more understood, the medical community is also expanding which deceases they are hoping to cure using CRISPR. The main challenge in the desire to ensure the safety of a given gRNA. This is currently done by testing the guide on living cells [9], however this is time consuming and expensive.

As these cyber-physical technologies are increasingly trending towards relying on computers for automation and verification of the gene editing process, so too does the need for automatic verification and validations. This has spawned a new field of research referred to as *cyberbiosecurity*. The devastating effects of cyberbiosecurity threats, or unverified gRNA's being used are obvious given the importance of certain applications, e.g., human cells, diagnostics and crops. Unlike other domains, the consequences of an unverified guide directly impact a human's way of living and even lives themselves.

The field of machine learning is growing and provides solutions to modern day problems such as facial recognition and fraud detection. Neural networks in particular, deep neural networks (DNN), are a key component of machine learning that can be adapted to understand a given domain. In this work, we apply a special kind of DNN called a convolutional neural network (CNN) to biological modeling, and show that machine learning can be used to predict and categorize in this domain.

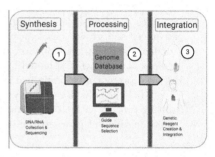

Fig. 1. A general model for genetic modification. The process on the user end may start with the collection of the DNA or RNA (1), or pulling down the information from a database (2) if the sequence has already been cataloged. This solution protects the boundaries between (1) and (2) as well as (2) and (3).

In this paper, we enumerate targets in the current state-of-the-art of gene editing techniques, in which most of the components are implemented manually in labs, as well as a proposed implementation of machine learning to identify and remove risks. The contributions of this paper are:

- We outline finding in the CRISPR space, specifically aspects of the domain that demonstrate features of data set.
- We present **Guide-Guard**, a convolutional neural network based solution for predicting the safety of a given guide sequence.
- We outline challenges in the data-mining domain faced in creating **Guide-Guard**, as well as how we over came them.
- We validate Guide-Guard's effectiveness on a real world CRISPR Cas13 database [7].

2   Background and Related Work

2.1   CRISPR

CRISPR stands for Clustered Regularly Interspaced Short Palindromic Repeats, and comes in many varieties. The most famous and used are CRISPR Cas9[1] and Cas13 [8] that are two proteins with different functionalities. This work focuses on the latter which modifies RNA in a similar fashion to Cas9 in figure Fig. 2. Another key difference is that Cas13 does not require a marker, Protospacer Adjacent Motif (PAM), to index the regions that can be edited. Save for those, the process of using various kinds of CRISPR is the same.

3   A General Model for Gene Editing

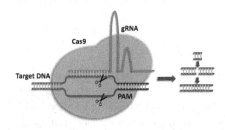

Fig. 2. A pictorial representation of CRISPR Cas9 in action. Note, CRISPR Cas13 functions nearly identically, but on RNA instead.

Figure 1 shows a high-level view of the major stages in planning for and executing a gene editing experiment. The biologist begins with a target DNA/RNA being identified and sequenced [3]. DNA/RNA sequencing is the process of determining the nucleic acid sequence and the order of nucleotides in DNA/RNA. It includes sensors used to determine the order of the four bases in a given DNA or RNA molecule. Using a variety of cyber-physical sensing and processing platforms. If the sequence has already been processed, then the previous step may be skipped entirely.

Target sequences are usually searched for off-target binding potential using web tools such as CRISPR Design Tool[2] or CCTop[3] and scored based on the potential for unintended binding. This information is forwarded onto the gene editing stage that includes obtaining synthesized primers and guide RNA (as synthesized sgRNA or as cr/tracr RNAs). The bench top protocols are performed in the lab, and the gene-edited target is sequenced and verified by comparison to the sequence.

---

[1] Cas stands for CRISPR-associated and is a protein required for DNA/RNA editing.

[2] http://crispr.mit.edu.

[3] https://crispr.cos.uni-heidelberg.de.

4   Methodology

For the following experiments, we used the data set from [7]. We chose this data set because it included multiple different types of intentionally induced mismatches, which allows us to study what happens if an accidentally happens in practice. Also, it has multiple genes within it, allowing for a better understanding of the generalizability of the these findings. Lastly, it is a standard to benchmark our solution in an objective measure, given the best sequences are known.

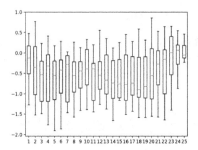

Fig. 3. A histogram of the binding potential where three mismatches occurs next to each other. The index given if the first mismatch.

4.1   Mismatch Location

From the data, we observed that for a single mismatch, there is a clear impact based on the location Fig. 4. From the histogram, we can see that the 18th nucleotide has a high importance in the CRISPR Cas13 binding energy. Additionally, the genes directly next to it also seem to have strong effect that tapers off the further away from the 18th nucleotide. This creates a normal distribution, with norm around the 18th.

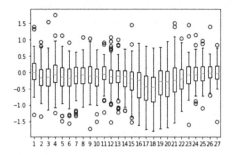

Fig. 4. A histogram detailing the binding potential as related to where a single mismatch occurs in the guide sequence.

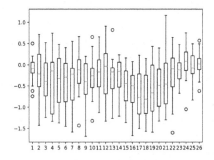

Fig. 5. The binding potential as related to where two mismatches occurs next to each other. The index is of the first mismatch.

This trend is continued into the triple consecutive mismatch histogram Fig. 3. When there are three consecutive mismatches present in the guide sequence, the bimodal distribution is more clear. The peak at the 5th is more pronounced, and the peak at the 18th persists. Note as well that generally all of the binding energies decrease, which is what is expected with further deviation from the perfect match guide.

The previous three experiments have been of consecutive mismatches. In the next Fig. 6, we examine what happens when two different mismatches occur, but not necessarily consecutively as was the case in Fig. 5. In this heat map, only the upper triangle is populated as the lower triangle would simply be a reflection. Also, the darker the cell is, the higher the activation energy. Note that the lowest activations are associated with the 5th and 18th nucleotides, consistent with our previous findings.

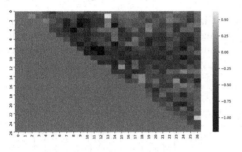

Fig. 6. A heat map detailing the binding potential as related to where two mismatches occur within a guide sequence, but may not be next to each other.

All of these values demonstrate the behavior of the guide sequence mismatch. These findings accord with the tertiary and secondary structure, which are the geometric structure of the gRNA. We used this information to put influence onto the channels associated with the 5th and 18th nucleotides. When we did this, we saw a **5.4% boost to accuracy** compared to when we did not place emphasis.

4.2 Nucleotide Replaced

Another noteworthy aspect of the data is that the nucleotide replaced, that is the original nucleotide in the target sequence which is mismatched in the guide sequence.

As can be seen from Fig. 7 Fig. 8 Fig. 10 Fig. 9 that U has the least effect when replaced, and G and C have higher effect. As such, in our set up of the model, we gave U a higher input strength to start with to provide the model with a deeper understanding of the space. When this was done, we saw a 3.8% increase to accuracy when compared to fully even input values.

4.3 Guide-Guard

In order to increase the dependability from the malicious actors in either the synthesis phase to the processing phase (1 to 2 in Fig. 1 and from the processing phase to the integration phase (1 to 3 in Fig. 1), or from user error, we designed Guide-Guard, using deep learning, to classify malicious gRNA that can be applied before use of gRNA. Wessels et al. [7] compiled a data set of RNA that can be analyzed to find their knockdown efficacy. The data set is used for predictions of guide RNA on target RNA, which is the same problem as finding malicious guides. From this data set, we make use of 3 transcriptomes, CD46, CD55, and CD71. There are about 5,000 guides to apply to each of these 3 transcriptomes in separate libraries for each.

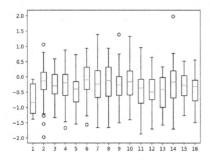

Fig. 7. A histogram detailing the binding potential as related to where a single mismatch occurs in the guide sequence, given it was originally an A.

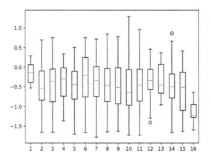

Fig. 8. A histogram detailing the binding potential as related to where a single mismatch occurs in the guide sequence, given it was originally an C.

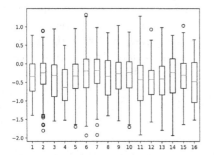

Fig. 9. The potential where a mismatch occurs in the guide sequence, given it was originally U. Although the dataset uses T, it should be U as Cas13 is for RNA.

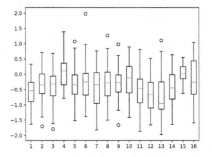

Fig. 10. A histogram detailing the binding potential as related to where a single mismatch occurs in the guide sequence, given it was originally an G.

4.4   Data Preparation

We begin with feature extraction by one-hot encoding our nucleotides into a computationally comprehensible format for our network. Leveraging the results from the survey, we weigh the 18th and to a lesser extent 5th nucleotide and G/C's more within the encoding. This is done for both the proposed guide sequence as well as the reverse-complement of target sequence. These are then zipped together, as apposed to concatenated, so that there are 46 encoded nucleotides (23 from the guides and 23 from the targets) being presented to the model. When we tested con-

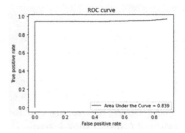

Fig. 11. The ROC representing our results in correctly identifying malicious guide sequences.

catenating the two sequences one after the other, we found that this out preformed zipping the sequences together when using a fully connected network, however when using a CNN, as we did here, we found that zipping them had an over all higher accuracy. This is a departure from previous works that assisted our results.

We then equally divide the data set by activation energy into 8. We chose 8 classes as this was the closes to having the edge of the top class be the cut off value for effective guide sequences found in [4]. There are 8 classes we assign to the data set, with 1 class (the top ones having the highest activation energy) being for positive inputs and 7 of them representing negative inputs. The classes are evenly distributed by number of data points to have an unbiased data set despite a large amount of negative inputs. We then train and validate using this data on our neural network, which is comprised of primarily dense layers and outputs one of these 8 classes. The highest of the 8 is a positive and secure input. The other 8 lack the knockdown efficacy to be safe for use. By having 8 equally distributed classes, we achieved the highest accuracy we could our of the tested division assignments as seen in Fig. 11.

4.5   Network Design

Guide-guard is a convolutional neural network, or CNN, based application. This was chosen due to CNN's ability to extract features from locality sensitive data, as such it makes a good choice for guide sequence selection. Its input layer takes in 46 values for the 23 nucleotides from each of the guide and target, with a kernel of 3 so the model is looking at a nucleotide and its neighbors. This is then put into another convolutional layer, which reduces the values down futher. These value are fed to a max pooling layer, which is finally flattened. These features are then piped to a fully connected network that goes from 400 to 200 to 100 to 50 to 25 and then finally 8 nodes. All of the activation functions utilized were ReLU, except the final layer, which utilizes softmax for classification. Since our goal was not to predict the exact activation energies, but rather to classify if a guide is sufficient for a given target, we utilized the categorical crossentropy loss function, and the Adam optimizer with a learning rate of .001.

5   Results and Discussion

It should be noted that for our testing we accept non-perfect matches for guide sequences for non-malicious inputs. These results can be seen in Table 1. We rate according to knockdown efficacy and thus take into account these non-perfect matches, as those appear quite regularly without any malicious intent. This is why the neural network is needed, since it is not as easily distinguishable as perfect or non-perfect match. The knockdown efficacy overall must be taken into account in order to correctly classify the harmful and secure inputs.

Our solution correctly classified about 84% of the inputs when tested against this data. To do this we use a 20-fold cross validation of our network on the data set to ensure proper results. Figure 11 represents the performance of our deep neural network at different classification thresholds. The receiver operating

Table 1. This table takes a more fine grained look of guide-guard. These are the accuracies and performance measures of the solution on guide sequences that are perfect matches as well as off targets.

	Perfect matches	Mismatch
Accuracy	85.51%	77.50%
True Positive Rate	98.87%	98.44%
True Negative Rate	79.48%	66.92%

curve shows our true positive rate in comparison to our true negative rate with the rating being the area under the curve. Our area under the curve was 0.839 as seen in Fig. 11, which is quite high. Our security checkpoint, while accurate, is also computationally efficient with a runtime of only 0.00055 s on average for a singular input. This average was found when run on 1 fold, or 1/20th, of the total data set.

As can be seen in Table 1, guide-guard has a higher accuracy on perfect matches than on mismatch guides. Perfect matches are guides which are the exact same sequence as the target being changed. Mismatch guides, on the other hand, are guide sequences which differ slightly from the target in some way. This discrepancy in accuracies could be explained with the observation that mismatches have a much higher variation in values, some differences in guides having no effect to binding potential, while others having dramatically different values. This would make it harder to classify. Even with this, guide-guard does better than the current best methods, which only look at perfect matches.

Our proposed solution can be used for anyone who needs to clarify the target sequence. This should be implemented before submitting to a database to secure against poisoning or any sort of errors that may occur.

Additionally, users should make use of this before applying RNA they have not developed themselves. This will add an extra layer of trust to the guide. With these implementations before Processing and Integration, as seen in Fig. 1, users can have a more secure and dependable experience.

This also applies to an automated approach with relative ease as well as the time required to verify a sample is only 0.00055 s on average when ran on a 2011 Macbook Pro 13". This means that even in higher scaled cases, the time required would increase by only about 5.5 s for every 10,000 inputs. When considering the potential dangers, or even just the potential time delays, from a single dangerous input, this would be remedied in 5.5 s.

When dangerous input warnings would appear as a result of our security program, the inputs can be checked and verified manually or discarded immediately. This allows automated and unverified processes to run smoother and more efficiently, along with adding security to any individual's CRISPR use. Databases and experiments can be secured through the implementation of our screening process before inputs are made on either one, steps 2 and 3 of Fig. 1.

6   Conclusion

The almost uninhibited accessibility to CRISPR has facilitated a new era of genome editing that is rapidly increasing the ceiling of innovation in biological design. However, the accessibility to such powerful technology must be done in a trustworthy way to mitigate threats of off-target behavior. In this paper, we propose Guide-guard as a solution to identify poor guides to a CRISPR Cas13 user's work, which can be used to stop any serious dangers to the end users.

References

1. The CRISPR-CAS immune system: biology, mechanisms and applications. Biochimie **117**, 119–128 (2015). https://doi.org/10.1016/j.biochi.2015.03.025, special Issue: Regulatory RNAs
2. Blackburn, P.R., Campbell, J.M., Clark, K.J., Ekker, S.C.: The CRISPR system–keeping zebrafish gene targeting fresh (2013)
3. Heather, J.M., Chain, B.: The sequence of sequencers: the history of sequencing DNA. Genomics (1), 1–8 (01). https://doi.org/10.1016/j.ygeno.2015.11.003
4. Metsky, H.C., et al.: Efficient design of maximally active and specific nucleic acid diagnostics for thousands of viruses (2020). https://doi.org/10.1101/2020.11.28.401877
5. Pertea, M.: Genes. https://doi.org/10.3390/genes3030344
6. Wang, D., et al.: Optimized CRISPR guide RNA design for two high-fidelity cas9 variants by deep learning. Nature Commun. (1), 4284. https://doi.org/10.1038/s41467-019-12281-8
7. Wessels, H.H., Méndez-Mancilla, A., Guo, X., Legut, M., Daniloski, Z., Sanjana, N.: Massively parallel cas13 screens reveal principles for guide RNA design. Nat. Biotechnol. **38** (2020). https://doi.org/10.1038/s41587-020-0456-9
8. Xu, D., et al.: A CRISPR/cas13-based approach demonstrates biological relevance of vlinc class of long non-coding RNAs in anticancer drug response. Sci. Rep. (1) (1794). https://doi.org/10.1038/s41598-020-58104-5
9. You, Y., Ramachandra, S.G., Jin, T.: A CRISPR-based method for testing the essentiality of a gene. Sci. Rep. **10**(1), 14779 (2020). https://doi.org/10.1038/s41598-020-71690-8

Topological Analysis of Credit Data: Preliminary Findings

James Cooper[1] , Peter Mitic[2,3](✉) , Gesine Reinert[4] ,
and Tadas Temčinas[4]

[1] Santander US, Boston, MA 02109, USA
[2] Department of Computer Science, University College London, London, UK
p.mitic@ucl.ac.uk
[3] Santander UK, London, UK
[4] Department of Statistics, University of Oxford, Oxford, UK

Abstract. Intuitively, similar customers should have similar credit risk. Capturing this similarity is often attempted using Euclidean distances between customer features and predicting credit default via logistic regression. Here we explore the use of topological data analysis for describing this similarity. In particular, persistent homology algorithms provide summaries of point clouds which relate to their topology. This approach has been shown to be useful in many applications but to the best of our knowledge, applying topological data analysis to prediction of credit risk is novel. We develop a pipeline which is based on the topological analysis of neighbourhoods of customers, with the neighbourhoods determined by a geometric network construction. We find a modest signal using three data sets from the *Lending Club*, and the *Japan Credit Screening* data set. The *Cleveland oncological* data set is used to validate the pipeline. The results have high variance, but they indicate that including such topological features could improve credit risk prediction when used as additional explanatory variable in a logistic regression.

Keywords: Credit risk · Topological data analysis · Barcode · Landscape · Logistic regression

1 Introduction

The context for this paper is bank unsecured lending. The bank makes its lending decisions using customer details (or *features*) such as the customer's employment status, income, expenditure, current total unsecured debt, previous credit record, and other third party and derived features (such as debt-to-income ratio). No single factor, nor combination of factors, is a sure-fire predictor of 'success' in repaying the loan. Bank predictions mostly succeed, but there is room for improvement. A better understanding of the probability of default can make even risky customers profitable. The heuristic behind credit risk prediction is that customers with "similar" features should be associated with similar risk of default. Finding useful measures of similarity is an ongoing issue in credit risk forecasting.

© The Author(s), under exclusive license to Springer Nature Switzerland AG 2022
H. Yin et al. (Eds.): IDEAL 2022, LNCS 13756, pp. 432–442, 2022.
https://doi.org/10.1007/978-3-031-21753-1_42

Here we investigate the use of Topological Data Analysis (*TDA*), which embraces the topological relationship between a single datum and other data. As such, there is a direct expression in topological terms of the phrase "If customer A paid back a loan, and customer B looks like customer A, then customer B should also pay back a loan". There are many applications of *TDA* in a financial context (e.g. [9]) but to the best of our knowledge, it has not been applied to credit risk modelling. Outside finance, there have been successes of classifiers based on *TDA* in many areas such as neuron spike data [15], cancer prognosis [4], and image classification [1]. This work follows previous discussions on financial credit-worthiness using data from the Lending Club (*LC* - https://www.lendingclub.com/), a US-based personal loan organisation, and the *Japan Credit Screening* data set[1].

We have formulated a procedural *pipeline* to apply *TDA* to credit data. To validate the pipeline, in the absence of related work for credit scoring, we apply and compare the pipeline to oncology data [7][2]. Using that data set, Wu and Hargreaves [17] carried out a related study, using logistic regression enhanced with *TDA* features. We set a high bar. For a training set, we select a data subset which achieves very high accuracy based on logistic regression alone, rather than a random sample. Hence, achieving similar results when including topological information, is promising.

The main contributions of this paper are, first, the finding that there can be a topological signal in credit risk data, and second, a pipeline for including topological summaries in credit scoring.

Nomenclature: In this paper the acronyms *LR* and *TDA* refer, respectively, to the *logistic regression* curve-fitting method, and the application of topological concepts to data analysis. We use the acronym *LR+TDA* to refer to a logistic regression calculation that incorporates topological components.

2 Literature Review: Credit Scoring

We concentrate on a summary of previous work on credit analysis. Work on *TDA* will be discussed in Sect. 3. Credit scoring was developed in the early 1940s in an attempt to control credit risk. In 1941 Durand (Chap. 4 in [8]) developed a 'scorecard' formula based on factors such as a customer's salary, age, sex, credit history, and occupation. Variants on that system are still used today. More factors have been introduced, and the weights attached to each factor have matured. A notable example is the U.S. *FICO* credit score[3], originally formulated in 1956 by the *Fair Isaac Corporation*. The idea of *nearest neighbours*, a key *TDA* component, was first applied by Chatterjee in 1970 [6]. In 1980, Ohlson [12] applied multivariate discriminator analysis (*MDA*) in an early probabilistic model of corporate bankruptcy prediction; similar probabilistic models have

---

[1] https://archive.ics.uci.edu/ml/datasets/Japanese+Credit+Screening.

[2] https://archive.ics.uci.edu/ml/datasets/heart+disease.

[3] https://www.fico.com/.

since been applied to retail contexts. An early application of a Logistic Regression (*logit*) model was formulated by Wigginton in 1980 [16]. *Logit* models were an advance on *MDA* models by removing the *MDA* requirement for characteristics to be drawn from a multivariate normal distribution. Further techniques, including genetic algorithms, neural networks and decision trees are discussed in [10]. These methods may miss subtle multivariate signals. Hence it is of interest to develop topological measures for credit scoring, and we address this issue here.

3   Concepts in Topological Data Analysis

Here we briefly introduce some concepts from topological data analysis which are useful for this paper. More detailed introductions can be found, for example, in [13] and [5]. Following [5], we start with a set of points P in N-dimensional Euclidean space $\mathbb{R}^N$ with a Euclidean norm $\|\cdot\|_2$. For each point $p \in P$ we define its ϵ-neighbourhood $N_\epsilon(p)$ as the open ball $B_\epsilon(p)$ of radius ϵ around p.

3.1   The Vietoris-Rips Filtration

For increasing ϵ, the balls around points will increasingly overlap, and eventually there will be no holes left between the balls. The process of increasing ϵ is used to construct characteristics of the point cloud P. First we create a sequence of so-called *Vietoris-Rips simplicial complexes* $VR_\epsilon(P)$. We start with a graph having vertices P and edges (p_0, p_1) for all pairs of points p_0, p_1 in P with distance $\|p_0 - p_1\|_2 \leq \epsilon$. We then set

$$VR_\epsilon(P) = \bigcup_{l \geq 0} VR_\epsilon(P)_l, \quad VR_\epsilon(P)_l = \{(p_0, \ldots, p_l) \mid \|p_i - p_j\|_2 \leq \epsilon \text{ for all } i, j\}.$$

Here $VR_\epsilon(P)_l$ can be viewed as a list of all l-simplices of the complex $VR_\epsilon(P)$. This construction is called *Vietoris-Rips filtration*, and we use it to calculate so-called *homology groups*. We shall only use the first two homology groups, which are easy to describe intuitively: The dimension-0 homology counts the number of connected components, and the dimension-1 homology counts the number of holes. The process as ϵ increases is illustrated in Fig. 1, reproduced from [5].

Fig. 1. An example of a Vietoris-Rips filtration. Adapted from Fig. 1 of [5].

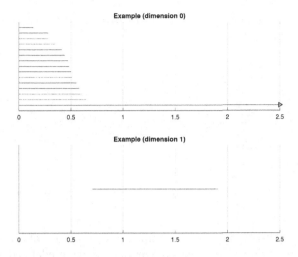

Fig. 2. The corresponding barcodes in dimension 0 and dimension 1 of the Vietoris-Rips filtration in Fig. 1. Adapted from Fig. 1 of [5].

3.2 Barcode

A *barcode* is a visual representation of point cloud connectivity as ϵ increases. A non-zero element (connected component for dimension 0, and hole for dimension 1) that first appears at $\epsilon = \epsilon_b$ and vanishes at $\epsilon = \epsilon_d$ is represented by the interval (or *bar*) $(\epsilon_b, \epsilon_d]$. Figure 2 shows the dimension 0 and 1 barcodes for the Vietoris-Rips filtration in Fig. 1.

The process described in Subsect. 3.1 takes a finite set of points $P \subseteq \mathbb{R}^N$ as an input and for every homological dimension $k \leq N$ outputs a barcode $\mathcal{B}(P, k)$, which can be represented as a multiset of intervals of the form $\left\{ (\epsilon_b(i), \epsilon_d(i)] \right\}_{i=1}^{m_k}$. An important feature of barcodes is that any two are directly comparable by many different stable metrics, in the sense that a small perturbation in the input point-cloud leads to only a small perturbation of the barcode, as measured by the metric. For an overview of stability results, see [14, Ch. 3].

3.3 Landscape

An alternative to the barcode encoding of topological information is a *persistence landscape* [3]. A *landscape* is derived from the set of m_k birth-death points at dimension k: $\left\{ (\epsilon_b(i), \epsilon_d(i)] \right\}_{i=1}^{m_k}$ of a barcode. With $\epsilon_b(i)$ plotted against $\epsilon_d(i)$, a piecewise continuous function, termed a *landscape*, models the extremities of the plot. In practice, we use a discretised form of this function, at some predefined resolution. Since we use dimension 0 and 1 homologies, we get two landscapes per point cloud. The precise mathematical definition of *persistence landscapes* is beyond the scope of this paper, and we refer the interested reader to [3].

Figure 3 shows an example of a typical *barcode* and the corresponding *landscapes* used in our analysis.

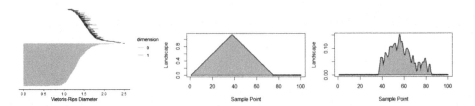

Fig. 3. Left to right: typical LC Barcode, with Landscapes at dimensions 0 and 1.

We would like to highlight a few important properties of landscapes. They are effectively vectorised expressions of barcodes at the predefined resolution (we used 500 reference points). All landscapes for a point cloud can be expressed in terms of the same resolution. That makes them easy to use in existing statistical and machine learning techniques. Second, they are also stable: a small perturbation of the point cloud will only result in a small perturbation of the landscape [3, Section 5]. Third, as real-valued functions, we can leverage different statistical techniques (such as adding, averaging) when analysing them.

An alternative to the *Landscape* metric is the *Wasserstein* metric, which measures distance between barcodes. We have found that *Wasserstein* calculations are very slow compared to *Landscape* calculations. Furthermore, they rarely improve predictive success compared to omitting *TDA* elements completely.

4   Application Pipeline to Credit Data

Topological data analysis focuses on structures, not individual nodes. Thus, as a preliminary step, we impose a structure on each node by defining a list of neighbouring nodes. Here we represent credit data as a network between customers, with two customers connected by an edge if their feature vectors are close in Euclidean space. With such a network representation we calculate the barcodes and landscapes for each individual neighbourhood. We use these topological summaries as additional measures of similarity in a logistic regression.

In more detail, we assume that a credit data set is available with N customers, and each customer has n features associated with it. The features are normalised to $[0, 1]$. A *node* with index i, n_i, comprises a set of n normalised features x_{ij}, and a binary outcome Y_i representing successful repayment of a loan or not (1 or 0 respectively); $n_i = \left\{ \{x_{i,1}, x_{i,2}, \ldots, x_{i,n}\}, Y_i \right\}$ with $x_i \in [0, 1], Y_i \in \{0, 1\}$. A collection of N nodes, $C_N = \{n_i\}_{i=1}^{N}$, constitutes the *point cloud* which is the basis of the credit risk *TDA* analysis. The neighbourhood of a target node n_i is a set of other nodes $\{n_j, j \neq i\}$ that are 'close to' n_i, according to some 'closeness' criterion. To construct edges in this network, we used a *k-nearest-neighbours* (*kNN*) algorithm with a Euclidean distance metric. The *kNN* approach was very slow to calculate, but provided marginally improved results compared to approximate methods such as *kd-Tree*, and a bespoke *feature filtering* method in which 'distant' nodes were filtered out using nodal feature-by-feature comparisons.

Once the network is constructed, each node (customer) n_i is assigned its neighbourhood as the subgraph induced by the nodes $n_j, j \neq i$ such that there is an edge between n_i and n_j in the network.

4.1 Homologies for Neighbourhoods

The central step when applying *TDA* in any context is the calculation of persistent homologies for each node. Persistent homologies are calculated using the neighbourhoods at dimensions 0 and 1. The dimension 2 components of the barcodes for the data used are often minor, and we assume that they have a negligible effect on *TDA* credit predictions. The homologies are then used to calculate outcome predictors. The *Landscape* predictor is derived from a persistence diagram *Landscape* construct [2]. It reports the proportion of Training nodes in the Landscape neighbourhoods who repay the loan, again separately for dimensions 0 and 1. With a *Landscape* predictor for each dimension, there is a total of two predictors. Finally, the predictors are used as a classifier for a set of Test nodes. The predicted outcomes are compared to the actual test outcomes.

4.2 Detailed *LR+TDA* Credit Risk algorithm

The *LR+TDA* pipeline details are described in the algorithm in this section. The first step, generation of a single Training set plus multiple Test sets, was designed to identify an "optimal" training set using the original data only. Specifically, this training set yields maximum *Accuracy* in multiple trials. The *high accuracy* training set presents a severe test for the case where the original data are augmented by *TDA* predictors. Consequently, results for *TDA*-augmented data that are close to those of the original data are sought.

Algorithm LR+TDA

1. Sample data
   (a) Draw 100 random samples (balanced 50:50 between defaulted and not), partition each into Training/Test sets in a ratio 2/3:1/3
   (b) Run *LR* calculations for each Training/Test pair, choose the training set that yields the highest *LR* Accuracy. Generate further similar Test sets
   (c) For small ($\leq$1000 nodes) data sets, only one Training/Test pair is generated.
2. Calculate neighbourhoods using *kNN*, such that the resulting number of neighbours is at least the square root of the number of Training nodes. The neighbourhood for each Training and Test node is derived from all other Training nodes only (Test data is not used to define any neighbourhood)

3. Calculate Training homologies, relative to Training nodes, separately for outcomes $Y = 0$ and $Y = 1$
4. Calculate Test homologies, also relative to Training nodes, separately for outcomes $Y = 0$ and $Y = 1$
5. Find least squares distances between Landscapes of neighbourhoods
    (a) Calculate landscapes for all nodes
    (b) Calculate least squares distance metrics for all Training landscapes, relative to other Training landscapes, at dimensions 0 and 1. Select the M shortest (we used $M = 25$)
    (c) Calculate least squares distance metrics for all Test landscapes, relative to all Training landscapes, at dimensions 0 and 1. Select the M shortest
6. Calculate Landscape predictors
    (a) For all Training Landscape neighbourhoods, calculate the proportion of the corresponding successful predictions at dimensions 0 and 1
    (b) For each Test node, calculate the proportion of Training nodes in its Landscape neighbourhood with Outcome $Y = 1$
7. Logistic regressions
    (a) Augment the original data with additional features
    (b) Do the LR calculations with and without the additional features.
8. Assess signal detection on Test data using a *Strength* algorithm:
    (a) If Landscape predictors for dimensions 0 and 1 agree, accept the common prediction.
    (b) If not, accept the predictor that is nearer to either 0 or 1 (i.e. the 'stronger' predictor)
    (c) Calculate the % of correct predictions, (we seek a % that exceeds the % of successful outcomes in the data)

The *kNN* calculations impose a severe restriction on the use of *TDA* with large data sets. The time taken to do them makes it infeasible to process more than 1000 nodes. Therefore we draw random samples from the data, and repeat each complete $LR+TDA$ calculation 10 times. We have found that using more than 10 repetitions reduces standard deviations of the results only minimally. The results were insensitive to the number of nearest neighbours used within the range 15–150, corresponding to approximately 5–50% of the Training sample size of $\frac{2}{3} \times 500$. The results are for 25 nearest neighbours.

5   Results

5.1   Data

Three data sets, sourced from the *Lending Club*, give details of unsecured loan applications, either defaulted or not, for the period 2007–14. The first, *LC-A*, has 42500 nodes and 25 features, with a default rate 15.1%. The second, *LC-B*, has 188000 nodes and 56 features. Its default rate is 15.7%, and predictions for it have proved to be particularly difficult [11].[4] Both default rates are high for a

---

[4] Neither is now available from the *Lending Club* website.

European bank. The third, *LC-C* (8900 nodes, 20 features) is used for teaching at *University College London*. The *LC* results are compared with the results of two alternatives. The first, *Japan Credit Screening Data*[5] is a small credit data set (690 nodes with 13 features), sourced from the *UCI*[6] repository. The second is the *Cleveland oncology* data [7] (300 nodes, 13 features), also sourced from the *UCI* repository. It is included to validate our results with that of an independent *TDA* study on the same data [17], as there is no available comparable study on credit risk data. All implementations were programmed in *R* on an Intel i7 quad core processor with 64 GB RAM, using, in particular, packages *TDA* and *TDAStats*.

5.2 Exploratory Classification Analysis

First we assess whether there is a topological signal in the data when not including logistic regression (step 7 of Algorithm *LR+TDA*). The *LC* data sets use balanced samples (50% in either outcome). We therefore seek a signal 'strength' of greater than 50% (i.e. a random guess). The *Japan* data set is slightly unbalanced, and we seek a signal 'strength' in excess of 55.5%. The *Cleveland* data set is also slightly unbalanced, and we seek a signal 'strength' in excess of 54.1%. No sampling was required for the *Japan* and *Cleveland* data. Hence, no standard deviations are reported. The results are collated in Table 1. Given these outcome balances, all signals exceed their 'random' levels. Thus, using *TDA* alone we detect a signal in all *LC* data sets. Although the signal is not very strong, even a small increase in precision may improve credit risk assessment.

Table 1. Classification success signal indicators, *LC*, *Japan* and *Cleveland* data, exploratory analysis. The figures in bold indicate signal strengths that exceed random levels, given extent of outcome balance.

		Accuracy	Sensitivity	Specificity	Precision
LC-A	Mean	**52.2**	0.51	0.53	0.52
	SD	3.88	0.05	0.07	0.04
LC-B	Mean	**51.2**	0.5	0.53	0.51
	SD	4.67	0.08	0.04	0.05
LC-C	Mean	**51.7**	0.53	0.5	0.51
	SD	5.48	0.1	0.05	0.05
Japan	Mean	**57.4**	0.63	0.5	0.61
Cleveland	Mean	**60.4**	0.67	0.5	0.67

5.3 Sampling Results

Step 1 of Algorithm 4.2 gave details of how the *LC* Training set and Test sets were generated. A total of ten Test sets were generated, and *LR+TDA* calculations

---

[5] https://archive.ics.uci.edu/ml/datasets/Japanese+Credit+Screening.
[6] University of California, Irvine.

were carried out on all Training-Test combinations. With samples of size 500 (333 Training, 167 Test), each complete $LR+TDA$ calculation took about 50 min to complete. Using a larger sample size (1000, 1500) did not improve accuracy. The *Japan* and *Oncology* Training data sets were selected in the same way, except that all data were used, so that only one test set is necessary. The results shown in Table 2 are a comparison of *LRs* with the original data only (column *No TDA*), and when the original data are augmented by *TDA* predictors.

Table 2. *TDA* Landscape results using *kNN* Success Indicators, *LC-A*, *LC-B* and *LC-C*, sample size 500, 10 runs each: *Japan, Cleveland*, all data, 1 run. Table entries in bold indicate improvements due to *TDA*: either non-decreased mean, or decreased standard deviation.

Data	Metric	Mean		SD	
		No TDA	Landscape	No TDA	Landscape
LC-A	*% Accuracy*	63.05	62.58	2.92	**2.79**
	Sensitivity	0.69	**0.70**	0.08	**0.08**
	Specificity	0.58	0.55	0.09	**0.08**
	Precision	0.62	0.61	0.03	**0.03**
	AUC	0.37	**0.37**	0.05	**0.04**
LC-B	*% Accuracy*	57.01	**57.13**	4.72	**4.47**
	Sensitivity	0.77	**0.79**	0.16	**0.15**
	Specificity	0.37	0.36	0.20	**0.18**
	Precision	0.55	**0.55**	0.04	**0.04**
	AUC	0.63	**0.63**	0.03	**0.03**
LC-C	*% Accuracy*	66.53	66.47	4.50	4.75
	Sensitivity	0.67	**0.67**	0.06	**0.06**
	Specificity	0.66	**0.66**	0.07	**0.07**
	Precision	0.66	**0.66**	0.05	**0.05**
	AUC	0.71	**0.71**	0.06	**0.06**
Japan	*% Accuracy*	90.00	**90.00**	n/a	n/a
	Sensitivity	0.87	**0.87**		
	Specificity	0.93	**0.93**		
	Precision	0.94	**0.94**		
	AUC	0.96	**0.96**		
Cleveland	*% Accuracy*	92.08	91.09	n/a	n/a
	Sensitivity	0.93	0.92		
	Specificity	0.90	**0.90**		
	Precision	0.93	**0.93**		
	AUC	0.97	**0.97**		

The results with *TDA* indicate parity with those without, with some modest improvements in the means of some metrics. It is encouraging that improvements (i.e. increases) in success metric means are usually matched by improvements (i.e. decreases) in success metric standard deviations. Comparison of results with those from the *Cleveland Oncology* and *Japan Credit* data provide further evidence that results including *TDA* features are comparable to those without. In particular, the independent study [17] using the same oncology data confirms our 90% accuracy figure, and also our figures for sensitivity, specificity and precision. Our result comparisons with and without *TDA*-augmented data are stringent because the training sets are optimised for maximum *LR Accuracy*. *LR+TDA* can compete, just!

6 Discussion

Here we have explored features which arise from *TDA*. A *TDA* aspiration is to capture features that a linear model cannot detect. We find indications for a subtle *TDA* signal in our data. We have tried minor modifications, namely weighting importance of neighbour scores by proximity; combining dimensions of *TDA* results in different ways, and including *Wasserstein*-derived predictors, but without significant change of predictive power. The *TDA* 'success' criterion used was stringent: a Training set optimised for *LR* accuracy. *LR* is the *de facto* method used in banking credit analysis. Therefore, *TDA* results that are a minor improvement over *LR* results for noisy credit data are encouraging.

A key consideration when implementing a credit decisioning model is that it should be explainable (i.e. reasons for the decision are apparent). Usually this means that the decision can be expressed in terms of a scorecard: customers get points based on their application credentials, and the loan is granted if they have enough points. The *TDA* pipeline is fully deterministic, so a decision is, in theory, explainable. However, more work is needed to express *TDA* results in terms of a score.

In our view, in order to achieve substantial improvements a more fundamental change would be needed. We propose a focus on how to expand the number of records we may use to train models, as well as introducing novel features whose structure is not revealed by linear methods. The next steps are for us are to see if combining *TDA* predictions based on different training samples can provide more insight, and to find if filtration-derived neighbourhoods provide a viable way forward.

Moreover, the high standard deviations in the *LR+TDA* predictions may be an indication of substantial heterogeneity in the data which is not modelled. Similar to personalised medicine, we shall try to identify subgroups of customers for which predictions can be reached with substantially smaller variance than for the entire population. Finally, we flag that the data sets here were fairly balanced, which was achieved for some of them via sampling. Highly unbalanced data sets may provide additional challenges.

Acknowledgements. GR was supported in part by EPSRC grants EP/T018445/1 and EP/R018472/1. TT acknowledges funding from EPSRC studentship 2275810.

References

1. Bernstein, A., Burnaev, E., Sharaev, M., Kondrateva, E., Kachan, O.: Topological data analysis in computer vision. In: Twelfth International Conference on Machine Vision (ICMV 2019), vol. 11433, pp. 673–679. SPIE (2020)
2. Bubenik, P., Dłotko, P.: A persistence landscapes toolbox for topological statistics. J. Symb. Comput. **78**, 91–114 (2017)
3. Bubenik, P., et al.: Statistical topological data analysis using persistence landscapes. J. Mach. Learn. Res. **16**(1), 77–102 (2015)
4. Bukkuri, A., Andor, N., Darcy, I.K. Applications of topological data analysis in oncology. Front. Artif. Intell. **38** (2021)
5. Byrne, H.M., Harrington, H.A., Muschel, R., Reinert, G., Stolz-Pretzer, B., Tillmann, U.: Topology characterises tumour vasculature. Math. Today (2019)
6. Chatterjee, S., Barcun, S.: A nonparametric approach to credit screening. J. Am. Stat. Assoc. **65**(329), 150–154 (1970)
7. Detrano, R. Heart Disease Data Set. V.A. Medical Center, Long Beach and Cleveland Clinic. UCI Machine Learning Repository (1988)
8. Durand, D.: Risk Elements in Consumer Instalment Financing. National Bureau of Economic Research (1941)
9. Gidea, M., Katz, Y.: Topological data analysis of financial time series: landscapes of crashes. Physica A **491**, 820–834 (2018)
10. Henley, W.E.: Statistical aspects of credit scoring, Ph.D., Open University (1995)
11. Mitic, P.: A metric framework for quantifying data concentration. In: Yin, H., Camacho, D., Tino, P., Tallón-Ballesteros, A.J., Menezes, R., Allmendinger, R. (eds.) IDEAL 2019. LNCS, vol. 11872, pp. 181–190. Springer, Cham (2019). https://doi.org/10.1007/978-3-030-33617-2_20
12. Ohlson, J.A.: Financial ratios and the probabilistic prediction of bankruptcy. J. Account. Res. **18**(1), 109–131 (1980)
13. Otter, N., Porter, M.A., Tillmann, U., Grindrod, P., Harrington, H.A.: A roadmap for the computation of persistent homology. EPJ Data Sci. **6**, 1–38 (2017)
14. Oudot, S.Y.: Persistence Theory: From Quiver Representations to Data Analysis, vol. 209. American Mathematical Society (2017)
15. Riihimäki, H., Chachólski, W., Theorell, J., Hillert, J., Ramanujam, R.: A topological data analysis based classification method for multiple measurements. BMC Bioinform. **21**(1), 1–18 (2020)
16. Wiginton, J.: A note on the comparison of logit and discriminant models of consumer credit behavior. J. Fin. Quant. Anal. **15**(3), 757–770 (1980)
17. Wu, C., Hargreaves, C.: Topological machine learning for mixed numeric and categorical data. Int. J. Artif. Intell. Tools **30**, 1–18 (2021)

A Comparative Study of LAD, CNN and DNN for Detecting Intrusions

Sneha Chauhan[1,2]([✉]), Loreen Mahmoud[1], Sugata Gangopadhyay[1], and Aditi Kar Gangopadhyay[3]([✉])

[1] Department of Computer Science and Engineering, IIT Roorkee, Roorkee, India
{schauhan1,loreen_fm,sugata.gangopadhyay}@cs.iitr.ac.in
[2] Department of Computer Science and Engineering,
NIT Uttarakhand, Srinagar, India
[3] Department of Mathematics, IIT Roorkee, Roorkee, India
aditi.gangopadhyay@ma.iitr.ac.in

Abstract. In recent years, with the growth of the Internet and network devices, a significant amount of information has been exposed to the attackers and intruders. Due to vulnerabilities in the system, the adversaries plan new ways of network intrusions. Many Intrusion Detection Systems (IDSs) are developed to protect the networks from malicious attacks and ensure reliability and availability within the organizations. IDSs built using various machine learning and data mining techniques are effective in detecting attacks. However, their performance decreases with an increase in the size of data. In this paper, we focus on developing an IDS model using Logical Analysis of Data (LAD). It is a supervised learning technique where patterns are generated using partially defined Boolean functions (pdBf), which can detect attacks based on certain features of the data. We compare the performance of LAD model with Deep Neural Network (DNN) and Convolutional Neural Network (CNN) IDS models. UNSW-NB15 and CSE-CIC-IDS2018 datasets are used for training and testing our proposed model. The results show that the performance of LAD model is competitive to CNN, DNN and other existing IDS models based on accuracy, precision, recall and F1 score.

Keywords: Logical analysis of data · Intrusion detection system · Deep neural network · Convolutional neural network

1 Introduction

With rapid development of cloud, big data and IoT domains, their usage has also become unavoidable. A lot of information is available online related to finance, government, military etc. which could be used in a wrong way by the hackers or adversaries. Information security and data privacy has now become a major challenge. Attackers not only launch denial of service attack but also inject virus, worms, spams to exploit the vulnerabilities of the system leading to exposure of the sensitive information stored in machines [19]. Thus, unauthorized

© The Author(s), under exclusive license to Springer Nature Switzerland AG 2022
H. Yin et al. (Eds.): IDEAL 2022, LNCS 13756, pp. 443–455, 2022.
https://doi.org/10.1007/978-3-031-21753-1_43

access, modification, destruction of data needs to be protected and confidentiality, integrity and availability of data must be considered in all the systems. Denning in her paper [9] highlighted the use of audit records to generate a rule based matching system. The system monitors the standard operations and any deviation from normal behaviour was reported.

Intrusion Detection System (IDS) is a crucial part of the computer network security that monitors the network traffic to detect intrusions. An IDS works on two different types of detection mechanism: anomaly detection and misuse detection. Anomaly detection based IDS looks for suspicious activities by matching the attacker's behaviour with that of normal user's behaviour. Any deviation from the normal behaviour triggers an alarm to alert the system administrator. Misuse detection or signature based IDS monitors the traffic by matching the known attack patterns or signatures with the traffic flowing into the network. This IDS has high detection rate if the attack is known but new or unknown attacks are difficult to detect compared to anomaly based IDS [16].

Recently, various machine learning and deep learning models are widely used in developing IDS as they provide efficient capability for intrusion detection. But there are certain complications like complex decision making, accuracy, competence which make IDS unsuitable for deployment in a real time scenario with large volume of traffic. In order to train an IDS model, large data and high computing power is required which is not always possible. Therefore, an IDS that can detect abnormal activity in real time or near real time is required. IDS can detect new attacks only when it has knowledge of normal activities.

To develop an IDS system, it is necessary to generate signatures or patterns that can identify normal and abnormal activities. This task can be achieved by the application of Logical Analysis of Data (LAD). LAD was introduced by Peter L. Hammer [11] as a technique to generate patterns or rules from the past data which helps in the classification of the new observations [5,6]. The main aim of LAD is to find patterns that can help in understanding a phenomena, what factors are responsible for it and find out ways which may affect its development.

Our research focuses on using LAD to detect intrusions by monitoring the network traffic. It labels the new observations as normal or attack based on the patterns generated by LAD. We have applied LAD to design an anomaly based IDS that can detect intrusions in near real time and with less computing power. Through the patterns, we can suggest the features which are more vulnerable for attack. UNSW-NB15 and CSE-CIC-IDS2018 datasets has been used for our experiment. CNN and DNN based IDS models are implemented for comparing the performance of our LAD based IDS model.

DNN is an Artificial Neural Network that consists of multiple hidden layers, and has an ability to model the non-linear relationships, as in each layer a non-linear transformation is executed on the layer input and finally we get a statistical model after finishing the process of training. CNN is a type of the Feed Forward Neural Networks that has an important part which is the feature extractor, for each convolution layer their is a convolution kernel which learns to reach the most convenient weights and it reduces the connections number and the over fitting possibility [18].

The research paper has following sections. Section 2 provides the summary of related works. Section 3 describes the implementation of LAD and the proposed IDS model. Experimental results and discussions are presented in Sect. 4. Finally, the paper is concluded in Sect. 5.

2   Related Work

In [13], the authors used Genetic Algorithm along with Logistic Regression as a feature selection method over the UNSW-NB15 and KDDCup99 dataset. Weka tool was used for simulation. Only 20 features out of 42 features in UNSW dataset were selected using Genetic Algorithm - Logistic Regression in conjunction with Decision Tree (DT) classifier. The accuracy achieved was 81.42%. A feature extraction technique based on Particle Swarm Optimization (PSO), Firefly Optimization (FO), Genetic Algorithm (GA) and Grey Wolf Optimization (GO) were applied on UNSW-NB15 dataset iteratively by the researchers to obtain an optimal subset of features [2]. The classification process was performed using SVM and J48 Tree based models. The SVM model attained an accuracy of 90.119% whereas J48 model achieved an accuracy of 90.48%. The authors in [12] explored the use of XGBoost algorithm for feature selection process along with various machine learning techniques such as ANN, DT, kNN, LR and SVM to develop an IDS model. The experiments were carried out on UNSW-NB15 dataset. 19 features were selected by the XGBoost algorithm. The accuracy obtained by ANN, kNN, DT, LR and SVM over the test data were 84.39%, 83.18%, 88.13%, 79.59% and 62.42% respectively. In [10], an ensemble model using logistics regression, decision trees, and gradient boosting was developed to built an IDS model. It was tested on CSE-CIC-IDS2018 dataset achieving an accuracy of 98.8%. Spearman's rank correlation coefficient was used for feature selection. In [17], the authors provide a 1D CNN-based IDS, they used NSL-KDD dataset as a benchmark, and they found that 1D-CNN based IDS detects malicious traffic better than SVM and Naive Bayes techniques, and they got accuracy of 99.56%. A CNN model was proposed by the authors in [15], and they have transformed the CICIDS2018 dataset (which was used as a benchmark) into images, and they got high accuracy of detecting intrusions using CNN comparing to other techniques including RNN.

3   Logical Analysis of Data and Its Implementation

LAD is a technique that is used to analyse and classify the data by applying combinatorial and optimization models. A brief description of LAD is given here. Consider a dataset Ω having two disjoint subsets Ω^+ and Ω^- consisting of positive and negative observations respectively. Each observation is an $n + 1$-dimensional vector denoting n features with the last bit for the class label. We can represent this as $\Omega^+ \cup \Omega^- \subseteq \{0,1\}^n$. All the observations in Ω can be represented by a partially defined Boolean function (pdBf) ϕ which maps each observation to either 0 or 1 i.e. $\Omega \rightarrow \{0,1\}$. In LAD, the goal is to determine an extension f of ϕ such that it can classify the unknown observations correctly. But this is practically unachievable, hence an approximate extension f' of f is obtained based on certain criteria [5,7]. LAD involves 4 steps that are explained in the below sections.

3.1   Binarization

In real life problems, the data is complex and has numerical as well as categorical values. In order to apply LAD on such data, the real data has to be converted into binary format. This process is termed as binarization.

In many fields of human endeavor, binarization of ordered attributes is a frequent technique. Blood pressure, body temperature and a variety of other medical characteristics are examples of what is referred to as normal or abnormal based on their values lying within or out of a certain range. Similar to these examples, binarization of numerical attributes is carried out by using cut-points. Several binary attributes are associated to each numeric attribute corresponding to the number of cut-points obtained for that attribute. The binary variable fall into two categories: level and interval variable. A level variable is formed when the value of the variable α is compared to a cut-point c_p as shown in Eq. 1. An interval variable is obtained by Eq. 2.

$$b(\alpha, c_p) = \begin{cases} 1, \text{if } \alpha \geq c_p. \\ 0, \text{otherwise} \end{cases} \tag{1}$$

$$b(\alpha, c_p^i, c_p^j) = \begin{cases} 1, \text{if } c_p^i \leq \alpha < c_p^j. \\ 0, \text{otherwise} \end{cases} \tag{2}$$

For generation of cut-points, the values of the numeric variable (say X) is sorted in descending order. Every two consecutive values (X_i, X_{i+1} where $1 \leq i \leq n$) are chosen such that both values belong to different class labels. This is done to maintain the disjoint nature of the positive and negative subset of the dataset. The cut-point is given by the average $\frac{X_i + X_{i+1}}{2}$.

When dealing with large datasets, a variable may produce large number of cut-points which increase the number of level and interval variables in the binarized output. This may lead to high memory usage. Also a variable having large number of cut-points may not influence the classification result so much. Hence, a threshold is set up after several rounds of analysis. A feature having cut-points less than the threshold is considered for level and interval variables otherwise it is discarded [8].

3.2   Support Set Generation

In LAD, feature selection is termed as support set generation. It is possible that the dataset created by binarizing the numerical attributes contains a number of redundant attributes that need to be removed. Support Set refers to a set of attributes which are able to keep the set of positive and negative observations disjoint, meaning that each observation is either positive or negative. A support set is said to be irredundant or minimal if removing an attribute from it causes a collision of positive and negative observations, i.e. there exists an observation in the dataset that is both positive and negative. A set covering problem can be used to solve the difficulty of getting a minimal support set for a binary dataset. The papers [3] and [6] describe few algorithms to solve the set cover problem. To obtain a minimal support set for our binary dataset, we have used Mutual Information Greedy Algorithm in our implementation which was proposed in [3]. In this algorithm, each attribute is given a score based on its entropy. The attribute having minimum best score is added to the support set.

3.3 Pattern Generation

Let us first have a look at some Boolean terminologies which are used in pattern generation process. A literal is a Boolean variable or its complement in Boolean algebra. A group of literals is called a term. The degree of a term is the number of literals present in it. A point $p \in \{0, 1\}^n$ is covered by a term T only if $T(p) = 1$. A positive (negative) pattern is a term that covers only positive (negative) observations. Pattern generation is an important phase in LAD technique as patterns give an exact interpretation of the problem. Patterns are Boolean attributes combined in a specific order to classify the positive and negative observations. Pattern generation can be performed using two approaches: Top down and Bottom up approach. In top down approach, the characteristic term of each observation is considered as a pattern i.e. combination of all attributes that makes that observation a positive one. It is possible that even after removing few literals, the term still covers observations and hence it is a pattern. Thus, in top down scenario, literals are removed one by one from the characteristic term until it becomes a prime pattern. Bottom up approach starts with one literal that covers some observations. If only positive observations are covered then it becomes a pattern, else we add literals to it one by one until a pattern is formed. We have used a breadth first enumerative technique which involves both approaches. Firstly the patterns of small degrees are generated from bottom up approach and then the remaining observations are covered by patterns generated by top down approach [5].

3.4 Classifier Design and Validation

The patterns obtained are then converted into rules which are used to build the classifier. Each pattern is written using if else statements. The if else structure is combined to form a classifier. If the condition given by a positive pattern is satisfied by the new observation then it is labelled as positive else it is labelled as negative. Positive and negative patterns can be combined to build a hybrid classifier.

4   Performance Evaluation

4.1   Datasets

There are many datasets available for validation of IDS models such as KDD-CUP'99, NSL-KDD, UNSW-NB15 and CSE-CIC-IDS2018 datasets. For our experiment, we have chosen UNSW-NB15 and CSE-CIC-IDS2018 datasets. These datasets have modern cyberattack types included, also they are the most recent datasets in IDS domain [1, 20].

The UNSW-NB15 dataset consists of instances that belong to nine different types of cyberattacks namely Analysis, DoS, Exploits, Backdoors, Shellcode, Reconnaissance, Fuzzers, Worms and Generic and rest are normal instances. Since our model is for binary classification, we have combined all the attack instances together and labeled them as Attack or 1 and normal instances are labeled as Normal or 0. The dataset has total of 42 features in which 3 are categorical features and 39 are numeric features. The training set of the dataset has 82,332 observations and the testing set includes 175,341 observations.

CSE-CIC-IDS2018 dataset was created to train anomaly based intrusion detection models to detect attacks in network traffic. The dataset is divided into 10 csv files with nine files having 79 features and one file has 83 features. For our experiment we have removed the Timestamp and IP addresses features so that each file has 78 features. The files having same type of attack are merged so that each file has different type of attacks. The results are given as per the attack categories. For LAD based IDS, we labeled the attack observations as 1 and normal observations as 0.

4.2   Experimental Setup

LAD-Based IDS. Before applying the LAD technique, the UNSW-NB15 dataset is preprocessed. The duplicate rows in the dataset are removed and the resulting training set has 53,951 observations and 43 attributes including the class label. For the CSE-CIC-IDS2018 dataset, each file contains rows indicating a specific type of attack. So we considered two different sets of data to build two LAD classifiers. File1 consists of FTP/SSH BruteForce and file5 has DDOS attack observations along with normal observations. These files are treated as two datasets and two LAD classifiers are built using these two data files. We did this to evaluate how our classifier responds to unknown attacks and also to limit the size of the training dataset. We have used only 30% of the data to build the LAD classifier. For better understanding of results, we will denote the LAD classifier developed using File1 data as LAD1 and other LAD classifier as LAD2.

During the binarization phase, the number of binary variables produced is directly proportional to number of cut-points. Hence, with large number of cut-points for a feature, the number of binary variables produced will also increase exponentially. This will result in increase of memory requirement and also such a feature may not have a great impact on classification of observations. So in order to limit the binary variables, we only generated level variables. Also a feature having cut-points more than 175 is discarded. We came to this conclusion after thorough analysis using the training data.

Another criteria which is incorporated in our implementation is the support of a pattern. Support of a positive (negative) pattern, is the number of positive (negative) observations covered by that pattern, provided that it does not cover any negative (positive) observation. For our experiment, we chose support to be 50 and 1 by empirical analysis for UNSW-NB15 and CSE-CIC-IDS2018 dataset.

As we have discussed earlier, IDS can be categorized as signature based and anomaly based IDS. Our model can be used in both ways i.e. if we generate positive patterns then it can work as signature based and if we generate negative patterns then it is anomaly based IDS. We can also design a hybrid classifier using both positive and negative patterns. In our study, attack observations are considered positive and normal observation as negative. Here, we have produced negative patterns observing the normal observations.

Deep Learning-Based IDS. We have proposed two deep learning-based IDS approaches- Convolutional Neural Network(CNN) based IDS and the Deep Neural Network(DNN) based IDS. We use different architectures of CNN and DNN for each dataset. For implementation we used Keras library and TensorFlow as backend for the development of the model and trained it on Google Colaboratory.

Preprocessing: Before using the data we have searched for any missing data or infinite values in order to get rid of them or replace them, the non numerical data has been transformed into numerical, the most effective features have been selected. We have scaled the data, as we have determined the maximum and minimum possible value of data in each column according to its data type, and finally the data has been split for training, validation, and testing sets.

CNN Model for CSE-CIC-IDS2018

- We used three 1D convolution layers, each layer has 256 filters.
- Kernel size of 5 in the first layer, and 7 for the other two layers.
- Each convolution layer is followed by batch normalization layer and a 1D max pooling layer, then we have a fully connected layer which ends with only one neuron (as we have only two classes).
- ReLu activation function is used in all the layers of the fully connected layer except the final one in which sigmoid activation function is used.
- Adam optimizer has been used with learning rate of 0.0001.
- Batch size of 128, and 500 epochs of training were there with early stopping with patience of 10 to stop the training when there is no improvement in training.
- The dataset was split to 10% for testing and 90% for training, and 20% out of the training set was taken as a validation set.
- The correlation between features and f_classif technique were used to select the features that affect the results the most, and 23 features have been selected.

DNN Model for CSE-CIC-IDS2018

- Three fully connected Dense layers were used, the first one has 600 neurons, the second one has 450 neurons, and the last one has 100 neurons.
- These layers are followed by a layer that contains only one neuron as we will classify that data into two classes.
- ReLu activation function is used in all the layers of the fully connected layer except the final one in which sigmoid activation function is used.
- The rest of the parameters (the optimizer, learning rate, Batch size, number of training epochs, early stopping patience, splitting of the data, feature selecting, and number of features selected) are the same that have been mentioned for CNN model.

CNN Model for UNSW-NB15

- Two 1D convolution layers with 150 filters and kernel size of 3 for each one.
- The second convolution layer is followed by 1D max pooling layer, and then there is a fully connected layer ends with one neuron
- Sigmoid activation function was used in the last layer, while we used ReLu activation function in all other layers.
- Adam optimizer is used with learning rate of 0.001.
- 100 of training epochs and early stopping with patience of 10.
- We used 20% of the training data set as a validation set, while training set and testing set are available in separate files from the beginning.
- Only 5 features have been selected out of the dataset features after computing the correlation between each feature and the two labels.

DNN Model for UNSW-NB15

- Three fully connected layers with 1024 neurons for the first one, 768 neurons for the second one, and 512 neurons for the third one, then we have the output layer with one neuron.
- Sigmoid activation function was used in the last layer, while we used ReLu activation function in all other layers.
- The optimizer, learning rate, number of training epochs, data splitting, and features selected, are chosen the same as in CNN model that we used for UNSW-NB15 dataset.

4.3   Experimental Results for UNSW-NB15 Dataset

The outcome of each step of LAD on UNSW-NB15 dataset is given below:

1. Binarization: During this step, 661 binary variables are generated. Now the size of the binary dataset becomes 53,951 × 662 including the class label.
2. Support Set Generation: 51 features are selected based on their discriminating power to classify the observations.
3. Pattern Generation: 17 negative patterns of degree 2, 3 and 4 are obtained during the pattern generation process.

4. Classifier Design: The rule based classifier is built using these 17 patterns. This LAD based IDS is validated using the UNSW-NB15 testing dataset.

The Performance has been evaluated using the following parameters: Accuracy, Precision, Recall and F1-score.

The results of the IDS model based on LAD, CNN and DNN for the UNSW-NB15 dataset are presented in the Table 1. The accuracy achieved by our LAD model is 93.16% which is higher than the other two models. The 99.75% recall shows that most of the attacks are correctly classified by LAD based IDS model. The precision of CNN and DNN is better than LAD model which indicates that there were less false positives in both the models.

Table 1. Result of IDS models on UNSW-NB15 Dataset.

IDS	Accuracy	Precision	Recall	F1-score
LAD-IDS	0.9316	0.9105	0.9975	0.9520
CNN-IDS	0.8589	0.9777	0.8116	0.8867
DNN-IDS	0.8644	0.9689	0.8273	0.8925

For performance analysis of our IDS models on UNSW-NB15 dataset, we have compared the results with the existing classifiers described in the literature review. It can be observed from the Table 2 that LAD based IDS has a better performance than the other methods. The LAD based IDS has an accuracy of 93.16% which is higher as compared to the classifiers like SVM (90.11%), J48 (90.48%) [2], GAA-ADS (91.8%), GAA-principal components (92.8%) [22] etc. The detection rate or recall of our model is 99.75% which is higher than other

Table 2. Comparison results of implemented classifiers with other ML methods on UNSW-NB15 Dataset.

Methods	Accuracy (%)	Precision (%)	Recall (%)	F1-score (%)
SVM [2]	90.11	–	–	–
J48 [2]	90.48	–	–	–
DT [12]	90.85	80.33	98.38	88.45
GAA-ADS-original features ($K = 10$) [22]	91.8	–	91.0	–
GAA-principal-components ($K = 10$) [22]	92.8	–	91.3	–
GALR-DT [13]	81.42	–	–	–
DT [21]	85.56	–	–	–
CNN-IDS	85.89	97.77	81.16	88.67
DNN-IDS	86.44	96.89	82.73	89.25
LAD-IDS	93.16	91.05	99.75	95.20

techniques given in [12]. The Precision and F1-Score of our model is 91.05% and 95.20% respectively.

4.4   Experimental Results for CSE-CIC-IDS2018 Dataset

Two LAD classifiers (LAD1 and LAD2) were built on the two CSE-CIC-IDS2018 dataset files. LAD1 classifier was developed using 1117 binary variables, 17 support set features and 4 negative (normal) patterns. Similarly, LAD2 classifier used 505 binary variables, 11 features in the support set and 6 negative (normal) patterns to develop its rules. We built the LAD1 classifier on the SSH/FTP Bruteforce attack. This classifier is tested on the other files except file 4 and 5. This classifier performed well to detect Bruteforce and DOS attack but could not detect Infiltration, SQL Injection and Bot attacks. The accuracy achieved for Bruteforce attack, DoS attacks-Golden Eye and Slowloris, DoS attack-Hulk and SlowHTTPTest are 99.93%, 99.63% and 99.47%. For the said attacks, the recall values are 100%, 94.29% and 99.78% which shows that most of the attacks were correctly classified. The LAD2 classifier was trained using the dataset of File 5 and it is tested using the data of file 4 and 5. This classifier detected the DDOS attack-HOIC and LOIC-UDP with an accuracy of 99.99% and has a recall of 100%. But the DDOS attacks-LOIC HTTP are not classified correctly as the recall and precision value is very low. All the results are tabulated in the Table 3.

From the Table 3 we can conclude that the LAD based IDS model could not generalise well. We tried to develop LAD classifier using only a part of the dataset in file 1 and 5. It was tested on the entire dataset. It was able to detect DoS attacks and Bruteforce attacks of other dataset files but LAD based IDS did not give good results for DDoS, infiltration and Bot attacks as these were not present in the training set. For some category of Bruteforce attack, DDOS attack and for all DOS attacks the classifier performed similar to CNN and DNN classifiers that were trained on the entire dataset. The DNN and CNN models have been able to detect all types of DoS and DDoS and few Bruteforce attacks very well with a recall of 100% approximately. But these models could not detect infiltration and SQL injection attacks.

Table 4 shows the overall performance of all the three classifiers on the entire dataset. The LAD based IDS has an accuracy of 96% which is less compared to Ensemble learning and HRCNN but here we have used very less amount of training dataset. Also the LAD-IDS is able to detect some of the DOS and Bruteforce attacks which were not present in the training set. The DNN-IDS has higher precision of 99.45% than the other methods.

Table 3. Results of IDS models on CSE-CIC-IDS2018 Dataset.

Attack	DNN-IDS				CNN-IDS				LAD-IDS			
	Accuracy	Precision	F1score	Recall	Accuracy	Precision	F1score	Recall	Accuracy	Precision	F1score	Recall
SSH-Bruteforce FTP-BruteForce	0.9996	0.9997	0.9996	0.9996	0.9996	0.9997	0.9997	0.9996	0.9993	0.9982	0.9991	1.0
DoS attacks-GoldenEye DoS attacks-Slowloris	0.9995	1.0	0.9994	0.9988	0.9993	0.9994	0.9991	0.9988	0.9963	0.9839	0.9630	0.9429
DoS attacks-Hulk DoS attacks-SlowHTTPTest	0.9992	0.9990	0.9995	1.0	0.9992	0.9990	0.9995	1.0	0.9947	0.9931	0.9954	0.9978
DDoS attacks-LOIC-HTTP	0.9985	0.9986	0.9985	0.9984	0.9983	0.9967	0.9983	1.0	0.8768	0.0848	0.077	0.0715
DDOS attack-HOIC DDOS attack-LOIC-UDP	0.9994	0.9993	0.9996	1.0	0.9994	0.9993	0.9996	1.0	0.9999	0.9999	0.9999	1.0
Brute Force -Web Brute Force -XSS SQL Injection	0.9970	1.0	0.6447	0.4757	0.9970	1.0	0.6447	0.4757	0.9990	0.0	0.0	0.0
Infiltration	0.6875	0.6895	0.8108	0.9839	0.6898	0.6914	0.8118	0.9832	0.8845	0.1515	0.0115	0.0060
BOT	0.9926	0.9996	0.9946	0.9896	0.9923	0.9996	0.9943	0.9891	0.8236	0.0	0.0	0.0

Table 4. Comparison of proposed IDS models with existing classifiers on CSE-CIC-IDS2018.

Classifier	Accuracy (%)	Precision (%)	Recall (%)
Deep Learning IDS [4]	94.40	–	–
Ensemble [10]	98.8	98.80	97.10
HRCNN [14]	97.6	96.33	97.12
CNN-IDS	95.11	95	95.4
DNN-IDS	95.1	99.45	91.69
LAD-IDS	96.02	98.95	80.79

5   Conclusion

In this paper, we have proposed 3 different approaches of intrusion detection system: LAD-IDS, CNN-IDS and DNN-IDS. The UNSW-NB15 and CSE-CIC-IDS2018 datasets are used as benchmark. For the UNSW-NB15 dataset, the LAD-IDS performs better than the other methods. The LAD-IDS has misclassified more normal instances as compared to CNN and DNN based IDS, that may be reduced by using a hybrid LAD classifier developed using both positive and negative patterns. We can also conclude that LAD performs well with small size dataset. For the CSE-CIC-IDS2018 dataset LAD-IDS is trained on a smaller dataset and tested for new unknown attacks. The results show that LAD-IDS is able to detect DOS, DDOS and SSH/FTP Bruteforce attacks effectively, while we can notice that CNN and DNN can detect all the types of attacks with very good accuracy except the infiltration attack because of the shortage of infiltration records. The positive aspect of LAD is that it can run on low performance hardware. Also LAD results are interpretable compared to other techniques. We are getting promising results for LAD-IDS, which can be further improved if we can reduce dataset size using undersampling techniques so that LAD can process it efficiently. Further, the overall performance of CNN and DNN may be improved by increasing the number of attacks' records that are less in the original dataset using oversampling techniques.

Authors' Contributions. Conceptualization and Supervision: Sugata Gangopadhyay and Aditi Kar Gangopadhyay. Investigation, Software Implementation of LAD: Sneha Chauhan. Investigation, Software Implementation of CNN and DNN: Loreen Mahmoud. Writing Original Draft: Sneha Chauhan and Loreen Mahmoud.

References

1. CSE-CIC-IDS-2018 dataset from university of newbrunswick. https://www.unb.ca/cic/datasets/ids-2018.html
2. Almomani, O.: A feature selection model for network intrusion detection system based on PSO, GWO, FFA and GA algorithms. Symmetry **12**(6), 1046 (2020). https://doi.org/10.3390/sym12061046
3. Almuallim, H., Dietterich, T.G.: Learning Boolean concepts in the presence of many irrelevant features. Artif. Intell. **69**(1–2), 279–305 (1994). https://doi.org/10.1016/0004-3702(94)90084-1
4. Basnet, R., Shash, R., Johnson, C., Walgren, L., Doleck, T.: Towards detecting and classifying network intrusion traffic using deep learning frameworks (2019). https://doi.org/10.22667/JISIS.2019.11.30.001
5. Boros, E., Hammer, P.L., Ibaraki, T., Kogan, A., Mayoraz, E., Muchnik, I.: An implementation of logical analysis of data. IEEE Trans. Knowl. Data Eng. **12**(2), 292–306 (2000). https://doi.org/10.1109/69.842268
6. Crama, Y., Hammer, P.L., Ibaraki, T.: Cause-effect relationships and partially defined Boolean functions. Ann. Oper. Res. **16**(1), 299–325 (1988). https://doi.org/10.1007/BF02283750
7. Das, T.K., Adepu, S., Zhou, J.: Anomaly detection in industrial control systems using logical analysis of data. Comput. Secur. **96**, 101935 (2020). https://doi.org/10.1016/j.cose.2020.101935
8. Das, T.K., Gangopadhyay, S., Zhou, J.: SSIDS: semi-supervised intrusion detection system by extending the logical analysis of data. CoRR (2020). arXiv:2007.10608
9. Denning, D.: An intrusion-detection model. IEEE Trans. Softw. Eng. **SE-13**(2), 222–232 (1987). https://doi.org/10.1109/TSE.1987.232894
10. Fitni, Q.R.S., Ramli, K.: Implementation of ensemble learning and feature selection for performance improvements in anomaly-based intrusion detection systems. In: 2020 IEEE International Conference on Industry 4.0, Artificial Intelligence, and Communications Technology (IAICT), pp. 118–124 (2020)
11. Hammer, P.L.: Partially defined Boolean functions and cause-effect relationships. In: Proceedings of the International Conference on Multi-Attribute Decision Making via OR-Based Expert Systems. University of Passau (1986)
12. Kasongo, S.M., Sun, Y.: Performance analysis of intrusion detection systems using a feature selection method on the UNSW-NB15 dataset. J. Big Data **7**(1), 1–20 (2020). https://doi.org/10.1186/s40537-020-00379-6
13. Khammassi, C., Krichen, S.: A GA-LR wrapper approach for feature selection in network intrusion detection. Comput. Secur. **70**, 255–277 (2017). https://doi.org/10.1016/j.cose.2017.06.005
14. Khan, M.A.: HCRNNIDS: hybrid convolutional recurrent neural network-based network intrusion detection system. Processes **9**(5) (2021). https://doi.org/10.3390/pr9050834

15. Kim, J., Shin, Y., Choi, E.: An intrusion detection model based on a convolutional neural network. J. Multimedia Inf. Syst. **6**, 165–172 (2019). https://doi.org/10. 33851/JMIS.2019.6.4.165

16. Kim, K., Aminanto, M.E., Tanuwidjaja, H.C.: Network Intrusion Detection Using Deep Learning: A Feature Learning Approach. Springer, Singapore (2018). https:// doi.org/10.1007/978-981-13-1444-5

17. Krishnan, A., Mithra, S.: A modified 1D-CNN based network intrusion detection system. Int. J. Res. Eng. Sci. Manag. **4**(6), 291–294 (2021)

18. Mahmoud, L., Praveen, R.: Artificial neural networks for detecting intrusions: a survey. In: Fifth International Conference on Research in Computational Intelligence and Communication Networks (ICRCICN), pp. 41–48. IEEE (2020)

19. Mishra, P., Varadharajan, V., Tupakula, U., Pilli, E.S.: A detailed investigation and analysis of using machine learning techniques for intrusion detection. IEEE Commun. Surv. Tutor. **21**(1), 686–728 (2018). https://doi.org/10.1109/COMST. 2018.2847722

20. Moustafa, N., Slay, J.: UNSW-NB15: a comprehensive data set for network intrusion detection systems (UNSW-NB15 network data set). In: 2015 Military Communications and Information Systems Conference (MilCIS), pp. 1–6. IEEE (2015). https://doi.org/10.1109/MilCIS.2015.7348942

21. Moustafa, N., Slay, J.: The evaluation of network anomaly detection systems: statistical analysis of the UNSW-NB15 data set and the comparison with the KDD99 data set. Inf. Secur. J. Glob. Perspect. **25**(1–3), 18–31 (2016). https://doi.org/10. 1080/19393555.2015.1125974

22. Moustafa, N., Slay, J., Creech, G.: Novel geometric area analysis technique for anomaly detection using trapezoidal area estimation on large-scale networks. IEEE Trans. Big Data **5**(4), 481–494 (2019). https://doi.org/10.1109/TBDATA.2017. 2715166

Effective Prevention of Semantic Drift in Continual Deep Learning

Khouloud Saadi[1] and Muhammad Taimoor Khan[2]($^\boxtimes$) ⓘ

[1] Islamabad, Pakistan
[2] National University of Computer and Emerging Sciences, Islamabad, Pakistan
`taimoor.muhammad@gmail.com`

Abstract. Lifelong machine learning or continual learning models attempt to learn incrementally by accumulating knowledge across a sequence of tasks. Therefore, these models learn better and faster. They are used in various intelligent systems that have to interact with humans or any dynamic environment. Dynamically expandable networks are continual deep learning models that allow its architecture to expand with a sequence of tasks. The model retains knowledge from the previous tasks that results in high performance on newer tasks. The existing models use Minkowski distance measures to separate nodes of the current network, resulting in higher catastrophic forgetting. These measures are susceptible to high dimensional sparse vectors, resulting in sub-optimum performance. We propose ang-DEN, as a dynamically expanding continual learning architecture that use angular distance metric. It addresses semantic drift through better separation of nodes achieving 97% average accuracy with an improvement of 1.3% across all tasks on MNIST variant datasets.

Keywords: Continual deep learning · Catastrophic forgetting · Semantic drift · ang-DEN

1 Introduction

Machine learning has become more and more important in today's industries showing considerable achievements in various domains [3]. The current dominant machine learning approach consists of training a model in isolation that generally require more data. In isolated learning, the model is taken out of service and retrained each time there is new data available. For many tasks the data and computational resources are not available in sufficient supply to train complex models. Moreover, in case of multiple tasks, the isolated learning mechanism does not share the common features across those tasks and rather each time learn it from scratch.

K. Saadi—Independent Researcher.
This work was supported by Fatima Al-Fihri predoctoral fellowship program (https:// fatimafellowship.com/).

© The Author(s), under exclusive license to Springer Nature Switzerland AG 2022
H. Yin et al. (Eds.): IDEAL 2022, LNCS 13756, pp. 456–464, 2022.
https://doi.org/10.1007/978-3-031-21753-1_44

In contrast, lifelong machine learning (LML) or continual learning model attempts to mimic the way we humans learn [7]. It learns incrementally by building on top of the knowledge acquired from the tasks performed in the past. The knowledge across multiple tasks accumulates to enhance learning of newer tasks without dropping its performance on the previously performed tasks. In order to perform a task T_{N+1}, the model accumulates its underlying dataset D_{N+1} and relevant knowledge from the previous T_1 to T_N tasks. The updated model is available for the application to perform any of the T_1 to T_{N+1} tasks. Generally, basic features are expected to be common across multiple tasks where more overlap is expected as the number of tasks increase. Continual learning empowers to have reduced computational cost, and have lesser dependency on the quantity and quality of underlying data for a task. Most importantly it helps to avoid relearning repeated patterns from scratch in performing different tasks [15].

Continual learning overshadows the existing machine learning paradigms by supporting incremental task adaptation with lesser semantic drift. The phenomenon of forgetting previously acquired information in order to learn new information is termed as catastrophic forgetting or semantic drift. In continual learning, the learning is persistent and long term. It facilitates in learning newer tasks with fewer training samples and without degrading performance on the previously learned tasks [18]. Continual learning models attempt to counter the semantic drift and rather utilize the previous information to facilitate learning newer tasks. Its significance increases many-folds in case of tasks with limited or noisy data.

Dynamically expandable networks (DEN) are expandable networks that are more often used with deep learning. They are also called continual deep learning models. The architecture of these models is revisited after performing each task in order to add the new information into the network [17]. Therefore, the model architecture updates and expands after each task. This is a memoryless approach which suits deep neural networks better due to the implicit nature of feature representation within the weights of the network. The other type of LML is memory-based where an external memory base is maintained. In memoryless approaches, the existing network architecture is separated into freeze, partially regularize and duplicate categories. The new information from the current task is accommodated into these categories based on its similarity with the previously learned tasks. In case of high similarity with the previous tasks, majority of the nodes are frozen otherwise duplicated. When the process of expanding for the current task completes, the model is capable of performing all the previous and current tasks. The existing research in continual deep learning use magnitude based distance measures as objective function to evaluate semantic drift or catastrophic forgetting in a node represented by a vector of its parameters [19]. In this paper we argue that measures such as euclidean and manhattan distance, as a family of Minkowski metrics, are not the optimal choice for this task. They are sensitive to scale and sparse where all the points with high dimensional feature space become uniformly distant [1]. We have used angular distance between the weight vectors of a node i.e., W_1 and W_2 to measure the semantic drift in a node as $angDis(W_1, W_2) = \arccos(\frac{W_1.W_2}{||W_1|| \times ||W_2||}) \times (180/\pi)$. With known intervals angular distance is better for applying thresholds to separate nodes to freeze, partially regularize and duplicate. Our proposed ang-DEN outperforms existing approaches by maintaining higher average accuracy above 97% with an improvement of 1.3% on standard MNIST variant datasets.

2   Related Work

Continual learning systems can function in a dynamic open environment where they can discover new problems which form new tasks. These new tasks need to be learned in real-time without forgetting the old tasks [3,4]. This is called learning on the job as it attempts to achieve artificial general intelligence by optimizing across multiple tasks [8]. The problem of catastrophic forgetting or semantic drift can be observed in transfer learning based models where once the weights of the model are updated according to the target task, the performance on the source task is expected to degrade. Continual learning addresses this problem by modifying the model architecture such that it learns newer tasks but also retains its learning of the previous tasks [19]. However, updating the network architecture more with a task helps achieve better results on the current task but degrades performance on the previously learned tasks. Thus, the model has high plasticity but low stability. On the other hand, lesser update in the network architecture with a task helps achieve high performance on the previous tasks but compromises learning of new tasks. Thus, the model has high stability but low plasticity. Finding a balance between the two has been a challenge is known as the stability-plasticity dilemma [4].

The continual learning approaches can be separated into two types i.e., memoryless and memory-bases. The memoryless approaches involve dynamic architectures that grows and expands with each task in a sequence. Progressive networks freeze the already trained nodes, while adding new nodes with each incoming task [17]. They often suffer from rapid growth incurring higher computational and storage costs. Dynamically expanding network duplicates the nodes that need to be retrained [19]. It allows partial regularization of th already trained nodes to make them usable across multiple tasks, which results in slower network growth. The existing approaches make use of magnitude distances from the Minkowski measures that do not provide better separation of nodes having large-sparse vectors [19]. These measures are too sensitive to feature scales, data dimensionality and their sparsity. Despite the optimal thresholds used, these measures present all points uniformly distant. Memory-based continual learning models maintain an extra memory unit to store refined information across all previously performed tasks. In case of episodic memory, several examples are stored from past tasks and used on minimizing negative backward transfer while learning new tasks [14]. In memory-based LML approaches exhaust memory very quickly and has limited learning capability. Therefore, the memoryless approach is more common with continual deep learning. However, in memoryless continual deep learning, nodes of the trained network are grouped using distance measure which may not be a better choice with sparse, high-dimensional vectors.

In order to understand the limitation of distance measure, consider an example. Let W_1 be the weight of a given node in the task t_1 and W_2 be the expected updated value of the corresponding nodes to accommodate task t_2. If $W_1 = [10, 10, 10, 10, 10]$ and $W_2 = [13, 13, 13, 13, 13]$, then the $dist(W_1, W_2) = \|(W_1 - W_2)\|_2^2 = (10-13)^2 + (10-13)^2 + (10-13)^2 + (10-13)^2 + (10-13)^2 = 45$. It is a high value but in actual the weights have increased in the same direction indicating a similar trend. It indicates that the distance measures are not providing optimum separation of nodes.

3   Proposed Approach

Our proposed ang-DEN approach provides a better balance between the plasticity and stability of the model. It is an improvement on the hybrid memoryless approach for continual deep learning that allows the model to regularize and expand as in [19]. It may also be referred to as the partial expansion method that ensures maximum utilization of the learned weights and expanding dynamically for newer high-level features. Since the sequence of tasks are expected to be independent of each other, therefore, the model requires an automated mechanism to decide which nodes to freeze, regularize under constraint or expand through duplication. The model's performance heavily relies on objective function used to categorize nodes into three categories. There are mainly three steps to learn a task. First, a selective retraining estimate similarity of the current task with previous tasks [19]. Second, the drift in each node with respect to the current task is evaluated using threshold in order to decide whether to freeze, partially regularize or duplicate the node [11, 19, 20]. Third, the freeze, partially regularize or duplicate action is performed on the task.

Algorithm 1. ang-DEN Algorithm

Input: Task Sequence $T = T_1, T_2, ..., T_N$, Threshold σ
Output: W_N

1: **for** $T_i \in T$ **do**
2:     **if** W_i is ϕ **then**
3:         Train the model as first task, W_1 using Eq. (1)
4:     **else**
5:         Load the trained model, W_{i-1}
6:         Generate W_i by retraining for T_i
7:         **for** all hidden nodes j **do**
8:             $\rho_j^{T_i} = \arccos(\frac{W_{j,T_i} . W_{j,T_{i-1}}}{||W_{j,T_i}|| \times ||W_{j,T_{i-1}}||}) \times (180/\pi)$
9:             **if** $\rho_j^{T_i} > \sigma$ **then**
10:                Duplicate the node j as j' and train for W_i
11:            **end if**
12:        **end for**
13:    **end if**
14: **end for**

Working of the ang-DEN is represented in Algorithm 1. It takes a sequence of N tasks i.e., from T_1 to T_N along with the threshold σ for separating nodes.

The output of the model is the weight matrix W_N that supports all N tasks. The lines 1 to 14 represent the processing of tasks in a sequence as a continuous process. Line 2 checks if this is the first task that the model is performing. In that case, the model is trained from scratch with the default architecture on line 3, using equation [19];

$$\min_{W_{T_i}} \mathcal{L}(W_1, T_1) + \mu \|W_1\|_1 \qquad (1)$$

where W_1 are the network parameters after training on task T_1, while μ is the regularization parameter. In lines 4 to 13, the model updates its architecture for each task from T_2 till T_N. It first loads the parameter weights till the last task i.e., W_{i-1} on line 5. At this stage the model is not trained for the current task T_i, however, the parameter weights W_i is generated using a small fraction of the dataset for task T_i using the same architecture, in line 6. It helps to evaluate the relevance between the current task and the previously performed tasks along with underlying semantic drift with respect to each node. The semantic drift is evaluated in lines 7–12 as the difference between W_i and W_{i-1} using the following equation;

$$angDis(W_i, W_{i-1}) = \arccos(\frac{\sum_{j=1}^{J} W_{1,j} \times W_{i-1,j}}{\sum_{j=1}^{J} W_{1,j} \times \sum_{j=1}^{J} W_{i-1,j}}) \times \frac{180}{\pi} \qquad (2)$$

as angular distance between all the corresponding nodes from $j = 1$ to J. line 8 provides $\rho_j^{T_i}$ as the semantic drift in the node j for the task T_i. In other words, the value of node j will deviate by a margin of $\rho_j^{T_1}$ from its previous value if the new task is to be accommodated within the same architecture. In lines 9 to 11, this value is compared with the drift threshold σ. A value above the threshold indicates that changing the parameters of node j for the current task T_i would result in degrading the model's performance on the previously learnt tasks. Therefore, such nodes are duplicated in line 10 as j'. It allows the model's architecture to expand and fit in the specific features of all learned tasks. The nodes with almost no semantic drift can be restrained from update to save computational cost. The updated architecture is fully trained for the task T_i to get W_i as;

$$\min_{W_i} \mathcal{L}(W_i; T_i) + \lambda \|W_i - W_{i-1}\|_2^2 \qquad (3)$$

The network weight matrix W_i is obtained after modifying the model architecture and fully training the dataset for the task T_i. The weight matrix W_{i-1} represents the parameter weights for the tasks T_1 till T_{i-1}. The λ represents the penalization parameter that controls the distance between W_i and W_{i-1}. To prevent catastrophic forgetting, a higher value for λ enforce W_i to be close in value to W_{i-1}. But if the sequential tasks are largely independent of each other, by using this regularization method, the model settles for a sub-optimal results. Therefore, the duplicate of such nodes are created and added adjacent to the corresponding nodes allowing better generalization across all tasks. Moreover,

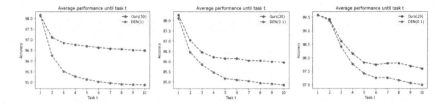

Fig. 1. Average accuracy on Permuted-MNIST and Permuted-Fashion-MNIST and Rotated-MNIST across 10 tasks in a sequence.

in the high dimensional feature space of deep neural networks, the angular distance provides better separation of nodes as compared to the magnitude based distance measures. It is robust against feature scales, sparsity and dimensionality, therefore, results in higher average accuracy across all tasks.

4 Experimental Setup and Results

All models have a default architecture of two hidden layers having 312 and 128 units respectively. The other settings include ReLu as activation function, Adam as optimiser while learning rate, batch-size and iterations are set to 0.001, 256 and 4300, respectively. In the experiments, three datasets i.e., Permuted-MNIST, Permuted-Fashion-MNIST, and Rotated-MNIST are used as in existing literature. Each dataset is treated as a sequence of 10 one-vs-rest tasks [10].

In Fig. 1, ang-DEN has a higher performance compared to the DEN approach respectively on the three datasets, Permuted-Fashion-MNIST, Permuted-MNIST, and Rotated-MNIST. The above graphs are generated with the split thresholds that gave the best results among the three datasets for both the DEN approach and ang-DEN. On the Permuted-MNIST dataset, ang-DEN has a higher accuracy compared to the DEN approach starting from task 2. On the Permuted-Fashion-MNIST dataset, our approach outperforms the DEN approach for all the tasks. On the Rotated-MNIST dataset, our approach outperforms the DEN approach starting from task 2.

We have compared ang-DEN with the latest continual deep learning models on the Permuted-MNIST dataset, as shown in Fig. 2 and Table 2. Figure 2 shows that it has outperformed all the existing models by a large margin i.e., dynamically expendable networks (DEN), elastic weight consolidation (EWC), regularized weight consolidation (RWC) and regularize, expand and compress (REC) [20]. Among these models, RWC [13], REC [20], and EWC [9] are fixed-size deep learning methods based on regularization to prevent semantic drift. While, DEN [19] and ang-DEN are partial expansion approaches. Thus, the group of the fixed-size approaches has low model complexity compared to the expandable networks approaches. When a single network was trained with the same setting for all the tasks, it resulted in an accuracy of 17.4%. The proposed approach has improved accuracy by 0.55% and 1.36% as compared to second and third best models, respectively. The ang-DEN has better objective function

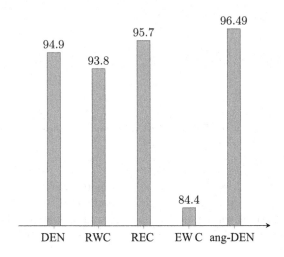

Fig. 2. Average accuracy on Permuted-MNIST dataset after 10 tasks.

Table 1. Model sizes on permuted-MNIST dataset.

Model	Dataset	Model size
ang-DEN	P-MNIST	0.27 M
DEN	P-MNIST	0.18 M
EWC	P-MNIST	0.01 M
REC	P-MNIST	0.01 M
RWC	P-MNIST	0.01 M

as angular distance, for separating nodes and adopts well to the other datasets as well.

We further report the size of each model on each of the datasets. Table 1 shows sizes of the model on permuted-MNIST dataset. The model size is represented by the total number of parameters of the final model over all tasks. There is a difference of 10K parameters between DEN and ang-DEN approaches. DEN has a low computational complexity compared to ang-DEN. In ang-DEN model, a better metric to compute similarity between nodes achieves higher accuracy while using more or higher parameters.

DEN and ang-DEN belong to the category of partially expanding models that duplicate some nodes while partially regularize others. These models are well suited for tasks with higher diversity. The other approaches that do not expand, settle for sub-optimal solutions. In Table 2, we report the competitive results of DEN and ang-DEN on the three datasets using the previously detailed setup. The hyper-parameter λ is evaluated for different values, where only the ones resulting in highest accuracy are reported. This threshold is responsible for the stability-plasticity balance in dynamically expanding networks. The λ values presented are scalars and angles for DEN and ang-DEN, respectively.

Table 2. Average accuracy and parameters used with (λ) threshold.

Model	Dataset	λ	Avg-accuracy	Parameters
DEN	P-MNIST	1	95.19	401018
ang-DEN		30	**96.49**	543394
DEN	F-MNIST	0.1	84.58	394555
ang-DEN		30	**86.01**	535378
DEN	R-MNIST	0.1	96.99	372342
ang-DEN		20	**97.59**	496651

ang-DEN approach has outperformed DEN on all the three datasets. For the Permuted-MNIST dataset, ang-DEN with a hyper-parameter λ of 30° yields the best performance. On the other hand, The best performance of the DEN is obtained with a hyper-parameter λ of 1. The ang-DEN model yields on average 1.3% improvement in accuracy as compared to the DEN approach. For the Permuted-Fashion-MNIST dataset, our approach with a hyper-parameter λ of 30° has the best performance over the rest and yields on average 1.43% higher accuracy in comparison to the DEN approach. For the Rotated-MNIST dataset, our approach with a hyper-parameter λ of 20° yields the best performance again and gives on average 1.17% accuracy improvement compared to the DEN approach. Table 2 shows that overall ang-DEN achieves around 1.3% improvement in accuracy as compared to the prior method.

5   Conclusion

Continual deep learning models allows incremental addition of tasks through expanding the network architecture. Angular distance measures provide a better mechanism for categorizing the network nodes based on their similarities or dissimilarities with the current task. It proves to be more effective towards balancing retention of past information while learning new information. ang-DEN has outperformed the existing approaches across the standard continual deep learning datasets by providing a balance between the stability and plasticity of the model.

In future, some external memory can be used to plug and play the already trained layers for different tasks based on their similarity. A better mechanism for finding thresholds and trisecting of nodes can further improve the utilization of past information into newer tasks.

References

1. Aggarwal, C.C., Hinneburg, A., Keim, D.A.: On the surprising behavior of distance metrics in high dimensional space. In: Van den Bussche, J., Vianu, V. (eds.) ICDT 2001. LNCS, vol. 1973, pp. 420–434. Springer, Heidelberg (2001). https://doi.org/10.1007/3-540-44503-X_27

2. Caruana, R.: Multitask learning. Mach. Learn. **28**(1), 41–75 (1997)
3. Chen, Z., Liu, B.: Lifelong machine learning. Synth. Lect. Artif. Intell. Mach. Learn. **12**(3), 1–207 (2018)
4. d'Autume, C.d.M., Ruder, S., Kong, L., Yogatama, D.: Episodic memory in lifelong language learning. arXiv preprint arXiv:1906.01076 (2019)
5. Fernando, C., et al.: PathNet: evolution channels gradient descent in super neural networks. arXiv preprint arXiv:1701.08734 (2017)
6. Fontenla-Romero, Ó., Guijarro-Berdiñas, B., Martinez-Rego, D., Pérez-Sánchez, B., Peteiro-Barral, D.: Online machine learning. In: Efficiency and Scalability Methods for Computational Intellect, pp. 27–54. IGI Global (2013)
7. Khan, M.T., Azam, N., Khalid, S., Yao, J.: A three-way approach for learning rules in automatic knowledge-based topic models. Int. J. Approx. Reason. **82**, 210–226 (2017)
8. Khan, M.T., Durrani, M., Khalid, S., Aziz, F.: Online knowledge-based model for big data topic extraction. In: Computational Intelligence and Neuroscience 2016 (2016)
9. Kirkpatrick, J., et al.: Overcoming catastrophic forgetting in neural networks. Proc. Natl. Acad. Sci. **114**(13), 3521–3526 (2017)
10. LeCun, Y., Cortes, C., Burges, C.: MNIST Handwritten Digit Database, vol. 2. ATT Labs (2010). http://yann.lecun.com/exdb/mnist
11. Li, X., Zhou, Y., Wu, T., Socher, R., Xiong, C.: Learn to grow: a continual structure learning framework for overcoming catastrophic forgetting. In: International Conference on Machine Learning, pp. 3925–3934. PMLR (2019)
12. Li, Z., Hoiem, D.: Learning without forgetting. IEEE Trans. Pattern Anal. Mach. Intell. **40**(12), 2935–2947 (2017)
13. Liu, X., Masana, M., Herranz, L., Van de Weijer, J., Lopez, A.M., Bagdanov, A.D.: Rotate your networks: better weight consolidation and less catastrophic forgetting. In: 2018 24th International Conference on Pattern Recognition (ICPR), pp. 2262–2268. IEEE (2018)
14. Lopez-Paz, D., Ranzato, M.: Gradient episodic memory for continual learning. Adv. Neural. Inf. Process. Syst. **30**, 6467–6476 (2017)
15. Parisi, G.I., Kemker, R., Part, J.L., Kanan, C., Wermter, S.: Continual lifelong learning with neural networks: A review. Neural Netw. **113**, 54–71 (2019)
16. Rebuffi, S.A., Kolesnikov, A., Sperl, G., Lampert, C.H.: ICARL: incremental classifier and representation learning. In: Proceedings of the IEEE Conference on Computer Vision and Pattern Recognition, pp. 2001–2010 (2017)
17. Rusu, A.A., et al.: Progressive neural networks. arXiv preprint arXiv:1606.04671 (2016)
18. Wu, Y., et al.: Large scale incremental learning. In: Proceedings of the IEEE/CVF Conference on Computer Vision and Pattern Recognition, pp. 374–382 (2019)
19. Yoon, J., Yang, E., Lee, J., Hwang, S.J.: Lifelong learning with dynamically expandable networks. arXiv preprint arXiv:1708.01547 (2017)
20. Zhang, J., Zhang, J., Ghosh, S., Li, D., Zhu, J., Zhang, H., Wang, Y.: Regularize, expand and compress: nonexpansive continual learning. In: Proceedings of the IEEE/CVF Winter Conference on Applications of Computer Vision, pp. 854–862 (2020)

A Sequence to Sequence Long Short-Term Memory Network for Footwear Sales Forecasting

Luís Santos[1], Luís Miguel Matos[4], Luís Ferreira[1], Pedro Alves[2], Mário Viana[3], André Pilastri[1], and Paulo Cortez[4(✉)]

[1] EPMQ, CCG ZGDV Institute, Guimarães, Portugal
{luis.santos,luis.ferreira,andre.pilastri}@ccg.pt
[2] KYAIA - SOLUÇÕES INFORMÁTICAS, LDA, Guimarães, Portugal
pedro.alves@ksi.pt
[3] OVERCUBE, S.A., Guimarães, Portugal
marioviana@m360.com.br
[4] ALGORITMI Research Centre/LASI, Department of Information Systems, University of Minho, Guimarães, Portugal
{luis.matos,pcortez}@dsi.uminho.pt

Abstract. Footwear sales forecasting is a critical task for supporting product managerial decisions, such as the management of footwear stocks and production levels. In this paper, we explore a recently proposed Sequence to Sequence (Seq2Seq) Long Short-Term Memory (LSTM) deep learning architecture for multi-step ahead footwear sales Time Series Forecasting (TSF). The analyzed Seq2Seq LSTM neural network is compared with two popular TSF methods, namely ARIMA and Prophet. Using real-world data from a Portuguese footwear company, several computational experiments were held. Focusing on daily sales, we analyze data recently collected during a 3-year period (2019–2021) and related with seven types of products (e.g., sandals). The evaluation assumed a robust and realistic rolling window scheme that considers 28 training and testing iterations, each related with one week of multi-step ahead predictions. Overall, competitive predictions were obtained by the proposed LSTM model, resulting in a weekly Normalized Mean Absolute Error (NMAE) that ranges from 5% to 11%.

Keywords: Time series forecasting · ARIMA · Prophet · Deep learning

1 Introduction

The accurate projection of sales is a crucial element to support inventory management systems. Indeed, inventory excesses or shortages are often the result of expectations not being met, which have an immediate detrimental effect on the company's profitability and competitiveness.

This work focuses on a Portuguese footwear company online store that sells several footwear products (e.g., Shoes, Sneakers) across Europe. By adopting

© The Author(s), under exclusive license to Springer Nature Switzerland AG 2022
H. Yin et al. (Eds.): IDEAL 2022, LNCS 13756, pp. 465–473, 2022.
https://doi.org/10.1007/978-3-031-21753-1_45

a Time Series Forecasting (TSF) approach, there is a potential to better support the inventory management system of the analyzed company (e.g., reducing stock costs). Moreover, in recent years there has been a growing interest in the usage of deep learning architectures to perform TSF tasks, such as Long Short-Term Memory (LSTM) networks. In this paper, we focus on a recently proposed Sequence (Seq2Seq) LSTM neural network [5], aiming to predict footwear sales. The adopted LSTM is compared with two popular Time Series Forecasting (TSF) methods, namely the Auto-Regressive Integrated Moving Average (ARIMA) and Prophet. Using real-world daily data from the analyzed company, related with a three-year period and seven types of products (e.g., sandals), we execute a realistic rolling window evaluation scheme that considers from 1 to $H = 7$ daily ahead predictions (up to one week) and several training and testing evaluations.

2   Related Work

TSF is widely adopted in several application domains (e.g., Finance, Production, Sales). Due to its importance, there is a wide range of methods that can perform TSF tasks. While proposed in the 1970s, the AutoRegressive Integrated Moving Average (ARIMA) methodology [1] is still a popular approach, including its Seasonal ARIMA (SARIMA) variant, which is capable of modeling trend and seasonal effect. Another popular method is Prophet, which was introduced by the Facebook company in 2018 [6]. Prophet is based on a additive regression approach that is capable of modeling trends, seasonal patterns and even outliers associated with weekends or specific events (e.g., holidays) [4]. In recent years, there has also been a growing interest in using deep learning methods, including LSTM recurrent networks, to perform TSF tasks [5,8,13].

In terms of the sales application domain, several studies have adopted TSF approaches. For instance, ARIMA was used to perform retail forecasting projections related with five categories of women's footwear [10]. Also, the SARIMA was adopted in [7] to forecast the monthly number of car sales in South Africa. In [3], ARIMA and Support Vector Machines were explored to predict the foot traffic of a retail store. As for the LSTM model, it was proposed in [13] to forecast sales of 66 different products over 45 weeks. In another study, the LSTM neural architecture obtained the best predictive results when predicting pharmaceutical sales, outperforming the Prophet model [8].

Recently, we have compared three distinct LSTM architectures to perform multi-step ahead predictions of a different application domain: movements of industrial workers [5]. The best results were obtained by a Seq2Seq LSTM, which assumes a encoder-decoder architecture. Following on these good results, in this paper we explore the Seq2Seq LSTM architecture for footwear sales prediction, comparing it with two other TSF method (ARIMA and Prophet).

3   Materials and Methods

3.1   Footwear Sales Data

The data was extracted from a database concerning the business management and sales of fashion products made available at the *Overcube* online platform over a three-year time span, from 2019 to 2021. The raw dataset was comprised of thirty-one product features (e.g., identification, size). The online platform also contains sales data related with 78 countries (e.g., Germany, Spain, Portugal).

In this work, we analyze seven footwear product categories (accessories, ankle boots, boots, sandals, shoe care, shoes and sneakers) that were sold at Portugal. The sales were aggregated into daily values. Then, the empty entries (corresponding to no sales) were replaced by zero values, resulting in a total of 1,095 daily records for all seven time series. Figure 1 shows an example of the accessories sales time series (left graph) and its respective autocorrelation values (right plot).

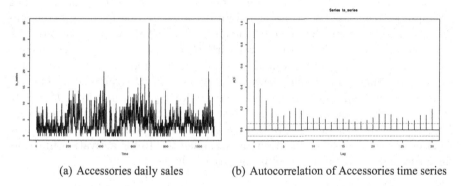

(a) Accessories daily sales   (b) Autocorrelation of Accessories time series

Fig. 1. Daily accessories sales time series (right, x-axis denotes the time period, y-axis the number of daily sales) and its autocorrelation values (left, x-axis denotes the time lags and y-axis the autocorrelations).

3.2 Modeling

An autoregressive time series model predicts a value for current time t based on past observations $\hat{y}_t = f(y_{t-k_1}, ..., y_{t-k_I})$, where f is the forecasting function and the k_i values denote the past time lags (assuming a total of I inputs). Daily series often present a weekly seasonal period ($K = 7$), which is the case of the analyzed footwear sales series. For example, the right of Fig. 1 shows higher autocorrelation values for the multiples of $K = 7$, confirming a weekly seasonal pattern. Therefore, the three time series methods (LSTM, ARIMA and Prophet) are set to built weekly forecasts, by computing from $h = 1$ up to $h = H = 7$ multi-step ahead daily predictions, where h denotes the ahead time in which the forecast is executed and $H = 7$ denote the maximum horizon value (up to one week).

The LSTM is a popular deep learning model to process temporal data. In effect, the LSTM is a special type of Recurrent Neural Network (RNN) that resolves the "short-term memory" problem by using a mechanism of gates that regulate the flow of information [2]. This type of RNN is capable of learning the order dependence in a sequential prediction problem, including time series. The

LSTM architecture is composed of a set of cells, where each cell includes three control gates: the "forget gate" that defines whether the information is relevant (1) or not (0); the "memory gate" that decides the new data that should be stored and modified in the cell; and the "output gate" that controls what is produced in each cell [11].

There are several LSTM variants. In this work, we explore a recently proposed Seq2Seq LSTM architecture that outperformed other two LSTM variants (standard LSTM and a stacked LSTM with two hidden layers) when predicting the shoulder angular movements of industrial workers [5]. Since the model is capable of memorizing temporal sequences, the Seq2Seq LSTM is only fed with one time lag y_{t-1}. The Seq2Seq LSTM model assumes a sequence to sequence architecture (encoder-decoder) that includes one model for reading and encoding the input sequence and a second model for decoding and performing predictions (Fig. 2). The Seq2Seq LSTM model includes two LSTM layers (each one with $L = 100$ cells), one repeat vector layer with $H = 7$ nodes, to repeat the incoming inputs for up to H times, and one time distributed layer, to process the output from the LSTM hidden layer and generate the H sequential output values. Thus, the Seq2Seq LSTM model is capable of performing from $H = 1$ to $H = 7$ multi-step ahead predictions when fed with the last known series value. The deep learning model was trained with the Adam optimizer, using the Mean Squared Error (MSE) loss function and assumed the ReLU activation function, since the sales data can not contain negative values.

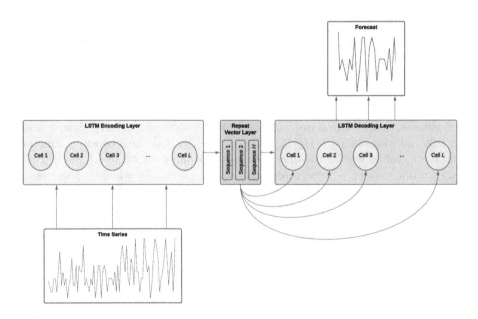

Fig. 2. Architecture of the Seq2Seq LSTM model.

As a baseline comparison, the proposed Seq2Seq LSTM model is compared with two popular seasonal TSF methods [6]: SARIMA and Prophet. Both methods were set using a weekly time period ($K = 7$).

3.3   Evaluation

The TSF methods are evaluated using a robust rolling window validation [9, 12], which simulates a real usage of a forecasting model through time, with several training and test updates (Fig. 3). The training set assumes a fixed window length with W examples. In the first iteration ($u = 1$), the model is adjusted to a training window with the W oldest values, and then predicts up to T test ahead predictions (in this paper, $T = H = 7$). Next, in the second iteration ($u = 2$), the training data is updated with S newer examples, allowing to fit a TSF model with W values and perform newer T predictions, and so on. In total, this produces $U = \frac{D-(W+T)}{S}$ model updates, where D is the data length (number of time series examples). After consulting with the company experts, we opted to use the realistic values of $W = 881$ and $H = 7$ with a $S = 7$, thus resulting in $U = 28$ model fitting and testing updates for each TSF method (LSTM, SARIMA and Prophet).

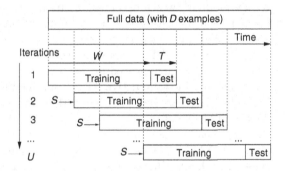

Fig. 3. Schematic of the rolling window procedure.

To measure the quality of the forecasts, we adopt the Normalized Mean Absolute Error (NMAE). The NMAE measure normalizes the popular Mean Absolute Error (MAE) by the output target range on the test set, thus resulting in a scale-independent percentage that is easier to understand and is represented by the following formula [9]:

$$NMAE = \frac{MAE}{(y_{max}-y_{min})} \qquad (1)$$

where the y_{max} and the y_{min} represent the highest and the lowest values of the target series. Note that the lower NMAE values, the more accurate is the model (the perfect value is 0%).

For each footwear product, the adopted rolling window scheme produces $U = 28$ sets of predictions, each set with $H = 7$ multi-step ahead forecasts (from $h = 1$ to $h = 7$). Following the procedure used in [3], we compute vertical and horizontal forecasting errors. The former assumes the NMAE value for a fixed h value, thus computed with $U = 28$ observations. The latter, termed here as Multi-Step Ahead Aggregation (MSAA), works by first computing the NMAE value using all multi-step ahead forecasts (7 values, from $h = 1$ to $h = 7$ for a targeted week). Then, the distinct $U = 28$ NMAE values (one for each tested week) are aggregated by computing the median values, which is less sensitive to outliers when compared with the average.

4   Results

All computational experiments were conducted using code written in the Python programming language. The Seq2Seq LSTM neural network implementation is based on the `tensorflow` API structure[1]. Each rolling window iteration assumes an initial Seq2Seq LSTM network with random generated connection weights. Then, the Adam optimizer is run, assuming an early stopping procedure (10% of the most recent training data is used as the validation set) and a maximum of 100 learning epochs. Regarding the SARIMA, we adopted the `auto.arima` function of the `pmdarima` python module[2], which executes an automatic SARIMA model identification and fit for each u-th iteration of the rolling window scheme. As for the Prophet, the `prophet` python package[3] was adopted. Similarly to the previous methods, the Prophet model is refit using the training data available in each iteration of the rolling window procedure.

Table 1 summarizes the obtained predictive results obtained by the ARIMA, Prophet and LSTM models. In terms of the vertical NMAE values (for a fixed h), the proposed Seq2Seq LSTM model produces competitive results. In effect, for almost all $h \in \{1, 2, ..., 7\}$ ahead ranges and footwear products, the LSTM model obtains the lowest vertical NMAE values. In 49 NMAE comparisons, there are only four cases in which Prophet produces similar (sneakers and $h = 7$) or better results (e.g., ankle boots and $h = 3$). Moreover, the average vertical NMAE values, considering all products (last row of Table 1) favors the deep learning method when compared with the two baseline methods (SARIMA and Prophet). Turning to the horizontal multi-step ahead forecasting results (column **MSAA**), the computed median NMAE values also position the Seq2Seq LSTM model at the first place, producing the lowest values for all seven footwear products, ranging from 4.8% (accessories) to 10.6% (sneakers). On average (considering all products), the NMAE median value is 7.07%, which is around 2% points better when compared with SARIMA and Prophet.

For demonstration purposes, Fig. 4 shows the last rolling window iteration weekly ahead forecasts ($U = 28$, h from 1 to 7) for the sandals product sales.

---

[1] https://www.tensorflow.org/.

[2] https://alkaline-ml.com/pmdarima/0.9.0/index.html.

[3] https://github.com/facebook/prophet.

Table 1. Predictive results (NMAE values for a fixed h, in %; median NMAE values for MSAA, in %; the best values are highlighted by using a **boldface** font).

Model	Product	$h = 1$	$h = 2$	$h = 3$	$h = 4$	$h = 5$	$h = 6$	$h = 7$	MSAA
SARIMA	Accessories	5.94	4.85	6.19	6.04	6.35	6.09	4.11	6.32
	Ankle Boots	5.71	3.78	3.87	6.13	4.62	4.04	4.35	6.33
	Boots	5.23	3.36	5.50	5.21	4.71	6.85	4.27	7.34
	Sandals	9.59	9.27	10.75	12.22	12.68	13.24	13.49	10.75
	Shoe Care	9.94	6.76	7.77	10.36	11.03	8.52	7.94	10.42
	Shoes	10.70	7.44	10.67	9.56	8.41	8.85	8.34	8.95
	Sneakers	10.54	12.04	11.46	10.09	16.45	12.98	14.04	13.71
Average		8.24	6.79	8.03	8.52	9.18	8.65	8.08	9.12
Prophet	Accessories	6.88	5.84	5.93	6.88	7.29	6.64	5.49	7.09
	Ankle Boots	5.79	3.89	**3.76**	5.87	4.78	3.39	4.31	5.82
	Boots	5.46	3.87	6.05	5.61	5.52	6.43	4.70	7.45
	Sandals	9.69	10.40	9.41	10.75	10.22	10.26	12.42	8.90
	Shoe Care	9.54	8.09	7.59	11.53	12.23	8.75	**6.82**	10.70
	Shoes	11.14	8.03	10.71	9.61	8.67	**6.39**	8.08	9.31
	Sneakers	11.87	13.24	11.18	11.37	17.35	13.67	**13.90**	14.11
Average		8.62	7.62	7.80	8.80	9.44	7.93	7.96	9.05
Seq2Seq LSTM	Accessories	**4.54**	**3.97**	**4.91**	**5.02**	**4.42**	**4.63**	**3.36**	**4.80**
	Ankle Boots	**4.94**	**3.49**	4.02	**5.37**	**4.31**	**3.10**	**2.60**	**4.75**
	Boots	**3.70**	**2.43**	**3.30**	**2.95**	**2.58**	4.32	**2.96**	**5.40**
	Sandals	**9.04**	**8.11**	**7.63**	**8.72**	**9.92**	**9.05**	10.53	**7.79**
	Shoe Care	**8.46**	**6.06**	**6.80**	**9.66**	**9.93**	**7.67**	6.92	**9.18**
	Shoes	**8.34**	**5.61**	**8.08**	**7.33**	**6.44**	7.08	**6.49**	**6.95**
	Sneakers	**9.61**	**6.55**	10.20	**6.88**	11.40	12.10	**13.90**	10.64
Average		**6.95**	**5.17**	**6.42**	**6.56**	**7.00**	**6.85**	**6.68**	**7.07**

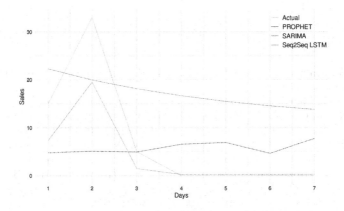

Fig. 4. Last rolling window weekly forecasts for the three TSF methods (x-axis denotes the horizon, y-axis the number of daily sales) for the sandals product.

In this example, the best forecasts are provided by the Seq2Seq LSTM model, which correctly identify a peak for $h = 2$ and also detect a stagnation "no sale" period from $h = 4$ to $h = 7$.

5   Conclusions

In this paper, we explored a recently proposed Seq2Seq LSTM deep learning method to forecast the number of daily sales of seven products from a Portuguese footwear company. Using real-world data, collected during a three-year time period and a robust rolling windows evaluation scheme, several computational experiments were conducted, comparing the proposed deep learning model with two popular TSF methods (ARIMA and Prophet). Overall, when considering both a fixed ahead forecast (vertical analysis) and multi-step ahead weekly forecasts (horizontal analysis), competitive results were obtained by the proposed Seq2Seq LSTM method. In particular, interestingly low multi-step ahead NMAE values were achieved by the deep learning method, ranging from 5% (accessories sales) to 11% (sneakers sales).

The obtained results were shown to the footwear company, which found them interesting and valuable to support the management of its footwear stocks. Indeed, in future work, we intend to deploy the proposed Seq2Seq LSTM model into a friendly decision support system, aiming to provide forecasting insights for the inventory management system currently adopted by the company. We also plan to adapt the forecasting methods to other time scales (e.g., monthly), explore multivariate forecasting methods and explore other recent forecasting methods (e.g., Temporal Convolutional Network).

Acknowledgments. This work was financed by the project "GreenShoes 4.0 - Calçado, Marroquinaria e Tecnologias Avançadas de Materiais, Equipamentos e Software" (N° POCI-01-0247-FEDER-046082), supported by COMPETE 2020, under the PORTUGAL 2020 Partnership Agreement, through the European Regional Development Fund (ERDF).

References

1. Box, G., Jenkins, G.: Time Series Analysis: Forecasting and Control. Holden Day, San Francisco (1976)
2. Chopra, S., Meindl, P.: Supply chain management. strategy, planning & operation. In: Boersch, C., Elschen, R. (eds.) Das Summa Summarum Des Management, pp. 265–275. Springer, Cham (2007). https://doi.org/10.1007/978-3-8349-9320-5_22
3. Cortez, P., Matos, L.M., Pereira, P.J., Santos, N., Duque, D.: Forecasting store foot traffic using facial recognition, time series and support vector machines. In: Graña, M., López-Guede, J.M., Etxaniz, O., Herrero, Á., Quintián, H., Corchado, E. (eds.) SOCO/CISIS/ICEUTE -2016. AISC, vol. 527, pp. 267–276. Springer, Cham (2017). https://doi.org/10.1007/978-3-319-47364-2_26
4. Ensafi, Y., Amin, S.H., Zhang, G., Shah, B.: Time-series forecasting of seasonal items sales using machine learning-a comparative analysis. Int. J. Inf. Manag. Data Insights **2**(1):100058 (2022)

5. Fernandes, C., et al.: A deep learning approach to prevent problematic movements of industrial workers based on inertial sensors. In International Joint Conference on Neural Networks, IJCNN 2022, Padua, Italy, 18–23 July 2022. IEEE (2022)

6. Hyndman, R.J., Athanasopoulos, G.: Forecasting: Principles and Practice, 3rd edn. O Texts (2021)

7. Makatjane, K., Moroke, N.: Comparative study of holt-winters triple exponential smoothing and seasonal arima: forecasting short term seasonal car sales in south africa. Risk Gov. Control Financ. Markets Institutions **6** (2016)

8. Meng, J., Yang, X., Yang, C., Liu, Y.: Comparative analysis of prophet and LSTM model in drug sales forecasting. **1910** (2021). IOP Publishing

9. Oliveira, N., Cortez, P., Areal, N.: The impact of microblogging data for stock market prediction: Using twitter to predict returns, volatility, trading volume and survey sentiment indices. Expert Syst. Appl. **73**, 125–144 (2017)

10. Ramos, P., Santos, N., Rebelo, R.: Performance of state space and ARIMa models for consumer retail sales forecasting. Rob. Comput.-Integrat. Manuf. **34**, 151–163 (2015)

11. Siami-Namini, S., Tavakoli, N., Namin, A.S.: A comparison of ARIMA and LSTM in forecasting time series. In: Arif Wani, M., Kantardzic, M.M., Mouchaweh, M.S., Gama, J., Lughofer, E. (eds.) 17th IEEE International Conference on Machine Learning and Applications, ICMLA 2018, Orlando, FL, USA, 17–20 December 2018, pp. 1394–1401. IEEE (2018)

12. Tashman, L.J.: Out-of-sample tests of forecasting accuracy: an analysis and review. Int. Forecast. J. **16**(4), 437–450 (2000)

13. Yu, Q., Wang, K., Strandhagen, J.O., Wang, Y.: Application of long short-term memory neural network to sales forecasting in retail—a case study. In: Wang, K., Wang, Y., Strandhagen, J.O., Yu, T. (eds.) IWAMA 2017. LNEE, vol. 451, pp. 11–17. Springer, Singapore (2018). https://doi.org/10.1007/978-981-10-5768-7_2

EfficientNet Architecture Family Analysis on Railway Track Defects

Jon Rengel[1(✉)], Matilde Santos[2], and Ravi Pandit[3]

[1] ETSI. Informática, Universidad Nacional de Educación a Distancia, 28040 Madrid, Spain
jrengel10@alumno.uned.es
[2] Institute of Knowledge Technology, University Complutense of Madrid, 28040 Madrid, Spain
msantos@ucm.es
[3] School of Aerospace, Transport and Manufacturing,
Cranfield University, Cranfield 43 0AL, MK, UK

Abstract. Keeping railway tracks in correct conditions is of paramount importance to ensure adequate performance of trains, and to avoid any issues that might end up in accidents. The development of computational technology together with artificial vision techniques have boosted the use of video cameras for inspecting railway tracks. Some deep learning techniques have been applied to the day, being Convolutional Neural Networks the most popular ones. This paper presents the use of EfficientNet architecture on a railway track fault detection as a novel technique in this field. The paper compares B0 to B7 network families as well as studies the effect of input image resolution on a small dataset. On the one hand, results show that image resolution has an impact on validation accuracy and in fact, specific networks developed for particular image size are not always the best option for that image. B7 family has outperformed the rest of networks, reaching validation accuracy of 89.1%, while B2 scored 87.5% of accuracy, being the best in computational cost and performance ratio. In average, B7 and B2 have been proved to be the better solutions, with 84.6% and 84.2% accuracy respectively.

Keywords: EfficientNet · Convolutional neural network · Failure detection · Railway

1 Introduction

Since the invention of the first steam engine, railways have become one of the most popular and safe land transportation for goods and people. According to the International Energy Agency (IEA), the global passenger rail activity is expected to double (116%) from present levels in 2050. With the increasing transport demand and current capacity bottlenecks, it is estimated that the required rail networks need to be extended by more than 430 000 track-kilometers through to 2050, a 27% increase from 2016 [1].

The large number of already existing rail tracks together with the expected increase of it, highlights the need of regularly inspecting the railway track for maintaining safe and reliable train operations. Historically, these regular inspections were manually performed

© The Author(s), under exclusive license to Springer Nature Switzerland AG 2022
H. Yin et al. (Eds.): IDEAL 2022, LNCS 13756, pp. 474–481, 2022.
https://doi.org/10.1007/978-3-031-21753-1_46

using a railway cart. However, more effective methods are required to ensure a better performance and to avoid human errors.

The development in computational technology and artificial vision techniques have lately boosted the use of video cameras for inspecting the tracks [2, 3]. Among the available techniques, image classification with deep learning methods have become popular, especially those involving Convolutional Neural Networks (CNN) [4–6].

From all the available architectures, EfficientNet has been selected for their outstanding performance on this images, and for the novelty of their application on this field. This neural network family presents the highest accuracy rate compared to the rest of the state-of-the-art techniques, as shown in Fig. 1 [7]. Moreover, it presents a relatively small number of parameters and FLOPS, meaning less computational cost to process the model.

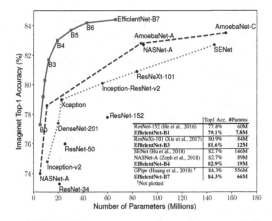

Fig. 1. Comparison of EfficientNet accuracy vs other architectures on ImageNet dataset [7].

Since the development of this architecture, it has been widely applied in multiple disciplines. The literature shows that it is especially popular on medicine, where it has been used for all sort of analyses, from detecting brain tumours [8] to Covid-19 diagnosis [9–12].

The structure of the paper is as follows. Section 2 details the methodology and all the scenarios used for experiments. In Sect. 3, a brief description of the dataset and image pre-processing is presented. Obtained results are discussed in Sect. 4. The paper conclusions and future work are summarized in Sect. 5.

2   Methodology

EfficientNet architecture is formed of eight different families, from B0 to B7. The higher the number of the family, the deeper the neural network becomes and therefore, the more computationally expensive it is. However, going to a higher EfficientNet family does not always imply getting better results [7].

In this study, EfficientNet B0 to B7 architectures have been modelled and analysed. The aim of the paper is to analyse the whole family in order to perform a comparison of them so as to obtain the most suitable architecture for this specific problem.

Each of the EfficientNet family architecture was developed for a specific image resolution. This resolution is called base resolution, as shown in Table 1, from B0 to B7. Besides, results are also obtained for a fixed resolution of the images, 300, that for some networks means a reduction and for other an increment of the size (Table 1). This way it is possible to see the effect of changing the resolution of the image when a specific network is applied.

Table 1. Analyzed scenarios on EfficientNet architecture

Family	Resolution	
	Base	Fixed
EfficientNetB0	224	300
EfficientNetB1	240	300
EfficientNetB2	260	300
EfficientNetB3	300	300
EfficientNetB4	380	300
EfficientNetB5	456	300
EfficientNetB6	528	300
EfficientNetB7	600	300

For each model, training images have been randomly picked up into batches, so simulation of the configurations listed in Table 1 has been run multiple times.

2.1 Model Set-Up

The model has been built as a binary image classification problem, where with the original architecture a pooling layer and two dense layers have been added.

For the training process, a first iteration has been performed using already existing pre-trained weights on ImageNet. With the aim of improving accuracy, a fine-tuning step has been included.

In the fine tuning, the learning rate of the optimizer has been reduced to 10^{-6} and the first block layers of each family have been defined as trainable. This has been the case for every layer in the first block except the BatchNormalization one. It is important to highlight that in every EfficientNet architecture, the number of layers of the first block varies, as every family architecture is slightly different.

3   Dataset Analysis and Pre-processing

The dataset for this work has been extracted from Kaggle [13], an open source online community of data scientists and machine learning practitioners which is a subsidiary of GoogleResults.

The dataset consists of a total of 300 railway track images which have already been classified into 'Defective' or 'Non defective'. Some of the annotated defects are fasteners, bolts, dents, cracks and misalignments. Out of those 300, 87 contain defective information, this is a 29% of the images. This dataset has been split into 70% for training, 20% for validation and 10% for testing.

Deep learning image classification dataset usually consist of thousands or tens of thousands of images, which makes the current dataset relatively small and complicates the correct training process. To minimize the effect of having such a small dataset, many different data augmentation techniques have been applied to extend it. Particularly, images were shifted, flipped, rotated, zoomed and sheared.

Initial image resolution of the dataset is 4000×3000. To match these resolutions to the ones of Table 1, images have been cropped so that they are now square and distortion do not affect the results. It is assumed that all of the relevant data of the images are centered and, therefore, the 3000 pixels of the center will be kept to minimize information loss. Afterwards, images have been resized to the original resolutions.

4   Results

Table 2 shows the average and best validation and testing accuracy results for EfficientNet architectures B0 to B7, for both image dimensions, base and fixed resolution. Moreover, it includes the number of parameters and FLOPS that corresponds to each model.

As expected, a strong correlation exists between number of parameters and FLOPS. The larger the number of parameters, the higher the FLOPS. In other words, the more complex the architecture, the higher the required computation cost to solve it.

On the other hand, attending to the FLOPS difference column, it can be seen that in the base resolution cases, it is above the reduced one from architectures B4 on. For those architectures, the base resolution image size is larger than the so-called fixed one, meaning that it works with a larger number of parameters. Similarly, the number of parameters for architectures B0 to B2 with fixed resolution images is larger because in these cases the base resolution is smaller than the fixed one (300). This is why the FLOPS diff. Column is negative for architectures B0 to B2.

In terms of validation accuracy, results are shown in Fig. 2. It can be seen that values drop when going down the EfficientNet architecture from B7 to B3. This is as expected, and similar to results obtained with the ImageNet Dataset on the original analysis shown in Fig. 1. However, when moving to architectures below B3, it is found that the average validation accuracy no longer drops, but increases. This was also stated in Ref. [7], where authors run the architectures with different datasets and found that lower level architectures provide sometimes higher accuracies than the higher level ones. Nevertheless, in this case, the efficiency obtained with B0 to B2 can be compared to B6 and B7.

Table 2. Results of the analyzed scenarios

Arch	Score	Param	Base resolution			Fixed resolution			FLOPS diff
			Val. Acc.	Test Acc.	FLOPS	Val. Acc.	Test Acc.	FLOPS	
B0	Avg	8.4E+06	82.2%	78.7%	1.5E+09	79.7%	74.1%	3.0E+09	−47%
	Top 1		84.4%	85.7%		81.3%	76.2%		
B1	Avg	1.3E+07	82.8%	76.8%	2.8E+09	81.2%	77.1%	4.4E+09	−37%
	Top 1		84.4%	**85.7%**		84.4%	81.0%		
B2	Avg	1.5E+07	**84.2%**	84.7%	4.0E+09	81.9%	81.7%	5.1E+09	−21%
	Top 1		**87.5%**	85.7%		82.8%	**85.7%**		
B3	Avg	2.2E+07	78.3%	76.8%	7.4E+09	78.3%	76.8%	7.4E+09	0%
	Top 1		82.8%	**85.7%**		82.8%	**85.7%**		
B4	Avg	3.5E+07	81.8%	79.7%	1.7E+10	79.0%	67.4%	1.1E+10	54%
	Top 1		85.9%	**85.7%**		81.3%	71.4%		
B5	Avg	5.7E+07	82.5%	74.4%	4.1E+10	79.3%	75.1%	1.8E+10	129%
	Top 1		84.4%	85.7%		82.8%	81.0%		
B6	Avg	8.2E+07	83.7%	75.3%	7.7E+10	84.3%	74.7%	2.6E+10	198%
	Top 1		85.9%	85.7%		87.5%	81.0%		
B7	Avg	1.2E+08	84.6%	75.1%	1.5E+11	**85.9%**	69.2%	4.0E+10	283%
	Top 1		**87.5%**	81.0%		**89.1%**	76.2%		

Regarding the best accuracy, Fig. 2 shows the obtained validation accuracy for the base and reduced resolution images. In this case, a peak value is obtained with the reduced resolution and the B7 architecture, with a 89.1% value.

Similar to the average validation accuracy behavior, base resolution best accuracy is always above the fixed one except for B6 and B7 architectures. It is remarkable that the top accuracy achieved by the B2 architecture on the base resolution images exceeds the value obtained by the B7 with the base resolution images, and it matches the B6 with the fixed one with a value of 87.5% accuracy.

Even if the validation dataset accuracy is already a good indicator of the performance of the trained model, a small dataset of 21 images in total (considering both, defective and non-defective cases) were put aside to run some further analyses.

The expected results should be similar regarding validation, but this is not the case. The highest average test accuracy was obtained with the B2 architecture (84.7%). For the best test accuracy, architectures B0 to B6 obtained a value of 85.7%.

The difference between the validation and test accuracies are explained by the used of a small dataset. The main risk of having scarce samples is that in a randomly distributed case, datasets might not generalize well enough.

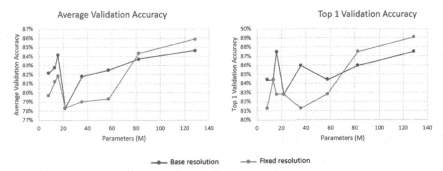

Fig. 2. Average and best validation accuracy for B0 to B7 EfficientNet architectures and different image resolution

4.1 Fault Detection

To ensure that the models had been trained correctly, an object detection algorithm based on image pyramid and sliding windows has been developed. This method allows to detect the area with highest probability of having the defect. The detected fault will be framed with a square. To avoid overlapping squares, non-maxima suppression (NMS) has been applied and afterwards, the highest probability area has been kept.

This process is illustrated in in Fig. 3 for an image that returned a prediction of 94.7% of being faulty, whose defect was a screw and bolt missing. Image (a) is the original

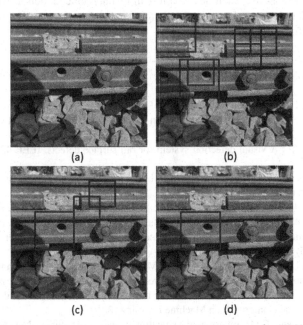

Fig. 3. (a) Original image; (b) areas with a prediction >85%; (c) areas after NMS with prediction >85%; (d) highest probability area with a 92.7% confidence

window, (b) the areas with a prediction above 85% probability, (c) areas with predictions above 85% after NMS, and (d) highest probability area with a 92.7% probability of being faulty.

Results show that the model has been satisfactorily trained, as it is capable of detecting the most promising areas with defects, in this case the area with a missing bolt.

5 Conclusions and Future Work

The paper has analysed the use of the EfficientNet CNN architecture to classify railway track fault defects. The validation accuracy of up to 89.1% show that the technique is suitable for such application. In the performed comparison between different network families, B7 always outperformed the rest. However, B2 showed to be the best in computer cost vs accuracy ratio.

Regarding future works, the performed fine tuning process only made the first block of each family trainable. Even if only 10% of the cases improved during this step, it is considered that further improvement is possible by making more blocks trainable. This comes at the cost of computational power.

Attending to the cases explored, standard deviations of 8.9% and 8.5% resulted from B5 and B6 architectures for base resolution, respectively. Deviations of the rest of the families are considered good enough as they remained below 6%. These deviations values show that additional iteration might be needed to obtain more stable results.

Other future work may include to explore more backbone models as well as applying morphological transformations on the images and different networks that have been proved useful for fault detection [14, 15].

References

1. International Energy Agency: The Future of Rail (2019)
2. Li, Q., Ren, S.: A real-time visual inspection system for discrete surface defects of rail heads. IEEE **61**(8), 2189–2199 (2012)
3. Li, Q., Ren, S.: A visual detection system for rail surface defects. IEEE **42**(6), 1531–1542 (2012)
4. James, A., et al.: TrackNet - a deep learning based fault detection for railway track inspection. IEEE **1**, 1–5 (2018)
5. Yaman, O., Karaköse, M., Ak, E., Ayd, I.: Ray Yüzeyi için Görüntü İş leme Tabanl ı Ar ı za Tespit Yakla ş ı m ı Image Processing Based Fault Detection Approach for Rail Surface. In: 2015 23nd Signal Processing and Communications Applications Conference (SIU), p. 4. IEEE (2015)
6. Wei, X., Wei, D., Suo, D.A., Jia, L., Li, Y.: Multi-target defect identification for railway track line based on image processing and improved YOLOv3 model. IEEE Access **8**, 1 (2020)
7. Tan, M., Le, Q.V.: EfficientNet : rethinking model scaling for convolutional neural network. In: International Conference on Machine Learning (2019)
8. Tumor, B., Using, C.: Brain tumor classification using dense Efficient-Net. Axioms **11**(1), 13 (2022)

9. Ebenezer, A.S., Kanmani, S.D., Sivakumar, M., Priya, S.J.: Materials today: proceedings effect of image transformation on EfficientNet model for COVID-19 CT image classification. Mater. Today Proc. **51**, 2512–2519 (2022)
10. Marques, G., Agarwal, D., De, I., Díez, T.: Automated medical diagnosis of COVID-19 through EfficientNet convolutional neural network. Appl. Soft Comput. J. **96**, 106691 (2020)
11. Abdou, P., Karou, K.: Accurate detection of COVID-19 using K-EfficientNet deep learning image classifier and K-COVID chest X-ray images dataset. In: 2020 IEEE 6th International Conference on Computer and Communications (ICCC), pp. 1527–1531 (2021)
12. Müftüoglu, Z., Ayyüce Kizrak, M., Yildirim, T.: Differential privacy practice on diagnosis of COVID-19 radiology imaging using EfficientNet. Cornell University Library, pp. 1–6 (2020)
13. Kaggle, Kaggle. [Online]. https://www.kaggle.com/. Accessed 1 May 2022
14. López-Estrada, F.R., Méndez-López, A., Santos-Ruiz, I., Valencia-Palomo, G., Escobar-Gómez, E.: Fault detection in unmanned aerial vehicles via orientation signals and machine learning. Rev. Iberoam. Autom. Inf. Indust. **18**(3), 254–264 (2021). https://doi.org/10.4995/riai.2020.14031
15. Márquez-Vera, M.A., López-Ortega, O., Ramos-Velasco, L.E., Ortega-Mendoza, R.M., Fernández-Neri, B.J., Zúñiga-Peña, N.S.: Fault diagnosis in industrial process by using LSTM and an elastic net. Rev. Iberoam. Autom. Inf. Indust. **18**(2), 160–171 (2021). https://doi.org/10.4995/riai.2020.13611

Challenging Mitosis Detection Algorithms: Global Labels Allow Centroid Localization

Claudio Fernandez-Martín[1](✉), Umay Kiraz[2,3], Julio Silva-Rodríguez[4],
Sandra Morales[1], Emiel A.M. Janssen[2,3], and Valery Naranjo[1]

[1] Institute of Research and Innovation in Bioengineering, Universitat Politècnica de València, Valencia, Spain
clferma1@i3b.upv.es
[2] Department of Pathology, Stavanger University Hospital, Stavanger, Norway
[3] Department of Chemistry, Bioscience and Environmental Engineering, University of Stavanger, Stavanger, Norway
[4] Institute of Transport and Territory, Universitat Politècnica de València, Valencia, Spain

Abstract. Mitotic activity is a crucial proliferation biomarker for the diagnosis and prognosis of different types of cancers. Nevertheless, mitosis counting is a cumbersome process for pathologists, prone to low reproducibility, due to the large size of augmented biopsy slides, the low density of mitotic cells, and pattern heterogeneity. To improve reproducibility, deep learning methods have been proposed in the last years using convolutional neural networks. However, these methods have been hindered by the process of data labelling, which usually solely consist of the mitosis centroids. Therefore, current literature proposes complex algorithms with multiple stages to refine the labels at pixel level, and to reduce the number of false positives. In this work, we propose to avoid complex scenarios, and we perform the localization task in a weakly supervised manner, using only image-level labels on patches. The results obtained on the publicly available TUPAC16 dataset are competitive with state-of-the-art methods, using only one training phase. Our method achieves an F1-score of 0.729 and challenges the efficiency of previous methods, which required multiple stages and strong mitosis location information.

Keywords: Mitosis detection · Weak labels · Histology · Digital pathology

The work of C. Fernández-Martín and U. Kiraz was funded from the Horizon 2020 of European Union research and innovation programme under the Marie Sklodowska Curie grant agreement No 860627 (CLARIFY Project). The work of Sandra Morales has been co-funded by the Universitat Politècnica de València through the program PAID-10-20. This work was partially funded by GVA through project PROMETEO/2019/109.

© The Author(s), under exclusive license to Springer Nature Switzerland AG 2022
H. Yin et al. (Eds.): IDEAL 2022, LNCS 13756, pp. 482–490, 2022.
https://doi.org/10.1007/978-3-031-21753-1_47

1   Introduction

In digital pathology, mitosis counting is one of the most important tasks in the histopathological clinical practice. In the case of breast cancer, the mitotic activity index (MAI) is considered one of the strongest proliferation-associated prognostic factors [1]. However, mitosis counting is a laborious and time-consuming task due to the large size of the Hematoxylin and Eosin (H&E) slides under a microscope, and the low occurrence of mitotic figures. In addition, the large heterogeneity of patterns and similarity between mitotic and non-mitotic cells (see Fig. 1) makes this task highly variable among clinical experts [2], which hinders its reproducibility.

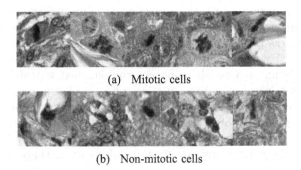

(a)   Mitotic cells

(b)   Non-mitotic cells

Fig. 1. Visual illustration of the morphological heterogeneity and the challenge of differentiating patterns between mitotic and non-mitotic cells, extracted from TUPAC16 [3].

In the last years, the advent of modern deep learning algorithms has emerged as a possible solution to bring objectivity and reproducibility to the challenge of mitosis localization. Deep learning using convolutional neural networks (CNNs) has reached remarkable results in a wide range of applications under the supervised learning paradigm. Nevertheless, it requires a reasonable amount of carefully-labeled data to perform properly. In the case of mitosis localization, this is a tedious process, which usually is repeated by different pathologists to reach consensus labels. Since delineating at pixel-level individual mitotic cells is an unfeasible task, the reference datasets normally contain centroid-based labels [3] or inexact pixel-level annotations [4]. Because of this, previous works to automate the mitosis localization process have struggled to match the available labels to the use of segmentation or object detection CNNs, which are typically used localization tasks. Contrary to this line of work, we propose to make use of inherent spatial localization capacity of CNNs in image-level classification tasks [5], without the need to resort to an exact localization of the mitotic cell inside the region of interest. Our main contributions are summarized as follows:

- A CNN for weakly supervised segmentation of mitotic figures on H&E patches using image-level labels.

- Concretely, training is driven by maximum aggregation of instance-level predictions.
- Comprehensive experiments demonstrate the competitive performance on the popular TUPAC16 dataset, using a single-phase pipeline without requiring the exact localization information for training our model.

2   Related Work

2.1   Mitosis Detection

Mitosis localization algorithms using CNNs deal with labels in the form of centroid annotations, or inexact pixel-level delineation of the mitotic cell. In that sense, Li *et al.* [6] propose a novel concentric loss to move from centroid labels to pixel-level segmentation using the pixels surrounding certain radius of the centroid. Other works focus on leveraging cell-level predictions using multi-phase pipelines [7–12]. For instance, Sohail *et al.* [12] propose a complex multi-phase pipeline that includes pseudolabeling centroid-labelled mitosis via previously trained Mask R-CNNs. Also, Nateghi *et al.* use multiple training stages to refine the false positive detection using hard-negative mining via stain priors, or prediction uncertainty. In contrast to these works, we study how training a CNN at the image-level for a classification task also allows the precise location of mitotic cells without using any localization information, shape or stain priors, or multi-phase refinement pipelines.

2.2   Weakly Supervised Segmentation

Weakly supervised segmentation (WSS) aims to leverage pixel-level localization using global (a.k.a image-level) labels during training. According to [13], WSS methods use fully-convolutional CNNs with an aggregation function that merges all the spatial information into one value, that serves as global prediction [14]. This output is then used to compute the loss function, and drives the network optimization. Different strategies include the aggregation of spatial features (embedding-based) or pixel-level predictions (instance-based). Finally, the probability maps before aggregation operation are used as segmentation predictions. Lately, these segmentation maps are refined to incorporate self-supervised learning pipelines [15] or uncertainty proxies [16], among others.

3   Methods

An overview of our proposed method is depicted in Fig. 2. In the following, we describe the problem formulation, and each of the proposed components.

Problem Formulation. In the paradigm of weakly supervised segmentation (WSS), the training set is composed of images $\{x_n\}_{n=1}^N$, whose binary label $\{Y_n^k\}_{n=1}^N$, such that, $Y_n^k = \{0, 1\}$ is known, and defines if a category k is present

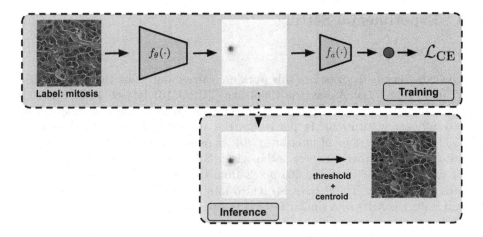

Fig. 2. Overview of the proposed method for mitosis localization.

within the image. Also, each positive image has pixel-level labels $y_{n,i}$ for each i pixel in the image, but they remain unknown during training. Further, we denote Y_n^k as Y_n for simplicity, since one unique class is taken into account, and we assume image index n.

Instance-Based WSS. In this work, we aim to train a CNN capable of locating positive mitosis during inference, while being trained only with image-level labels. To do so, we make use of an instance-based weakly supervised learning strategy. Let us denote a CNN model, $f_\theta(\cdot) : \mathcal{X} \to \mathcal{H}^K$, parameterized by θ, which processes instances $x \in \mathcal{X}$ to output sigmoid-activated instance-level probabilities, h_i, such that $h_i \in [0, 1]$. Also, we use a parameter free aggregation function, $f_a(\cdot)$, in charge of combining the pixel-level scores into one global output H, such that $H = f_a(f_\theta(x))$. Then, the optimization of θ is driven by the minimization of cross entropy loss between reference and predicted image-level score.

$$\mathcal{L}_{ce} = Y log(H) + (1 - Y) log(1 - H) \tag{1}$$

In this work, we propose to use the maximum operation as aggregation function, $f_a(\cdot)$. Although this aggregation only backpropagates gradients through the maximum-activated spatial regions, this effect produces that only very discriminative cells will be classified as mitosis, which avoids false positive predictions.

Inference. During inference, pixel-level predictions are inferred using the pixel-level predictions given by the trained CNN, $y_i = f_\theta(x)$. The probability maps are resized to the original image dimensions by bi-linear interpolation. Then, sigmoid scores are converted to a binary mask by applying a threshold to the probability maps. Concretely, the threshold is obtained from the operative point of the ROC curve between image-level predictions and references. Finally, a centroid is assigned to each element in the mask, to be located as a mitosis.

4   Experimental Setting

4.1   Datasets

The experiments described in this work are carried out using the popular 2016 TUmor Proliferation Assessment Challenge (TUPAC16) dataset [3]. TUPAC16 is publically available and is composed of 73 breast cancer whole slide images from two different institutions. In particular, the auxiliary mitosis dataset contains 1552 processed regions of interest at 40× magnification, with centroid-labelled mitosis by consensus of expert pathologists. Following relevant literature in [6], we extracted patches of size 500 pixels from the regions of interest for computational efficiency. The dataset is divided into patient-level training, validation, and testing cohorts in a similar fashion to prior literature [6].

4.2   Metrics

We use standard metrics for mitosis localization evaluation. First, the model is optimized using only global image-level labels, by means of the accuracy, AUC, and F1-score. Then, the comparison with state-of-the-art methods on mitosis detection is assessed using the standard criteria of mitosis detection contests [12]. A detected mitosis is considered true if it is located at most 30 pixels from an annotated mitosis. Under this criteria, precision, recall and F1-score are computed.

4.3   Implementation Details

The proposed method is trained using ResNet-18 [17] convolutional blocks as a backbone. Concretely, the first 3 blocks pre-trained on ImageNet are used as feature extractor, which are retrained for the mitosis detection task. We trained this architecture during 40 epochs to optimize Eq. 1 using a batch size of 32 images and a learning rate of 0.0001. In order to deal with class imbalance, the images are sampled homogeneously according to its class in each epoch. Also, color normalization and augmentation techniques are employed to increase robustness against stain variations and artifacts in the digitized slides. Images are color-normalized using the stain normalization method of Macenko *et al.* [18], and data augmentation is included during training using spatial translations, rotations, and blurring.

5   Results

5.1   Comparison to Literature

The quantitative results obtained by the proposed method for mitosis localization on the test cohort are presented in Table 1. Also, we include results reported in previous literature on the TUPAC16 dataset. The proposed weakly-supervised method reaches an F1-score value of 0.729, which is comparable to prior literature without accessing to any supervision regarding the exact location of the mitosis in the image. It should be noted that, in addition, the best previous methods use additional training data, and require multiple stages of label refinement. In contrast, the proposed method uses only one training cycle. Moreover, the proposed approach obtains the best precision on mitosis localization that only use one training phase. This could be due to maximum aggregation, which propagates gradients only in those regions that are highly discriminating.

Table 1. Performance comparison of the proposed model with existing methods on test subset of TUPAC16 auxiliary dataset.

Method	Precision	Recall	F-score	Multiple phases	Location supervision	External data
Paeng *et al.* (2017) [7]	-	-	0.652	×	×	
Zerhouni *et al.* (2017) [8]	0.675	0.623	0.648	×	×	
Akram *et al.* (2018) [9]	0.613	0.671	0.640	×	×	×
Li *et al.* (2019) [6]	0.64	0.70	0.669		×	
Wahab *et al.* (2019) [10]	0.770	0.660	0.713	×	×	
Mahmood *et al.* (2020) [19]	0.641	0.642	0.642	×		
Nateghi *et al.* (2021) [11]	0.764	0.714	0.738	×	×	
Sohail *et al.* (2021) [12]	0.710	0.760	0.750	×	×	×
Proposed	0.739	0.720	0.729			

5.2 Ablation Experiments

In the following, we depict ablation experiments to motivate the choice of the different components of the proposed method.

Weakly Supervised Setting. First, we study the configuration of the WSS model architecture. To do so, we explore the most outstanding configurations. First, embedding-based approaches that aggregate spatial features before the classification layers, and instance-based approaches that apply the classification layer spatially. Also, we use different aggregation methods, such as mean and max operations, and the trainable attentionMIL mechanism [13]. Results are presented in Table 2. The figures of merit show that, although all methods reach similar results at image-level, only the instance-based with maximum aggregation performs properly on mitosis localization, since it is the only method that penalizes false positive localization during training.

Table 2. Performance comparison of the different configurations of the WSS proposed model, in terms of aggregation strategies. Results are presented for mitosis localization and image-level classification.

Configuration	F-score image-level	F-score localization
Embedding - mean	0.762	0.134
Embedding - max	**0.772**	0.234
AttentionMIL [13]	0.768	0.014
Instance - mean	0.753	0.004
Instance - max	0.761	**0.729**

On the Importance of the Feature Complexity. Convolutional neural networks combine stacked convolutional and pooling operations, which merge spatial information. Thus, later layers in CNNs extract high-level features with complex shapes, and low spatial resolution. Although CNNs for classification tasks usually benefit from deep structures, we observed that spatial resolution and low-level features are vital for mitosis localization, as shown in Fig. 3. For that reason, we used only 3 residual blocks of ResNet-18 architecture for the proposed method.

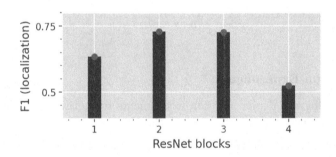

Fig. 3. Ablation study on the number of residual blocks used for feature extraction. Metric presented for mitosis localization.

5.3 Qualitative Evaluation

Finally, we present visual results of the proposed method performance on the test subset in Fig. 4. In particular, correct detections of mitotic cells (true positives), cells wrongly classified as mitosis (false positives) and non-detected mitosis (false negatives) are shown in green, yellow and blue colors, respectively. Visual results show a promising performance of the proposed method, with false positive classifications occur with irregularly-shaped non-mitotic cells.

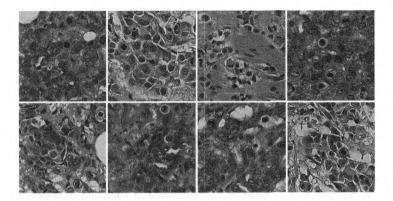

Fig. 4. Qualitative evaluation of the proposed method for mitosis localization. Green: true positive; Blue: false negative; Yellow: false positive. (Color figure online)

6   Conclusions

In this work, we have presented a deep learning model for weakly supervised mitosis location on H&E histology images. In particular, the model is composed of a narrow CNN backbone that leverages pixel-level predictions. Then, those predictions are grouped into an image-level score using maximum aggregation, that serves as proxy for CNN training via global labels. Thanks to the maximum operation, that only focus on very discriminative cells, obtained results have very few false positive predictions, and reaching a precision of 0.739 and an F-score of 0.729 on TUPAC16 dataset. The proposed approach, yet simple, reaches competitive performance in comparison to previous literature, without requiring any information of mitosis localization in the image during training. This calls into question the efficiency of other approaches, which require this location information, and resort to multiple phases of training to refine centroid-based labels and to alleviate false positive predictions. Further research could complement the proposed setting to take into account uncertainties on predicted mitoses, and to incorporate location information using a soft, constrained formulation.

References

1. Baak, J.P.A., et al.: Prospective multicenter validation of the independent prognostic value of the mitotic activity index in lymph node-negative breast cancer patients younger than 55 years. J. Clin. Oncol. **23**, 25 (2005)
2. Elmore, J.G.J.G., et al.: Diagnostic concordance among pathologists interpreting breast biopsy specimens. JAMA. **313**(11), 1122–1132 (2015)
3. Veta, M., et al.: Predicting breast tumor proliferation from whole-slide images: the TUPAC16 challenge. Med. Image Anal. **54**, 111–121 (2019)
4. Roux, L., et al.: Mitosis detection in breast cancer histological images an ICPR 2012, contest. J. Pathol. Inform. **4**(1), 1–7 (2013)

5. Oquab, M., Bottou, L., Laptev, I., Sivic, J.: Learning and transferring mid-level image representations using convolutional neural networks. In: Proceedings of the IEEE Computer Society Conference on Computer Vision and Pattern Recognition (CVPR) (2014)

6. Li, C., Wang, X., Liu, W., Latecki, L.J., Wang, B., Huang, J.: Weakly supervised mitosis detection in breast histopathology images using concentric loss. Med. Image Anal. **53**, 165–178 (2019)

7. Paeng, K., Hwang, S., Park, S., Kim, M.: A unified framework for tumor proliferation score prediction in breast histopathology (2016)

8. Zerhouni, E., Lanyi, D., Viana, M., Gabrani, M.: Wide residual networks for mitosis detection. In: 2017 IEEE 14th International Symposium on Biomedical Imaging (ISBI 2017) (2017)

9. Akram, S.U., Qaiser, T., Graham, S., Kannala, J., Heikkilä, J., Rajpoot, N.: Leveraging unlabeled whole-slide-images for mitosis detection (2018)

10. Wahab, N., Khan, A., Lee, Y.S.: Transfer learning based deep CNN for segmentation and detection of mitoses in breast cancer histopathological images. Microscopy **68**(3), 216–233 (2019)

11. Nateghi, R., Danyali, H., Helfroush, M.S.: A deep learning approach for mitosis detection: Application in tumor proliferation prediction from whole slide images. Artif. Intell. Med. **114**, 102048 (2021)

12. Sohail, A., Khan, A., Wahab, N., Zameer, A., Khan, S.: A multi-phase deep CNN based mitosis detection framework for breast cancer histopathological images. Sci. Rep. **11**(1), 1–8 (2021)

13. Ilse, M., Tomczak, J.M., Welling, M.: Attention-based deep multiple instance learning. In: 35th International Conference on Machine Learning (ICML) (2018)

14. Silva-Rodríguez, J., Colomer, A., Naranjo, V.: WegleNet: a weakly-supervised convolutional neural network for the semantic segmentation of Gleason grades in prostate histology images. Comput. Med. Imaging Graph. **88**, 101846 (2021)

15. Wang, Y., Zhang, J., Kan, M., Shan, S., Chen, X.: Self-supervised equivariant attention mechanism for weakly supervised semantic segmentation. In: Proceedings of the IEEE Computer Society Conference on Computer Vision and Pattern Recognition (CVPR) (2020)

16. Belharbi, S., Rony, J., Dolz, J., Ayed, I.B., Mccaffrey, L., Granger, E.: Deep interpretable classification and weakly-supervised segmentation of histology images via max-min uncertainty. IEEE Trans. Med. Imaging **41**(3), 702–714 (2022)

17. He, K., Zhang, X., Ren, S., Sun, J.: Deep residual learning for image recognition. In: Proceedings of the Conference on Computer Vision and Pattern Recognition (CVPR) (2016)

18. Macenko, M., et al.: A method for normalizing histology slides for quantitative analysis. In: 2009 IEEE International Symposium on Biomedical Imaging: From Nano to Macro, pp. 1107–1110 (2009)

19. Mahmood, T., Arsalan, M., Owais, M., Lee, M.B., Park, K.R.: Artificial intelligence-based mitosis detection in breast cancer histopathology images using faster R-CNN and deep CNNs. J. Clin. Med. **9**(3), 749 (2020)

Go-Around Prediction in Non-Stabilized Approach Scenarios Through a Regression Machine-Learning Model Trained from Pilots' Expertise

Jesús Cantero[1], Adrián Colomer[1(✉)], Laëtitia Launet[1], Alexandre Duchevet[2], Théo De La Hogue[2], Jean-Paul Imbert[2], and Valery Naranjo[1]

[1] Institute of Research and Innovation in Bioengineering, Universitat Politècnica de València, Valencia, Spain
{jacanram,adcogra,lmlaunet,vnaranjo}@i3b.upv.es

[2] ENAC - Ecole Nationale de l'Aviation Civile, 7 Avenue Edouard Belin CS 54005, 31055 Toulouse, France
{alexandre.duchevet,theo.de-la-hogue,jean-paul.imbert}@enac.fr

Abstract. To face the pilots' shortage that could hit the aeronautical world in the future, single-pilot operations are envisaged as a solution. Even if the single pilot will have to assume the tasks that are today done by two pilots in the cockpits, the safety level of the flight should remain the same compared to standard operations. The challenge will be significant during high workload phases when complex decisions are to be made. We focused our work on the final approach phase during which the decision to perform a go-around or to land must be taken. This decision reveals to be particularly difficult as studies show that 97% of the time, the choice to land is made whereas a go-around was needed. In this paper, we propose a machine learning regression algorithm based on expert users classification to predict the need for a go-around during the final approach. An average precision of 0.96, a sensitivity of 0.84 and an F1-score of 0.88 shows a promising behaviour for the GA automatic identification task.

Keywords: Go-around prediction · Non-stabilized approach · Pilot expertise · Regression neural network · Aircraft parameters

1 Introduction

With more than 600,000 pilots to train over the next 20 years [1] and the perspective of a qualified pilots shortage, the aeronautical industry is today pushing towards Single Pilot Operations (SPO). To ensure an adequate safety level and

This work has received funding from the Clean Sky 2 Joint Undertaking (JU) under grant agreement No 831884. The DGX A100 used for this research has been funded by the European Union within the operating Program ERDF of the Valencian Community 2014–2020 with the grant number IDIFEDER/2020/030.

© The Author(s), under exclusive license to Springer Nature Switzerland AG 2022
H. Yin et al. (Eds.): IDEAL 2022, LNCS 13756, pp. 491–499, 2022.
https://doi.org/10.1007/978-3-031-21753-1_48

compensate the absence of a second pilot in the cockpit, several solutions are envisaged. First, a ground operator or a harbor pilot could support and relieve the on-board pilot from some tasks during workload-heavy phases [2,3]. New systems could also monitor pilots' physiological state and adapt their inter-faces accordingly to enable a more efficient work [4]. Finally, advancement in automation and artificial intelligence could bring a support to decision-making, particularly in situations that necessitate a quick reaction from the pilot [5,6]. During the final approach phase for example, the decision to perform a go around or not is critical and could be difficult to be made by a ground operator as it requires a quick reaction and a great situation awareness. The final approach is the last segment of a flight before landing, when the alignment and the descent to the runway are made. This phase is workload heavy for pilots as any errors may quickly lead to fatal consequences. Therefore, it usually requires the full attention of two pilots, one in charge of piloting the aircraft and another one in charge of monitoring and alerting in case of deviations according to the airline stabilisation policy. The objectives of pilots during this phase is to stabilise the aircraft for landing. Stabilising the Aircraft means bringing the Aircraft in the right configuration at the right speed at the stabilisation gate to perform a safe landing. In order to reduce their workload, manufacturers provide guidance to help the operators to define their own stabilisation policy, based on the mon-itoring of the most relevant flight parameters [7]. This way, an approach will be considered stabilised if these parameters do not go over certain thresholds. However, according to [8], 97% of unstable approaches are continued to landing, suggesting that the decision to perform a go around is complex to take. Indeed, IATA [9] identifies several factors that governs the go-around decision like over confidence, inadequate awareness of wind conditions or the lack of operator pol-icy, organisational culture and training to support the decision.

The advances of artificial intelligence in the last decade have led to the inclu-sion of this technology into the aerospace sector. In this sense, during the last two years, a few research works explored data-driven approaches for GA prediction. In particular in [10], the authors proposed an online prediction tree-based model to estimate the landing speed. The authors defend that is the most relevant parameter that affects for critical landing detection and they report promising results predicting landing speed. However, a go-around manoeuvre is a more complex situation in which other factors are included (unstable approach, visi-bility, meteorological conditions, or obstacle in the runway). The authors of [11] present a macro and microscopic data-driven models to predict GA occurrences. The reported results show a 50% detection rate of GAs with false positive rate below 7%. Finally, the most recent contribution from Dhief et al. [12] propose a novel safety metric based on machine learning techniques that may assist tower controllers in detecting and predicting go-around events. First, a data-driven model is developed for labelling go-around events. Then, features are engineered for a tree-based learning model to predict go-around events. For the full data set the model is able to detect 33% of the go-around with a 25% false alert rate.

Taking all this into account, in this paper we propose a machine learning methodology based on expert users classification to predict go-arounds during the final approach. To the best of the authors' knowledge, this is the first time in which a data-driven model mimicking the expertise of qualified pilots is trained. Soft labels are computed taking into account the experience of pilots and a tailored regression neural network is trained to estimate go-around in non stabilised scenarios. This work is the first step towards the future digital assistant in the cockpits to support pilots in difficult decision-making in single-pilot operations.

2 Materials

2.1 Data Acquisition

For building the data base, a total of 71 different approach scenarios on the Paris-Orly airport were simulated, based on the data obtained from historical data of real approaches. The last 7 nautical miles of stabilised and unstabilised approaches were performed on an online Flight simulator, as shown in Fig. 1 using the meteorological conditions of the real flights. During simulation, pilots were asked to evaluate the difficulty of the approach all along the flight using as inputs video data and the flight parameters, differentiating among three different difficulty levels :

- Standard: "I'm feeling at ease with the situation. The crew should be able to perform a stabilised approach and a safe landing without problem."
- Difficult: "The situation looks complex and the crew may have some difficulties to perform a stabilised approach and a safe landing."
- Very difficult: "The situation looks very complex and I have serious doubts about the capability of the crew to perform a stabilised approach and a safe landing."

In the case the pilots considered that a safe landing was not possible, they could also call for a go-around, finishing the scenario simulation. If no go-around was called, all scenarios end at 200 ft Above Ground Level (AGL). At the end of each simulation, the level of confidence in the decision taken was also registered.

A total of 33 pilots participated in the evaluation, with an average experience of 6125 flight hours on commercial aircraft. A total of 668 simulations were evaluated by the pilots.

2.2 Data Description

The available data is mainly of two different types:

1. On the one hand, for each one of the 71 different scenarios, the values of the 21 different parameters defining the state of the aircraft at each timestep (given by the sampling frequency) were recorded and stored in the corresponding csv file, identified with the corresponding scenario's filename.
2. On the other hand, one csv file collecting all the annotations of the different pilots throughout the whole duration of each simulation was available. Each pilot's experience in number of hours was also collected.

Fig. 1. Simulation environment for expertise data gathering.

2.3   Data Conditioning and Pre-processing

From the 21 initial flight parameters, only the 9 most relevant features were selected: Altitude, Vertical Speed, Ground Speed, Indicated Air Speed, Pitch, Roll, Localizer Deviation, Glide Slope Deviation and Distance to Runway Threshold. Note that the temporal information was used for assigning to each timestep its corresponding label, obtained from the pilots' annotations. From the pilots' evaluation data file, the scenario file name, the level of confidence about the decision and difficulty level at each timestep were used.

The different csv files were converted to pandas data frames and saved in python lists in order to preprocess the data before feeding it to the designed machine learning algorithm. Note that from the 71 different scenarios available, 57 were reserved for training the model, which corresponds to a total of 50360 timesteps. The remaining 14 flights were used to evaluate the performance and ability of generalization of the predictive model.

For the data preprocessing, the labels were obtained from the pilots' annotations at timestep level by linear interpolation of the weighted mean of all pilot's annotations (using as weights the level of confidence and pilot's experience). A visual representation of this transformation is shown in Fig. 2. Note that 0 label refers to a standard situation, 1 to difficult scenario, 2 a very difficult one and 3 that a go-around manoeuvre is required.

Finally, the labels were filtered to obtain a smoother curve and both the input and label lists were converted to numpy arrays and normalized before feeding the machine learning algorithm.

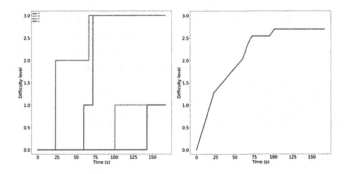

Fig. 2. Regression labels obtained by weighted mean and linear interpolation (right) from pilots' annotations (left). 0=standard situation, 1=difficult scenario, 2=very difficult and 3=go-around.

3 Methodology

3.1 Go-Around AI Prediction Module

The Go-Around prediction is performed by a Multi Layer Perceptron (MLP) neural network. As it can be observed in Fig. 3, the architecture consists of one input layer with 9 input units, one for each of the flight parameters considered, one hidden layer of 4 units with rectified linear unit (ReLU) activation and one output unit with sigmoid activation for difficulty level regression.

A droupout layer with 0.2 dropout rate and a batch normalization layer were added for optimising the architecture configutarion. For further reducing overfitting, a weight of 0.003 is fixed for the kernel regularizer and a value of 0.01 is chosen for both the bias and activity regularizers.

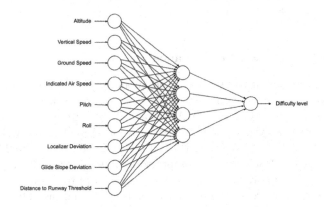

Fig. 3. Diagram of the MLP regression model architecture.

The model was trained for 80 epochs with a decay rate of 0.001 per epoch. A 20% of the training set was reserved for validation purposes. The batch size was set to 256 samples per batch, using Adam as optimizer and the Mean Squared Error (MSE) as a loss function. Note that hyperparameter tunning was performed via KerasTuner tool [13].

4   Results

4.1   Training Learning Curves

As it can be appreciated in the learning curve in Fig. 4, both the training and validation losses stabilise at a MSE value of less than 0.1, so it can be deduced that the model is learning with not overfitting, proving that it can successfully generalise to new samples.

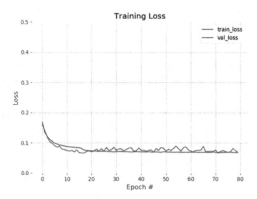

Fig. 4. MSE loss evolution per epoch for the validation set (in blue) and the training set (in red). (Color figure online)

4.2   Model Prediction Performance

To evaluate the similarity between the model's predictions and the labels obtained from the pilots' annotations for each timestep of the test scenarios, the MSE, Mean Absolute Error (MAE) and R2 score were chosen as figures of merit. Average values of several figures of merit for regression task were obtained. In particular a MSE of 0.056, a MAE of 0.172 and a R2 score equals to 0.374.

In addition to these metrics, comparisons between the difficulty level predicted by the model and the weighted labels extracted from pilots' annotations for the test scenarios were obtained for a qualitative analysis. These results are shown in Fig. 5(a) and Fig. 5(b) for two representative scenarios (i.e., Go-around and not Go-around examples).

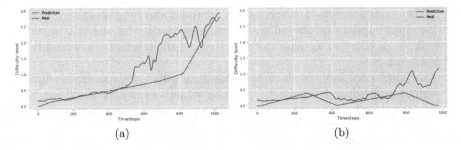

(a)   (b)

Fig. 5. Comparison between the difficulty predicted by the model and the labels obtained from the pilots' annotations. (a) Go-around scenario and (b) Non Go-around scenario.

To evaluate the final classification between Go-around (GA) and not Go-around (not GA) at scenario level, the precision, recall and f1-score were used as figures of merit. The obtained results are shown in Table 1.

Table 1. Model prediction quantitative metrics.

Classes	Precision	Recall	F1-score	N° of scenarios
Not GA	0.92	1.00	0.96	11
GA	1.00	0.67	0.80	3
Average	0.96	0.84	0.88	-

5 Discussion

Experimentations were conducted with seven Commercial Air transport pilots to test the Non Stabilised Approach assistant and consequently the go-around prediction algorithm presented in previous parts. Based on the experimentation results, it can be concluded that despite the fact that the MLP regression model produces quite noisy predictions, it is able to call for a go-around decision when needed. When calling a go-around, the call was indeed justified by the situation in 75% of the time. On the contrary, the algorithm was not able to recognise go-around situations in some cases. To explain these mixed results, four aspects should be considered:

- With only 71 simulated scenarios as input data to create the model, it is evident that only a small part of all possible situations were exposed to the algorithm. Therefore, some remote situations could not be recognised by the assistant as they were never exposed to it.
- Another interesting aspect to take into consideration for this problematic is the fact that the difficulty level is a subjective value, to the point that one same pilot may estimate a different difficulty value in different simulation runs

for the same scenario. For this reason, it would also be useful to incorporate the rules imposed by the different airline companies for the GA prediction task, in an effort to increase the objectivity of the labels, i.e. availability of a gold standard.

- The pilots pool was also heterogeneous with different experiences, different airlines and cultural differences, resulting sometimes in contrasted results for go-around calls during the same approach scenario. Some pilots seeing no problem whereas others calling a go-around rapidly.
- Last but not least, another valid approach consists in developing models that consider the experience of the pilot. This could be accomplished by splitting the training data into different sets according to the pilots' experience ranges (measured in flight hours) and training an AI model for each of them. In this sense, a wider dataset should be gathered.

The objective of involving experts users in the process of model creation and assess to what extent it is feasible has been achieved. This methods revealed to be resource heavy in terms of approach scenarios classification and produced mixed results. As this can be explained by the diversity of pilots that participated to the study, one solution could be to focus on a specific group of pilots, for example pilots that are Captain in a specific airline with a significant number of flight hours. This way we can assure that their profiles are similar and suppose that their approach scenarios classifications will be homogeneous, thus facilitating the model creation. The assistant based on this model could this way be an image of the experienced captains of this airline and be relevant to be used in operation by this airline as it would be coherent with its operational policy. This matching between the airline and the model of the assistant can be seen as a benefit. Other benefits are to be found and study. As experts users were involved in the classification process, to what extent algorithm outputs are better understood by users in operations? Does this expert users involvement has an impact on explainability of the outputs?

6   Conclusion

In this work, we presented a non stabilised approach assistant based on AI trained by gathering the expertise of tens of pilots. This was achieved by designing a total of 71 different approach scenarios and by asking 33 pilots to evaluate the difficulty of the approach all along the flight simulations. This way, using as labels the linear interpolation of the weighted mean of the difficulty level perceived during scenario simulation by expert pilots, a Multi Layer Perceptron regressor was trained.

Finally, the prediction outputs obtained from the AI module are used for announcing the parameters deviations and suggesting corrective actions to the pilot. Moreover, the assistant provides go-around decision making support, suggesting the pilot to interrupt the approach if the estimated difficulty level surpasses a certain threshold for a determined period of time.

References

1. Boeing, Pilot and Technician Outlook 2021–2040. https://www.boeing.com/commercial/market/pilot-technician-outlook/. Accessed 24 Feb 2022
2. Lachter, J., Brandt, S.L., Battiste, V., Ligda, S. V., Matessa, M., Johnson, W.W.: Toward single pilot operations: developing a ground station. In Proceedings of the international conference on human-computer interaction in aerospace, pp. 1–8, July 2014
3. Koltz, M.T., et al.: An investigation of the harbor pilot concept for single pilot operations. Proc. Manuf. **3**, 2937–2944 (2015). https://doi.org/10.1016/j.promfg.2015.07.948
4. Liu, J., Gardi, A., Ramasamy, S., Lim, Y., Sabatini, R.: Cognitive pilot-aircraft interface for single-pilot operations, knowledge-based systems, vol. 112, pp. 37–53 (2016). ISSN 0950-7051, https://doi.org/10.1016/j.knosys.2016.08.031
5. Duchevet, A., et al.: Toward a Non Stabilized Approach assistant based on human expertise. (2020) In: 1st International Conference on Cognitive Aircraft Systems - ICCAS 2020, 18 March 2020–19 March 2020 (Toulouse, France). (unpublished)
6. Núñez, J., et al.: Aircraft Dynamic Rerouting Support. In: 1st International Conference on Cognitive Aircraft Systems - ICCAS 2020, 18 March 2020–19 March 2020 (Toulouse, France) (2020). (unpublished)
7. Airbus, October 2020. https://safetyfirst.airbus.com/prevention-of-unstable-approaches/. Accessed 24 Feb 2022
8. Smith, J.M., Jamieson, D.W., Curtis, W.F.: Why are go-around policies ineffective? The psychology of decision making during unstable approaches. In: 65th Annual FSF International Air Safety Seminar Santiago, Chile (2012)
9. International Air Transport Association. Unstable Approaches: Risk Mitigation Policies, Procedures and Best Practices, 2nd Edition ISBN 978-92-9229-317-8 (2016)
10. Puranik, T.G., Rodriguez, N., Mavris, D.N.: Towards online prediction of safety-critical landing metrics in aviation using supervised machine learning. Transp. Res. Part C Emerg. Technol. **120**, 102819 (2020)
11. Figuet, B., Monstein, R., Waltert, M., Barry, S.: Predicting airplane go-arounds using machine learning and open-source data. In: Proceedings, vol. 59, p. 6 (2020). https://doi.org/10.3390/proceedings2020059006
12. Dhief, I., Alam S., Chan Chea Mean, C.C., Lilith, N.: Tree-based machine learning model for go-around detection and prediction. In: Proceedings 11th SESAR Innovation Days (2021)
13. O'Malley, T., et al.: KerasTuner (2019). https://github.com/keras-team/keras-tuner

Special Session on Intelligent Techniques for Real-World Applications of Renewable Energy and Green Transport

Identification of Variables of a Floating Wind Turbine Prototype

Juan Tecedor Roa[1], Carlos Serrano[1], Matilde Santos[2],
and J. Enrique Sierra-García[3(✉)]

[1] Facultad de Informática, Universidad Complutense de Madrid,
28040 Madrid, Spain
[2] Institute of Knowledge Technology, University Complutense of Madrid,
28040 Madrid, Spain
[3] Department of Electromechanical Engineering, University of Burgos,
09006 Burgos, Spain
jesierra@ubu.es

Abstract. In this paper, using real data from a low scale prototype of a wind turbine, different models have been obtained based on machine learning techniques. These models have been shown to be useful to forecast some key statistical metrics of the dynamics of the wind turbine. The models are dependent on the wind speed and the blade pitch angle. These models can be used to develop a digital twin of the wind turbine and predict its behavior, even for wind speed and pitch angles outside the ranges used for training the system.

Keywords: Wind turbine · Neural networks · Dynamics identification

1 Introduction

Renewable energies are raising in importance and relevance due to climate change, among other reasons. Wind turbines are a type of renewable energy generators that extract the energy from the wind that passes though its blades and is converted into electricity [1,4].

More concretely, offshore wind turbines and, specifically, floating ones (FOWT), are a type of wind turbines that offer several advantages over traditional wind turbines (onshore or offshore bottom-fixed counterparts), due to stronger and more stable wind speeds, larger sizes, wider spaces,... and thus increased power generation [9]. They are however, a very new technology still in development that benefits from modern techniques such as machine learning to increase their efficiency and reliability while decreasing costs [6,7].

Scale models allow to perform experiments with reduced costs and less restrictions due to space limitations. For example, a real world, modern wind turbine might span 100 m in diameter [10], which is unfeasible for most experimental labs. Low scale models have much smaller dimensions, for instance the one used in this study is under 50 cm in length. Among other advantages, these models

© The Author(s), under exclusive license to Springer Nature Switzerland AG 2022
H. Yin et al. (Eds.): IDEAL 2022, LNCS 13756, pp. 503–512, 2022.
https://doi.org/10.1007/978-3-031-21753-1_49

allow us to test control strategies before their deployment in the real systems, measure and extract relevant signal data, and analyze the relationships between the system variables [3]. This is specially interesting in the WT field due to the data obtained in real WTs are normally confidential and they are not available for the scientific community. In the case of FOWT, these data would help to understand better the relation between the wind velocity, the pitch angle, and the acceleration experimented by the WT, with the ultimate purpose of designing multiobjective controllers to stabilize the power and reduce vibrations [5].

To study these interrelated variables in this work, we have worked with real signal data that is measured from a prototype of a FOWT described in the following section. Different machine learning techniques have been applied to obtain regression models that identify several output variables of the wind turbine considering the wind velocity and the pitch angle as inputs. Although the scaling is an issue when working with small replicas, the methodology here proposed is a step forward in the development of digital twins of these complex systems [8].

The remaining part of the paper is structured as follows. The scale model and experiment set up are described in Sect. 2. Section 3 details the processing of the experimental data. The regression models and the results are discussed in Sect. 4. Finally the paper ends with the conclusions and the future works.

2   Small Replica and Experiment Set up

The scale model of the wind turbine is shown in Fig. 1. It was built at the System Engineering and Automatic Control Group of the University Complutense of Madrid, Spain. The sensor and microcontroller used were MPU6050 and Arduino Mega, respectively. To emulate the performance of the wind turbine, specific blades have been designed and built with a 3D-printer. A symmetric aerodynamic profile, the NACA0012 (National Advisory Committee for Aeronautics) has been selected to capture more wind. These blades have more inertia than RC commercial blades, adding more realism to the behavior of the prototype. As pitch actuator, an SG90 RC servomotor is used with a two-blade variable pitch mechanism. The blades are 10 cm long and 2 cm width; the platform has dimensions 16×24 cm. The model has an IMU (Inertial Measurement Unit) that allows instantaneous measurements of the acceleration $[g]$, angular velocity $[°s^{-1}]$ and magnetic flux density $[\mu T]$.

The axes of the wind turbine variables are also shown in Fig. 1. All axes are perpendicular. The Z axis is perpendicular to the ground plane, pointing up; the Y axis is parallel to the axis of the turbine, pointing towards the back, and the X axis is pointing right. Note some of the springs are clearly visible at the right of the photograph with blue circles in their top terminations. The microcontroller can be seen at the left and the IMU is behind it, partially obscured by it.

Experiments were carried out with this small prototype in a 6-m long wind tunnel, with the following calibrated wind speeds: 8.5, 9.3, 10.1, 11, 11.6, 12.4, 13.1, and 13.8 $[ms^{-1}]$.

The low scale model of the turbine allows the control of the blade pitch angle with the following rotations: -30, -20, -10, 10, 20 and 30 [°]. These are the two input variables of the regression models we have obtained.

Fig. 1. Front view of the wind turbine with labeled axes.

In addition, to implement the barge-type floating wind turbine, a set of springs along the perimeter of its platform have been placed, so to emulate the oscillation caused by the waves and the wind.

3   Experimental Data

The combination of wind speed (8 values) and blade pitch angles (6 values) gives 48 combinations that will be used for the experiments. Each experiment lasts 10 s, and measurements are taken every 100 ms, thus, we have 99 samples per experiment. Table 1 shows some of the data obtained from an experiment. Note the three signals measured: acceleration (Acc), angular velocity (Gyro), and magnetic flux density (Mag), all them along axes x, y and z. The first column is the time and it is possible to see the 100 ms separation between samples. To illustrate the shape of these signals they are also shown in Fig. 2.

Table 1. Sample of data at $8.5\,\mathrm{ms}^{-1}$ and $10°$.

Time [0.1 s]	Acc x [g]	Acc y [g]	Acc z [g]	Gyro x [°s^{-1}]	Gyro y [°s^{-1}]	Gyro z [°s^{-1}]	Mag x [μT]	Mag y [μT]	Mag z [μT]
1	$-0{,}001062012$	$0{,}000982666$	$0{,}978607178$	$-0{,}002361081$	$-0{,}010571808$	$-0{,}010266725$	$-26{,}67796875$	$-6{,}1425$	$-9{,}7734375$
2	$-0{,}010552979$	$-0{,}00043335$	$1{,}0246521$	$0{,}041690331$	$0{,}010730982$	$-0{,}008038288$	$-26{,}16773438$	$-5{,}338125$	$-9{,}7734375$
⋮	⋮	⋮	⋮	⋮	⋮	⋮	⋮	⋮	⋮
99	$-0{,}00824585$	$-0{,}018615723$	$1{,}048773193$	$0{,}004443608$	$-0{,}017681582$	$-0{,}026210657$	$-26{,}31351563$	$-5{,}630625$	$-9{,}7734375$

Some analysis and pre-processing have been carried out on the data. First, as the measured signals are vectorial a new feature is added to the dataset; the modulus of each vector is calculated and added as a new column. This is obtained by taking the square root of the sum of the squares of the components of each vector. This new feature was included as it was considered to be very informative and easier to interpret than the metrics along a single axis. Table 2 shows the newly added columns along some statistical figures of the variables:

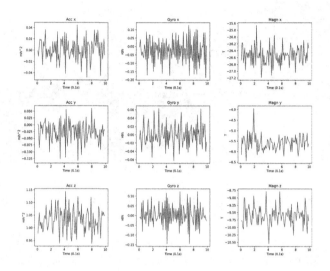

Fig. 2. Signals at a wind speed of 8.5 ms^{-1} and a pitch angle of 10°.

Table 2. Data at a wind speed of 13.8 ms^{-1} and 10° pitch angle.

	Time [0.1 s]	Acc x [g]	Acc y [g]	Acc z [g]	Gyro x [°s^{-1}]	Gyro y [°s^{-1}]	Gyro z [°s^{-1}]	Mag x [μT]	Mag y [μT]	Mag z [μT]	len Acc	len Gyro	len Magn
count	99.000000	99.000000	99.000000	99.000000	99.000000	99.000000	99.000000	99.000000	99.000000	99.000000	99.000000	99.000000	99.000000
mean	50.000000	0.027833	−0.030619	1.100040	−0.041565	0.004803	0.001589	−25.751742	−5.556761	−8.289062	1.113743	0.159670	27.623939
std	28.722813	0.093319	0.140280	0.263003	0.105575	0.062644	0.114191	0.316227	0.396620	0.426263	0.261994	0.064237	0.303083
min	1.000000	−0.203320	−0.346960	0.684369	−0.306463	−0.131929	−0.215256	−26.459297	−7.458750	−10.195312	0.698817	0.038939	26.800792
25%	25.500000	−0.031522	−0.118387	0.926773	−0.125084	−0.040072	−0.089940	−25.985508	−5.703750	−8.507812	0.941567	0.115168	27.418520
50%	50.000000	0.022394	−0.037952	1.013049	−0.046399	0.009802	0.004165	−25.803281	−5.557500	−8.296875	1.025440	0.152568	27.638048
75%	74.500000	0.078992	0.059320	1.198538	0.037200	0.051533	0.079461	−25.548164	−5.338125	−8.015625	1.208403	0.194613	27.837288
max	99.000000	0.322821	0.334290	1.694971	0.210176	0.129700	0.341136	−24.782813	−4.680000	−7.101562	1.746060	0.347568	28.295480

Finally, the tables for the different experiments were integrated into a single table, to have all information we used to feed the supervised models. In addition, 10 different statistical metrics were obtained for each variable: median, mean, var (variance), ptp ($ptp = range = max - min$), max (maximum), min (minimum), std (standard deviation) and median_abs, max_abs and min_abs (variations of median, max and min, where the absolute value is taken before applying the function).

These statistical metrics are the target variables to be predicted. That is, taking as inputs wind speed and the blade angle, the machine learning models will learn to predict the mean, or the median, ... of Acc x, or Gyro y, ... as it is possible to see summarized in Table 3. As there are 12 measured signals (9 directly measured + 3 modulus), a total of 120 different target outputs will be considered. Therefore Table 3 has 48 rows (6 angles × 8 wind speeds) and 122 columns: 2 (angle and wind speed) + (10 metrics × 12 signals) with a total of 5.856 cells.

Table 3. Grouped data set columns.

angle	windspeed	median Acc x	mean Acc x	var Acc x	ptp Acc x	amax Acc x	amin Acc x	std Acc x	...	min_abs len Magn
...	...	...	...	...	...	...	...	...	...	...

Before training the models, the data were scaled using a Standard Scaler, as it was found that the data were mostly normally distributed. This scaler subtracted the mean of the data and then divided by the standard deviation [2]. The variables were represented in time, to observe their evolution. In addition, as an example, Fig. 3 shows a 3D graphical representation of the standard deviation of the $Gyro_y$ regarding the angles and wind speeds. This is one of the statistical metrics that we have predicted with the regression models.

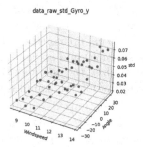

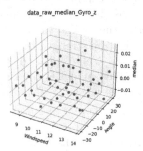

Fig. 3. Standard deviation of the angular velocity in the y axis.

Fig. 4. Median angular velocity in the z axis.

Some of the statistical metrics show a nonlinear or even almost random behavior, that makes some of them more difficult to learn than others. Figure 4 is one example of seemingly randomly distributed data.

To further study the dynamics of the wind turbine, a periodicity analysis was performed using the FFT (Fast Fourier Transform), in order to find the frequency at which the turbine vibrates. A prior theoretical analysis showed that the turbine would vibrate faster at large wind speeds and that the main rotation would be in the pitch axis (with the top moving backwards and forwards) due to the direction of the wind and the exerted forces. The results of the FFT confirmed this statement.

4   Supervised Regression Models

As explained, our objective is to obtain different models that receive the wind velocity and the pitch angle as inputs and compute the statistical metrics listed in the previous section. Thus the problem trying to be solved is a regression problem. To do it, six different techniques to obtain supervised regression models were tested: Linear regression, Polynomial regression, Ridge regressor, Huber regressor, Gaussian regressor and Neural Networks (MLPRegressor). All these techniques were implemented using the library Scikit-learn in Python [2]. The hyperparameter tuning, where applicable, was performed using Scikit-learn's GridSearch.

The following list details the models and the motivation for using them [2]:

- **Linear** Simplest model. Ordinary least squares linear model with coefficients $w = (w_1, \ldots, w_p)$. No changes on default parameters, defaults used: *fit_intercept = True*, *normalize = False* and *positive = False*.
- **Polynomial** Same Linear model as previous with polynomial features added to the pipeline. More complex than linear (potentially better fit but also higher change of overfitting). Degrees 2 to 4 were found to be the best performers (*degree = 2* chosen for already good performance and lower chance of overfit), *interaction_only = False*, *include_bias = True* and *order = C*.
- **Ridge** This regressor is useful when the independent variables are highly correlated. It minimizes the following objective function: $||y - \theta x||_2^2 + \alpha ||\theta||_2^2$, where α is the regularization term, θ are the weights, x is the training data and y is the target values. No changes on the parameters, defaults used. Regularization strength $\alpha = 1.0$, *fit_intercept = True*, *normalize = False*, *max_iter = None*, *tol = 10^{-3}* (precision of the solution) and *solver = auto*.
- **Huber** This model is robust to outliers. It optimizes the squared loss for the samples where $|\frac{y - \theta x}{\sigma}| < \epsilon$ and the absolute loss for the samples where $|\frac{y - \theta x}{\sigma}| > \epsilon$, where σ and θ are the parameters to be optimized. σ makes sure that if y is scaled, ϵ does not need to be rescaled too (this archives the same robustness). θ is the hypothesis, x is the training data and y is the target values. Default parameters: Number of samples that should be classified as outliers $\varepsilon = 1.35$, *max_iter = 100*, regularization parameter $\alpha = 0.0001$, *warm_start = False*, *fit_intercept = True* and *tol = 10^{-5}*.
- **Gaussian** This is a non-parametric Bayesian regressor, it features a very different nature to the previous models and thus, different behavior. No tuning, defaults used: *kernel = None*, value added to the diagonal of the kernel matrix during fitting $\alpha = 10^{-10}$, *optimizer = fmin_l_bfgs_b*, *n_restarts_optimizer = 0*, *normalize_y = False*, and *random_state = None*.
- **MLPRegressor** Optimizes the squared error using gradient descent of a neural network. Changes: *hidden_layer_sizes = (2, 4, 2)*. The rest, were default parameters: *activation = relu*, *solver = lbfgs*, L2 penalty (regularization term) $\alpha = 0.0001$, *batch_size = auto = min(200, n_samples)*, *max_iter = 1000*, *random_ = None*, *tol = 10^{-4}*, *warm_start = False* and *max_fun = 15000*. The default *hidden_layer_size* was (100), considered to be excessive. The solver *lbfgs* was vital for acceptable performance.

5   Results

Figure 5 and Fig. 6 show an example of the prediction errors, more specifically they depict the absolute prediction error of the mean of the acceleration modulus when the polynomial model is used. As the prediction accuracy varies with the wind and the angle, instead of showing a 3D curve, two different figures have been included, in Fig. 5 each line represents a different angle and in Fig. 6 each line represents a different wind speed.

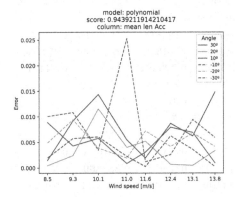

Fig. 5. Absolute error at different wind speeds and angles. Each line represents a different angle.

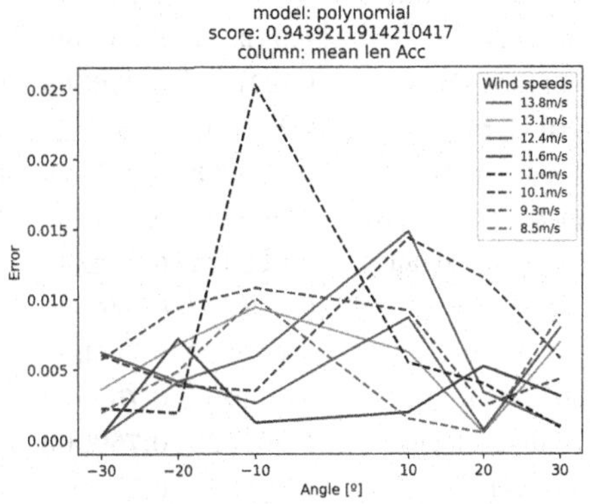

Fig. 6. Absolute error at different wind speeds and angles. Each line represents a different wind speed.

All errors are very low, this means that the predictions are very close to the real value. These figures do not show any clear growing trend with the wind neither the angle. Interestingly the worst predicted value is in the middle of the range, exactly the pair $(11\,\mathrm{ms}^{-1}, -10°)$. The modulus of the acceleration has been selected as it is probably the most important metric to predict together with the gyroscope values.

The best performing models had an error of $\approx 1\%$. The error amount was calculated using the following equation: $E = \frac{|r-p|}{|p|} \times 100$ with $r = real\_value$ and $p = prediction$. In addition, the R^2 score was used (also known as the coefficient of determination) to evaluate the performance of the models:

$$R^2 = 1 - \frac{SS_{res}}{SS_{tot}} = 1 - \frac{\sum_i (y_i - f_i)^2}{\sum_i (y_i - \bar{y})^2} \tag{1}$$

where SS_{res} is the residual sum of squares, SS_{tot} is the total sum of squares, f_i is the predicted value of the i-th sample and y_i is the corresponding real value for the same sample and $\bar{y}$ (the mean of the observed data) is: $\bar{y} = \frac{1}{n}\sum_{i=1}^{n} y_i$[2].

Table 4 shows the models for which any technique obtained a score greater than 0.75, that is the best predicted metrics. The techniques with a score over

0.75 have been boldfaced. Of the 120 metrics only 15 were obtained with a score higher than 0.75. It is observable how in general the metrics related with the z-axis and y-axis are better predicted than the others.

Table 4. Each row has a trained model that has at least a R^2 score over 0.75. Scores over 0.75 are in bold.

function\model	linear	polynomial	ridge	huber	gaussian	mlpr
median len Magn	0.569974	**0.880292**	0.525195	−0.517531	−5.590902	−0.009269
median_abs Magn z	0.538011	**0.791167**	**0.894096**	0.371187	0.563975	−0.066978
mean Acc z	**0.922887**	**0.934596**	**0.934883**	**0.901910**	−0.715218	−0.067811
min_abs Gyro y	0.737107	**0.756568**	0.564034	0.623832	−11.222342	0.286432
ptp Gyro y	0.096708	0.456102	**0.843081**	0.661692	−22.541098	0.482977
mean len Acc	**0.906163**	**0.957825**	**0.810409**	**0.829352**	0.486996	**0.844713**
median_abs Gyro y	0.361724	0.482616	0.676426	**0.913542**	−2.896763	−0.081325
mean len Magn	**0.820942**	0.645844	0.723610	0.614641	0.573424	−0.069494
std len Acc	0.663692	0.609930	**0.759360**	**0.766275**	−3.079215	−0.685077
var Acc z	0.482386	**0.795998**	0.485798	0.190519	−12.515135	−0.054823
std Acc z	0.618925	**0.756320**	0.273930	0.691286	−16.100500	0.682967
median Magn z	**0.862423**	**0.750962**	0.727603	0.690615	**0.884369**	0.574318
std Gyro y	0.730612	0.700493	0.743868	**0.834661**	−2.084311	−0.052904
var Gyro y	0.685186	**0.849338**	**0.752464**	0.478595	−2.848960	0.329246
mean Magn z	**0.765588**	**0.848840**	**0.818878**	**0.798631**	**0.918513**	−0.067585

In addition the Table 5 shows per each technique the number of metrics predicted with a score R^2 over a specified threshold. As it is possible to see, the conclusions from Table 5 are the following. The Gaussian model has the worst performance of the 6 techniques that have been compared. Nonetheless, later on, this model did perform well in some limited scenarios. Neural Networks (MLPR) outperformed the Gaussian model but still only worked for half of what the other four models did. Although the best models are Linear, Polynomial, and Hubber regressors, with 2 models with a R^2 score over 0.90, the Polynomial got an score over 0.95. This means a very accurate prediction.

Table 5. Number of models that have a R^2 score over a certain threshold.

Model	Scores over 0.0	Scores over 0.25	Scores over 0.5	Scores over 0.75	Scores over 0.9	Scores over 0.95
ridge	82	47	27	7	1	0
huber	77	56	29	6	2	0
mlpr	36	26	11	1	0	0
linear	72	47	22	5	2	0
polynomial	78	57	33	10	2	1
gaussian	9	8	4	2	1	0

6 Conclusions and Future Works

In this work, some real data from a wind turbine prototype have been studied. Although the measurements may not be very accurate, different machine techniques have been applied to those data for different combinations of wind speed and pitch angle in order to predict some metrics of the main variables. Some of the supervised regression models were able to predict some output variables with great accuracy. Other variables were, however, very hard to predict.

The trained models learnt how to predict values of the variables for new conditions and even for scenarios outside the experiment range, allowing to explore new configurations. As future works, more experimental data should be used to improve the training of the models, and other variables as the vibration of the wind turbines could be predicted. The prototype of the turbine will be also tested in a water pond and real data will be measured. In addition, it would be interesting to discuss the use of regression vs. ARIMA models.

Acknowledgments. This work is partially supported by the Spanish Ministry of Science and Innovation under the project MCI/AEI/FEDER number RTI2018-094902-B-C21.

References

1. Mikati, M., Santos, M., Armenta, C.: Electric grid dependence on the configuration of a small-scale wind and solar power hybrid system. Renew. Energy **57**, 587–593 (2013)
2. Pedregosa, F., et al.: Scikit-learn: machine learning in Python. J. Mach. Learn. Res. **12**, 2825–2830 (2011)
3. Rico-Azagra, J., Gil-Martínez, M., Rico, R., Nájera, S., Elvira, C.: Benchmark de control de la orientación de un multirrotor en una estructura de rotación con tres grados de libertad. Revista Iberoamericana de Automática e Informática industrial **18**(3), 265–276 (2021)
4. Seo, Y.H., Ryu, M.S., Oh, K.Y.: Dynamic characteristics of an offshore wind turbine with tripod suction buckets via full-scale testing. Complexity **2020** (2020)
5. Serrano, C., Santos, M., Sierra-García, J.: Intelligent hybrid control of individual blade pitch for load mitigation. In: Brito Palma, L., Neves-Silva, R., Gomes, L. (eds.) CONTROLO 2022. LNEE, pp. 599–608. Springer, Cham (2022). https://doi.org/10.1007/978-3-031-10047-5_53
6. Sierra-García, J., Santos, M.: Redes neuronales y aprendizaje por refuerzo en el control de turbinas eólicas. Revista Iberoamericana de Automática e Informática industrial **18**(4), 327–335 (2021)
7. Sierra-García, J.E., Santos, M.: Lookup table and neural network hybrid strategy for wind turbine pitch control. Sustainability **13**(6), 3235 (2021)
8. Tajadura, I., Sierra-García, J.E., Santos, M.: Communication library to implement digital twins based on matlab and IEC61131. In: Brito Palma, L., Neves-Silva, R., Gomes, L. (eds.) CONTROLO 2022. LNEE, pp. 262–271. Springer, Cham (2022). https://doi.org/10.1007/978-3-031-10047-5_23

9. Tomás-Rodríguez, M., Santos, M.: Modelado y control de turbinas eólicas marinas flotantes. Revista Iberoamericana de Automática e Informática Industrial **16**(4), 381–390 (2019)
10. Veers, P.S., et al.: Trends in the design, manufacture and evaluation of wind turbine blades. Wind Energy: Int. J. Progress Appl. Wind Power Convers. Technol. **6**(3), 245–259 (2003)

Dynamic Optimization of Energy Hubs with Evolutionary Algorithms Using Adaptive Time Segments and Varying Resolution

Rafael Poppenborg[(✉)], Hatem Khalloof, Malte Chlosta, Tim Hofferberth, Clemens Düpmeier, and Veit Hagenmeyer

Institute of Automation and Applied Informatics (IAI),
Karlsruhe Institute of Technology (KIT), Karlsruhe, Germany
{rafael.poppenborg,hatem.khalloof,malte.chlosta,clemens.duepmeier,
veit.hagenmeyer}@kit.edu, tim.hofferberth@student.kit.edu

Abstract. Many Renewable Energy Sources (RESs) need to be installed and integrated into the existing grid infrastructure for further developing the energy system towards a 100% RESs supply. To provide the required flexibility for actively compensating local volatile power fluctuations caused e.g. by RES, scheduling of controllable Distributed Energy Resources (DERs) using the concept of Energy Hubs (EHs) and their integration into power networks is a promising approach. The complexity of optimized operation of such EHs while simultaneously providing the required flexibility can be faced by using Evolutionary Algorithms (EAs) for calculating the schedules of internal components of the EHs. To focus the computational effort of the used EA on more important time segments of the problem space, a dynamic optimization method with adaptive time intervals and time resolution is presented. The method improves the quality of the optimization in form of better approximation to a given target value an EH can follow. The proposed approach leads to an average reduction of 11% in the Root Mean Squared Error (RMSE) between target value and measurement by appropriate increased calculation time.

Keywords: Multi-timescale optimization · Evolutionary algorithm · Dynamic optimization · Scheduling

1 Introduction

In the Paris Agreement[1], 196 parties committed to the goal of reducing greenhouse gas emissions. To achieve this, a wide range of Renewable Energy Sources (RESs) needs to be installed and integrated into the existing grid infrastructure. With a high penetration of such RESs in the energy system, new challenges in

---

[1] https://unfccc.int/sites/default/files/english_paris_agreement.pdf visited 14.04.2022.

© The Author(s), under exclusive license to Springer Nature Switzerland AG 2022
H. Yin et al. (Eds.): IDEAL 2022, LNCS 13756, pp. 513–524, 2022.
https://doi.org/10.1007/978-3-031-21753-1_50

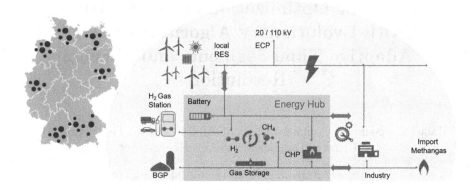

Fig. 1. Energy Hub concept.

energy security are raised for fair energy provision and environmental sustainability. Especially, fluctuation of energy generation by RESs causes problems in grid stability. Therefore, flexible solutions for the energy system are needed to compensate for the deficiencies of RESs, and the concept of Energy Hubs (EHs) providing flexibility for grid operation appears to be a promising approach [5–7] and [14]. An EH combines Distributed Energy Resources (DERs) including, e.g. RESs, energy storage and energy conversion systems as depicted in Fig. 1 to support the electricity grid by providing controllable flexibility for balancing load and supply. While deploying more EHs with various energy carriers into the electric power grid is generally feasible, the task of optimizing the scheduling of the internal DERs within an EH is not an easy task. The main reasons for this complexity can be summarized as follows: (i) The use of RESs is inherent uncertain, since it depends on weather conditions in the energy generation, which are, in turn, uncertain. (ii) Also the variations in load to be compensated are related to several factors like uncertain human behaviour, in different timescales. (iii) A wide range of additional criteria need to be fulfilled by an optimization solution, leading to non-linearities and complex boundary conditions, multi-objectives and increase of dimensionalities.

For scheduling of DERs within an EH an Energy Management System (EMS) is needed [4]. By using a multi-objective optimization algorithm like Evolutionary Algorithms (EAs), the EMS calculates appropriate schedules for the sub-plants of the EH that can meet a third-party control demand of e.g. a power network operator and additionally respect internal optimization criteria and boundary conditions. It then uses these schedules to coordinate the internal powerflow between dispatchable generation units and loads. In the current concepts of EHs, several problems regarding finding a near optimal solution for internal schedules that also meets the external control demand are unsolved. For example, and as stated in [16], simulation results show that an EH using an EA has difficulties optimizing schedules that cover a load with high variability. Moreover, analyzing the flexibility provision in those cases shows that the used EA optimization solution could not reduce the relative power variation of the specified load in frequencies above $5\,h^{-1}$ and therefore provided no flexibility in higher frequencies. One promising solution is dynamically adjusting the time resolution used for optimizing schedule parts in intervals with high fluctuation and adapting the time horizon for the optimization.

In the present paper, such a dynamic optimization approach with adaptive time segments and varying time resolution is introduced. One challenge in using a higher time resolution to improve the optimization result is the increase in calculation time. To face this challenge, only time segments that show a high deviation in meeting the target value should be optimized with higher resolution. Thus, identifying the right time segments has an important role in finding a good trade-off between the improvement of the optimization results and the increase of the total calculation time.

The present work is structured as follows: Sect. 2 provides an overview of the related work and outlines the contribution of the present work. Section 3 illustrates the method of the dynamic optimization approach and is divided into two parts. First, the impact of various time segments and resolution for the optimization algorithm is analyzed. Second, the new method for optimization is presented. The evaluation of our approach is described in Sect. 4. Finally, in Sect. 5, a conclusion is given and an outlook for future work is provided.

2   Related Work

In current research, several concepts using more than one timescale are discussed, e.g., multi-timescale coordinated optimized scheduling in [18], multi-timescale rolling optimal dispatch in [17], timescale adaptive dispatch in [13] and multi-timescale model predictive control in [3], which are grouped in our paper under the term multi-timescale scheduling. The basic idea is using multiple timescales for calculating the optimal unit commitment of various energy resources [11]. Multi-timescales are mainly used in literature for solving problems regarding uncertainties in the optimization, e.g. in [1,17] and [13], separating optimization for economic and operational factors e.g. in [19,20] and [3], improving the control in energy systems or shifting deferrable loads in [15]. In this context a rolling optimization approach is often used allowing different look-ahead periods to combine short-term and long-term benefits [12]. In the following, different multi-timescale scheduling approaches are described to distinguish them from present work.

In [18], a hierarchical optimization approach with three different timescales is used to schedule a combined system of RES, thermal generator, hydro pumped storage, and batteries. The first timescale is a day-ahead scheduling for thermal units based on a 24 h ahead forecast. For optimizing the dispatch of hydro-pumped unit power outputs, a second scheduling is proposed in which a day-ahead schedule as well as a 1 h ahead forecast are used. On the smallest time scale of 15 min, the aforementioned schedules are taken into account to obtain an optimal battery system schedule.

In [1], an approach for an integrated multi-timescale optimization of a coupled multi-type energy supply similar to an EH is presented. Regarding multi-timescale operation, a day-ahead schedule takes the uncertainty of RES generation into account. Additionally, real-time dispatch for storage, combined cooling and heat power, and ice storage air conditioners is used to react to fluctuations in RES and demand.

In [17], a multi-timescale rolling optimal dispatch framework is developed to cope with the impact of uncertainty in load on hybrid micro-grids at timescales of day-ahead and intraday. For the day-ahead scheduling, a distributed robust optimization model considering uncertainties in source-load power is used. While performing the intraday rolling optimization, a relaxed penalty cost for the final state of charge is added to ensure cyclic regulation of the energy storage.

In [13], the selection of the timescale is based on a threshold which is defined by a confidence interval. The optimization addresses scheduling of a RES on an island and has to deal with very high uncertainties. When the prediction error exceeds the available reserve, the energy system can become unstable. Therefore, the timescale is dynamically selected to stay within the confidence interval.

A multi-timescale economic scheduling strategy for a virtual power plant is presented in [20]. Their goal is to unlock the potential of a large quantity of deferrable load (DL) by participating in the wholesale energy and reserve market. A day-ahead bidding and real-time operation are used for the multi-timescale scheduling. With the proposed strategy, efficient management of a large number of DL can be realized while reducing energy management complexity and increasing overall cost-effectiveness.

A multi-timescale coordinated optimization of an EH is proposed in [3]. It includes a global optimization of day-ahead economic dispatch, a local intraday model predictive control with 15 min timescale and a minimization of the total adjustment amount of all controllable devices every 5 min.

The authors of [19] propose a multi-timescale model for regional integrated energy systems. The multi-timescale aspect is used on two levels in the model. On the first level, it is applied to differentiate between day-ahead and intraday scheduling. For day-ahead scheduling, the objective is to minimize costs in scheduling the energy system. The intraday scheduling uses a rolling optimization which is divided into three different control sub-layers. They achieve a balance of supply and demand in the system and can also restrain the fluctuation of renewable energy and load in the intraday scheduling.

In [21], the main goal is to coordinate the substation on-load tap changer operation on an hourly time scale in a power network with photovoltaic (PV) inverters and battery storage operations on a 15 min basis in the context of smart distribution grids.

The usage of multi-timescale scheduling in the context of deferrable appliances in a smart home is proposed in [15]. The appliances are categorized into two groups. While one group can be shifted on a hourly scale within a day, the second group can be additionally shifted between days.

With the main focus on managing uncertainties in the demand and power generation of RES, multiple methods are applied. In [1], several fixed timescales are used for the optimization, and specific facilities are assigned to one of these. Additionally, in [18] and [17] the timescales are optimized in hierarchical order so that lower timescales use the result of a higher timescale optimization as an input. In [13], the timescales are dynamically chosen. Another approach is presented in [19,20] and [3] using a day-ahead timescale with focus on economic and intraday timescales for operational aspects. [19] introduces –as a multi-timescale approach– an intraday rolling optimization for control, and is applied with a special focus on battery systems in [21]. Finally, [15] applies different timescales to shift deferrable loads.

However, in the aforementioned related work, there is no approach that uses different timescales to apply adaptive time segments for scheduling flexibility provided by Distributed Energy Resources (DERs) on a day-ahead scale. The method presented in Sect. 3 uses different timescales within the same schedule to solve the problem of optimizing schedules for an EH to follow a target value with different DERs included. In other words, all DERs are considered in the day-ahead scheduling without distinction of specifically related timescales. Timescales are therefore applied using adaptive time segments which are chosen based on the deviation to the target value. Furthermore, the EA implemented for evaluation of the present method relies on [2,9] and [10].

3   Method

The goal of the present method is to improve the total power exchange of the EH at its electrical connection point to provide flexibility to the grid. The total power exchange results from the superposition of the calculated schedules provided for each subsystem to meet a third-party target. The main task of the optimizer, such as EA – beside other objectives like economic cost – is to minimize the difference between the power output of the EH $P_{EH}(t)$ and the third-party target value $P_{target}(t)$ for all time segments according to Eq. 1.

$$min\, d_{rmse} = \sqrt{\frac{\sum_t^n \left(P_{EH}(t) - P_{target}(t)\right)^2}{n}} \qquad (1)$$

The general idea of our method for solving the problem described in Sect. 1 is to concentrate the computational effort of the EA on specific time segments where there is high variability in the given target value. The present method allows an EA to generate schedules with more control point changes for those segments than for others, resulting in a better approximation. The computational effort to find a solution in a specific time segment is scaled according to the time resolution considered for this segment, and focus is to shift the computational effort to time segments which are more difficult to optimize.

More general, this approach can also be used vice versa. If there is little to no fluctuation within a time segment, there is no need to provide more than one schedule entry for this segment. The computational effort is adaptable for each time segment. Time resolution determines the number of entries in the schedule and, therefore, also the number of possible solutions to approximate the target value for the power output of an EH. Optimizing a complete schedule with an EA in higher time resolution but with the same computational effort does not necessarily lead to an improvement of the optimization as only a larger search space is inspected.

The time resolution in the optimized schedules can be inspected from two perspectives, the use case and the optimization algorithm, namely the EA in combination with the EMS. In the perspective of the use case, the time resolution determines how fast an EH can respond to changes in the target value. For example, if the demand changes in 15 min intervals and the schedule only provides control steps every 30 min, the EH is not able to respond to changes below 30 min. The use case benefits only from a higher time resolution as long as it is not more granular than the input data. Looking at the EA calculation time or the number of populations analyzed is another perspective. An EA uses genetic operators which are based on random changes in the solutions to find new and better solutions over multiple generations. In general, scheduling Distributed Energy Resources (DERs) is an NP-hard problem and therefore when applying an EA, the goal could not be to find the best possible solution but only a good approximation in timely manner using the available computing power. The more fine-granular the time scale for schedules is defined, the larger the search space which directly results in higher computation time for finding an appropriate solution. To understand the proposed solution, it is important first to understand that the EA itself does not work with absolute numbers. The value of the power output of a subsystem in a certain scheduling interval, as well as the start time and duration are not interpreted by the EA as absolute values. In the used EA, a schedule (chromosome) is represented by a chain of set points (genes) which can be interpreted as a schedule with a certain number of time intervals. The set point value is a unitless number between -1 and 1. "Duration" and "Start Time" can be interpreted as relative values, whereas the "Start Time" is the index of the corresponding schedule segment. It is one main duty of the EMS to transform the contextless information within the chromosomes to their technical interpretation context. Thereby, translating the unitless and relative values to their absolute values, e.g. the real power output of a technical sub-system will be calculated by the EMS based on the relative value given by the optimizer and configuration parameter about how to interpret this number for the given subsystem. The same transformation will be done for the scheduling interval by translating index values into start times and duration values into time intervals. A second important duty of the EMS after this translation is then to calculate objective functions and to verify the boundary conditions.

The complete optimization process, including the implemented use case, is depicted in Fig. 2. First, the current state of the system under concern is determined via the system simulation (1) to initialize a new optimization run with low time resolution (2). Then, the first optimization is run (red box, 3–8). The optimizer performs a loop where the EA first generates new populations (3) by applying genetic operators to the (initial) populations (8). The schedules are

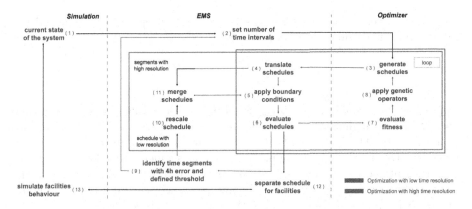

Fig. 2. Optimization process overview.

then send to the EMS which translates them into the technical context (4), and then evaluates the schedules (6) by first calculating and checking the boundary conditions (5) and then calculating the objective functions. With the results from the evaluation from the EMS a new fitness value is calculated in (7) by the optimizer. If the fitness value is too low, the optimizer performs the next generation cycle. If the fitness threshold configured is reached, the optimization run ends and the found schedules are translated into the technical context and are then send to the simulation. The simulation will apply the schedules to the system model, which calculates the resulting behavior of the EH at its connection point to the external network. These values can then be compared to the given target value curve to calculate how well the internal schedules approximate the target values. In this first optimization run using a low time resolution, the EA has less options to respond to changes in the target value. Giving only a set point for one larger time interval that also approximates the given target values is unlikely. Hence, after the first optimization run, an analysis of the approximation quality is performed where time segments are identified, which does not provide a good approximation (9). Then, a new optimization run (green box) is initialized which uses a higher time resolution for the determined time segments. When this optimization run finishes, the new schedule parts with higher time resolution are rescaled (10) and then merged (11) with the schedules for the lower-resolution time segments.

3.1 Identifying Time Segments for Adaptive Time Resolution

To apply this approach, a method is needed to identify time segments that could not be adequately approximated with a lower time resolution. The time resolution is then defined for the segments according to their variability and approximation error. One approach is by sequentially executing two optimization processes as depicted in Fig. 2. Suitable criteria for classifying time segments for a second optimization need to be defined. As one possible criterion, the absolute error d_{abs} of certain time segments s between the proposed schedule and the target values is calculated according to Eq. 2 in our case study.

$$d_{abs}(s) = \sum_{t_s} |P^{EH}_{target}(t_s) - P^{EH}(t_s)| \tag{2}$$

Depending on a configurable threshold of d_{abs} the identified segments are then optimized with higher time resolution and combined with the low time resolution segments. The EMS builds the actual time segments using the generated schedules. During the combining process the time segments with low time resolution are replaced by the new time segments. The evaluation will be done for the complete schedule and not only for the new generated schedules. The combining process is explained in more detail in the following Subsect. 3.2.

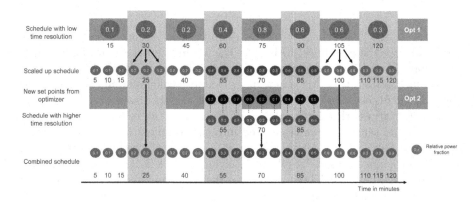

Fig. 3. Combining schedules with different time resolution.

3.2   Combining Schedules

The process of combining and interpreting time segments with different time resolutions is an important step for the subsequent evaluation of a schedule. Having only one schedule including different time segments makes the evaluation more efficient than evaluating the time segments separately. The evaluation of schedules includes applying the respective boundary conditions. Evaluating each time segment separately would result in executing a full optimization process for each time segment and reduce the time horizon of the optimization to the length of the time segment. But the goal of the optimization is finding a near optimal schedule for the entire time horizon which means all time segments combined. Therefore, the entire time horizon must also be evaluated in the optimization process with adaptive time segments and varying time resolution. By combining and interpreting different time resolutions as one schedule, the entire time horizon can still be evaluated as one while using the advantage of the different time resolutions as depicted in Fig. 3.

4   Evaluation

To assess whether the previously presented approach is a viable method to improve the ability of an EH to follow a target value, an evaluation is done using a concrete use case. The evaluation setup includes a specification of the environment for the evaluation with a test scenario, as well as an implementation of the approach. Furthermore, a precise definition of the evaluation criteria used to analyze the test results is needed.

4.1   Specification of Evaluation Environment and Criteria

The evaluation environment is based on previous work described in [9, 10, 16] and [2]. The co-simulation environment from [16] is used with the combination of Distributed Energy Resources (DERs) depicted in Fig. 1. The EA used for scheduling the EH is the General Learning Evolutionary Algorithm and Method (GLEAM) from [2] in combination with the Energy Management System (EMS) from [16]. The decisive criterion is the Root Mean Squared Error (RMSE) d_{dev} calculated according to Eq. 1. Furthermore, the calculation time for the complete process is evaluated and compared.

The settings for the EA are as follows: The generation amount of 100 is considered as termination criterion and the population size of 50 is experimentally defined in previous work. A drastically reduced population size produces low quality solutions, whereas a population size bigger than 50 does not lead to significant better results. Further configurations, e.g. mutation and crossover, are adopted from [8]. In the first optimization process, the simulation and optimization time resolution is 15 min, which corresponds to 96 intervals in the schedule. The simulation and optimization time resolution in the second optimization process is both 5 min. The number of intervals for optimization is then calculated after the time segments are identified. The optimization time horizon is one day and seven days are inspected. The number of optimizations that can be tested in the second optimization process depends on which time segments can be identified for each day. The criterion used to identify time segments is the absolute deviation according to Eq. 2 for a rolling period of 4 h between the target value and the proposed schedule.

4.2   Results

The results of the first optimization runs for an exemplary day of the evaluated week are shown in Fig. 4. According to the criterion described above, the time segment identified for a second optimization with higher time resolution is marked by the solid blue line between 9.75 h and 17.25 h. The RMSE after first simulation run accumulates to 2.0 MWh with a total calculation time of 17 min. For the identified time segment of 7.5 h a second optimization with 90 intervals is conducted. The combined schedule is shown in Fig. 5. Blue dotted lines mark the time segment where the second schedule is inserted. By evaluating the merged schedule, only solutions with continuous transition at the edges of the identified time segments are valid. In the example shown in Fig. 5, the RMSE is reduced by 15.5% to 1.69 MWh. The average improvement for the evaluated week is 11%, while the computation time increases up to three times.

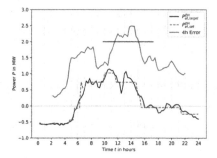

Fig. 4. Optimization result with low time resolution for an example day.

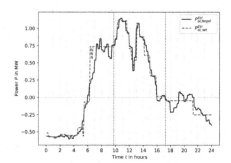

Fig. 5. Combined schedule for an example day.

5   Conclusion and Outlook

For supporting the integration of Renewable Energy Sources (RESs) into power networks by using the balancing flexibility of Energy Hubs (EHs), the performance of a respective optimization method is crucial for scheduling Distributed Energy Resources (DERs) as sub-systems of an EH regarding multiple objectives while also respecting complex boundary conditions. In this context, the present paper analyzes the impact and usage of applying different time resolutions to the optimization of schedules. It then presents a new approach by using adaptable time segments for EA based schedule optimization and a corresponding method to identify time segments that should be optimized with different time resolutions. Using this method, a schedule is first calculated with low time resolution, and then the results are used to identify time segments that should be optimized with higher time resolution. Then, the identified parts of the schedule are recalculated with higher time resolution. It is also explained how time segments of different time resolutions are combined and interpreted to build one schedule for the EH. For the evaluation of the method, a concrete test setup is implemented. Tests regarding different time resolutions for the optimization process, as well as the new optimization approach, are executed and the results are described. The analysis shows that the use of different time resolutions for the full schedule of an optimization process has not necessarily a positive impact on the performance. However, applying the developed method shows that using a higher time resolution for only specific time segments reduces the average deviation between the electrical power output of an EH and the target value by 11%. The analysis also shows that the approach can achieve similar results regarding the deviation and calculation time using different parameter settings of the EA. Therefore, the present research shows that scheduling an EH with a dynamic optimization approach using adaptive time segments and varying time resolution can enhance the flexibility provided by an EH. The presented research focuses on developing a dynamic optimization method to improve the ability of an EH to follow a target value. In future work, a detailed evaluation of the test setup implemented should be performed. The sequential optimization process presented increases the computation time. Future work in this area should focus first on the detection of time segments that are hard to optimize for EAs, so that these are identified in advance to save one optimization run. For this purpose, artificial neural networks offer great opportunities to characterize time series data. Second, computational performance of the used EA should be investigated and possibilities for parallelizing the algorithm should be used.

Acknowledgment. The authors gratefully acknowledge funding by the German Federal Ministry of Education and Research (BMBF) within the Kopernikus Project ENSURE 'New ENergy grid StructURes for the German Energiewende'.

References

1. Bao, Z., Zhou, Q., Yang, Z., Yang, Q., Xu, L., Wu, T.: A multi time-scale and multi energy-type coordinated microgrid scheduling solution-part i: model and methodology. IEEE Trans. Power Syst. **30**, 2257–2266 (2015). https://doi.org/10.1109/TPWRS.2014.2367127
2. Blume, C., Jakob, W.: Gleam - general learning evolutionary algorithm and method: Ein evolutionärer algorithmus und seine anwendungen. Schriftenreihe des Instituts für Angewandte Informatik - Automatisierungstechnik, Universität Karlsruhe (TH), vol. 32. KIT Scientific Publishing (2009). https://doi.org/10.5445/KSP/1000013553
3. Cheng, S., Wang, R., Xu, J., Wei, Z.: Multi-time scale coordinated optimization of an energy hub in the integrated energy system with multi-type energy storage systems. Sustain. Energy Technol. Assess. **47**, 101327 (2021). https://doi.org/10.1016/j.seta.2021.101327
4. Fiorini, L., Aiello, M.: Energy management for user's thermal and power needs: a survey. Energy Rep. **5**, 1048–1076 (2019). https://doi.org/10.1016/j.egyr.2019.08.003
5. Geidl, M., Andersson, G.: A modeling and optimization approach for multiple energy carrier power flow, pp. 1–7 (2005). https://doi.org/10.1109/PTC.2005.4524640
6. Geidl, M., Andersson, G.: Optimal power flow of multiple energy carriers. IEEE Trans. Power Syst. **22**(1), 145–155 (2007). https://doi.org/10.1109/TPWRS.2006.888988
7. Geidl, M., Koeppel, G., Favre-Perrod, P., Klockl, B., Andersson, G., Frohlich, K.: Energy hubs for the future. IEEE Power Energ. Mag. **5**(1), 24–30 (2007). https://doi.org/10.1109/MPAE.2007.264850
8. Jakob, W., Quinte, A., Stucky, K.-U., Süß, W.: Fast multi-objective scheduling of jobs to constrained resources using a hybrid evolutionary algorithm. In: Rudolph, G., Jansen, T., Beume, N., Lucas, S., Poloni, C. (eds.) PPSN 2008. LNCS, vol. 5199, pp. 1031–1040. Springer, Heidelberg (2008). https://doi.org/10.1007/978-3-540-87700-4_102

9. Khalloof, H., et al.: A generic distributed microservices and container based framework for metaheuristic optimization. In: Proceedings of the Genetic and Evolutionary Conference Companion, Kyoto, Japan, 15–19 July 2018, pp. 1363–1370. Association for Computing Machinery (ACM) (2018). https://doi.org/10.1145/3205651.3208253

10. Khalloof, H., Jakob, W., Shahoud, S., Duepmeier, C., Hagenmeyer, V.: A generic scalable method for scheduling distributed energy resources using parallelized population-based metaheuristics. In: Arai, K., Kapoor, S., Bhatia, R. (eds.) FTC 2020. AISC, vol. 1289, pp. 1–21. Springer, Cham (2021). https://doi.org/10.1007/978-3-030-63089-8_1

11. Kurita, A., et al.: Multiple time-scale power system dynamic simulation. IEEE Trans. Power Syst. **8**, 216–223 (1993). https://doi.org/10.1109/59.221237

12. Le, K.D., Day, J.T.: Rolling horizon method: a new optimization technique for generation expansion studies. **PAS-101**, 3112–3116 (1982). https://doi.org/10.1109/TPAS.1982.317523

13. Li, C., et al.: A time-scale adaptive dispatch method for renewable energy power supply systems on islands. IEEE Trans. Smart Grid **7**, 1069–1078 (2016). https://doi.org/10.1109/TSG.2015.2485664

14. Maroufmashat, A., Taqvi, S.T., Miragha, A., Fowler, M., Elkamel, A.: Modeling and optimization of energy hubs: a comprehensive review. Inventions **4**, 50 (2019)

15. Mehdi, R.A.: Scheduling deferrable appliances and energy resources of a smart home applying multi-time scale stochastic model predictive control. Sustain. Cities Soc. **32**, 338–347 (2017). https://doi.org/10.1016/j.scs.2017.04.006

16. Poppenborg, R., et al.: Energy hub gas: a multi-domain system modelling and co-simulation approach. In: Proceedings of the 9th Workshop on Modeling and Simulation of Cyber-Physical Energy Systems, no. 12. Association for Computing Machinery (2021). https://doi.org/10.1145/3470481.3472712

17. Qiu, H., Gu, W., Xu, Y., Zhao, B.: Multi-time-scale rolling optimal dispatch for ac/dc hybrid microgrids with day-ahead distributionally robust scheduling. IEEE Trans. Sustain. Energy **10**, 1653–1663 (2019). https://doi.org/10.1109/TSTE.2018.2868548

18. Xia, S., Ding, Z., Du, T., Zhang, D., Shahidehpour, M., Ding, T.: Multitime scale coordinated scheduling for the combined system of wind power, photovoltaic, thermal generator, hydro pumped storage, and batteries. IEEE Trans. Ind. Appl. **56**(3), 2227–2237 (2020). https://doi.org/10.1109/TIA.2020.2974426

19. Yang, H., Li, M., Jiang, Z., Zhang, P.: Multi-time scale optimal scheduling of regional integrated energy systems considering integrated demand response. IEEE Access **8**, 5080–5090 (2020). https://doi.org/10.1109/ACCESS.2019.2963463

20. Yi, Z., Xu, Y., Gu, W., Wu, W.: A multi-time-scale economic scheduling strategy for virtual power plant based on deferrable loads aggregation and disaggregation. IEEE Trans. Sustain. Energy **11**, 1332–1346 (2020). https://doi.org/10.1109/TSTE.2019.2924936

21. Zafar, R., Ravishankar, J., Fletcher, J.E., Pota, H.R.: Multi-timescale model predictive control of battery energy storage system using conic relaxation in smart distribution grids. IEEE Trans. Power Syst. **33**, 7152–7161 (2018). https://doi.org/10.1109/TPWRS.2018.2847400

Special Session on Computational Intelligence for Imbalanced Classification

Solving Multi-class Imbalance Problems Using Improved Tabular GANs

Zakarya Farou[1]([✉]) [iD], Liudmila Kopeikina[1], and Tomáš Horváth[1,2] [iD]

[1] Department of Data Science and Engineering, Institute of Industry - Academia Innovation, ELTE Eötvös Loránd University, Pázmány Péter sétány 1/C, 1117 Budapest, Hungary
{zakaryafarou,hasx2y,tomas.horvath}@inf.elte.hu
[2] Institute of Computer Science, Pavol Jozef Šafárik University, Jesenná 5, 040 01 Košice, Slovakia

Abstract. Multi-class imbalance problems are non-standard derivative data science problems. These problems are associated with the skewness in the data underlying distribution, which, in turn, raises numerous issues for conventional machine learning techniques. To address the lack of data in imbalance problems, we can either collect new data or oversample the underrepresented classes by synthesizing artificial data from original instances. This paper focuses on the latter and introduces two novel tabular GAN variants to handle multi-class imbalance problems. Empirical results on three datasets from the UCI repository demonstrated that the suggested approaches that use our proposed filtering algorithm based on neighboring rules improved the ability of the decision tree classification model to recognize underrepresented class instances, decreased the bias toward the majority class, and enhanced its generalization ability.

Keywords: Imbalanced learning · Generative adversarial networks · Data augmentation · Data filtering · Multi-class classification

1 Introduction

Nowadays, class imbalance problems have drawn much attention due to the classification difficulties caused by uneven class distributions. In many real-world scenarios, the data distribution among dataset classes is not uniform. At least one class, also known as the minority class, has fewer samples than the others (majority classes). Standard machine learning algorithms are usually biased towards the majority classes in imbalanced problems. In this case, classifiers perform well in majority classes but poorly in minority classes, although the minority class may have more valuable knowledge.

To handle class imbalance problems, we could under-sample the majority classes, collect and annotate data, or synthesize new ones from the existing data, which will over-sample the minority classes. As shown in Fig. 1, gathering and labeling the data needs domain experts, also known as raters that examine

© The Author(s), under exclusive license to Springer Nature Switzerland AG 2022
H. Yin et al. (Eds.): IDEAL 2022, LNCS 13756, pp. 527–539, 2022.
https://doi.org/10.1007/978-3-031-21753-1_51

and annotate the data. Each rater gives his judgment (a label) about a given data. If a disagreement occurs between raters, i.e., $C = False$, then an expert rater will give his judgment and resolve the disagreement.

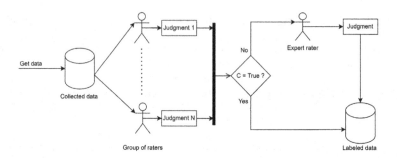

Fig. 1. Data annotation workflow.

This solution is time-consuming, costly, and sometimes difficult due to privacy constraints. As an alternative, many data scientists proposed several data augmentation algorithms that create artificial data solely by using the existing data. These solutions are speedy and do not require any new data to be collected or annotated. However, data augmentation would only be worthwhile if the generated data have new patterns pertinent to the task and have not yet been seen in pre-training [15], which is not the case for many SMOTE alike over samplers. They try to overpopulate underrepresented classes without considering the original data distribution. As a result, they generate noisy samples, borderline examples, and overlapping regions while solving the lack of data problem. Also, most efforts so far are only focused on binary, i.e., two-class imbalance problems. In contrast, there are unsolved issues in a multi-class domain, which are increasingly encountered in real-world applications.

Moreover, even though the problem of imbalanced learning is mature enough, the growing number of new real-world cases dealing with imbalanced data makes the domain relevant and of high interest. Specifically, there is a need to explore the actual distribution aligned within the original data and examine the methods for multi-class cases.

Thus, the paper aims to investigate the usability of generative adversarial networks (GANs) as over samplers, i.e., generating minority class samples as similar as possible to the original multi-class data. Previous studies [7,9] have shown that using GANs as a data augmentation technique has some drawbacks, leading to bad instances creation. In this turn, we aim to develop, implement, and experimentally evaluate an approach that could tackle both problematics. That is, create artificial data consistent with the existing ones and eliminate the issues of the bad instances, also known as noisy instances and borderline samples.

The paper is structured as follows: Sect. 2 explains various concepts and solutions related to multi-class imbalance learning and highlights previous works related to tabular GANs and their usability. Section 3 describes the proposed methodology to tackle the previously mentioned problematics. The description regarding the used dataset, performed experiments, results, and discussion are incorporated in Sect. 4. Lastly, Sect. 5 outlines the conclusion about the conducted research and the potential research direction to further improve learning and classification with class imbalance problems.

2 Multi-class Imbalance Learning

Multi-class imbalance is one of the most challenging problems within the machine learning domain which is present by nature in many real-world application such as fault prediction [1] and fraud detection [16], cancer detection [7], text classification [5], particles identification [8], and many more [12]. However, most existing methods aim to address the binary class imbalance problems, as the multi-class imbalance problem is more complicated since it involves dealing with many classes and their complex relationships.

The multi-class imbalanced problem extends the traditional binary problem where the number of classes in the dataset is enlarged to k classes instead of two ($k > 2$). While imbalance exists in the binary class imbalance problem when one class severely outnumbers the other class, extended to multiple classes, the effects of imbalance are even more problematic. The reason is that, given k classes, there are various ways for class imbalance to manifest itself in the dataset. There are two significant reasons why the multi-class imbalance problem is of extreme interest. First of all, as we already mentioned, most algorithms do not deal with the wide variety of challenges multi-class imbalance presents. Secondly, several classifiers do not easily extend to the multi-class domain.

2.1 Handling Multi-class Imbalanced Problems

Diverse techniques have been proposed to overcome the problem mentioned above [7]. These techniques aim to enhance supervised models sensitivity toward data by inferring rules that form a reasonable decision boundary that segregates well negative C_i^- from positive C_j^+ classes. We can divide the techniques into decomposition-based and resampling-based approaches. In decomposition-based approaches, the original problem is reduced to two-class subsets. Then it can be directly solved by any existing technique for the binary imbalanced scenarios. We distinguish two types of decomposition techniques, one-versus-all (OVA) and one-versus-one (OVO). OVA [10] tries to set $F - 1$ classifiers for L classes. While OVO [10] aims to build $F \times (F - 1)/2$ binary classifiers to classify samples from L classes. However, resampling-based approaches preprocess the training data by adding more minority observations or reducing the size of the majority class. Resampling methods can be further categorized into random sampling approaches and augmentation approaches. In random sampling, we balance class distribution through the random replication of minority class examples or the random dismissal of majority class instances. While in augmentation approaches, the choice of samples to add/eliminate is decided based on specific criteria rather than randomly chosen. Data augmentation is also known as directed sampling.

2.2 Data Augmentation Using Tabular GANs

The GAN, introduced in 2014, is applied in various domains such as language, speech processing, and text generation. However, the most successful domain is computer vision. Although researches on GANs for tabular data are still sparse compared to other data modalities, there are already several GAN-based approaches [7,14,19] for creating valuable synthetic structured tabular data.

In [14], the authors proposed a table-GAN adapted from deep convolutional GAN (DC-GAN). It converts every row of tabular data into a square two-dimensional matrix and then processes it by two-dimensional convolutions. However, the values of columns can be heterogeneous. That is why the proposed approach tends to be inappropriate for structured data. Also, in [2], the authors achieved high performance on tabular data by replacing the Vanilla GAN loss with a Wasserstein GAN Gradient Penalty (WGANGP) loss and demonstrated that loss functions used in image generation could also be promising for tabular data generation. Furthermore, in [13], the authors applied a GAN to generate passenger name record data. They proposed an architecture that explicitly uses cross layers in both the generator and the discriminator to compute feature interactions. To sum up, recent studies show that GANs are scalable and could be used with tabular data. That is why it is worth investigating the potential of tabular GANs on imbalanced datasets.

3   Methodology

The current work proposes the approach displayed in Fig. 2 to solve the multi-class imbalanced classification problems. We assume that each imbalanced dataset X has N samples and M features and that each instance is assigned to one of the S classes $C_1 \cdots C_s$. The number of samples in every class C_i is different so that C_i can be either a majority or a minority class. In the subsequent sections, we will explain in detail each step of the pipeline.

3.1   Data Pre-processing

The prepossessing steps for the input data are transparent and unambiguous. Numerical variables are scaled using standardization, while categorical values are replaced by the count of the observations that show that category in the dataset. Moreover, we split the data into 30:70 ratio for X_{test} and X_{train} sets respectively. We preserve X_{test} from further modifications as we will use it for the evaluation phase. However, X_{train} will be augmented by the mean of CTWGAN and CTGAN discussed in the following subsection.

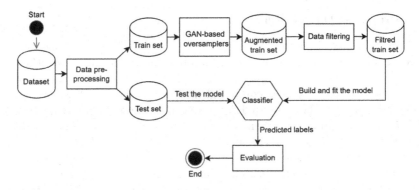

Fig. 2. Overall pipeline.

3.2 Data Augmentation

For balancing class distributions and improving model performance on the imbalance dataset, the current study examines different oversampling methods to increase the weight of the minority class. Mainly, we used five over-samplers that are known to be useful for multi-class imbalance learning as baselines, namely: synthetic minority oversampling (SMOTE) [10], random walk oversampling (RWO) [11], adaptive neighbor synthetic minority oversampling (ANS) [20], K-means SMOTE [18], and geometric SMOTE (G-SMOTE) [4]. These baselines are compared with four GANs, namely conditional tabular GAN (CTGAN) [19], gene expression generator [7] that we will refer to it as a conditional tabular Wasserstein GAN (CTWGAN), and our proposed models FTWA, and FTA described in Sect. 3.3. Compared to other GAN oversamplers, CTWGAN and CTGAN are recent and dedicated only to tabular data, which explains our choice of choosing these two algorithms.

Conditional Tabular Wasserstein GAN. CTWGAN [7] is adapted from [17] to use tabular data samples of a single class as an input to generate more artificial instances referring to the same class. CTWGAN's primary feature is that Wasserstein distance is used as a loss function to reflect the similarity between the original and generated data distribution. In [3], the authors gave some theoretical insights about Wasserstein distance which is the most practicable loss function for GAN's discriminator. Compared to the traditional cross-entropy loss function, Wasserstein distance considers the distribution of each variable in the real and generated samples. It determines how far apart the distributions are for real and generated data by providing criticism to the so-called critic.

Conditional Tabular GAN. Compared to standard TGANs, CTGAN [19] consists of three key elements: the conditional vector, the generator loss, and the training-by-sampling to deal with the imbalanced discrete columns. It also applies the mode-specific normalization to overcome the non-Gaussian and multimodal distribution.

3.3 Proposed Data Filter

After augmenting the data with $CTWGAN$ and $CTGAN$, we use our proposed Algorithm 1 to filter out the generated data. The filter uses the training set T, where $T = \{\langle T_1, L_1 \rangle, \ldots, \langle T_p, L_p \rangle\}$, the original error rate per class ER (the ER of each class is computed as 1 - Accuracy of that class), where $ER = \{C_1, \ldots, C_s\}$, the generated dataset G from the previous step, where $G = \{\langle G_1, L_1 \rangle, \ldots, \langle G_n, L_n \rangle\}$, the number of neighbors K, and a distance measure D as input and outputs the filtered dataset F, where $F = \{\langle F_1, L_1 \rangle, \ldots, \langle F_m, L_m \rangle\}$, $m < n$. Such filtering allows us to keep as much generated data as possible while reducing ER, omitting noisy instances from G, and optimizing the purity of the generated data from the training set. The reason behind this filter is that both GANs have a shared common drawback. While synthesizing new instances for a specific class, the GAN's generator is fed only with single class instances. As a result, GAN assumes only the data distribution of that class without considering other class instances. Such a process helps avoid annotating the data as all generated data will have the same label as the generator input data. However, it will yield overlapping regions and noisy examples. That is why using a filter is highly advised. The idea is to evaluate the neighborhood of the generated data and remove those generated instances in which the class differs from the class of its neighbors and does not improve class recognition. Thus, we preserve good generated data points from the safe regions and use them for model training.

As a result, we built two new tabular GAN variants that use CTWGAN and CTGAN as data augmentation techniques and Algorithm 1 as a filter. For simplicity, the filtered CTWGAN data augmentor is abbreviated as FTWA while the filtered CTGAN augmentor is abbreviated as FTA.

Algorithm 1. Proposed filtering algorithm based on neighboring rules

---

1: $i \leftarrow 0$
2: $F \leftarrow \emptyset$
3: **while** ($i < n$) **do**
4:      generated_data_neighbors = get_neighbors(G[i], T, K, D)
5:      **if** same_class(G[i], generated_data_neighbors) **then**
6:          $F \leftarrow F + G[i]$
7:      **else**
8:          $ER' = calculate\_new\_error\_rate(T + G[i])$
9:          **if** $ER'[i] < ER[i]$ **then**
10:             $F \leftarrow F + G[i]$
11:         **end if**
12:     **end if**
13:     $i \leftarrow i + 1$
14: **end while**
15: **return** F

Given the set of generated data G of length n, we aim to determine whether the i-th instance $\langle G_i, L_i \rangle$ is a noisy sample or not. For every generated sample G_i, we identify its K nearest neighbors from the training set T while considering the distance measure D (K and D are hyperparameters). We set $D = Euclidean$ $distance$ as a distance measure and utilized $NearestNeighbors$ implementation from sklearn.neighbors to locate the K nearest neighbors of a given G_i. Furthermore, we used a *Ball Tree* algorithm to transform input training data into a fast indexing structure. As soon as the K neighbors are found, we identify the representative class by the majority rule. If the label of G_i is the same as the class of all its nearest neighbors, we consider G_i as data generated in the safe area, then we add G_i to the list of filtered data F. Otherwise, we calculate the new error rate ER' on the $T + G_i$ set. If the ER' is smaller than the initial error rate ER, we append G_i to F. Otherwise, we omit it.

3.4 Classification Using Decision Trees

Once we obtain F, we append it with T, resulting in an augmented training set $(F \cup T)$ where $p < p+m$. In other words, we oversampled T with F, and will use $F \cup T$ to train a decision tree classifier (DT)[1]. Compared to black-box models like neural networks, tree-based learning algorithms are considered one of the best and mostly used supervised learning approaches for multi-class as well as binary problems as they are accurate and easy to comprehend.

4 Experimental Results

4.1 Datasets Description

For the experimental part, we used three multi-class imbalanced datasets from the UCI machine learning repository [6], namely, page blocks, internet firewall, and shuttle datasets. The datasets belong to distinct application domains and have different characteristics (Table 1). The source code of conducted experiments, and proposed filter are publicly available on GitHub repository[2].

As previously mentioned, in multi-class imbalanced classification, a dataset has k classes. Therefore, we consider the class with the highest number of instances as a majority class, while the remaining ones as minority classes.

Table 1. Description of datasets used for the experiments.

Dataset	# Train	# Test	# Features	# Classes
Page blocks	3666	1806	10	5
Internet firewall	43906	21626	11	4
Shuttle	43500	14500	9	7

---

[1] We used the sklearn implementation of *DecisionTreeClassifier()* with a GridSearch() to optimize its hyper parameters.

[2] https://github.com/LyudaK/msc_thesis_imblearn.

4.2 Evaluation Metrics

Cohen's Kappa (K): is a powerful metric that has very well performance on imbalanced class and multi-class tasks, it measures the ratio of agreement between the real and predicted values. It ranges between $[-1,1]$ where 0 means random agreement, and 1 means a perfect agreement. In the multi-class case, we compute K based on the following formula:

$$K = \frac{c \times s - \sum_k^K p_i \times t_k}{s^2 - \sum_k^K p_i \times t_k} \tag{1}$$

where:

- $c = \sum_k^K C_{k,k}$ the total number of elements correctly predicted,
- $s = \sum_i^K \sum_j^K C_{i,j}$ the total number of elements,
- $p_k = \sum_i^K C_{k,i}$ the number of times that class k was predicted (column total),
- $t_k = \sum_i^K C_{i,k}$ the number of times that class k truly occurred (row total).

F1-score: is another well-known metric to assess model's performance for both binary and multi-class classification problems. In short, it consider all classes one by one and aggregates their *Precision* and *Recall* measures under the concept of harmonic mean.

$$F1\text{-}Score = 2 \times \frac{precision \times recall}{precision + recall} \tag{2}$$

Geometric mean (G-mean): is strongly advised to measure performance for classification purposes. It aims to maximize the accuracy of each class. For multi-class cases, we compute G-mean as the following:

$$G\text{-}mean = \sqrt[n]{sen \times spe} \tag{3}$$

where, *sen* is the sensitivity, *sep* is the specificity, and n is the number of classes.

4.3 Classification Results

The decision tree was selected as a classifier for all experiments, and we set up each data augmentation method to equalize the number of instances in each class (the number of instances in each class is equal to the number of samples in the majority class). For FTWA and FTA, and more precisely during the filtering stage, the number of nearest neighbors was equal to 5, and the euclidian distance was selected as a distance metric. Hyperparameters are optimized with the help of an exhaustive search over specified parameter values for an estimator (GridSearchCV from the sklearn package).

Table 2. Classification results on Shuttle, Page Blocks and Internet Firewall datasets.

Oversampler	None	SMOTE	G-SMOTE	RWO	K-means SMOTE	ANS	CTWGAN	CTGAN	FTWA	FTA
Metrics	Shuttle									
G-mean	0.84	0.78	0.80	0.75	0.78	0.78	0.85	0.85	0.88	**0.89**
Cohen's Kappa	0.57	0.37	0.39	0.33	0.40	0.40	0.59	0.59	0.78	**0.81**
F1-score	0.85	0.75	0.76	0.71	0.80	0.80	0.85	0.85	0.91	**0.92**
Metrics	Page blocks									
G-mean	0.85	0.83	0.82	0.84	0.82	0.82	0.62	0.64	0.66	**0.87**
Cohen's Kappa	0.61	0.43	0.37	0.45	0.57	0.58	0.39	0.39	0.53	**0.67**
F1-score	0.92	0.87	0.85	0.88	0.92	0.88	0.89	0.88	0.92	**0.93**
Metrics	Internet firewall									
G-mean	0.88	0.77	0.72	0.88	0.72	0.87	0.72	0.72	0.83	**0.88**
Cohen's Kappa	0.79	0.72	0.94	0.83	0.94	0.72	**0.99**	0.98	**0.99**	0.99
F1-score	0.91	0.88	0.95	0.94	0.98	0.97	**0.99**	0.98	**0.99**	0.99

Table 2 summarizes the results of the comparison between data augmentation techniques based on Cohen's Kappa, G-mean, and F-1 scores. Additionally, we examined how GAN-based methods (including the proposed ones) influence the error rate per each class compared to SMOTE (Table 3). Note that class 1 in Table 3 refers to the majority class, while the remaining classes are minority classes (the one we generated data for).

Table 3. Error rate per class for Shuttle, Page Blocks, and Internet Firewall datasets.

Class	Baseline	SMOTE	CTWGAN	CTGAN	FTWA	FTA
Shuttle						
1	0.12	0.40	0.33	0.13	0.07	**0.06**
2	0.32	0	0.12	0.58	0.02	0.28
3	0.41	0.30	0.58	0.30	0.29	**0.18**
4	0.23	0.28	0.23	0.15	0.04	**0.02**
5	0.05	0.19	0	0	0.02	0
6	0.50	0.25	0	0.5	0	0.5
7	0.50	0.50	0.2	0	0	0
Page blocks						
1	0.05	0.14	**0.03**	**0.03**	0.07	0.06
2	0.21	0.46	0.56	0.41	**0.03**	0.06
3	0.66	**0.22**	0.44	0.46	0.41	**0.22**
4	0.31	0.66	0.82	**0.17**	0.46	**0.17**
5	0.82	0.43	0.31	0.17	0.17	**0.10**
Internet firewall						
1	0.11	0.08	0.008	0.07	0	0
2	0.22	0.02	0.008	0.008	**0.007**	**0.007**
3	0.02	0	0	0	0	0
4	1	0.88	0.75	0.5	**0.33**	0.38

Table 2 shows that GAN-based approaches have slightly better predictions of minority class instances than the baselines. However, the proposed techniques, FTWA and FTA, that use the given filtering approach perform better than the original implementations. Furthermore, the experimental results indicate that FTA is the best data augmentation method for the datasets used in the experimental part. Based on the good results obtained, we assume that the data filter could reduce the effect of conditional GANs drawbacks, which helped improve the performance of the DT classifier on the chosen datasets. The results for each dataset can be summarized as follows:

- **Shuttle dataset:** we used DT classifier with entropy criterion, max_depth= 13, and log2 as max_features for classifying shuttle test data. Training the classifier with the augmented data by SMOTE and its variants have worse results than the trained model with the initial training set. The reason is that the class distribution is highly skewed, and the number of samples in one class dramatically exceeds the number of instances in the minority classes. As a result, SMOTE and its variants added noise samples and may have created overlapping regions that led to decreased performances. However, GAN-based methods achieved better performances on the minority classes, and according to Table 3, the application of a filter helps to improve the overall performance and reduces all classes ER except one making the proposed FTA the best.
- **Page Blocks dataset:** for classifying the samples from the dataset, we used a decision tree with entropy criterion, max_depth= 14, and sqrt as max_features. Table 2) shows that the classifier achieved high performance even on the initial set of data (up to 92% of F1-score) without any augmentation. However, DT makes false predictions for 1/3 samples from minority classes on the test set. Furthermore, the proposed filtering improved the classification performance and finally achieved 93% in the F1-score (Table 3). FTA gave the best results, from which the proposed filtering removed noise and boundary samples. The model started to distinguish underrepresented classes 3 and 5 better and decreased the ER for classes 2 and 4.
- **Internet Firewall dataset:** for classifying the samples from the Internet Firewall dataset, we used a decision tree with entropy as a criterion, max_depth= 11, and auto as max_features. For the Internet Firewall dataset, the smallest minority class has 2% of the whole dataset, whereas a sufficient number of instances represents the rest classes. The baseline model shows promising results, but Table 2 shows that it misclassifies all instances from the 4th class. Both SMOTE and GAN variants decreased the ER for the 2nd and 3rd classes up to 0%. However, SMOTE did not improve the model for the 4th class, contrary to FTWA and FTA, which could decrease the ER to 0.33%. To sum up, according to Table 3, the proposed method allowed to achieve 99% in F1-score and decrease the ER for underrepresented class 4 from 100% to 33%, meaning that 11 out 18 samples from the named class were appropriately distinguished. The best results were obtained when we trained the classifier on the filtered data generated by FTWA.

5   Conclusion

Dealing with non-standard derivative datasets that are imbalanced by nature is a common problem in classification tasks. To solve such problems, we can either gather new data for the underrepresented classes or synthesize artificial data from the existing ones. In this paper, we demonstrated the power of generative adversarial networks on multi-class imbalanced datasets that could be used as an alternative to data collection. We affirmed that GANs are efficient directed sampling methods regardless of the data modality. To further enhance their performances and reduce their drawbacks, we proposed a filtering algorithm based on neighboring rules that added two new tabular GANs variants to the literature: FTWA and FTA. Experimental results showed that augmenting the training set using FTWA and FTA added new patterns pertinent to the task and helped improve the classification performance of decision tree classifiers. We can conclude that both FTWA and FTA produced synthetic samples with the same data distribution as real samples for the datasets that we used. Furthermore, they increased the ability of the classification model to identify rare events, underrepresented class instances, reduced the bias toward the majority class, and improved the generalization ability of the trained models.

As future directives, we would focus on high-dimensional and other multi-class imbalanced datasets to inspect the performance of the proposed approaches at a broader range. Using FTWA or FTA on high-dimensional data may take long-running considering every feature dimension. We can run a pre-processing activity in these cases to pull out the most valuable attributes and generate new samples from a subset of essential features. Also, as most of the attributes in the used datasets were numeric, we will investigate several options for extending the proposed pipeline to handle categorical and mixed-type attributes. Last but not least, it would be interesting to analyze how different classifiers affect the performance of the proposed models.

Acknowledgements. This research is supported by the ÚNKP-21-3 New National Excellence Program of the Ministry for Innovation and Technology from the source of the National Research, Development, and Innovation Fund.

References

1. Balaram, A., Vasundra, S.: Prediction of software fault-prone classes using ensemble random forest with adaptive synthetic sampling algorithm. Autom. Softw. Eng. **29**(1), 1–21 (2022)
2. Baowaly, M.K., Lin, C.C., Liu, C.L., Chen, K.T.: Synthesizing electronic health records using improved generative adversarial networks. J. Am. Med. Inform. Assoc. **26**(3), 228–241 (2019)
3. Biau, G., Sangnier, M., Tanielian, U.: Some theoretical insights into Wasserstein GANs. J. Mach. Learn. Res. **22**, 1–45 (2021)

4. Camacho, L., Douzas, G., Bacao, F.: Geometric SMOTE for regression. Expert Syst. Appl. **193**, 116387 (2022)
5. Dogra, V., Verma, S., Jhanjhi, N., Ghosh, U., Le, D.N., et al.: A comparative analysis of machine learning models for banking news extraction by multiclass classification with imbalanced datasets of financial news: challenges and solutions. Int. J. Interact. Multimedia Artif. Intell. **7**(3), 35–53 (2022)
6. Dua, D., Graff, C.: UCI machine learning repository (2019). http://archive.ics.uci.edu/ml
7. Farou, Z., Mouhoub, N., Horváth, T.: Data generation using gene expression generator. In: Analide, C., Novais, P., Camacho, D., Yin, H. (eds.) IDEAL 2020. LNCS, vol. 12490, pp. 54–65. Springer, Cham (2020). https://doi.org/10.1007/978-3-030-62365-4_6
8. Farou, Z., Ouaari, S., Domian, B., Horváth, T.: Directed undersampling using active learning for particle identification. In: Singh, P.K., Singh, Y., Chhabra, J.K., Illés, Z., Verma, C. (eds.) Recent Innovations in Computing. Lecture Notes in Electrical Engineering, vol. 855, pp. 149–162. Springer, Singapore (2022). https://doi.org/10.1007/978-981-16-8892-8_12
9. Feng, Q., Guo, C., Benitez-Quiroz, F., Martinez, A.M.: When do GANs replicate? On the choice of dataset size. In: Proceedings of the IEEE/CVF International Conference on Computer Vision (ICCV), pp. 6701–6710, October 2021
10. Fernández, A., García, S., Galar, M., Prati, R.C., Krawczyk, B., Herrera, F.: Learning from Imbalanced Data Sets, vol. 11. Springer, Cham (2018). https://doi.org/10.1007/978-3-319-98074-4
11. Kong, J., Rios, T., Kowalczyk, W., Menzel, S., Bäck, T.: On the performance of oversampling techniques for class imbalance problems. In: Lauw, H.W., Wong, R.C.-W., Ntoulas, A., Lim, E.-P., Ng, S.-K., Pan, S.J. (eds.) PAKDD 2020. LNCS (LNAI), vol. 12085, pp. 84–96. Springer, Cham (2020). https://doi.org/10.1007/978-3-030-47436-2_7
12. Lango, M., Stefanowski, J.: What makes multi-class imbalanced problems difficult? An experimental study. Expert Syst. Appl. **199**, 116962 (2022)
13. Mottini, A., Lheritier, A., Acuna-Agost, R.: Airline passenger name record generation using generative adversarial networks. arXiv preprint arXiv:1807.06657 (2018)
14. Park, N., Mohammadi, M., Gorde, K., Jajodia, S., Park, H., Kim, Y.: Data synthesis based on generative adversarial networks. arXiv preprint arXiv:1806.03384 (2018)
15. Saha, P.K., Logofatu, D.: Efficient approaches for data augmentation by using generative adversarial networks. In: Iliadis, L., Jayne, C., Tefas, A., Pimenidis, E. (eds.) EANN 2022. CCIS, vol. 1600, pp. 386–399. Springer, Cham (2022). https://doi.org/10.1007/978-3-031-08223-8_32
16. Singh, A., Ranjan, R.K., Tiwari, A.: Credit card fraud detection under extreme imbalanced data: a comparative study of data-level algorithms. J. Exp. Theor. Artif. Intell. **34**(4), 571–598 (2022)
17. Wang, Q., et al.: WGAN-based synthetic minority over-sampling technique: improving semantic fine-grained classification for lung nodules in CT images. IEEE Access **7**, 18450–18463 (2019)
18. Wu, T., Fan, H., Zhu, H., You, C., Zhou, H., Huang, X.: Intrusion detection system combined enhanced random forest with smote algorithm. EURASIP J. Adv. Signal Process. **2022**(1), 1–20 (2022)

19. Xu, L., Skoularidou, M., Cuesta-Infante, A., Veeramachaneni, K.: Modeling tabular data using conditional GAN. In: Advances in Neural Information Processing Systems 32 (2019)
20. Yi, X., Xu, Y., Hu, Q., Krishnamoorthy, S., Li, W., Tang, Z.: ASN-SMOTE: a synthetic minority oversampling method with adaptive qualified synthesizer selection. Complex Intell. Syst. **8**, 2247–2272 (2022)

Convolutional Neural Network Approach for Multiple Sclerosis Lesion Segmentation

Nada Haj Messaoud[1,3]([✉]) [ID], Asma Mansour[1] [ID], Rim Ayari[1] [ID],
Asma Ben Abdallah[1] [ID], Mouna Aissi[2] [ID], Mahbouba Frih[2],
and Mohamed Hedi Bedoui[1] [ID]

[1] Laboratory of Technology and Medical Imaging, Faculty of Medicine, University of Monastir,
Monastir, Tunisia
Hajmessaoud.nada@gmail.com
[2] Neurology Department, Fatouma Bouguiba Hospital, Monastir, Tunisia
[3] Faculty of Sciences of Monastir (FSM), University of Monastir, Monastir, Tunisia

Abstract. Nowadays Deep Learning (DL) based automatic segmentation has outperformed traditional methods. In the present paper, we are interested in automatic MS lesion segmentation of 2D images based on DL techniques. The main challenge consists in proposing a new model that takes advantage of referenced CNN models: U-Net, ResNet, and DenseNet with a reduced number of parameters and a shorter execution time. To evaluate the proposed approach named "Concat-U-Net", we compared its performance to those of three implemented models, namely U-Net, U-ResNet, and Dense-U-Net. Furthermore, we employed just one modality (FLAIR) from the public ISBI dataset to segment MS lesions accurately. The best Dice value obtained was 0.73, which outperformed those reported in the literature. Our approach reduced the elapsed execution time from 48 s to 7 s. By reducing the number of parameters, an 85.42% time gain was achieved.

Keywords: Deep Learning · Multiple sclerosis segmentation · MRI · Data augmentation

1 Introduction

Multiple Sclerosis (MS) is an autoimmune disease of the central nervous system that induces lesions in the brain tissues, more precisely in the white matter [1]. These lesions are detected using Magnetic Resonance Imaging (MRI), notably the T2-FLAIR, which reveals and specifies the edges of periventricular and juxtacortical lesions, which is an important criterion in MS diagnosis [2]. The variation of lesion charge acquired in the different MRI sequences is a crucial criterion in determining disease progression in volume and location. Given the variability of MS lesions in terms of volume, location, shape, subjects, and texture, precise segmentation becomes a crucial task to help clinicians make an accurate diagnosis and characterize disease progression. Although manual annotation of lesions by experts in brain imaging is the gold standard approach and remains feasible in practice, it is time-consuming and prone to inter-observer variability. This

© The Author(s), under exclusive license to Springer Nature Switzerland AG 2022
H. Yin et al. (Eds.): IDEAL 2022, LNCS 13756, pp. 540–548, 2022.
https://doi.org/10.1007/978-3-031-21753-1_52

has resulted in poor reproducibility and has led to the development of several automatic lesion segmentation methods aimed at increasing segmentation accuracy and reliability. In this field, Machine Learning (ML) and Deep Learning (DL) based methods used in a wide range of applications such as identifying diseases and diagnosis, segmentation, classification, and prediction have shown promising results. Some studies used cross-sectional [9] and longitudinal lesions from MR images [12] and demonstrated better performance in 2D or 3D segmentation [15]. Current research and clinical applications are focused on white matter lesions (WML) visible in MRI sequences [3, 16]. However, some recent research works have been particularly interested in the case of small cortical lesions spread into the gray matter or originating from the cortex, as they are considered an important biomarker for this disease [1]. In the same context, several challenges have been organized for MS lesion segmentation [3, 15]. These challenges have offered datasets used in several methods to evaluate the proposed algorithms. Among these, the works summarized in Table 1.

Table 1. Summary of some Deep Learning works reported in the literature for the segmentation of MS lesions.

Works	Dataset	Modalities	Methods	Performance
Hashemi (2018) [10]	MICCAI 2016	MRI	3D patch-wise FC-dense-net	DSC = 69.9
	ISBI 2015			DSC = 65.64
Zhang (2018) [8]	Clinical studies	MRI	MS-GAN	DSC = 67.2 Sensitivity = 69.2 Precision = 72.4
Aslani (2019) [2]	Clinical studies	MRI	2D-CNN	DSC = 0.76
	ISBI 2015			DSC = 0.69
Hagiwara (2019) [12]	Clinical studies	MRI	Conditional GAN	PSNR = 35.9 NRMSE = 27
McKinley (2021) [14]	Clinical studies	MRI	nnU-Net	DSC = 56.2
	MSSEG			DSC = 74.5
Afzal (2021) [16]	MICCAI 2016	MRI	Two 2D-CNN	DSC = 67 Sen = 48 Pre = 90
	ISBI 2015			
M. Sadeghibakhi (2022) [17]	ISBI 2015	MRI FLAIR	Attention-based CNNs	DSC = 0.63 LTPR = 0.45

In the present paper, we focused on automatic MS lesion segmentation based on DL architectures from MR brain images. Our major objective is to propose a new robust DL architecture that can achieve precise segmentation performance while reducing the number of parameters in order to accelerate execution time and reduce the risk of overfitting, especially with the use of small-labelled databases, as is the case in medical imaging. The robustness of our approach is guaranteed by the inclusion of a preparation phase composed of (i) a pre-processing step. (ii) Patch extraction, followed by the collection of segmented patches and thresholding of the resulting image. The major additional contribution of this work is, the segmentation of lesions with different sizes, numbers, and shapes occurs either in the white matter or in the grey matter. The workflow of our proposed MS segmentation approach is shown in Fig. 1. The rest of the paper is organized as follows: Multiple Sclerosis Databases are described in Sect. 2. The proposed workflow is detailed in Sect. 3. A comparison of the performance obtained in terms of the number of parameters and execution time is provided in Sect. 4. Finally, Sect. 5 provides a discussion of the obtained results and a conclusion.

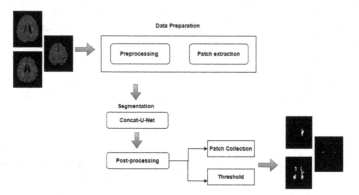

Fig. 1. Workflow of the proposed segmentation method

2 Multiple Sclerosis Data Base

ISBI 2015 Longitudinal MS Lesion Segmentation Challenge: included 19 subjects divided into two sets [15], 5 subjects for training and 14 for the test. In the current work, we are interested mainly in the training database because the ground truths have been made available to the public. The training set consists of 5 subjects with an average of 4.4 time points, which means that 21 data sets with their ground truth are available. Each volume consists of 182 slices (including black images), with a size of (181 × 217) and a 1-mm cubic voxel resolution. For each image, two ground truth annotations from two raters are given (Rater 1: 10 years of experience, Rater 2: 4 years of experience. After removing the black slices that influence the model, we obtained 1389 images from the first expert and 1557 from the second one.

3 Proposed Automatic MS Lesion Segmentation Method

3.1 Preprocessing

In T2-FLAIR MRI slices, the lesions appeared as hyper intense, so we needed to improve the contrast and eliminate the regions influencing the detection of these lesions by performing a preprocessing step (see Fig. 2). At this level, based on the literature [12, 16,

17], three operations were adopted. Gamma correction and intensity normalization was carried out for better image quality. All intensities were normalized to make the feature extraction step easier and more efficient. In addition, histogram equalization was carried out. To limit extremely unbalanced data and omit uninformative samples, only slices containing at least one pixel labeled as a lesion were selected. Figure 3 shows example of image before and after preprocessing.

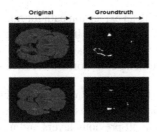

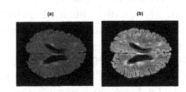

Fig. 2. Example of two T2 FLAIR slices.   **Fig. 3.** (a) Original image, (b) image after preprocessing.

3.2 Patch Extraction

A limited number of lesions of different shapes, sizes, and locations characterize MS lesions in MR images. Segmentation of an entire image using deep learning architectures generates inaccurate results. A patch extraction process was adopted to overcome this problem and improve system performance. In the current work, the size of the patch was chosen empirically while taking into account the fact that the patch size should be larger than the lesion size in order to produce a more accurate segmentation. Some works used small patches (7×7) or (9×9). Others showed that large patches produced more accurate results [2]. Based on the literature and given that the size of lesions can be larger than 30 pixels, we have chosen overlapping patches of size 48×48.

3.3 DL Segmentation Architectures

U-Net, ResNet and DenseNet networks are widely used in medical imaging, and have achieved very high segmentation performance.

- **U-Net**
  U-Net was proposed by [7] in 2015 for biomedical image segmentation. It consisted of a contraction path (downsampling) for feature extraction associated with an expansion path (upsampling) for building the segmented image. This architecture has shown better performance in image segmentation, but has a large number of parameters.
- **U-ResNet**
  ResNet [6] is a fully convolutional network, introduced by He et al. in 2016 and was initially dedicated to image classification. A significant benefit of ResNet is that the gradient can directly flow over the identity function from lower to higher layers. These

connections are known as Shortcut connections and have shown better performance in medical image segmentation [2]. This approach allows the network to fit the residual mapping instead of learning the underlying mapping. Instead of saying H(x), initial mapping, let the network fit (1), giving (2).

$$F(x) := H(x) - x \tag{1}$$

$$H(x) := F(x) + x \tag{2}$$

We improved the original architecture by adding an expansion path for finer information retrieval. We can say that the U-ResNet model looks like the U-Net model previously described, but it differs by the use of residual units as a basic block instead of simple neural units to build a rich network with long and short consecutive layers.

- **Dense-U-Net**
  DenseNets [5], proposed by Huang et al. in 2017, are based on dense blocks and pooling operations. This architecture can be considered an extension of ResNets. Each layer of a dense block receives feature maps from all previous layers and transmits its output to all subsequent layers. One of the greatest advantages of Dense Nets is the improved flow of information and gradients throughout the network, making them easy to train. We kept the U-shape of U-Net and we added the dense blocks in both contraction and expansion paths.
- **Concat-U-Net: proposed architecture for MS lesion segmentation**

The proposed model Concat-U-Net exploits the three architectures detailed previously. Therefore, we are particularly interested in skip connections used in U-Net, residual blocks in ResNet, and feature map concatenations in DenseNet. Compared to ResNet, DenseNet can improve information flow between layers and network efficiency by introducing direct connections (concatenations) from any layer to all subsequent layers rather than combining the identity function and layer outputs by summation, which can hinder information flow in the network [5]. To obtain finer results, it is very important to use low-level details while keeping high-level semantic information. Thus, the major contribution of U-Net is the use of skip connections between the contraction path and the expansion path (see Fig. 4). Our network architecture is similar to the U-Net but it is more convenient since it uses fewer parameters to reduce the risk of overfitting as detailed in [13]. We added connections between convolution layers by concatenating the feature maps. As in ResNet, we created a propagation path to allow information to propagate between low and high levels more easily. Therefore, our model has long skip connections between the convolution layers. As in DenseNet, we used feature maps concatenations to allow the model to be: (i) Deeper, (ii) Optimal by reducing the number of parameters. These features were concatenated to be used as inputs in the next layer.

3.4 Post-processing

Post-processing was carried out in order to merge the segmented patches and apply a threshold in order to generate a binary image of MS lesions. Therefore, all segmented patches in an image were collected. Then, these patches were copied in slicing order, depending on the size of the image used. Finally, an empirical threshold was used to enhance the resulting image.

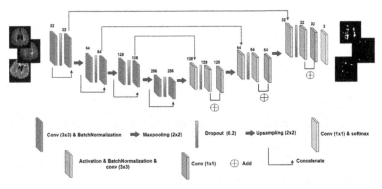

Fig. 4. Concat-U-Net architecture. In each block, the number of filters used is denoted

4 Results and Evaluation Metrics

For the evaluation metrics, we chose the most commonly used in MS lesion segmentation works cited in the literature (see Table 2.).

Table 2. Evaluation metrics used.

Dice coefficient	Precision value (PPV)	Sensitivity (True Positive Rate)
$\dfrac{2TP}{2TP+FP+FN}$	$\dfrac{TP}{TP+FP}$	$\dfrac{TP}{TP+FN}$

4.1 Implementation Details

The training phase requires establishing a set of parameters such as the optimizer, learning rate, number of epochs, and batch size. These are usually experimentally selected or based on recent studies with the aim of producing precise segmentation performance. The implementation is conducted on Intel Core i9-11900F @ 2.50 GHz, 32 Gb RAM and a Nvidia GeForce RTX 3090. The models were implemented in Python language using Keras with Tensorflow backend. After the experimental study, U-Net and Dense-U-Net were trained using an SGD optimizer with a 0.01 learning rate. However, for the other two methods, U-ResNet and Concat-U-Net - the "Adam" optimizer was used with a learning rate of 0.00001. We chose 32 as batch size, «categorical cross entropy» as loss function and 50 as several epochs. Figure 5 shows Learning Curve details.

4.2 Results on ISBI Database

We propose a 5-fold cross validation strategy, which entails dividing the MS images database (Rater 1) into five subsets in order to carry out five trials for each database, in order to effectively analyze the proposed method. These experiments aim to assess the robustness of the proposed method (see Table 3).

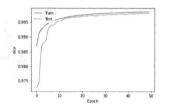 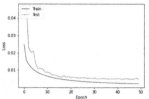

Fig. 5. (a) Dice coefficient over each iteration; (b) loss over each iteration.

Table 3. Average performance measure for 5-fold cross validation for the ISBI dataset. Average value of precision, sensitivity, Dice, PR-curve are given.

Method	Precision	Sensitivity	Dice	Test time (s)	#Parameters
U-Net	**0.87**	0.49	0.61	29 s	31,378,590
U-ResNet	0.83	0.62	0.66	10 s	8,214,386
Dense-U-Net	0.84	0.61	0.67	48 s	57,077,260
Concat-U-Net	0.82	**0.75**	**0.73**	**7 s**	**2,290,402**

As shown in Table 3 the lowest time elapsed was 7 s, obtained by Concat-U-Net. To better validate the segmentation results, we calculated the similarity value between the ground truths provided by the two experts (Rater 1 and Rater 2), and we obtained a Dice value of 71%, which was lower than the one obtained by our model (see Table 4). This demonstrates the efficiency of Concat-U-Net. It can be noticed that concatenations used in our approach allowed better detection of all types of lesions with a reduced number of parameters and a shorter processing time (see Fig. 6).

Table 4. Comparison of our method with other state-of-the-art methods tested on the ISBI dataset (Only images with available ground truths were considered).

Method	Rater 1	
	Dice	Sensitivity
Extra Tree forest [4]	0.70	0.53
3D CNN with shortcut connections [11]	0.68	0.74
Concat-U-Net	**0.73**	**0.75**

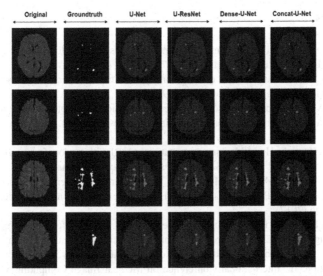

Fig. 6. Example of Output segmentation results of the different methods. On all images, true positives, false negatives, and false positives are colored in green, blue and red, respectively. (Color figure online)

5   Discussion and Conclusion

In this paper, we designed an automated pipeline for Multiple Sclerosis lesion segmentation from MRI data based on DL methods. Our main contribution is proposing an hybrid architecture that exploits three well-known architectures, U-Net, ResNet, and DenseNet, allowing for better segmentation performance with fewer parameters and a shorter processing time. The public ISBI dataset is used to test the performance of the proposed model. By analyzing the results detailed in the previous section, we can conclude that Concat-U-Net showed high improvements in MS lesion segmentation. This could have resulted from the use of "Shortcut connections" which facilitated information propagation and helped the network exploit more of the contextual information associated with the lesions. The proposed model is much more optimal than U-ResNet in terms of reducing the number of parameters to 2.29 million and accomplishing a short execution time. Compared to the other models, we achieved an 85.42% time gain. In conclusion, when interpreting the obtained results, we noticed that with a reduced number of training parameters, we obtained better segmentation precision and better performance with a more optimized execution time. Future research will make the use of the images segmented by our model to extract important and helpful MRI brain abnormalities, such as the location, size, number, and type of MS lesions, in order to analyze the relationship between these features, patient clinical data, and the expanded disability status scale (EDSS).

References

1. Thompson, A.J., et al.: Freedman and others, diagnosis of multiple sclerosis: 2017 revisions of the McDonald criteria. Lancet Neurol. **17**, 162–173 (2018)
2. Aslani, S., et al.: Multi-branch convolutional neural network for multiple sclerosis lesion segmentation. Neuroimage **196**, 1–15 (2019)
3. Shoeibi, A., et al.: Applications of deep learning techniques for automated multiple sclerosis detection using magnetic resonance imaging: a review. Comput. Biol. Med. **136**, 104697 (2021)
4. Maier, O., Handels, H.: MS lesion segmentation in MRI with random forests. In: Proceedings of 2015 Longitudinal Multiple Sclerosis Lesion Segmentation Challenge, pp. 1–2 (2015)
5. Huang, G., Liu, Z., Van Der Maaten, L., Weinberger, K.Q.: Densely connected convolutional networks. In: Proceedings of the IEEE Conference on Computer Vision and Pattern Recognition (2017)
6. He, K., Zhang, X., Ren, S., Sun, J.: Deep residual learning for image recognition. In: Proceedings of the IEEE Conference on Computer Vision and Pattern Recognition (2016)
7. Ronneberger, O., Fischer, P., Brox, T.: U-Net: convolutional networks for biomedical image segmentation. In: International Conference on Medical Image Computing and Computer-Assisted Intervention (2015)
8. Zhang, C., et al.: MS-GAN: GAN-based semantic segmentation of multiple sclerosis lesions in brain magnetic resonance imaging. In: 2018 Digital Image Computing: Techniques and Applications (DICTA), pp. 1–8. IEEE, December 2018
9. Moeskops, P., Viergever, M.A., Mendrik, A.M., De Vries, L.S., Benders, M.J.N.L., Išgum, I.: Automatic segmentation of MR brain images with a convolutional neural network. IEEE Trans. Med. Imaging **35**, 1252–1261 (2016)
10. Hashemi, S.R., Salehi, S.S.M., Erdogmus, D., Prabhu, S.P., Warfield, S.K., Gholipour, A.: Asymmetric loss functions and deep densely-connected networks for highly-imbalanced medical image segmentation: application to multiple sclerosis lesion detection. IEEE Access **7**, 1721–1735 (2018)
11. Brosch, T., Tang, L.Y.W., Yoo, Y., Li, D.K.B., Traboulsee, A., Tam, R.: Deep 3D convolutional encoder networks with shortcuts for multiscale feature integration applied to multiple sclerosis lesion segmentation. IEEE Trans. Med. Imaging **35**, 1229–1239 (2016)
12. Hagiwara, A., et al.: Improving the quality of synthetic FLAIR images with deep learning using a conditional generative adversarial network for pixel-by-pixel image translation. Am. J. Neuroradiol. **40**(2), 224–230 (2019)
13. Zeng, C., Gu, L., Liu, Z., Zhao, S.: Review of deep learning approaches for the segmentation of multiple sclerosis lesions on brain MRI. Front. Neuroinform. **14**, 55 (2020)
14. McKinley, R., Wepfer, R., Aschwanden, F., et al.: Simultaneous lesion and brain segmentation in multiple sclerosis using deep neural networks. Sci. Rep. **11**, 1087 (2021)
15. Carass, A., et al.: Longitudinal multiple sclerosis lesion segmentation: resource and challenge. NeuroImage **148**, 77–102 (2017)
16. Afzal, H.M., et al.: Automatic and robust segmentation of multiple sclerosis lesions with convolutional neural networks. CMC-Comput. Mater. Continua **66**(1), 977–991 (2021)
17. Sadeghibakhshi, M., Pourreza, H., Mahyar, H.: Multiple sclerosis lesions segmentation using attention-based CNNs in FLAIR images. IEEE J. Transl. Eng. Health Med. **10**, 1–11 (2022)

Author Index

Printed in the United States
by Baker & Taylor Publisher Services

Printed in the United States
by Baker & Taylor Publisher Services